HANDBOOK OF ENZYME BIOTECHNOLOGY

Second Edition

Editor:

ALAN WISEMAN, PhD, FRSC, MIBiol
Department of Biochemistry
University of Surrey, Guildford, England

ELLIS HORWOOD LIMITED
Publishers · Chichester

Halsted Press: a division of
JOHN WILEY & SONS
Chichester · New York · Ontario · Brisbane

First published in 1985
Reprinted 1987 by
ELLIS HORWOOD LIMITED
Market Cross House, Cooper Street, Chichester, West Sussex, PO19 1EB, England

The publisher's colophon is reproduced from James Gillison's drawing of the ancient Market Cross, Chichester.

Distributors.

Australia, New Zealand, South-east Asia:
Jacaranda-Wiley Ltd., Jacaranda Press,
JOHN WILEY & SONS INC.,
G.P.O. Box 859, Brisbane, Queensland 40001, Australia

Canada:
JOHN WILEY & SONS CANADA LIMITED
22 Worcester Road, Rexdale, Ontario, Canada.

Europe Africa:
JOHN WILEY & SONS LIMITED
Baffins Lane, Chichester, West Sussex, England.

North and South America and the rest of the world:
Halsted Press: a division of
JOHN WILEY & SONS
605 Third Avenue, New York, N.Y. 10016, U.S.A.

©1985 A. Wiseman/Ellis Horwood Limited

British Library Cataloguing in Publication Data
Handbook of enzyme biotechnology. – 2nd Ed.
(Ellis Horwood series in enzyme biotechnology)
1. Enzymes – Industrial applications
I. Wiseman, Alan
661'.8 TP248.E5

Library of Congress Card No. 84-25167

ISBN 0-85312-420-5 (Ellis Horwood Limited)
ISBN 0-420-20153 (Halsted Press)

Typeset by Ellis Horwood Limited
Printed in Great Britain by Butler & Tanner, Frome, Somerset

Table of Contents

Chapter 4   Principles of immobilization of enzymes

Professor J. F. KENNEDY, Research Laboratory for the Chemistry of
Bioactive Carbohydrate and Proteins, Department of Chemistry, Uni-
versity of Birmingham B15 2TT, England and The North East Wales
Institute, Deeside, Clwyd CH5 4BR, Wales and Dr. C. A. WHITE,
Vincent Kennedy Ltd., 47 Conchar Road, Sutton Coldfield, B72 1LL,
England

**Chapter 4   Data on techniques of enzyme immobilization and
                bioaffinity procedures**

Professor J. F. KENNEDY, Research Laboratory for the Chemistry of
Bioactive Carbohydrate and Proteins, Department of Chemistry, Uni-
versity of Birmingham B15 2TT, England and The North East Wales
Institute, Deeside, Clwyd CH5 4BR, Wales and Dr. C. A. WHITE,
Vincent Kennedy Ltd, 47 Conchar Road, Sutton Coldfield, B72 1LL,
England

Chapter 5   Enzymes in clinical analyses — data

Dr. B. J. GOULD, Department of Biochemistry, University of Surrey,
Guildford, Surrey GU2 5XH, England, and Dr. B. F. ROCKS, Depart-
ment of Pathology, The Royal Sussex Country Hospital, Brighton,
East Sussex, BN2 5BE, England

Table of Contents

PRINCIPLES OF INDUSTRIAL ENZYME ISOLATION AND UTILIZATION

CHAPTER 1

Introduction to principles

Dr. ALAN WISEMAN, Biochemistry Division, Department of Biochemistry, University of Surrey, Guildford, England

1.1 GENERAL INTRODUCTION

The first edition of *Handbook of Enzyme Biotechnology,* edited by Alan Wiseman, was published in 1975 by Ellis Horwood, and this most successful book was later translated into Japanese (1977) and Czech (1981). Part I of that book covered the principles of enzyme production and utilization, while Part 2 was a collection of data for use in industrial and other applications, of enzymes. Many specialist topics were mentioned in this source book, although only a few were subjected to detailed analysis. This led therefore to a series of books, *Topics in Enzyme and Fermentation Biotechnology,* published from 1977 by Ellis Horwood.

Topics 1 (1977) reviewed enzyme synthesis in continuous culture, foam separation of biological materials, aeration of culture fluids, enzymic modifications of antibiotics, patents, glucose isomerase, and cytochromes P-450. *Topics 2* (1978) reviewed enzymes immobilized on inorganic supports, enzyme electrodes and enzyme-based sensors, antibiotic-inactivating enzymes, biological treatment of aqueous wastes, and stabilization of enzymes. *Topics 3* (1979) reviewed the uses of oxyanions in enzyme equilibrium displacement, developments in microbial extra-cellular enzymes, rennets and cheese, scale-up of fermentation processes, and new and modified invertases and their applications. *Topics 4* (1980) reviewed enzymes in therapy, medical uses of proteolytic enzymes, solid substrate fermentation, measurement of process variables, and immobilized microbial cells. *Topics 5* (1981) reviewed immobilized coenzymes, large-scale enzyme extraction and recovery, aspects of Gramicidin S, papain, and alcohol dehydrogenases. *Topics 6* (1982) reviewed 4-hydroxycoumarin antibiotics, microbiological aspects of secondary metabolites, enzyme stabilization, beer fermentation, and microbial oxygenases. *Topics 7* (1983) reviewed immobilized plant and animal cells, disordering macromolecular structure for enzyme attack, microbial enymes in the biodegradation of sulphated surfactants, thermophilic, anaerobic, and cellulolytic bacteria, monoclonal antibodies, immobilized enzymes in water and air purification, and the limitations of fermentation processes for utilization of food wastes. *Topics 8* (1984) reviewed xylanses: function properties and applications, biological control of nitrogenous pollution in waste water, and computers and microprocessors in industrial fermentation. *Topics 9* (1984) reviewed the physiology of hydrocarbon-utilizing microorganisms, applications of reactive dyes in biotechnology and biochemistry, application of immobilized enzymes to fundamental studies on enzyme structure and function, and progress with design of enzymes and mimics. *Topics 10* (1985) is in press.

Many theoretical possibilities have come to fruition in the ten years since the first edition of the *Handbook of Enzyme Biotechnology* was published. Nevertheless, a vast number of such possibilities have not emerged at industrial level, so far as one can ascertain, perhaps because of a lack of the real scientific information required for success. Another factor is always the economics of a suggested process at any particular time.

Once again, the *Handbook* sets out to summarize in concise form the principles and practice associated with industrial enzymes in their widest sense. We have therefore extended the coverage to the range of enzymes used in clinical laboratories, where repeated use of these procedures gives rise to a part of the general 'industrial' requirement for enzymes.

1.2 INTRODUCTION TO PART A OF *HANDBOOK OF ENZYME BIO-TECHNOLOGY II*, PRINCIPLES OF INDUSTRIAL ENZYME ISOLATION AND UTILIZATION

The most important principles involved in enzyme utilization are becoming clearer with the move towards the use of immobilized enzymes in various forms. Much of classical enzyme kinetics has needed to be remoulded towards the particular general requirement of product formation — and indeed in every process the key features of enzymology have to be re-established. Stability and stabilization of enzymes is often of great importance in this context. (See Part A: sections 3.27 and 4.4 and review by Mozhaev & Martinek, (1984)).

Enzymes are bought and sold, by activity rather than by weight. There is no need to use a more highly purified, or modified, enzyme than is necessary for the particular process, as work done on the enzyme will be expensive. Nevertheless, the presence of inhibitors could make the prediction of the effect of the enzyme difficult, especially in enzyme kinetic terms. Deciding if the process will really work on a large scale may defy prediction.

Many of the important principles are associated with the use of a variety of immobilized enzyme reactors. This area of enzyme engineering is of great importance in assessing the practicalities of the particular application of the immobilized enzyme or immobilized cell (see Part A: Chapter 3). But first the enzyme must be successfully isolated (see Part A: Chapter 2) and immobilized (see Part A; Chapter 4). Some of the most sophisticated applications of enzymes, however, are to be found in clinical biochemistry, and here the use of antibodies has allowed the development of a variety of remarkable techniques such as enzyme immunoassay (see Part A: Chapter 5). The reader is referred to Part B of the book for data, practical details, and applications.

REFERENCES
Mozhaev, V. V. & Martinek, K. (1984) *Enzyme & Microbial Technology*, 6, 50–59.

ACKNOWLEDGEMENT
We acknowledge the kind cooperation of Surrey University Press (Blackie & Son Ltd., Glasgow) for the agreement to use some figures (as indicated) in Part A Chapter 3, from *Principles of Biotechnology* (Ed. Wiseman, A.) 1983.

CHAPTER 2

Large-scale extraction and purification of enzymes and other proteins

Dr. M. D. SCAWEN and **Professor J. MELLING,** PHLS Centre for Applied Microbiology and Research, Porton Down, Salisbury, England

2.1 INTRODUCTION

In the recent surge of interest in biotechnology considerable emphasis has been placed upon advances in the genetic aspects as well as in fermentation technology. Biotechnologists have, however, always appreciated that the techniques for extraction and purification of proteins from an inseparable part of the whole biotechnological process. There has, in the last few years, been an increasing need to ally protein purification techniques to large-scale fermentation, sometimes on a vast scale involving thousands of litres of products, to purify a range of proteins from microbial cultures. Proteins which are to be administered to patients pose special problems. These cannot adequately be covered in a chapter devoted to the wider problem of purification, but in general, very high standards of hygiene and process control are required. In addition, the use of techniques or materials which could result in any contamination of the final product must be carefully evaluated in terms of product safety and acceptability to the licensing authorities.

This chapter will deal mainly with the isolation and purification of proteins from microbial sources, but the techniques described can, and do, apply equally well to the extraction of materials from plant and animal tissue.

Bacterial protein can be conveniently classified into extracellular and bound and free intracellular proteins. Although the purification of intracellular proteins necessitates some cell breaking process, once this has been achieved then, except that the volumes involved may be smaller than for extracellular proteins, there is no fundamental difference in the purification procedures for those two groups of proteins.

Before describing in detail the individual methods involved in isolation and purification of proteins, it is worth considering the relationship between small-scale and large-scale processes. There are many techniques available, but not all are suitable for large-scale work. This may result from some inherent restriction in the available technique or apparatus or from the increase in time which usually accompanies the transition to a large-scale process. Effective bacterial disruption, for example, can be accomplished by sonication, freezing and thawing, grinding or solid shear. The scale-up of these techniques in many cases is not practical, and in others the result is inadequate. Similarly, the use of ultracentrifugation is a valuable laboratory technique which cannot be conveniently used for large-scale processes.

The definition of large-scale is a matter of opinion, but in the context of this chapter will refer to processes carried out with one or more kilograms of wet cell paste. The term 'extraction' will be used to define the methods by which microorgansims are disrupted, and the terms 'isolation' and 'purification' will relate to the other techniques involved in obtaining a pure, or specified, product (see previous review by Darbyshire (1981) and introductory review by Bucke (1983)).

2.2 EXTRACTION BY CHEMICAL METHODS

2.2.1 Alkali

This method has been used with considerable success in both small- and large-scale extractions of various bacteria. Wade (1968) isolated the therapeutic enzyme L-asparaginase (3,5,1,1-L-asparagine amido-hydrolase) by exposing bacteria to an alkaline pH between 11 and 12.5 for 20 minutes. Salton (1964) has reported the use of alkali treatment in the preparation of plant, fungal and bacterial cell walls, and the degree of hydrolysis was fairly extensive as membrane-bound components such as cytochrome C were also released. The success of alkali treatment is dependent on the stability of the required enzyme to high pH. It has also been suggested that this method may inactivate proteases should they be present, and reduce the possibility of pyrogen contamination of therapeutic enzyme preparations. Such methods have possible applications for rapid inactivation and lysis of genetically engineered mocroorganisms.

2.2.2 Lysozyme and EDTA

Lysozyme is an enzyme, produced commercially from hen egg white, which specifically catalyses the hydrolysis of β-1-4-glycosidic bonds in the mucopeptide moeity of bacterial cell walls. Gram-positive bacteria, which rely to a greater extent on the wall mucopeptide for rigidity than Gram-negative bacteria, are more susceptible to lysozyme than Gram-negative bacteria. However, final rupture of the cell envelope often depends upon the osmotic pressure of the suspending medium once the cell wall has been disrupted. In Gram-negative bacteria the breakdown of the cell wall is rarely achieved by lysozyme alone. The addition of EDTA, which has been shown to cause release of lipopolysaccharide from Gram-negative cell envelopes (Gray & Wilkinson 1965), is a necessary adjunct. It has been suggested that EDTA acts by chelation of divalent cations essential for wall stability, and that this then allows access of lysozyme to act on the mucopeptide layer (Edwards & Noller 1964).

This technique is rarely used for the large-scale extraction of bacterial enzymes because of the relatively high cost of lysozyme, although this is small in comparison with the value of a purified enzyme. Crude hen egg white is much cheaper and often equally effective, and experiments have been carried out with immobilized lysozyme (Dunnill & Lilly 1972). Lysozyme is a very gentle method of cell lysis and is to be favoured if the enzyme is susceptible to the temperature rise during physical lysis procedures. It has recently been used for the lysis of a *Pseudomonas fluorescens* prior to the purification of an aryl acyl hydrolase (Hammond *et al.* 1983).

2.2.3 Detergents
Detergents are either ionic, for example sodium lauryl sulphate (anionic) and cetyldiethylammonium bromide (cationic) or non-ionic, for example Tweens and Tritons. Under conditions of low ionic strength and appropriate pH, detergents will combine with lipoproteins to form micelles (Morton 1955). Therefore, the lipoprotein constituents of biological membranes can be solubilized, or the membranes made permeable. The mechanism of detergent/lipoprotein complex formation is still not fully understood, but is thought to involve electrostatic (salt links) and Van der Waals' forces (Klevens 1950). These complex formations appear to be critically dependent upon pH and temperature.

Ionic detergents are more reactive than non-ionic detergents and can lead to the dissociation of lipoproteins. This in turn can further lead to protein denaturation, precipitation, and, occasionally, to the hydrolysis of peptide bonds (Putnam 1948). For this reason, detergents are not desirable in the extraction of enzymes. Further, salt precipitation of proteins is made difficult in the presence of detergents. However, this can be overcome in many cases by the use of ion exchange chromatography (Tzagoloff & Penefsky 1971).

Nevertheless, detergents do have considerable use in some extraction processes. Morris & Darlow (1971) indicated their usefulness in the fractionation of virus particles; Triton X-100 was used for the large-scale release of cholesterol oxidase from *Nocardia* sp. (Buckland *et al.* 1974), and sodium cholate was used to solubilize pullulanase (pullulan-6-glucan hydrolase), a membrane-bound enzyme from intact cells of *Klebsiella pneumoniae* (Kroner *et al.* 1978). Useful tables giving some details of the extraction of enzymes by detergents can be found in *Methods in Enzymology* 22, p. 229 and in recent volumes of *Amino acids peptides and proteins,* published by the Royal Society of Chemistry as Specialist Periodical Reports (e.g. vol. 14, 1982).

2.2.4 Cold shock
The effect of cold shock (a rapid reduction from normal growth temperature to 0°C) was first observed by Sherman & Albus (1923). The literature on this phenomenon indicates that Gram-negative bacteria are more susceptible than Gram-positive organisms (Hagen 1971). The effect, which only occurs under

specialized conditions, results in a loss of viability and release into the surrounding medium of 260 nm adsorbing material, amino acids, and ATP (Strange & Dark 1962, Strange & Ness 1963, Strange 1964).

However, as a technique for bacterial disruption on a large-scale, cold shock would be difficult to employ. Two of the major limitations imposed by this technique are firstly that cold shock has little or no effect on cell suspensions with a density greater than 10^8/ml (Strange 1964), and secondly bacteria are considerably more susceptible to cold shock when in the exponential growth phase (Gorill & McNeil 1960). On the large scale the size of vessel and sophistication of equipment required to ensure success with this method would be unpractical.

2.2.5 Osmotic shock

Osmotic shock has been used in the extraction of hydrolytic enzymes and binding proteins from a number of Gram-negative bacteria, including *Salmonella typhimurium* and *E. coli* (Neu & Heppel 1967, Heppel 1967, 1969), and in the release of DNA from T (even) bacteriophages by rupture of the protein coat (Anderson 1949). The method involves washing the bacteria in buffer solution to free them from growth medium and then resuspending them in 20% buffered sucrose. The cells are then allowed to equilibrate, resulting in the loss of some internal water, and are then removed from suspension by centrifugation. The cell paste obtained is rapidly dispersed in water at approximately 4°C. It has been suggested that the sudden increase in the osmotic pressure inside the cell causes the release of certain cell constituents (Cedar & Schwartz 1967), but it is clear that this method is in some respects similar to that of cold shock (see above).

Only 4–7% of the total bacterial protein is released by osmotic shock, although in many cases viability is drastically reduced. If the required enzyme is located in the periplasmic region, osmotic shock and cold shock are both worth considering as a 14- to 20-fold increase in purification may be obtainable compared to other disruptive techniques (Charm & Matteo 1971). Osmotic shock has proved to be particularly useful for the extraction of periplasmic aminoglycoside inactivating enzyme, for example kanamicin acetyl transferase from *E. coli* (Haas & Dowding 1975).

It is also a valuable technique from the release of enzymes from marine bacteria, such as luciferase from *Photobacterium fischeri* (Gunsulas-Miguel *et al.* 1972). In the case of marine organisms it is only necessary to resuspend the cells in dilute buffer to achieve lysis as the growth medium contains 30 g/l of NaCl. Osmotic shock is not suitable for enzyme release from Gram-positive bacteria and, in part, this can be explained by the fact that several Gram-positive cocci have internal osmotic pressures in the region of 20 atmosphered (Mitchell & Moyle 1956). The integrity of the cell is maintained by a rigid mucopeptide layer in Gram-positive bacteria, and this may enable them to maintain such internal osmotic pressures, whereas Gram-negative organisms cannot.

2.3 EXTRACTION BY PHYSICAL METHODS

2.3.1 Sonication

The term ultrasonics is used to denote the frequencies above the range of human hearing; such frequencies are 20 kHz and above. When expressed in wavelength, ultrasonic waves in liquid range from 6 to 2.4×10^{-4} cm. This brings them into a similar order of magnitude to those of visible light.

The application of ultrasonics in liquid creates the phenomenon known as cavitation. Areas of compression and rarefaction occur, and cavities, which form in the areas of rarefaction, rapidly collapse as the area changes to one of compression. The bubbles produced in the cavities are thus compressed to several thousand atmsopheres. Subsequent to their collapse, shock waves are formed, and these shock waves are considered to be the destructive element in this procedure. The physical principles involved in viral disruption have been reviewed by Pollard (1953).

Ultrasonic oscillators have been successfully used in a number of extraction procedures. Salton (1953), Bosco (1956), and Ikawa & Snell (1960) have all employed this method for the preparation of bacterial cell walls. The effect on protozoa, *Infusoria,* and common water weed, *Elodea,* is one of nonselective destruction (Bergmann & Hatfield, 1938).

Hugo (1954), and Marr & Cota-Robles (1957) have reviewed this method of cell disintegration in relation to enzyme release. However, it should be noted that some bacteria, for example *Staphylococci* (Datta & Penefsky 1970), are very resistant to ultrasonic disruption. In addition, membrane-bound proteins, such as those related to electron transport systems, may require considerable exposure to achieve solubilization.

The efficiency of ultrasonic treatment depends upon various environmental parameters including pH, temperature (Datta & Penefsky 1970), and ionic strength (Penefsky & Tzagoloff 1971) of the suspending medium as well as the time of exposure. The selection of a set of conditions is essentially an empirical choice, and the conditions will vary with the particular organism as well as the required product.

Although ultrasonic treatment has proved to be a useful and versatile method for laboratory-scale work, its application to the disruption of quantities of bacteria is limited by the difficulty of transmitting sufficient power to large volumes of suspension.

2.3.2 Freezing and thawing

The effects of freezing and thawing on microorganisms are in some ways similar to those observed during cold and osmotic shock. However, in addition to rapid cooling and intracellular and extracellular solute concentration, the formation of intracellular and extracellular ice crystals result in further damage to the cell. Mazur (1969, 1970) reviewed the various physicochemical events leading to injury by freezing. The efficiency of freezing and thawing for general protein release is, however, limited, and less than 10% of the total soluble protein is normally liberated, even from Gram-negative bacteria, by a single operation. Proteins located in the periplasmic space can be released more readily, however, and up to 60% of an R-factor mediated Penicillinase from *E. coli* appeared in the supernatant after the cells had been frozen and thawed once (Melling, unpublished results). Freezing and thawing is often, perhaps unintentionally, an adjunct to other cell-breaking methods since bacterial pastes are often stored at −20°C prior to enzyme extraction and purification.

Freezing and thawing may result in a loss of enzyme activity, and Cowman & Speck (1969) have demonstrated reduced protease activity in various strains of *Streptococcus lactis* after freezing and thawing and subsequent storage at 30°C for ten days. Speck & Cowman (1969) demonstrated that this inactivation is related to changing enzyme structure. On storage at 22°C, after freezing and thawing, the membrane associated protease was found to be in the active monomer and dimer form, but after storage for one day at 30°C the enzyme aggregated to form inactive high molecular weight units. Hanafusa (1969) has demonstrated denaturation and inactivation of ATPase. Catalase, on the other hand, showed no apparent change in structure but considerable loss of activity. These results indicate that considerable problems may be encountered in using this method for the extraction of enzymes. Furthermore, unless bacterial pastes are frozen in small batches, thawing and subsequent resuspension present problems.

2.3.3 Solid shear

Buchner & Hahn (1903) were the first to use this method for breaking microorganisms to release enzymes. They mixed yeast paste and kieselguhr and subjected the mix to hydraulic pressure. The kieselguhr acted as an abrasive, and following application of pressure, zymase was released. A review of these early studies can be found in *Alcoholic fermentations* (Harden 1923).

Hughes (1951) developed the fly press. In this process bacterial paste was mixed with an abrasive, or frozen to −20°C in which case the ice crystals formed acted as an abrasive. The paste was placed in a cylindrical hole in a metal block and a tight-fitting metal plunger was then inserted. A pressure of some 10–15 tons per square inch (150–230 MPa) was applied to the top of this plunger. A very high percentage of the cells in a particulate suspension have been found to be ruptured using this method (Hughes 1951, Hugo 1954, Hunt *et al.* 1959, Hughes 1962). Fraser (1951) successfully obtained enzyme preparations from bacterial endospores by alternating the Hughes Press with freezing and thawing. Eaton (1962) modified the metal block such that it could be separated into two halves, and this facilitated almost complete recovery. However, even using a block with 2 and 3 plungers little more than 30 g of cell paste can be handled, and therefore large-scale work using this method is impossible (Hughes *et al.* 1971).

The X-press was developed which is capable of processing 100 g of *Saccharomyces cerevisiae* per minute (Edebo 1969). Frozen cells are disrupted by passage through a perforated disc, the outlet temperature being approximately −22°C. Rupture of the cells is thought to be caused by the shear forces exerted by the passage of the extruded paste through the small orifice, the shear being aided by crystal formation in the frozen paste. Solid shear cell disruption has proved to be an excellent method for obtaining enzyme protein and bacterial cell wall preparations.

A larger version of the solid-shear disintegrator has been established by Magnusson & Edebo (1976) which is capable of disrupting up to 10 kg yeast paste per hour. Such equipment is perhaps over-complicated for routine use, although the fact that cell breakage occurs at −20°C may be an advantage when heat labile enzymes are being extracted.

2.3.4 Grinding or agitation with abrasives
Booth & Green (1938) developed a grinding mill for the disruption of bacteria. This did not prove satisfactory as both *E. coli* and *B. subtilis* required 2 hours' treatment, and *Sarcina leutea* 4 hours for enzyme release.

Curran & Evans (1942), and King & Alexander (1948) studied the disintegration of bacteria by agitation in the presence of glass beads. King & Alexander (1948) showed the optimum conditions to be (a) beads of 0.13−0.26 mm diameter, (b) 300−500 strokes per minute, and (c) an amplitude of 6 cm.

Mickle (1948) developed an apparatus which has found wide application in the disruption of microorganisms. Its most popular use is in the preparation of microbial cell walls. Walls have been obtained from bacteria (Salton & Horne 1951), bacterial spores, mycobacteria, and fungi (Salton 1964). The apparatus comprises a glass cup with indentations suspended on the end of a variable length arm. Into the glass cup is placed 10 ml of bacterial suspension (10−20 mg of protein/ml) together with an equal volume of Ballatini beads (0.1−0.2 mm in diameter). The cups are then shaken at a frequency of 50 cycles/sec. Overheating can be avoided by working in a cold room; while the required degree of disruption has to be worked out on an empirical basis. Shockman *et al.* (1957) described a method similar in principle to that of Mickle (1948), which could cope with up to 6 g of material per run.

Although the above methods are only suitable for small quantities of bacteria, the same principle has been used for disruption of larger amounts. Rehacek *et al.* (1969) described an apparatus which employed intense stirring of cell suspensions with glass beads. Rehacek (1971) reported a modification to the original incorporating a bead separation system and enabling continuous operation to be used with a flow of bacterial suspension through a chamber filled with glass beads.

The disruptive effect was suggested (Rehacek *et al.* 1969) to result from shear forces produced between velocity gradients; such gradients being due to the stream lines. In addition, collisions between organisms and beads occur as well as grinding between rolling beads with different speeds. The degree of disintegration has been found to depend upon the rate of stirring, the concentration of microorganisms, the concentration and size of beads, and the contact time.

A commercially produced apparatus employing these principles (Dynomill, W. A. Bachofen, Switzerland) has produced a most efficient method for the disruption of some 'tough' bacteria including *Streptococcus mutans, Streptococcus haemolyticus* (Melling *et al.* 1973a), *Staphylococcus aureus,* and *Micrococcus lysodiekticus.* Up to 5 kg per hour of such bacterial paste could be treated using the laboratory scale machine with a 600 ml continuous-flow grinding container. The effect of various parameters on breaking of yeast cells has been discussed by Dunnill & Lilly (1972); and Marfy & Kula (1974) have studied the release of several enzymes from brewers' yeast under various conditions. Rehacek & Schaefer (1977) studied the disintegration of yeast cells in a novel type of agitator in which the agitator discs were placed alternately perpendicularly and obliquely on the drive-shaft, an arrangement which was claimed to be more efficient. Recently Woodrow & Quirk (1982) examined the conditions necessary for the optimum release of a number of bacterial enzymes using a Dyno-Mill fitted with a 600 ml continuous flow chamber. The Dyno-Mill type of cell disruptor has the significant advantage that it can be readily mounted in a safety cabinet when pathogenic or genetically engineered microorganisms are to be disrupted.

2.3.5 Liquid shear

Milner *et al.* (1950) used a liquid shear system for the disruption of chloroplasts. The suspension was passed through a needle valve at pressures up to 20 000 p.s.i. (137 MPa) and a flow rate of 10 ml/min. Under these conditions, rupture of *E. coli* was indicated by an increase in glutamic acid decarboxylase activity. Ribi *et al.* (1959) improved this procedure with an apparatus working at a loading of 40 000 p.s.i. (275 MPa). The outlet temperature was controlled at 15°C with gaseous CO_2, and the apparatus was capable of handling 300 mg of bacterial protein.

More recently the APV Manton Gaulin homogenizer has been used for disrupting bacteria on a continuous basis. The machine is a positive-displacement pump, incorporating an adjustable valve with a restricted orifice. The cells in liquid suspension are passed through the homogenizer at pressures up to 55 MPa. There would appear to be three distinct mechanisms causing disruption of the cells. The cells are pumped through a non-return valve after which they impinge on the homogenizing valve at the selected operating pressure. Subsequent to this, they pass through a narrow channel bounded by the homogenizing valve and an impact ring. It is at this point that the greatest shear forces are exerted. The pressure is then rapidly reduced to atmospheric pressure as the suspension passes into the outlet. This sudden reduction in pressure is the third force exerted on the suspension and having a disruptive effect. Charm & Matteo (1971) indicated the three ways in which the homogenizers can be used; single pass, batch recycle, and continuous recycle and bleed. They also gave equations for determining the number of whole cells remaining, and the overall processing rate of the methods. Protein release on disruption of yeast cells is described by the first order rate equation (Follows *et al.* 1971).

$$\text{Log}(Rm/Rm - R) = KNP^{2.9}$$

R = amount of soluble protein released in g per gm of packed cells
Rm = the maximum amount of soluble protein released
K = a temperature-dependent rate constant
N = the number of times the suspension is passed through the machine
P = operating pressure
2.9 = power variation

Protein release was found to be coincident with cell disruption, although some enzymes were released more slowly, and some more quickly than the mean protein release rate. These differences were not sufficient, however, to achieve significant purification. Although such rate equations may be used to describe protein release under well-defined conditions, in practice their value is limited, and conditions need to be empirically determined.

Gram-negative bacteria are, in general, more easily disrupted then Gram-positive organisms (Charm & Matteo 1971), and the history of the material including such factors as growth conditions (Atkinson 1973), and freezing and thawing, affect the protein release rate. In addition, the rate of liberation of some enzymes may differ from that of the majority of the cell protein. In particular, periplasmic enzymes are more readily liberated (Melling & Scott 1972), and recently a very mild homogenization treatment (3.5 MPa) has been found to remove surface appendages from *E. coli* without disrupting the bacteria.

2.4 ISOLATION AND PURIFICATION (for practical details see Part B, Chapter 2)

2.4.1 Nucleic acid removal
In the context of this chapter nucleic acids will be considered as contaminants, and the interest is in their removal from an extract while maintaining enzyme activity.

Factors such as shear (Davidson 1959, Trim 1959), high pH (Ralph & Bergquist 1967), low ionic strength (Marmur *et al.* 1963), and the presence of nucleases (Gierer 1957), lead to nucleic acid denaturation, and although in the purification of nucleic acids these conditions should be avoided, in enzyme purification such factors can be of positive help.

A second group of techniques which may be more correctly considered to remove nucleic acids are those involving precipitation. They appear to act by complex formation between the negative phosphate residues of the nucleic acid molecules and positively charged groups of the precipitant. The resulting complex may then be removed by centrifugation.

Cetyltrimethyl ammonium bromide. The isolation of bacterial nucleic acids from *Mycobacterium tuberculosis, Mycobacterium phlei,* and *Sarcina lutea* by precipitation with the cationic detergent cetylmethyl ammonium bromide (cetavlon) has been described by Jones (1953). Guerritore & Bellelli (1959) demonstrated that precipitation of a solution of RNA in water was a function of pH, salt concentration, and the nucleic acid/cetavlon ratio. Using ribonucleic acid at a concentration of 500 mg/ml in water at pH 7.0 and a temperature of 20°C they found that the optimum nucleic acid/cetavlon ratio was 1−2. Sodium chloride, sodium sulphate, and sodium citrate all increased the solubility of the nucleic acid at concentrations above 0.2 M; while glycine, glucose, and urea had little or no effect. In these salt solutions the precipitation remained unaffected within the pH range 5.0 to 9.0.

Streptomycin sulphate. Oxenburgh & Snoswell (1965) developed a technique for removing nucleic acids from bacterial extracts using streptomycin sulphate. This method was designed because of difficulty encountered in using other methods. Protamine sulphate and manganese chloride have been used (Heppel 1955) and often proved difficult to reproduce (Snoswell 1963) leading to a considerable loss of protein (Oxenburgh & Snoswell 1965).

The precipitation of nucleic acids is dependent on salt concentration (Moskowitz 1963, Donovic *et al.* 1948), and for this reason Oxenburgh & Snoswell (1965) found it necessary to dialyse their ammonium sulphate extracts thoroughly before adding streptomycin. They found that precipitation was dependent on pH and the ratio of streptomycin/protein. The optimum pH ranged from pH 6−8, and the best results were obtained with a ratio of streptomycin/protein of 1.0. With an extract of *Lactobacillus plantarum* and using a solution of 10 mg/ml protein, 10% (w/v) of streptomycin at pH 7.0, a conductivity of 0.38 mS and a streptomycin/protein ratio of 1.0, only 24% of the protein was lost. This compared favourably with other methods and was preferred because of its reproducibility.

Apart from loss of protein when using streptomycin, there are potential safety problems. Streptomycin itself is toxic, and the use of large quantities may lead to over-exposure of the operators. Also, continual use may encourage streptomycin-resistant bacteria which may cause difficulties if streptomycin needs to be used therapeutically.

Polyethyleneimine. Polyethyleneimine is a long-chain cationic polymer with a molecular weight of about 24 000, and it has been reported (Atkinson & Jack

1973) to be highly effective for the precipitation of nucleic acids. By adding polyethyleneimine to a cell extract following breaking, it was found possible to remove precipitated nucleic acids and cell debris simultaneously by centrifugation. Using a synthetic DNA–RNA mixture with a concentration of polyethyleneimine of 0.294% in the presence of 0.02 M NaCl at pH 7.0 and a temperature of 40°C, 89% of the RNA and 96% of the DNA was precipitated. Precipitation with polyethyleneimine has also been used to selectively purify the restriction endonuclease *Eco*RI (Bingham *et al.* 1977).

With an extract of *E. coli* EM 20031 it was shown that the effectiveness in precipitating nucleic acids decreased in the order polylysine > polyethyleneimine > Cetavlon > streptomycin sulphate > protamine sulphate > $MnCl_2$. Spermine, spermidine, tetramethylenediamine, and $MnCl_2$ were relatively inefficient and resulted in over 75% of the nucleic acids remaining in solution. Extracts of *Micrococcus lysodiekticus* treated with protamine sulphate showed a loss of catalase activity and ineffective removal of RNA. In contrast, treatment of a similar extract using polyethyleneimine resulted in 90% precipitation of the nucleic acid and 70% recovery of catalase and oxaloacetate decarboxylase.

Difficulties have been encountered using polyethyleneimine, and it was found that some enzymes of *Thermus aquaticus* were precipitated by polyethyleneimine, although this could be overcome by increasing the ionic strength. Triosephosphate isomerase from *B. stearothermophilus* has a high affinity for nucleic acids (Atkinson *et al.* 1972), and was co-precipitated with the nucleic acid. In addition, dihydrofolate reductase from *L. casei* complexed with polyethyleneimine causing enzyme inactivation. Activity could, however, be restored by extensive dailysis, during which time the polyethyleneimine was removed. Trim (1959) and Melling & Atkinson (1972) noted depolymerization of nucleic acids when bacterial extracts were prepared by liquid shear, and it was demonstrated that only 85% of nucleic acids were removed in this type of preparation compared with 95% when extracts were obtained by non-shear methods. Finally, one severe limitation on the use of polyethyleneimine is the fact that it may be carcinogenic owing to the presence of unreacted monomer.

Nuclease treatment. This has become the method of choice in recent years. It is well known that bovine pancreatic ribonuclease and deoxyribonuclease can be used to hydrolyse bacterial nucleic acids (Davidson 1965), and deoxyribonuclease treatment of disrupted bacterial suspensions has been reported (Burgess 1969, Burgess *et al.* 1969), although not investigated in detail.

Melling & Atkinson (1972) examined nuclease treatment as a method for the removal of nucleic acids from bacterial suspensions. It was clear that with the two strains of *E. coli* used, nuclease treatment was effective in depolymerizing nucleic acids and hence improving the recovery of supernatant after centrifugation to remove cell debris. However, at this stage there was a relatively high nucleotide content, amounting to some 15–20% of the total protein plus nucleotides. This could be reduced to a very low level by ammonium sulphate precipitation and subsequent dialysis of the precipitate. Since such steps are commonly used during an extraction procedure, it appeared to provide a useful means of reducing nucleic acid contamination.

It was apparent that an organism's own nucleases may provide an often unrecognized 'nuclease treatment', and the results with a ribonuclease-deficient strain of *E. coli* (MRE 600) emphasized the importance of ribonuclease as well as deoxyribonuclease treatment if removal of nucleic acids was to be effective.

The cost of nuclease treatment of 5 kg of bacterial paste was £1, compared with £30 for streptomycin sulphate and £350 for protamine sulphate.

2.4.2 Concentration by precipitation

Ammonium sulphate. Salting out of proteins has been employed for many years and has fulfilled the dual purpose of both purification and concentration of specified proteins. The salt most commonly used is ammonium sulphate, because of its high solubility, lack of toxicity to most enzymes, cheapness and, in some cases, its stabilizing effect on enzymes (Dixon & Webb 1979).

The theory of enzyme fractionation by salting out has been discussed by Green & Hughes (1955) and Dixon & Webb (1961). In concentrated electrolyte solutions the logarithm of the decrease in protein solubility is a linear function of increasing salt concentration (ionic strength). The general equation describing the salting out process is:

$$\log s = B^1 - K^1s \frac{\Gamma}{2}$$

where:

s = solubility of protein in g/l of solution
B^1 = intercept constant
K^1s = (slope of line) salting out constant
Γ = ionic strength (moles/l)

The value B^1 is dependent on the salt used, but will vary with pH, temperature and the nature of the protein in solution. K^1s, on the other hand, is independent of pH and temperature but will vary with the protein in solution and the salt used (Charm & Matteo 1971).

A protein solution of known concentration(s) will begin to precipitate at the ionic strength given by the following equation:

$$\frac{\Gamma}{2} = \frac{B^1 - \log s}{K^1s}.$$

As pointed out by Charm & Matteo (1971), reproducible results can only be obtained if pH, temperature, and protein concentration are kept constant, and under these conditions the salt concentration required to precipitate an enzyme will vary with the concentration of enzyme (Dixon & Webb 1961).

There are some useful points to consider when varying such parameters as pH, temperature, and type of electrolyte. At constant ionic strength the solubility of a protein increases when it is on the acid or alkaline side of its isoelectric point. This means that a reduced ionic strength will precipitate a protein if held at its isoelectric point. In dilute electrolyte solutions proteins are more soluble at temperatures above 0°C. However, in concentrated electrolyte solutions this effect may be reversed. Salts of a high valency produce higher ionic strengths than low valency salts and are therefore more efficient in the precipitation of proteins.

The use of ammonium sulphate in enzyme purification has been described by many authors. The extraction of penicillinase from *E. coli* strain W3310 (Melling & Scott, 1972) is a fairly typical example. Ammonium sulphate (20%, w/v) precipitation of unwanted protein improved the specific activity 4-fold, and increasing the ammonium sulphate concentration to 56% (w/v) to precipitate the penicillinase resulted in a 5-fold purification overall coupled with a reduction in volume from some 60 litres to a paste weighing 2.4 kg.

Organic solvents. The addition of organic solvents to aqueous solutions of proteins reduces the solubility of the proteins by reducing the dielectric constant of the medium. As increasing amounts of organic solvents are added, protein molecules tend to interact more with other protein molecules than with water. Complexing of oppositely charged protein molecules continues until a point is reached at which the protein is precipitated. This occurs as a result of the decrease in the dielectric constant, and hence an increase in the coulombic attractive forces between the unlike charges of protein molecules (Green & Hughes 1955). Addition of salt to an aqueous organic solvent increases the solubility of the protein. Temperatures above 4°C can lead to denaturation in the presence of organic solvents, and for this reason it is desirable to work at temperatures below 0°C, which is feasible because of the depression of the freezing point produced by organic solvents.

As pointed out by Green & Hughes (1955) the effects of organic solvents may be due to factors other than modification of the dielectric constant. It was suggested that the exchange of protein-bound water for molecules of the organic solvent may lead to solvation of the protein. Because the organic solvents used have to be water miscible it was also suggested that the protein might precipitate owing to dehydration. However, these effects are unpredictable and probably depend on the degree of ionization of the protein such that the presence of many charged groups would possibly lead to dehydration, and few to solvation.

Various organic solvents can be employed for the precipitation of proteins, with methanol, ethanol, and propan-2-ol being the most important, although acetone and even diethyl ether can be employed. These latter two solvents have the serious disadvantage of greatly increased flammability.

For large-scale work, the precipitation and concentration of enzymes with organic solvents has not been widely used. Although the degree of precision by which a protein can be precipitated using this method is high, the flammable nature of the materials together with a relatively high cost has made them less attractive than other methods, with the notable exception of serum fractionation. In this process, despite the introduction of other techniques, ethanol precipitation has reigned supreme since its introduction by Cohn et al. (1946). This process has been developed into a highly automated computer-controlled system (Foster & Watt 1980).

High molecular weight polymers. Other organic precipitants which can be used for the fractionation of proteins are the water soluble polymers, of which polyethylene glycol is the most widely used. This has the advantages of being non-toxic, non-flammable, and non-denaturing to proteins. Its mode of action and applications have been described by Kula et al. (1977) and Hao et al. (1980), but it is not widely used, except in the blood processing field.

2.4.3 Concentration by ultrafiltration

Although the technique of ultrafiltration is considered here as a method for concentration, it can nevertheless constitute a valuable means of achieving a purification of some desired proteins by exploiting differences in molecular weight (size) between the required protein and other unwanted proteins.

Ultrafiltration should be considered as one part of a spectrum of filtration techniques ranging from reverse osmosis (molecular weight cut-off up to about 250), ultrafiltration (molecular weight cut-offs between 500 and 300 000), micro-particle filters up to particle filtration, with some overlap between the ultrafiltration and micro-particle ranges.

There are two types of ultrafiltration membrane, the microporous ultrafilter and the diffusive ultrafilter. The microporous ultra-filter can best be described as similar to the conventional idea of a filter. The membrane is rigid with a series of extremely small random probes running through it with an average size of the order of 500–5000 Å (Michaels 1968). Thus, very small molecules will pass through the membrane whereas very large ones will be retained within the structure of the membrane. It is because these filters retain intermediate size molecules that they are susceptible to blockage. This leads to the smaller pores being blocked first, leaving only the larger pores for passage of solute and solvent. The result is that the filter becomes less retentive and does not discriminate so well between different solutes. However, the ultimate result is blockage to most molecular species except the very smallest, and thus a finer pore filter may be formed. To minimize blockage or fouling it is advisable to choose a membrane that has a mean pore size well below that of the solute to be retained. From this it can be deduced that the microporous membranes are useful for the concentration of solutes of high molecular weight ($<1 \times 10^6$). Microporous filters have, however, been largely superseded by membranes of the diffusive type, bacause as well as concentration they are capable of more selective molecular discrimination.

The diffusive ultrafilter is essentially an homogeneous hydrogel membrane through which solvents and solutes are transported by molecular diffusion (Michaels 1968) under the action of a concentration or activity gradient. The transportation of a molecule requires considerable kinetic energy and is a thermally activated process. Michaels (1968) considered that the energy required depends mainly on the dimensions of the diffusing molecule, the density of the membrane polymer matrix, as well as the forces of interaction between the diffusing molecule and the matrix, and the constraints upon free motion of the matrix polymer segments. Thus, for a membrane of high permeability, the conditions would be a highly hydrated gel matrix and a strong specific affinity between polymer and solvent. Conversely, for a membrane of low permeability the conditions and reduced affinity between solute and polymer.

The first recorded demonstration of ultrafiltration was by Schmidt in 1861. In this he partially retained gum arabic, using bovine pericardium as a membrane. The first artificial membrane was constructed by Martin (1896) which he used in the fractionation of snake venoms. Bechold (1907) introduced the term ultrafiltration and also the use of flat reinforced collodion membranes. A breakthrough due to the development of anisotropic diffusive membranes was made between 1961 and 1965 (Loeb *et al.* 1960, Loeb 1961, Blatt *et al.* 1965, Michaels 1965). It is largely owing to these developments that ultrafiltration membranes are such useful tools in the isolation and purification of enzymes on a large scale.

The anisotropic membrane consists of a highly consolidated, but very thin skin (0.1–5 μm) supported by a relatively thick (20 μm–1 mm) porous substructure which acts as a support. Because the active layer is so thin, such membranes exhibit extremely high flux rates, while the porosity of the support is such that any molecule passing through the membrane is not retained by the support (Loeb & Sourirajan 1964). The advantage of this type of membrane, which contains no pores in the conventional sense, is that it does not 'plug' or block within the membrane, and so there is no reduction in solvent permeability at constant pressure.

Solute transport through these membranes is largely governed by Fick's law of diffusion, within the membrane. Thus, solute flux is proportional to the solute concentration gradient which is determined by solute concentrations in the 'concentrate' and 'ultrafiltrate', and is independent of pressure. This means that as solvent flux is increased by an increase in pressure the solute flux is little altered, so that the retention efficiency of the membranes increases (Michaels 1968).

Although the diffusion coefficients of various solutes differ according to the shape and size of the molecules, they tend in general to vary exponentially with the square of the molecular diameter.

At relatively low pressures the solvent activity is nearly directly proportional to hydrostatic pressure. However, as already mentioned, solvent permeability is temperature dependent. This is because temperature influences the activation energy for the diffusion of solvent within the membrane. Temperature will also affect the fluidity of the solvent, but is not significant in this process.

Although high solvent flux rates are readily obtained by using anisotropic membranes, the flow of solvent may be drastically reduced by a phenomenon known as concentration polarization (Loeb & Sourirajan 1964). This arises as a result of the build-up of a layer of rejected solute at the membrane surface which impedes solvent flow. The resistance to flow due to this layer thus increases until a point is reached where the rate of diffusion of solute away from the polarization layer equals the rate of deposition. This phenomenon of concentration polarization also explains why the solvent flux often becomes invariant with an increasing transmembrane pressure drop. Although solvent flux through a membrane is a hydraulic pressure-activated process and should increase with increasing pressure, a point is reached where the rate of build-up of the solute polarization layer is such that flow is impeded and further increases in pressure have no effect.

The design of ultrafiltration equipment has been aimed at keeping such polarization layers to a minimum. In laboratory apparatus this has been achieved with a stirrer to increase the rate of removal of solute from the layer. In larger equipment high flow rates through hollow channels have been employed (Loeb & Sourirajan 1964).

The ultrafiltration characteristics of proteins may vary, depending upon a number of factors. Adsorption of proteins to the membranes has been observed by several workers, and this area has been reviewed by Van Oss (1970). Where this does occur it was suggested that a high solute concentration should be used, presumably to saturate the membrane sites without affecting the recovery too significantly. Melling & Westmacott (1972) studied the ultrafiltration characteristics of a partially purified preparation of penicillinase derived from *E. coli*. Using a 30 000 molecular weight cut-off (anisotropic) membrane the throughput of enzyme varied from 1.6% to 55% depending upon buffer pH and salt concentration. Maximum throughput was at pH 5.1–6.2 which is near the isoelectric point of the enzyme, where penicillinase–protein interactions might be expected to be minimal. At pH 8.0 the throughput could be increased from 1.6% to 20% by increasing the buffer concentration from 1 to 30 mM and was possibly due to some buffer-shielding effect on penicillinase–protein interactions. The addition of EDTA also increased enzyme throughput, while urea had the reverse effect.

Salt concentration, the nature of the salt used, and pH all affected the retention of a β-lactamases from *Staph aureus, B. cereus,* and *E. coli* as well as a β-N-acetylglucosaminidase from *B. subtilis* (Melling *et al.* 1973b, 1978). The case of β-N-acetylglucosaminidase is of particular interest since it was shown that the pattern of liberation of this enzyme from the bacterial cell with increasing NaCl concentration was reflected by the effect of NaCl on enzyme retention by an ultrafiltration membrane (Brewer *et al.* 1973). As well as effects related to pH and salt concentration, specific protein–protein interactions may affect the retention of a protein. Elongation factor Tu has a throughput of 45% with a 100 000 molecular weight cut-off membrane, but addition of factor Ts, which is known to form a complex with Tu reduced throughput to 8% (A. Atkinson, unpublished results).

By manipulating those factors which influence the ultrafiltration characteristics of proteins, considerable increases in purification may be obtained, for example 7-fold for *E. coli* penicillinase (Melling & Westmacott 1972). In view of this, and taking into account the relative ease of scale-up from laboratory processes, ultrafiltration should provide a powerful technique for large-scale work in enzyme purification as well as concentration. Besides its unquestioned role as a means of concentration, surveys of enzyme purification suggest that ultrafiltration for concentration is used in the majority of protein purifications. Ultrafiltration can also be used for the rapid removal of low molecular mass solutes, such as ethanol or water from protein solutions, a process which has been termed diafiltration. In this process the liquid lost as ultrafiltrate is continuously replaced by fresh buffer, thus maintaining a constant volume of solution in the ultrafiltration cell. It can be calculated (Nelsen 1978) that the removal of 99% of the solutes requires the throughput of 4.6 × the initial solution volume. If necessary the throughput volume can be reduced by an initial concentration step. By suitable choice of ultrafiltration apparatus this technique is applicable to any scale of operation ranging from a few millilitres to hundreds (or thousands) of litres. Finally, it is worth noting some of the characteristics which make ultrafiltration a useful method in enzyme work.

(1) A low temperature operation, if desired (thus, by ultrafiltration at a cold room temperature, $-5°C$ to $+5°C$, loss in activity due to exposure at high temperature is avoided and autodigestion can be minimized.
(2) No phase change required.
(3) Operation at low hydrostatic pressure.
(4) Gentle and non-destructive.
(5) No chemical reagents required.
(6) Simultaneous concentration and purification if desired.
(7) Maintenance of constant ionic strength and pH of the concentrate avoiding inactivation of the enzyme.
(8) Economical.

2.4.4 Concentration by freeze-drying

A number of authors (Merryman 1966, Mackenzie 1966, Longmore 1968, Rowe 1970) have reviewed the process of freeze-drying, and accordingly no detailed discussion of the basic process will be attempted here.

Although, in certain instances, freeze-drying may be employed for the concentration of proteins, its use is limited by the fact that unless the salt concentration in solutions is sufficiently reduced, eutectic mixtures may be formed. This may lead to incomplete drying or severe foaming and protein denaturation. Such problems can be avoided by careful control of the temperature of the frozen mass to maintain a temperature below that of the lowest partial eutectic. It may also be necessary to reduce the salt concentration before commencing the freeze-drying process.

In view of the other methods of concentration which are now available for protein solutions, freeze-drying may best be reserved for the preparation of materials for storage or transportation.

Many different proteins have been freeze-dried without denaturation or loss of activity (Schantz et al. 1972, Melling & Scott 1972). However, it is not uncommon to find that some proteins show denaturation and complete loss of biological activity. This appears to be especially true for these proteins which are composed of subunits, and the studies of catalase (Hanfusa 1969) and myosin (Tanford & Louvrien 1962) provide useful information in this respect. Rosenkranz & Scholtan (1971) found, by examination of ORD and CD spectra, that the structure of E. coli asparaginase was significantly increased by freeze-drying. Marlborough et al. (1975) and Hellman et al. (1983) found that freeze-drying (as opposed to freezing and thawing) of Erwinia carotovora asparaginase appears to perturb the tertiary and quaternary structure of the enzyme molecule, with the consequent weakening of the interaction between subunits. However, if the freeze-dried product was reconstituted in buffer solution within the pH range 5.0 to 7.0 these perturbations were reversed. Using buffers with pH values outside these limits resulted in loss of activity and denaturation.

In the light of these observations, the freeze-drying of enzymes should not only be carefully controlled, but it is also clear that it may be necessary to investigate the process in some detail for each enzyme. In particular where materials are for clinical use these requirements are especially important.

2.4.5 Gel chromatography

In this section, as in the others, it is the preparative, rather than the analytical, applications of the technique which will be emphasized. The interest in this technique is based on its ability to fractionate and purify enzymes. From this viewpoint, consideration will be given to the gross structure of gels, and to the general theory which accounts for the separation of biologically active compounds, according to their molecular size and shape.

As pointed out by Ackers (1970), confusion has arisen owing to the number of titles applied to this subject, and it might therefore be of use if these were listed:

(1) Gel filtration (Porath & Flodin 1959).
(2) Molecular sieve filtration (Fasold *et al.* 1961).
(3) Molecular sieve chromatography (Hjerten & Mosbach 1962).
(4) Exclusion chromatography (Pederson 1962).
(5) Restricted diffusion chromatography (Steers & Ackers 1962).
(6) Gel permeation chromatography (Moore 1964).
(7) Gel chromatography (Determann 1964)
(8) Steric chromatography (Haller 1968).

For simplicity and adequacy, Ackers (1970) preferred the title 'Gel chromatography' to describe this subject, and this will be used throughout this section. Should readers wish to review the field in detail, they are referred to Ackers (1970), where a number of useful references can be found.

The observations by Wheaton & Baumann (1953), and later Lathe & Ruthven (1956) suggested the principle of gel chromatography. In 1959 Porath & Flodin demonstrated the separation of proteins by using crosslinked dextrans in a chromatographic process. Commercial pressure then led to development of the Sephadex (Pharmacia, Uppsala, Sweden) series of gels as well as several other products.

The principles of gel chromatography can be illustrated by reference to the Sephadex gels, crosslinked with epichlorohydrin. The dextran used is a soluble polysaccharide of glucose and is a product of the microorganism *Leuconostoc mesenteroides*. The glucose residues are predominantly $\alpha-1-6$ linked glucosidic bonds, the dextran chains being crosslinked via glucose hydroxyl groups, resulting in glycerin ether bonds (Ackers 1970). The degree of crosslinking is controlled by the amount of epichlorohydrin used, which in turn controls the degree of hydration (water regain) within the gel matrix and hence the porosity. The greater the water regain per unit of dry weight of gel, the larger the molecular species which can be fractionated. Conversely, an increase in the amount of dextran matrix present per unit dry weight of gel, leads to a reduction in the size of the molecules which can be fractionated. Thus, if a mixture of high and low molecular weight material is placed on top of a gel column of low water regain (that is, highly crosslinked), the high molecular weight species will pass down the column in the mobile phase, and will be collected first. The low molecular weight species will be retarded owing to its passage through the stationary phase, which is inside the gel.

The process of gel chromatography is the diffusional partitioning of solute molecules between the readily mobile solvent phase and that confined in spaces within the porous particles that make up the stationary phase (Ackers 1970).

The mechanism of gel chromatography may be more easily understood by considering the gel as having a porous bead-like structure. The mobile phase is that liquid external to the gel particles, known as the void volume (V_o), while the stationary phase is that liquid which occupies the interior of the gel particles (V_i). Diffusional exchange of solute takes place between the stationary and mobile phases. The only volume unaccounted for in the total volume of the column (V_t) is that occupied by the gel matrix (V_g). Therefore:

$$V_t = V_o + V_i + V_g .$$

Taking Sephadex G75 as an example, the fractionating range of this material is from 3000 to 80 000 MW and will, to varying degrees, be affected by diffusional partitioning between V_o and V_i. Molecules within this range will be fractionated according to their molecular weight, shape, and size, giving each molecular species its own partition coefficient. This parameter represents the extent to which a given molecule penetrates the stationary phase. Mathematical calculations for partition coefficients have been given by Ackers (1970), but simply, the partition coefficient K_{av} can be obtained from the following equation:

$$K_{av} = (V_e - V_o)/(V_t - V_o) .$$

V_e = volume of solvent required to elute solute from column
V_o = void volume
V_t = total volume of column bed.

The value of K_{av} has been shown, empirically, to be inversely proportional to the log of the molecular weight (Andrews 1970, Brewer 1971). Figure 2.1 illustrates the relationship for globular proteins on Sephadex G75. From this curve it can be seen that three variants of gel chromatography are possible; A to B desalting, B to C molecular fractionation, and C to D exclusion.

We have briefly explained the mechanism of gel chromatography. However, it should be noted that there are four basic theories governing gel chromatography (Laurent et al. 1969), (1) The Exclusion Theory, (2) The Theory of Restricted Diffusion, (3) Partition due to Molecular Surface Forces, and (4) Partition due to the Osmotic Pressure in the Gel. As yet a unifying theory has not appeared.

In recent years a variety of materials other than crosslinked dextrans have been introduced as matrices for gel chromatography. These include polyacrylamide, agarose, polystyrene, crosslinked agarose, agarose–polyacrylamide mixtures, porous glass, and polymethylmethacrylate. Most of these materials offer significant advantages over the conventional gels, particularly when the scale-up of a chromatographic separation is contemplated. In particular, the particles are considerably more rigid, smaller, and uniform in size, resulting in improved resolution and flow properties.

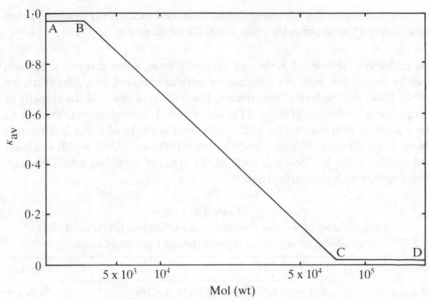

Fig. 2.1 – Relationship between K_{av} and molecular weight for globular proteins (Brewer 1971): by courtesy of *Process Biochemistry*.

2.4.6 Ion exchange chromatography

There are three types of ion exchanger available: the ion exchange resins, ion exchange celluloses, and ion exchange dextran and agarose gels. The latter types have found the widest application, but the ion exchange resins have proved valuable on a more limited scale.

Ion exchange resins. These may be generally described as water insoluble polymers containing either anionic or cationic groups. The nature of the functional group varies, but they are of the general form: cationic exchangers RH^+ and anionic exchangers ROH^- where R represents the resin polymer. Exchange of either H^+ or OH^- ions will take place for ions combined with some weaker acid or base, in this case the proteins, and thus the protein molecules become reversibly attached to the resin.

The functional groups found in cationic exchangers include $-SO_3H$, $-COOH$, and $-OH$ in decreasing order of acidity, while anionic exchangers often contain quaternary ammonium groups $-NH_3^+ Cl^-$ (or other anions).

Resins have the advantages of being stable to the physical stresses usually encountered, with a high capacity for protein adsorption, they sediment rapidly, and when packed into columns allow high liquid flow rates. The high capacity of these materials (up to 5 meq/g) is often disadvantageous when considering their use for protein purification, since the protein may not withstand the harsh conditions necessary for elution from the resins.

Ion exchange resins have nevertheless been used successfully in some protein purifications (Yamanaka & Okunuki 1963, Richardson *et al.* 1971).

Ion exchange celluloses. A variety of charged groups, either cationic or anionic, may be introduced into the cellulose by various chemical processes (Peterson 1970). However, the level of substitution is usually very low, and the capacity of ion exchange celluloses (Table 2.1) is about 1/10th that of resisns. This allows the elution of polyelectrolytes under mild conditions. In addition, substitution above 1 meq/g may lead to solubilization of the cellulose, which is clearly undesirable. Table 2.1 indicates some of the types of celluloses which are available, together with their properties.

Table 2.1
Characteristics of various cellulosic ion exchangers (Peterson 1970)
Reproduced by permission of North Holland Publishing Company

Anion exchangers	Ionizable group		meq/g
AE-cellulose	Aminoethyl-	$-OCH_2-CH_2-NH_2$	0.3−1.0
DEAE-cellulose	Diethylaminoethyl-	$-O-CH_2-CH_2-\overset{x}{\underset{\uparrow}{N}}(C_2H_5)_2$	0.1−1.1
TEAE- cellulose	Triethylaminoethyl-	$-O-CH_2-CH_2-N(C_2H_5)_3$	0.5−1.0
GE-cellulose	Guanidoethyl	$-O-CH_2-CH_2-NH-\overset{NH}{C}-NH_2$	0.2−0.5
PAB-cellulose	*p*-Aminobenzyl-	$-O-CH_2-(C_6H_4)-NH_2$	
ECTEOLA- cellulose	Triethanolamine coupled to cellulose through glycerol and polyglyceryl chains. Mixed groups		0.1−0.5
BD-cellulose	Benzoylated DEAE-cellulose		0.8
BND-cellulose	Benzoylated-naphthoylated DEAE-cellulose		0.8
PEI-cellulose	Polyethyleneimine adsorbed to cellulose or weakly phosphorylated cellulose		0.1
Cation exchangers			
CM-cellulose	Carboxymethyl-	$-O-CH_2-COOH$	0.5−1.0
P-cellulose	Phosphate	$-O-\overset{O}{\underset{OH}{P}}-OH$	0.7−7.4
SE-cellulose	Sulfoethyl-	$-O-CH_2-CH_2-\overset{O}{\underset{O}{S}}-OH$	0.2−0.3

Other ion exchange gels. Cellulose has been the traditional medium for the ion exchange chromatography of proteins. In recent years, however, a number of other ion exchange materials have been introduced, which generally have superior flow properties. Ion exchange Sephadex gels are prepared by introducing DEAE on CM groups into Sephadex G25 or G50 (Pharmacia). Those based on Sephadex G50 have a high capacity for proteins but unfortunately are rather soft and tend to shrink and swell when the pH or ionic strength of the buffer is varied.

The introduction of DEAE or CM groups into crosslinked agarose or acrylic copolymers gives ion exchange Sepharoses (Pharmacia) and Trisacryl (LKB). These two types of exchangers have a high capacity for proteins, and small, uniform bead shaped particles which support high flow rates. They do not deform under the pressures normally encountered in chromatography, and more importantly do not shrink or swell when the pH or ionic strength of the buffer is changed. As they do not even shrink or swell at extremes of pH they can be regenerated in the column, a considerable advantage in large-scale operations. One further advantage of these types of exchanger is that they can be sterilized repeatedly when they are being used for the purification of therapeutic products.

The one major disadvantage of these materials is their high cost, particularly when compared with the more traditional materials like cellulose. However, in many cases the higher flow rates obtainable means that the extra cost is recoverable by a greater production capacity.

Ion exchange chromatography can differentiate between components in a polyelectrolyte mixture on the basis of the charge carried by the individual polyelectrolytes. However, the net charge, charge density, and molecular size of the protein as well as the pH and ionic strength of the solution are all parameters which affect this chromatographic process. The partition coefficient is the expression which relates the number of molecules of a particular electrolyte in the solvent to the number attached to the stationary phase at equilibrium. Therefore, the greater the number of equilibrations a polyelectrolyte species is subjected to the greater is its resolution. Efficiency of separation is also related to an exchanger's capacity; the greater the capacity the larger the number of equilibration, which in turn reduces the differences in partition coefficient required to separate different electrolytes.

In the large-scale isolation of enzymes, however, column chromatographic are usually reserved for the final stages of purification, owing to the large volumes which may have to be handled at the beginning. However, ion exchangers can be of considerable use in the early stages of enzyme isolation when employed in a 'batch' technique. This involves binding of the polyelectrolyte to the exchanger such that at any given time the greater part of the polyelectrolyte is associated with the matrix. Thus it can be said that the polyelectrolyte is immobilized. The conditions generally required for the immobilization to occur are obtained when the electrostatic bonds formed are strong. This requires that the charge on the polyelectrolyte is high and opposite to that of the exchanger. Normally, basic polyelectrolytes (isoelectric point above pH 7.0) are adsorbed onto cationic exchangers (for example, CM cellulose) at a pH below the isoelectric point, and acidic polyelectrolytes (isoelectric point below pH 7.0) onto anionic exchangers (for example DEAE cellulose) at a pH above the isoelectric point. Therefore, it is important to know the isoelectric point of the particular polyelectrolyte. After choosing the exchanger to be used, the conditions of pH and ionic strength are of prime importance. The pH has to be of sufficient extreme to give a high charge on both exchanger and polyelectrolyte; while the ionic strength should be chosen such that the salt helps to preserve the activity and solubility of the enzyme, but is not high enough to compete effectively with the enzyme for sites on the exchanger. If the conditions are chosen correctly it is possible to adsorb the enzyme onto the exchanger, wash with buffer to free it of all non-bound proteins, and then elute the enzyme from the exchanger.

The eluting methods in common use are change of pH, increase in ionic strength, of a combination of both. By changing the pH of a suspension of a cationic exchanger, with an adsorbed basic enzyme, from acid to alkaline the charge of both the exchanger and the enzyme is reduced, thus the electrostatic bonds are broken and the enzyme released. When the ionic strength is increased, competition for the sites occupied by the enzyme is increased, until a point is reached when the enzyme is again released. A combination of both processes is useful when the enzyme is particularly sensitive to changes in ionic strength and pH. Fractional desorption can be advantageous when several polyelectrolytes are adsorbed to the same exchanger. This can often be controlled by varying both pH and ionic strength. Enzymes whose isoelectric points are well removed from neutral are generally more easily purified by using ion echange chromatography by the 'batch' method. With L-asparaginase (Wade 1968), which has an isoelectric point of 8.6, a 100-fold purification is possible using CM cellulose batch-wise.

The batch technique is also of use in removing contaminating polyelectrolytes such as nucleic acids. The increase in protein purification may be marginal, but later steps are often aided by the removal of these large, highly charged polyelectrolytes.

Batch adsorption and subsequent elution may appear to be crude and not to exploit the fractionating ability of the ion exchange technique to the full; however, it has proved a most useful technique which allows simultaneous purification and concentration to be achieved prior to the application of more sophisticated methods.

2.4.7 Affinity chromatography

Affinity chromatography can provide perhaps the most elegant solution to the problem of purifying individual proteins from a complex mixture, However, affinity chromatography, although frequently employed on a laboratory scale, is only rarely employed for the large-scale purification of proteins.

Affinity chromatography relies on the more or less specific interaction of a protein with an immobilized ligand. Such ligands fall into the two main groups:

(a) Those which are specific for the desired protein, in which case the ligand is generally an antibody, a substrate, or substrate analogue or an inhibitor.

(b) Those which can interact with a variety of proteins. Such general ligands may be specific for different classes of enzymes such as 5' AMP, 2',5' ADP, or NAD$^+$, or they interact with a wide range of proteins such as hydrocarbon chains or dyes.

Whatever the nature of the interaction it is essential that it be reversible. A wide variety of eluants can be employed to desorb a protein from an affinity ligand, ranging from nonspecific methods such as an increase in ionic strength or alteration in pH to specific methods using a cofactor or free ligand.

The matrix to which the ligand is attached should exhibit the following basic properties:

(a) Weak reactivity with proteins in general, to avoid nonspecific adsorptions.
(b) Show good solvent flow rates after coupling of ligand.
(c) Contain chemical groups that can be modified such that the matrix is not damaged.
(d) These chemical groups should be plentiful, leading to a high capacity after coupling.
(e) Must be mechanically and chemically stable under the various conditions required for coupling and for elution.
(f) Have a loose porous and hydrophilic network allowing free passage to large macromolecules, preferably be uniform, spherical, and rigid.

A large number of studies on the effects of different matrices on affinity chromatography have been carried out (for example Turkova 1978, Lowe 1979, Scouten 1980). To date the most frequently used matrix is 4% agarose in bead form, which satisfies most of the desired criteria and is readily available.

Coupling techniques. The most frequently employed coupling procedure employs cyanogen bromide under alkaline conditions (Porath *et al.* 1967, Axen *et al.* 1967) to generate a reactive imidocarbonate which can be subsequently reacted with a nucleophilic moiety on the ligand. Detailed studies on the chemistry and quantitation of the cyanogen bromide activation process have been carried out by Kohn & Wilchek (1981, 1982).

Bisoxiranes (bisepoxides) such as 1,4-bis-(2,3-epoxypropoxy-) butane form ether linkages with a variable length hydrophilic spacer between the matrix and ligand (Sundberg & Porath 1974). They can be used to couple ligands containing hydroxyl, amino, or thiol groups. Other coupling reagents have been proposed such as toulene-*p*-sulphonyl chloride (tosyl chloride) or 2,2,2 trifluoro ethane sulphonyl chloride (tresyl chloride) (Nilsson & Mosbach 1980, 1981), which from stable linkages between the matrix and amine-containing ligands. Carbonylating reagents such as 1,1'-carbonyl di-imidazole or 1,1'-carbonyldi-1,2,4-triazole can also be used for the rapid coupling of labile, amine containing ligands (Bethel *et al.* 1981).

Despite the introduction of numerous other coupling reagents, the cyanogen bromide method remains the most widely used, despite the toxicity of cyanogen bromide, because it is relatively simple and gives good coupling yields. The cyanogen bromide technique can also be used to introduce hydrophobic spacer groups into the matrix by coupling 1,6-diamino hexane or 6-amino hexanoic acid to a cyanogen bromide activated matrix.

These coupling techniques can be used to couple small molecules such as cofactors or substrates, or large molecules such as antibodies to a solid support.

The conditions necessary for absorption and elution of a protein from an affinity matrix are extremely variable and have been extensively reviewed (for example Lowe 1979, Scouten 1980).

In general terms absorption takes place under conditions of low ionic strength and neutral pH. Nonspecific elution can be effected by change in pH, ionic strength, or temperature. Biospecific elution, which will often give a higher degree of purification, can be effected by substrate or substrate analogue.

Despite the vast amount of literature relating to affinity chromatography and its undoubted impact on enzyme purification, reports of its use in large-scale purification are rare. Details of enzymes purified by large-scale affinity chromatography are given by Scawen *et al.* (1980) and Janson & Hedman (1982). The main reasons for this lack of acceptance are perhaps the expense of the matrix, the instability of many biological ligands, the complex chemistry involved in coupling them to a matrix, and the low capacity of many such ligands. However, under certain circumstances affinity chromatography can give excellent results, and Robinson *et al.* (1974) have described the large-scale automated purification of β-galactosidase from *E. coli* using immobilised *p*-aminophenyl-β-galacto-pyranoside as the ligand.

One type of affinity chromatography which does not suffer from the disadvantages listed above is dye affinity chromatography using immobilized reactive dyes. The uses of these dyes have been reviewed by Dean & Watson (1979), Lowe *et al.* (1980, 1982), Atkinson *et al.* (1982), Kopperschlager *et al.* (1982). Dye affinity chromatography has recently been employed for the large-scale purification of enzymes as diverse as alcohol dehydrogenase from horse liver (Roy & Nishikawa 1979), protein kinase from pig's liver (Baydoun *et al.* 1982), and glycerokinase from *Bacillus stearothermophilus* (Scawen *et al.* 1983).

The dyes act as pseudo-affinity ligands and can bind a wide range of proteins, often with a surprising degree of specificity. Bound proteins can be eluted non-specifically by increasing the ionic strength or biospecifically by using a substrate or cofactor.

From a practical point of view the reactive dyes offer numerous advantages. They are readily available in large quantities, are inexpensive, and can be coupled to a support under mildly alkaline conditions without the use of toxic chemicals. Once coupled, the resulting bond is stable and the capacity of the conjugate is high. Indeed the capacity of a dye ligand can be some 20 times greater than for a nucleotide ligand in the case of bacterial malate dehydrogenase (Scawen *et al.* 1982).

2.4.8 Non-specific adsorbents

Hydroxyapatite. Hydroxyapatite (HA) has been used in batch techniques for enzyme purification over a considerable period (Levin 1959), and column chromatography using HA was developed by Swingle & Tiselius (1951) The method has only slowly been accepted by protein chemists, and the following reasons for its slow acceptance were given by Bernardi (1971):

(a) Laborious preparation procedure of HA.
(b) The unknown mechanism of interaction of proteins with HA.
(c) The introduction of cellulose ion exchangers in 1956 by Peterson and Sober.

Laboratory-scale preparations of crystalline hydroxyapatite have been described by several authors (Tiselius *et al.* 1956, Main *et al.* 1959, Bernardi 1971). Atkinson *et al.* (1973) published a cheap and simple method for large-scale production of HA which has a uniform crystal size, showed good flow rates and protein binding capacity. Commercially available HA was criticized by these latter authors as being expensive and often unreliable in terms of its crystal structure, flow rate, and binding capacity.

A mechanism of interaction between proteins and HA has been described by Bernardi & Kawasaki (1968) including the influence of the secondary and tertiary structure of proteins in relation to their chromatographic behaviour. Parameters determining resolution in HA columns, and the theoretical basis of chromatography of structurally rigid macromolecules, have been dealt with by Kawasaki & Bernardi (1970a, b), Kawasaki (1970a, b). Bernardi (1971) considered that chromatography on HA should be used as an additional tool in the purification of proteins because the basis of separation is different from that of either ion exchange or gel filtration.

The preparation of hydroxyapatite (Atkinson *et al.* 1973) is based on the conversion of secondary calcium phosphate (brushite) by boiling under alkaline conditions.

$$0.5M\ Na_{1.5}H_{1.5}PO_4(pH6.8)\ +\ 0.5M\ CaCl_2 \xrightarrow{\text{ambient}}$$

$$CaH_2PO_42H_2O \xrightarrow[\ 100°C\ ]{NH_3} Ca_{10}(PO_4)_6(OH)_2$$

$$\text{Brushite} \qquad \text{Hydroxyapatite}$$

The mechanism by which HA separates macromolecules is not fully understood. However, it is considered that exposed Ca^{2+} and PO_4^{3-} sites on the surface of the HA crystals contribute to the fractionation of macromolecules. Acidic and neutral proteins are thought to bind to Ca^{2+} sites on HA; neither the binding nor the elution would appear to be affected by the presence of high concentrations of NaCl, KCl, or $CaCl_2$ (Bernardi 1971, Bernardi et al 1972). The elution of acidic and neutral proteins is generally with low concentration phosphate buffers (30–120 mM) at about pH 6.8. Basic proteins, however, bind to the PO_4^{3-} groups on the HA crystals, adsorption and elution being strongly affected by the presence of NaCl, KCl, and $CaCl_2$. Elution can generally be brought about by these salts and phosphate buffers ranging in concentration from 120 to 230 mM. There are a number of exceptions to these general rules. Phosphoproteins normally require high concentrations of phosphate for elution (Bernardi et al. 1972), while some basic proteins are not eluted by $CaCl_2$, even at very high concentration (Bernardi et al. 1972). Bernardi (1971) also indicated that basic proteins bind strongly at pH 6.8, whereas basic amino acids bind test at pH 7.5. Atkinson et al. (1973) pointed out the unpredictable elution properties of acidic proteins capable of binding phosphate groups, the example being tryptophanyl-tRNA synthetases obtained from B. stearothermophilus and E. coli. These acidic proteins, containing no phosphate groups, can be absorbed onto HA in 1.0 M NaCl and yet require 400 mM phosphate at pH 6.8, comparable to that required for very basic proteins, for elution. It was also noted that enzyme-bound cofactors changed the eluting pattern of dihydrofolate reductase on HA.

Using HA on the large scale for the isolation and purification of enzymes can give good results, indicating that it is useful to the protein chemist. Using a 16 litre column of HA loaded with 500 g of protein extracted from B. stearothermophilus, six enzymes were successfully separated: triose phosphate isomerase, glyceraldehyde-3-phosphate dehydrogenase, NADH dehydrogenase, valyl-, methionyl- and tryptophanyl-tRNA synthetases. Methyionyl-t-RNA from 20 kg of E. coli EM 20031 containing 3.6×10^6 units was also fractionated on HA. The yield was 85%, and a 12-fold increase in purification was obtained (Bruton et al. 1975). Valyl- and tyrosyl-tRNA synthetases, rhodanese, and glyceraldehyde-3-phosphate dehydrogenase from B. stearothermophilus could also be separated by HA chromatography (Atkinson et al. 1979).

Celite. Celite is a diatomaceous earth composed mainly of calcium carbonate with some silicate also present. This material has been used successfully in the purification of two β-lactamases from B. cereus (Davies et al. 1974). The enzymes were adsorbed directly from the culture supernatant at pH 7.0 and eluted at the same pH with a high ionic strength tris-zinc-citrate-NaCl buffer. Although both celite and powdered glass had been used previously (Miller et al. 1965, Kuwabarra 1970) only the β-lactamase I was recovered by these workers, possibly owing to the low pH used for adsorption (pH 4.5) or the higher pH for elution (pH 8.5).

Whether or not celite is applicable to other enzymes is not clear, but the 20-fold purification obtained with the penicillinases suggests that such a cheap and readily available material may bear further investigation.

2.4.9 Hydrophobic interaction chromatography

Hydrophobic chromatography was developed following the observation that certain proteins were unexpectedly retained on affinity gels containing hydrophobic spacer arms (Er-el et al. 1972, Shaltiel & Er-el 1973). This concept was extended, and families of hydrophobic adsorbents prepared using a homologous series of hydrocarbon chains of different lengths (Shaltiel 1974). In practice most proteins can be satisfactorily purified using C-8 or phenyl substituents.

Hydrophobic interactions are strongest at high ionic strength, so adsorption to a hydrophobic matrix can be conveniently performed after a salt precipitation or ion exchange chromatography step. Elution can be effected by altering the solvent ionic strength, pH, or composition or by the use of a dielectric constant modifier such as ethanediol.

The effectiveness of different salts in promoting hydrophobic interaction varies, and both anions and cations can be arranged in a soluphobic series depending on whether they promote hydrophobic interactions (salting-out effect) or disrupt the structure of water (chaotropic effect) and lead to a weakening of hydrophobic interactions (Hjerten et al. 1974) (see Table 2.2).

Table 2.2

Effect of ions on hydrophobic interactions

Increasing salting-out effect

←——————————————————

Cations NH_4^+, Rb^+, K^+, Na^+, Cs^+, Li^+, Mg^{2+}, Ca^{2+}, Ba^{2+}

Anions PO_4^{3-}, SO_4^{2-}, CH_3COO^-, Cl^-, Br^-, No_3^-, ClO_4^-, I^-, SCN^-

——————————————————→

Increasing chaotropic effect

2.4.10 High performance liquid chromatography

One of the most significant advances in chromatography in recent years has been the introduction of high performance liquid chromatography. Originally intended for the separation of small organic molecules that were soluble in non-aqueous solvents, the technique rapidly developed into a form suitable for the separation of larger, biological compounds soluble in aqueous solvents. In recent years, following the development of macroporous supports, the technique has become applicable to molecules as large as proteins and nucleic acids (Hancock & Sparrow 1983, Hearn 1982, Scoble & Brown 1983).

The high efficiency of HPLC columns is derived from the small particle size, with diameters in the range 5–50 μm, one-tenth the diameter of classical soft gels. This small particle size necessitates high pressures in order to give usable flow rates. Thus the particles must be extremely rigid. Originally this entailed the use of silica, but the surface of silica particles is detrimental to proteins. However, silica can be derivatized by reaction with monochloro- or monoalkoxysilanes to give hydrophilic surfaces which are more attractive to proteins. The chemistry of surface modified silica supports has been extensively reviewed in recent years (for example Majors 1981, Unger 1978).

A major disadvantage of silica is that it is unstable above pH 7–8; this has led to the development of rigid, polymeric supports which exhibit a much greater range of pH stability (for example Mono Q, Pharmacia; TSK-PW, Toyo Soda Co. Ltd) which have proved useful for high performance ion exchange chromatography in particular.

All of the mornal chromatographic techniques, gel filtration, ion exchange, hydrophobic and affinity chromatography are available in HPLC (Regnier & Gooding 1980, Regnier 1982, Pearson *et al.* 1982, Larsson *et al.* 1983). However, most of the applications of HPLC to protein purification described so far have been confined to the small scale. Thus μg quantities of angiotensins were separated and purified by high performance ion exchange chromatography (Dizdaroglu *et al.* 1982), and even ng amounts of luteinizing hormone releasing hormone from milk by reverse phase chromatography (Amarant *et al* 1982). However, mg quantities of brain hexokinase have been isolated by high performance ion exchange chromatography (Polakis & Wilson 1982), and lactate dehydrogenase has been purified by high performance liquid affinity chromatography, again on the mg scale (Small *et al.* 1983). HPLC has considerable potential for scale-up because in addition to the use of larger columns, the technique can be readily automated, so as to allow the repeated injection of sample onto the same column.

2.4.11 Electrophoretic techniques

Electrophoresis. In recent years the basic concepts and theories of electrophoresis have been reassessed (for example Hjelmeland & Chrambach 1981, 1982, Rosenfeld 1982).

The principles of electrophoresis are the same irrespective of the particular technique used, and can be simply stated as the acceleration of charged particles in an electric field; this is opposed by friction due to passage through the surrounding medium such that the particles move at a constant speed proportional to their charge. Cations move towards the cathode, and anions towards the anode. The passage of current through a solution is described by Ohm's law: $V = R \times I$ where V = voltage, R = resistance, and I = current, and in practice an increase in current can be effected by increasing the voltage. Since the ions carry the current, migration will be increased with increase in current.

The rate of movement of an ion is expressed as $Q \times N$ where Q = field strength and N = the net charge on the ion. The rate of movement is opposed by the frictional force encountered by the ions as they pass through the solution. Stokes' Law states that spherical particles of radius a, moving at velocity V in a medium with viscosity coefficient η, will be subjected to a frictional force F, such that:

$$F = 6 \pi \eta a V .$$

It can be seen that F will vary with changes in η which can be caused by a change in temperature. Thus the rate of movement of an ion will vary with temperature as well as applied current, molecular size, and shape. This does not imply, however, that molecules showing similar rates of migration are identical.

When an electrical current is passed through a volume element of unit dimensions, heat is generated according to the general formula:

$$w = \frac{xE^2}{A} .$$

w = heat generated
x = specific conductance
E = field of strength
A = mechanical heat equivalent.

The heat produced (which is dependent on the voltage applied) can lead to a number of undesirable effects in preparative as well as analytical techniques.

(a) Changes in density leading to convection currents.
(b) Decrease in viscosity causing increased diffusion, thus affecting resolving power.
(c) Increase in conductivity; however, this can be overcome by using a constant current source.
(d) Evaporation of volatile components leading to changes in pH, ionic strength, and conductivity. Evaporation can also cause water flow from the electrode vessels, resulting in accumulation of buffer salts near the centre of the electric field (only evident when using certain techniques).
(e) Denaturation of thermolabile enzymes.

Another problem is electro-osmosis. This is the transport of water from one electrode to another. It results from the support medium being charged. At neutral and alkaline pH the charge is usually negative, the charged medium therefore attempts to move towards the anode. To balance this, positively charged water molecules (H_3O^+) move to the cathode. This movement of water does not affect the fractionating capacity (Wieme 1965), but can cause loss of water from reservoirs, thus altering the homogeneity of the electrophoretic field.

In general terms, high ionic strength buffers result in a decrease in mobility compared to those of low ionic strength. The former, however, generally lead to more discrete bands or zones. pH has a dramatic effect on the mobility of enzymes in an electric field. Proteins are zwitterions, and as such can normally exist in three ionic forms. The form is dependent on the isoelectric point (pI) of the protein and the pH of the solution. A protein at its pI will be electrically neutral ($NH_3 + RCOO^-$) and will not migrate in an electric field. On the acid side of its pI the species will be positively charged ($NH_3 + RCOOH$) and in alkaline solution will be negatively charged (NH_2RCOO-).

The number of electrophoretic techniques available to the protein chemist for large-scale work are few. In most cases they are avoided because of equipment cost, time involved in fractionation, and that only exceptionally do they result in an improved purification over alternative methods. Of the preparative techniques available only one or two constitute truly large-scale methods.

There are several preparative electrophoretic techniques available, some of which will be mentioned. Using column polyacrylamide gel electrophoresis, up to 10 g of protein can be fractionated (Shaw 1969), while continuous curtain electrophoresis permits the application of 15–20 mg of protein per hour. Forced flow electrophoresis has been used in the purification and concentration of various veterinary vaccines of bacterial origin (Wieme 1965), with a projected capacity of approximately 9 litre/h. Continuous zone electrophoresis was introduced by Grassman & Hannig (1949), and a recent modification of the Philpot continuous preparative free electrophoresis technique (Philpot 1973) has been introduced by Thompson *et al.* (1980). This apparatus has a flow rate of 0.5–1.5 litre/h (50 mg/ml) over a working day, enzyme recoveries varying between 80 and 100%.

Isoelectric focusing. Isoelectric focusing in beds of granulated gel-Sephadex or agarose is an attractive proposition, but the heat generated when using thicker gels limits the protein loading to about 1 g, which is not sufficient for truly large-scale operation (Frey & Radola 1982).

Multimembrane electrodecantation. Although now rarely used in practice, the phenomenon of electrodecantation was observed some years ago by Pauli & Valko (1929), and later Gutfreund (1943) applied the technique to separation of proteins. Polson (1953) developed the multimembrane electrodecantation apparatus which can be used for continuous or batch fractionation of protein mixtures.

The method involves placing a protein solution into a cell containing a number of semipermeable membranes. The pH of the solution is adjusted to the isoelectric point of the protein to be purified. A current is passed through the solution by two sloping electrodes. Because the required protein is at its isoelectric point it remains stationary in the electric field; all other proteins, being charged, will move either towards the cathode or the anode. This movement will cease when the migrating proteins meet the semipermeable membranes. The accumulation of protein at the membrane surface will cause an increase in density, thus it will move to the bottom of the cell. To help this decantation process a temperature gradient is created by the sloping electrodes, which means that the viscosity of the buffer is greater at the bottom of the cell than at the top. Thus the unwanted protein accumulated at the base of the cell and can be drawn off, leaving the required component in the cell. Fresh buffer can be added to fill the cell, and the process can be repeated to give further purification. In the continuous process, a series of such cells are linked together which reduces the carry-over of immobile protein in the outlet stream.

Using this technique, tetanus toxin (Largier 1956), diptheria toxoid (Largier 1957), and human serum globulin (Polson 1953) have been fractionated. Mathies (1952) obtained a 3-fold purification of alkaline phosphatase; trypsin and deoxyribonuclease have been fractionated by Polson (1953, 1956). Considerable loss of enzyme activity was observed in each case. Fleetwood & Milne (1967) obtained antibody and globulin with an 80% recovery and a 650-fold increase in purification. Brummelhuis & Krijnen (1970) have prepared antilymphocyte globulin (ALG), an immunosuppressive agent, from horse antihuman lymphocyte sera. Winchester *et al.* (1971) have used this technique in the purification of bovine 1-D glucosidase, acetylglucosaminidase, and a bacterial protease. The first two showed an approximate 10-fold purification, and the protease proved to be homogeneous by polyacrylamide gel electrophoresis. They also quote a throughput of 0.5–1.0 litre of 2% (w/v) of crude enzyme protein per hour, the only limitation being blockage of the membranes after an unspecified period of continuous running.

Multimembrane electrodecantation appears to be a suitable method for the large-scale purification of enzymes. To date, the method has been used by only relatively few people. However, as information becomes available and the expertise required is more easily obtained, the method may, in the future, become more generally accepted in the large-scale fractionation of enzymes.

2.4.12 Chromatofocusing

Although employed as a chromatographic technique, chromatofocusing is in effect a form of isoelectric focusing carried out on a bed of specialized ion exchange resin eluted with amphoteric buffers (Sluyterman & Wijdnes 1978, Sluyterman & Elgersma 1978, Sluyterman 1982).

Chromatofocusing can combine the resolving power of isoelectric focusing with the convenience of chromatography, thus it is far more suitable to large-scale applications, although all the applications described so far are only small-scale. Chromatofocusing has been used for the fractionation of hexokinases from yeast (Koetzke & Entian 1982) and in the presence of 6M urea for the fractionation of bovine eye-lens crystalin (Bloemendal & Groenwoud 1981).

2.4.13 Aqueous two-phase separation

Aqueous two-phase systems can be created by mixing solutions of polyethylene glycol and dextran or polyethylene glycol and certain salts, especially ammonium sulphate or potassium phosphate. Proteins and cell debris will partition between the two phases. The exact location of a particular protein depends on such parameters as its molecular weight, the concentrations and molecular weights of the polymers, the temperature, pH and ionic strength of the mixture, and the presence of polyvalent salts (Kula 1979). Unfortunately the optimum conditions for favourable partition have to be found empirically for each protein. Two-phase systems can be effectively applied to remove debris from cell homogenates and at the same time achieve a certain degree of purification.

The two phases can in some cases be separated in a settling tank, but a more rapid and efficient separation can be achieved by using a centrifugal separation. As it is often easier to separate ligands of different density than solids from liquids, particularly on a large scale, it is claimed that this technique can be used to great advantage in large-scale enzyme preparations (Kroner et al. 1978). However, despite these apparent advantages aqueous phase partition is only rarely used, perhaps because of the non-recoverable cost of the dextran and polyethylene glycol employed, although a recent report suggests that it is possible to use crude dextran with a considerable saving in cost (Kroner et al. 1982).

Two-phase separation has been used for the separation and large-scale purification of pullulan-6-glucan hydrolase and 1,4-α-glucan phosphorylase from 5 kg Klebsiella pneumoniae cell paste (Hustedt et al. 1978), and RNA polymerase and glutamine synthetase from E. coli (Takahashi & Adachi 1982).

By coupling a ligand to polyethylene glycol it is possible to increase the partition coefficient of a protein in favour of the polyethylene glycol phase. Thus by using the dye Cibacron blue it was possible to achieve a 58-fold purification of yeast phosphofructokinase in two steps (Koperschlager & Johansson 1982).

REFERENCES

Ackers, G. K. (1970) Adv. in Protein Chemistry 23 343–446.
Amarant, T., Fridkin, M., & Koch, Y. (1982) Eur. J. Biochem. 127 647–650.
Anderson, T. F. (1949 Bot. Rev. 15 464–505.
Andrews, P. (1970) Meth. Biochem. Anal. 18, 1–53.
Atkinson, A. (1973) Process Biochem. 8 9–13.
Atkinson, A., Bradford, P. A. & Selmes, I. P. (1973) J. Appl. Chem. Biotechnol. 23 517–529.
Atkinson, A. & Jack, G. (1973) Biochim. Biophys. Acta 308 41–52.
Atkinson, A., Phillips, B. W., Callow, D. S., Stones, W. R. & Bradford, P. A. (1972) Biochem. J. 127 63–64.
Atkinson, A., Banks, G. T., Bruton, C. J., Comer, M. J., Jakes, R., Kamalagharon, T., Whitaker, A. R. & Winter, G. (1979) J. Appl. Biochem. 1 247–258.
Atkinson, A., McArdell, J. E., Scawen, M. D., Sherwood, R. F., Small, D. A. P., Lowe, C. R. & Bruton, C. J. (1982) In Affinity Chromatography and Related Techniques, pp. 399–410. Ed. by T. C. J. Gribnau, J. Visser and R. F. J. Nivard. Elsevier, Amsterdam.

Axen, R., Porath, J. & Ernbach, S. (1967) *Nature* **214** 1302–1304.
Baydoun, H., Hoppe, J., Freist, W. & Wagner, K. G. (1982) *J. Biol. Chem.* **257** 1032–1036.
Bechold, H. (1907) *Kolloid Z,* **2** 3 and 33.
Bergmann, L. & Hatfield, H. S. (1938) *Ultrasonics* Wiley, New York.
Bernardi, G. (1971) *Methods Enzymol.* **22** 325–339.
Bernardi, G. and Kawasaki, T. (1968) *Biochim. Biphys. Acta* **160**, 301–310.
Bernardi, G., Giro, M. G. & Gaillard, C. (1972) *Biochim. Biophys. Acta* **278** 409–420.
Bethel, G. S., Ayres, J. S., Hearn, M. T. W. & Hancock, W. S. (1981) *J. Chromatogr.* **219** 353–359.
Bingham, A. H. A., Sharman, a. F. & Atkinson, A. (1977) *FEBS Lett.* **2** 250–256.
Blatt, W. F., Feinberg, M. P., Hoppenberg, H. B. & Saravis, C. A. (1965) *Science* **150** 224–225.
Bloemendal, H. & Groenwoud, G. (1981) *Anal. Biochem.* **117** 327–329.
Booth, B. H. & Green, D. E. (1938) *Biochem. J.* **32** 855–861.
Bosco, G. (1956) *J. Inf. Dis.* **99** 270–274.
Brewer, J. (1971) *Process Biochem.* **6** 39–42.
Brewer, S. J., Berkeley, R. C. W. & Melling, J. (1973) *Biochem. Soc. Trans.* **1** 1093–1095.
Brummelhuis, H. G. J. & Krijnen, H. W. (1970) in *Inter. Symp. on Anti-Lymp. Serum. Versailles Symp. Series Immuno. Biol. Standard* vol. 16, pp. 145–152 Karger.
Bruton, C. J., Jakes, R. & Atkinson, A. (1975) *Eur. J. Biochem.* **59** 327–333.
Buchner, E. & Hahn, H. (1903) *Die Zymase Garung.* Oldenburg, Munich.
Bucke, C. (1983) in *Principles of Biotechnology* pp. 151–170. (Ed. by A. Wiseman), Surrey University Press.
Buckland, B. C., Richmond, W., Dunnill, P. & Lilly, M. D. (1974) In *Industrial Aspects of Biochemistry* FEBS vol. 30, pp. 65–79. Ed. by B. Spencer. North Holland Publishing Co., Amsterdam and London.
Burgess, R. R. (1969) *J. Biol. Chem.* **244** 6160–6167.
Burgess, R. R., Travers, A. A., Dunn, J. J. & Baurz, E. K. F. (1969) *Nature* **221** 43–46.
Cedar, H. & Schwartz, J. H. (1967) *J. Biol. Chem.* **242** 3753–3755.
Charm, S. E. & Matteo, C. C. (1971) *Methods Enzymol.* **22** 476–556.
Cohn, E. J., Strong, L. E., Hughes, W. L., Mulford, D. J., Ashworth, J. N., Metin, M. & Taylor, H. L. (1946) *J. Am. Chem. Soc.* **68** 459–475.
Cowman, S. E. & Speck, M. L. (1969) *Cryobiology* **5** 291–299.
Curran, H. R. & Evans, F. R. (1942) *J. Bacteriol.* **43** 125–138.
Darbyshire, J. (1981) In *Topics in Enzyme and Fermentation Biotechnology.* Vol. 5, pp. 147–186 Ed. by A. Wiseman. Ellis Horwood.
Davidson, J. N. (1965) In *The Biochemistry of the Nucleic Acids,* 5th Ed. Methuen & Co. Ltd.
Davidson, P. F. (1959) *Proc. Nat. Acad. Sci. U.S.A.* **45** 1560–1568.
Davies, R. B., Abraham, E. P. & Melling, J. (1974) *Biochem. J.* **143** 115–127.
Dean, P. D. G. & Watson, D. H. (1979) *J. Chromatogr.* **165** 301–319.
Determann, H. (1964) *Agnew. Chem.* **76** 635–644.
Dixon, M. & Webb, E. C. (1961) *Adv. in Protein Chem.* **16** 197–219.
Dixon, M. & Webb, E. C. (1979) in *Enzymes,* Longmans, London.
Dizdaroglu, M., Krutzsch, H. C. & Simic, M. G. (1982) *Anal. Biochem.* **123** 192–193.
Dinovick, R., Bayan, A. P., Canales, P. & Pansy, J. (1948) *J. Bacteriol.* **56** 125–137.
Dunnill, P. & Lilly, M. D. (1972) *Biotechnol. Bioeng. Symp.* No. 3, p. 103.
Dunnill, P. & Lilly, M. D. (1972) in *Enzyme Engineering,* pp. 101–113.
Eaton, N. R. (1962) *J. Bacteriol.* **83** 1359–1360.
Edebo, L. (1969) In *Fermentation Advances,* pp. 249–271. Ed. by D. Perlman. Academic Press, New York.
Edwards, C. D. & Noller, E. C. (1964) *Proc. Okla, Acad. Sci. U.S.A.* **44** 196–200.
Elsworth, R., Meaking, L. R. P., Pirt, S. J. & Capell, G. H. (1956) *J. Appl. Bact.* **19** 264–278.
Er-el, Z., Zaidenzaig, Y. & Shaltiel, S. (1972) *Biochem. Biophys. Res. Commun.* **49** 383–390.
Fasold, H., Gundlach, H. G. & Turba, F. (1961) In *Chromatography,* p. 406. Ed. by E. Heftman. Reinhold, New York.
Fleetwood, J. G. & Milne, G. R. (1967) *Protides of Biolg. Fluids* **15** 545–550.
Follows, M., Hetherington, P. J., Dunnill, P. & Lilly, M. D. (1971) *Biotechnol. Bioeng.* **XIII** 549–560.
Foster, P. & Watt, J. G. (1980) In *Methods of Plasma Protein Fractionation,* pp. 17–31. Ed. by J. M. Curling. Academic Press.
Fraser, D. (1951) *Nature* **167** 33–34.

Frey, M. D. and Radola, B. J. (1982) *Electrophor,* **3** 216–
Gierer, A. (1957) *Nature* **179** 1297–1299.
Gorill, R. H. & McNeil, E. M. (1960) *J. Gen. Microbiol.* **22** 437–442.
Gospodarnadez, D., Ge-Mung Lut & Cheng, J. (1982) *J. Biol. Chem.* **257** 12266–12276.
Grassman, W. & Hannig, K. (1949) DBP505399 May 24th.
Gray, G. W. & Wilkinson, S. G. (1965) *J. Gen. Microbiol.* **39** 385–399.
Green, A. A. and Hughes, W. L. (1955) *Methods Enzymol.* **1** 72–90.
Guerritore, D. & Bellelli, L. (1969) *Nature* **184** 1638.
Gunsalus-Miguel, A., Meighen, E. A., Nicoli, M. S. & Nealson, K. H. (1972) *J. Biol. Chem.* **247** 398–404.
Gurfreund, H. (1943) *Biochem. J.* **37** 186–188.
Haas, M. J. & Dowding, J. E. (1975) *Methods Enzymol.* **43** 611–628.
Hagen Per-Otto (1971) In *Inhibition and Destruction of the Microbiol. Cell,* pp. 39–70. Ed. by W. B. Hugo. Academic Press, London.
Haller, W. (1968) *J. Cromatog.* **32** 676–684.
Hammond, P. M., Price. C. P. & Scawen, M. D. (1983) *Eur. J. Biochem.* **132** 651–655.
Hanafusa, N. (1969) In *Freezing and Drying of Microorganisms,* pp. 117–129. Ed. by T. Nei. University of Tokio Press, Tokio.
Hancock, W. S. & Sparrow, J. T. (1983) In *High Performance Liquid Chromatography, Advances and Perspectives,* vol. 3, pp. 50–87. Ed. by C. Horvath. Academic Press, London and New York.
Hao, Y. L., Ingham, K. C. & Wickerhauser, M. (1980) In *Methods of Plasma Protein Fractionation,* pp. 57–74. Ed. by J. M. Curling. Academic Press, London and New York.
Harden, A. (1923) *Alcoholic Fermentation.* Longmans, Green and Co., London.
Hearn, M. T. W. (1982) In *Advances in Chromatography,* vol. 20, pp. 2–82. Ed. by J. C. Giddings. Marcel Dekken, New York and Basel.
Hellman, K., Miller, D. S. & Cammack, K. A. (1983) *Biochim. Biophys. Acta* **749,** 133–142.
Heppel, L. (1955) *Methods Enzymol.* **1** 137–138.
Heppel, L. A. (1967) *Science* **156** 1451–1455.
Heppel, L. A. (1969) *J. Gen. Physiol.* **54** 955.
Hjelmeland, L. M. & Chrambach, A. (1981) *Electrophor.* **2** 1–.
Hjelmeland, L. M. & Chrambach, A. (1982) *Electrophor.* **3** 9–.
Hjerten, S. & Mosbach, R. (1962) *Anal. Biochem.* **3** 109–118.
Hjerten, S., Rosengren, J. & Pahlman, S. (1974) *J. Chromatogr.* **101,** 281–.
Hughes, D. E. (1951) *Brit. J. Exptl. Path.* **32** 97–109.
Hughes, D. E. (1962) *J. Gen. Microbiol.* **29** 39–46.
Hughes, D. E., Wimpenny, J. W. T. & Lloyd, D. (1971) In *Methods in Microbiology,* vol. 5B, pp. 1–54. Ed. by J. R. Norris and D. W. Ribbons. Academic Press, London and New York.
Hughes, W. L. (1954) In *The Proteins* **2,** Part B. Ed, by H. Neurath and K. Baily. Academic Press, London.
Hugo, W. B. (1954) *Bacteriol. Revs.* **18** 87–105.
Hunt, A. L., Rodgers, A. & Hughes, D. E. (1959) *Biochim. Biphys. Acta* **34** 354–372.
Hustedt, H., Kroner, K. H., Stach, W. & Kula, M-R. (1978) *Biotechnol. Bioeng.* **20** 1989–2005.
Ikawa, M. & Snell, E. E. (1960) *J. Biol. Chem.* **235** 1376–1382.
Janson, J. C. & Hedman, P. (1982) *Adv. Biochem. Eng.* **25** 43–99.
Jones, A. S. (1953) *Biochim. Biophys. Acta* **10** 607–612.
Kawasaki, T. (1970a) *Biopolymers* **9** 277–289.
Kawasaki, T. (1970b) *Biopolymers* **9** 291–306.
Kawasaki, T. & Bernardi, G. (1970a) *Biopolymers* **9** 257–268.
Kawasaki, T. & Bernardi, G. (1970b) *Biopolymers* **9** 269–276.
King, H. K. & Alexander, H. (1948) *J. Gen. Microbiol.* **2** 315–324.
Klevens, H. B. (1950) *Chem. Revs.* **47** 1–74.
Kohn, J. & Wilchek, M. (1981) *Anal. Biochem.* **15** 375–382.
Kohn, J. & Wilchek, M. (1982) *Enzyme Microb. Technol.* **4** 161–163.
Kopperschlager, G. & Johansson, G. (1982) *Anal. Biochem.* **124** 117–124.
Kopperschlager, G., Bohme, H. J. & Hofman, E. (1982) *Adv. Biochem. Eng.* **25** 101–138.
Kopetzki, E. & Entian, K-D. (1982) *Anal. Biochem.* **121** 181–185.
Kroner, K. H., Hustedt, H., Granda, S. & Kula, M. R. (1978) *Biotechnol. Bioeng.* **20** 1967–1988.
Kroner, K. H., Hustedt, H. & Kula, M. R. (1982) *Biotechnol. Bioeng.* **24** 1015–1045.
Kula, M. R. (1979) In *Applied Biochemistry and Bioengineering.* Ed. by L. B. Wingard, E. Katzir-Katchalski, E. & L. Goldstein. Vol. 2, Academic Press, London and New

Kula, M. R., Honig, W. & Foellmer, H. (1977) In *Proceedings of International Workshop on Technology for protein Separation and Improvement of Blood Plasmid Fractionation*. pp. 361–371. Ed. by H. E. Sandberg. National Institutes of Health Publication No. 78–1422, Washington D.C.

Kuwabara, S. (1970) *Biochem. J.* **118** 457–465.

Largier, J. F. (1956) *Biochim. Biophys. Acta* **21** 433–438.

Largier, J. F. (1957) *J. Immunol.* **79** 181–186.

Larsson, P. O., Glad, M., Hansson, L., Mansson, M. O., Ohlson, S. & Mosbach, K. (1983) *Advances in Chromatography* **21** 41–84.

Lathe, G. H. & Ruthven, C. R. J. (1956) *Biochem. J.* **62** 665–674.

Laurent, T. C., Obrink, B., Hellsing, K. & Wasteson, A. (1969) *Prog. in Sep. and Purification* **2** 199–218.

Leloir, L. F. & Goldenberg, S. H. (1962) *Meth. in Enzymol.* **5**, 45–147.

Leloir, L. F., Rongine De Fekete, M. A. & Cardini, E. C. (1961) *J. Biol. Chem.* **236** 636–641.

Levin, O. (1959) *Methods Enzymol.* **5** 27–32.

Linko, M., Walliander, P. & Linko, Yu-yen (1973) *F.E.B.S. Abs.* No. 59. Dublin.

Linggood, F. V., Mattews, A. C., Pinfield, S., Pope, C. G. & Sharland, T. R. (1955) *Nature* **176** 1128.

Loeb, S. (1961) *UCLA Engineering Report,* 42–61.

Loeb, S., Sourirajan, S. & Yuster, S. T. (1960) *Chem. Eng. News* **38** (15) 64.

Loeb, S. & Sourirajan, S. (1964) US Patent 3, 133–132.

Longmore, A. P. (1968) *Proc. Fourth Int. Vac. Congr.* Pt. I, 79–82.

Lowe, C. R. (1979) *An Introduction to Affinity Chromatography, Laboratory Techniques in Biochemistry and Molecular Biology*. Ed. by T. S. Work & E. Work. North-Holland Publishing Co., Amsterdam.

Lowe, C. R., Small, D. A. P. & Atkinson, A. (1980) *Int. J. Biochem.* **13** 33–40.

Lowe, C. R., Clonis, Y. D., Goldfinch, M. J., Small, D. A. P. & Atkinson, A. (1982) In *Affinity Chromatography and Related Techniques*, pp. 389–398. Ed. by T. C. J. Gribnau, J. Visser & R. J. F. Nivard. Elsevier, Amsterdam.

Mackenzie, A. P. (1966) *Bull. Parental Drug. Ass.* **20** 101–129.

Magnusson, K. E. & Edebo, L. (1976) *Biotechnol. Bioeng.* **18** 975–986.

Main, R. K., Wilkins, M. J. & Cole, L. J. (1959) *Science* **129** 331–332.

Majors, R. E. (1981) In *High Performance Liquid Chromatography, Advances and Perspectives*. Vol. 1. Ed. by C. Horvath. pp. 00–00. Academic Press, London and New York.

Marfy, F. & Kula, M. R. (1974) *Biotechnol. Bioeng.* **16** 623–634.

Marlborough, D. I., Miller, D. S. & Cammack, K. A. (1975) *Biochim. Biophys. Acta* **386** 376–589.

Marmur, J., Roland, R. & Schildkraut, C. L. (1963) *Prog. Nucleic Acid Res.* **1** 231–300.

Marr, A. G. & Cota-Robles, E. H. (1957) *J. Bacteriol.* **74** 79–86.

Martin, C. J. (1896) *Proc. Roy. Soc. New South Wales,* April 5th.

Mathies, J. C. (1952) *Science* **115** 144–146.

Mazur, P. (1969) *Ann Rev. Plant Physiol.* **20** 419–448.

Mazur, P. (1970) *Science* **168** 939–949.

Melling, J. & Atkinson, A. (1972) *J. Appl. Chem. Biotechnol.* **22** 739–744.

Melling, J., Berkeley, R. C. W., Caulfield, M. P. & Cammack, K. A. (1978) in Protein: Structure Function and Industrial Applications. Ed. by E. Hoffman *et al.*, pp. 453–462. Proc. 12th FEBS meeting, Dresden.

Melling, J., Evans C. G. T., Harris-Smith, R. & Stratton, J. E. D. (1973a) *J. Gen. Microbiol.* **77** XVIII.

Melling, J., Downs, D. J. & Brewer, S. J. (1973b) *J. Appl. Chem. Biotechnol.* **23** 166.

Melling, J. & Scott, G. K. (1972) *Biochem. J.* **130** 55–66.

Melling, J. & Westmacott, D. (1972) *J. Appl. Chem. Biotechnol.* **22** 951–958.

Merryman, H. T. (1966) In *Cryobiology*, pp. 609–663. Ed. by H. T. Merryman. Academic Press, New York.

Michaels, A. S. (1965) *Ind. Eng. Chem.* **57** (10), 32–40.

Michaels, A. S. (1968) *Ultrafiltration*. In *Progress in Separation and Purification*. Ed. by E. S. Perry. Wiley, New York.

Mickle, H. (1948) *J. Roy. Microscop. Soc.* **68** 10–12.

Miller, G., Bach, G. & Markus, Z. (1965) *Biotechnol. Bioeng.* **7** 517–528.

Milner, H. W., Lawrence, N. S. & French, C. S. (1950) *Science* **111** 633–634.

Mitchell, P. & Moyle, J. (1956) *Soc. Gen. Microbiol. Symp. No. 6*, pp. 150–180. Cambridge University Press, London.

Moore, J. C. (1964) *J. Polymer Sci.* **B2** 835–843.
Morris, E. J. & Darlow, H. M. (1971) In *Inhibition and Destruction of the Microbial Cell*, pp. 687–710. Ed. by W. B. Hugo. Academic Press, London.
Morton, R. K. (1955) *Methods Enzymol.* **1** 25–51.
Moskowitz, M. (1963) *Nature* **200** 335–337.
Nelsen, L. (1978) In *Proceedings of the International Workshop on Technology for Protein Separation and Improvement of Blood Plasma Fractionation*, pp. 133–145. Ed. by H. E. Sandberg. National Institutes of Health, Publication No. 78–1422, Washington D.C.
Neu, H. C. & Heppel, L. A. (1967) *J. Biol. Chem.* **140**, 3685–3692.
Nilsson, K. & Mosbach, K. (1980) *Eur. J. Biochem.* **112** 397–402.
Nilsson, K. & Mosbach, K. (1981) *Biochim. Biophys. Res. Commun.* **120** 449–457.
Oxenburgh, M. S. & Snoswell, A. N. (1965) *Natire* **203** 1416–1417.
Pauli, W. & Valko, E. (1929) *Electrochemie der Kolloide*. Springer Verlag, Vienna.
Pearson, J. D., Lin, N. T. & Regnier, F. E. (1982) *Anal. Biochem.* **124** 217–230.
Pederson, K. O. (1962) *Arch. Biochem. Biophys. Suppl.* **1** 157–168.
Penefsky, H. S. & Tzagoloff, A. (1971) *Methods Enzymol.* **22** 204–219.
Peterson, E. A. (1970) *Laboratory Techniques in Biochem. and Molecular Biology*, pp. 228–397. Ed. by T. S. Work & E. Eork. North Holland Pub. Co., Amsterdam.
Philpot, J. S. L. (1973) In *Methdological Developments in Biochemistry*, Vol. 2, p. 81. Ed. by E. Reid. Longmans, London.
Polakis, P. G. & Wilson, J. E. 1982) *Biochem. Biophys. Res. Commun.* **107** 937–943.
Pollard, E. C. (1953) In *The Physics of Viruses*. Academic Press, New York.
Polson, A. (1953) *Biochim. Biophys. Acta* **11** 315–325.
Polson, A. (1956) *Biochim. Biophys, Acta* **22** 61–65.
Porath, J., Axen, R. & Ernback, S. (1967) *Nature* **215** 1491–1492.
Porath, J. & Flodin, P. (1959) *Nature* **1657–1659**.
Putnam, F. W. (1948) *Adv. in Protein Chemistry* **4** 80–122.
Ralph, R. K. & Bergquist, P. L. (1967) *Meth. in Virology* **11** 465–538.
Regnier, F. E. (1982) *Anal. Biochem.* **126** 1–7.
Regnier, F. E. & Gooding, K. M. (1980) *Anal. Biochem.* **103** 1–25.
Rehacek, J. (1971) *Experimentia* **27** Fasc 9.
Rehacek, J., Beran, K. & Bicik, V. (1969) *Appl. Microbiol.* **17** 412–466.
Rehacek, J. & Schaeffer, J. (1977) *Biotechnol. Bioeng.* **19** 1523–1534.
Ribi, E., Perrine, T., List, R., Brown, W. & Goode, G. (1959) *Proc. Soc. Exptl. Biol. Med.* **100** 647–649.
Richardson, M., Richardson, D., Ramshaw, J. A. M., Thompson, E. W. & Boulter, D. (1971) *J. Biochem.* (Tokyo) **69** 811–813.
Robinson, P. J., Wheatley, M. A., Johnson, J. Ch. & Dunnill, P. (1974) *Biotechnol. Bioeng.* **16** 1103–1112.
Rosenfeld, L. (1982) *Origins of Clinical Chemistry. The Evolution of Protein Analysis*. Academic Press, New York.
Rosenkranz, G. & Scholtan, W. (1971) Hoppe-Seyer's *Z. Physiol. Chem.* **352** 1081–1090.
Rowe, T. W. G. (1970 In *Current Trends in Cryobiology*, pp. 61–138. Ed. by A. U. Smith. Plenum Press, New York.
Roy, S. K. & Nishikawa, A. H. (1979) *Biotechnol. Bioeng.* **21** 775–785.
Salton, M. R. J. (1953) *J. Gen. Microbiol.* **9** 512–523.
Salton, M. R. J. (1964) In *The Bacterial Cell Wall*. Elsevier, Amsterdam.
Salton, M. R. J. & Horne, R. W. (1951) *Biochim. Biophys. Acta* **7** 177–179.
Scawen, M. D., Atkinson, A. & Darbyshire, J. (1980) In *Applied Protein Chemistry*, pp. 281–324. Ed. by R. A. Grant. Applied Science Publishers, Essex, U.K.
Scawen, M. D., Darbyshire, J., Harvey, M. J. & Atkinson, T. (1982) *Biochem. J.* **203** 699–705.
Scawen, M. D., Hammond, P. M., Comer, M. J. & Atkinson, T. (1983) *Anal. Biochem.* **132** 413–417.
Schantz, E. J., Ressler, W. G., Woodburn, M. J., Lunch, J. M., Jacoby, H. M., Scoble, H. A. & Brown, P. R. (1983) In *High Performance Liquid Chromatography, Advances and Perspectives*, vol. 3, pp. 2–49. Ed. by C. Horvath. Academic Press, London and New York.
Schmist, A. (1861) *Poggendorff Ann.* **114** 337.
Scoble, H. A. and Brown, P. R. (1983) In *High Performance Liquid Chromatography, Advances and Perspectives*, vol. 3, pp. 2–49. Ed. by C. Horvath. Academic Press, London and New York.
Scouten, W. H. (1980) *Affinity Chromatography, Bioselective Adsorption on Inert Matrices*.

John Wiley and Sons, New York.
Shaltiel, S. (1974) *Methods Enzymol.* **34** 126–.
Shaltiel, S. & Er-el, Z. (1973) *Proc. Nat. Acad. Sci. U.S.A.* **70** 778–781.
Shaw, D. J. (1969) *Electrophoresis.* Academic Press, London.
Sherman, J. M. & Albus, W. R. (1923) *J. Bact.* **8** 127–139.
Shockman, G. D., Kolb, J. J. & Toennies, G. (1957) *Biochim. Biophys. Acta* **24** 203–204.
Silverman, S. J., Gorman, J. C. & Spero, L. (1972) *Biochemistry* **11** 360–366.
Sluyterman, L. A. A. E. & Elgersma, O. (1978) *J. Chromatogr.* **150** 17–30.
Sluyterman, L. A. A. E. & Wijdnes, J. (1978) *J. Chromatogr.* **150** 31–44.
Sluyterman, L. A. A. E. (1982) *Trends. Biochem. Sci.* **7** 168–170.
Small, D. A. P., Atkinson, T. & Lowe, C. R. (1983) *J. Chromatogr.* **266** 151–156.
Snoswell, A. M. (1963) *Biochim. Biophys. Acta* **77** 7–19.
Speck, M. L. & Cowman, R. A. (1969) In *Freezing and Drying of Microorganisms,* pp.
    39–51. Ed. by Nei. University of Tokio Press, Tokio.
Steers, R. L. & Ackers, G. K. (1962) *Nature* **196** 457–476.
Strange, R. E. (1964) *Nature* **203** 1304–1305.
Strange, R. E. & Dark, F. A. (1962) *J. Gen. Microbiol.* **29** 719–730.
Strange, R. E. & Ness, A. G. (1963) *Nature* **197** 819.
Sundberg, L. & Porath, J. (1974) *J. Chromatogr.* **90** 87–98.
Swingle, S. M. & Tiselius, A. (1951) *Biochem. J.* **48** 171–174.
Takahashi, T. & Adachi, Y. (1982) *J. Biochem.* (Tokyo) **91** 1719–1724.
Tanford, C. & Louvrein, R. (1962) *J. Am. Chem. Soc.* **84** 1892–1896.
Thompson, A. R., Mattock, P. & Aitchison, G. F. (1980) In *Electrophoresis '79,* pp. 591–
    605. Ed. by B. J. Radola. Walter de Gryter, Berlin.
Tiselius, A., Hjerken, S. & Levin, O. (1956) *Arch. Biochem. Biophys.* **65** 132–155.
Trim, A. R. (1959) *Biochem. J.* **73** 298–304.
Turkova, J. (1978) *Affinity Chromatography, Journal of Chromatography Library,* vol. 12.
    Elsevier, Amsterdam.
Tzagoloff, A. & Penefsky, H. S. (1971) *Methods Enzymol.* **22** 219–230.
Unger, K. K. (1978) *Porous Silica: Its Properties and use of a Support in Column Liquid
    Chromatography.* Elsevier, Amsterdam.
Van Oss, C. J. (1970) *Prog. in Sep. and Purif.* **3** 97–128.
Wade, H. E. (1968) British Patent Appl. No. 40344/68.
Wheaton, R. M. & Baumann, W. L. (1953) *Ann. N.Y. Acad. Sci.* **57** 159–176.
Wieme, R. J. (1965) *Agar Gel Electrophoresis.* Elsevier, Amsterdam.
Winchester, B. G., Caffrey, M. & Robinson D. (1971) *Biochem. J.* **121** 161–168.
Woodrow, J. R. & Quirk, A. V. (1982) *Enzyme Microb. Technol.* **4** 385–389.
Yamanaka, T. & Okunuki, K. (1963) *Biochim. Biophys. Acta* **67** 379–393.

CHAPTER 3

Principles of industrial enzymology: Basis of utilization of soluble and immobilized enzymes in industrial processes

PETER S. J. CHEETHAM, Tate & Lyle plc, Group Research and Development, Philip Lyle Memorial Research Laboratory, PO Box 68, Reading, England

GLOSSARY OF SYMBOLS

ATP   adenosine triphosphate see also the di and monophosphates ADP and AMP

NAD^+ nicotinamide adenine dinucleotide, see also the reduced form NADH

$NADP^+$ nicotinamide adenine dinucleotide phosphate, see also the reduced form NADPH

CMP   cytosine monophosphate, see also the diphosphate CDP

CSTR   continuous stirred tank reactor

PFR   plug-flow reactor

ΔG free energy change

$\Delta G^{\#}$
or E free energy of high energy transition state (activation energy)

ΔH change of enthalpy during reaction

T absolute temperature

h Planck constant

ΔS change of entropy during reaction

R universal gas constant

M molarity

pH   a measure of the proton concentration of a solution

A Arrhenius constant

pO_2 partial pressure of oxygen

Re Reynolds number

V_{max} maximum velocity of enzyme catalysed reaction

V'_{max} maximum velocity of reaction of an enzyme working in reverse

V''_{max} maximum velocity of reaction of an immobilized enzyme

V_t reaction velocity after a period of incubation, t

V^s_{max} maximum rate of reaction per unit surface area

V^v_{max} maximum rate of reaction per unit volume of support or cell

K_m	Michaelis–Menten constant
K'_m	Michaelis–Menten constant of enzyme working in reverse
K''_m	Michaelis–Menten constant of an immobilized enzyme
k	rate constant (various subscripts used)
K	equilibrium constant
K_s	dissociation constant of E-S complex
$[I]$	inhibitor concentration
K_i	inhibitor constant
K_{ip}	product inhibitor constant
t	period for which an enzyme catalysed reaction proceeds
$[S]$	substrate concentration
$[S_b]$	substrate concentration in the bulk solution
$[S_s]$	substrate concentration present at the surface of a cell or enzyme
$[S_t]$	substrate concentration remaining after reactiion for a period, t.
$[S']$	substrate concentration external to a permease zone
$[S'']$	substrate concentration in a metabolic zone
$[S_{equ}]$	substrate concentration present at equilibrium
$[P]$	product concentration
$[P_{equ}]$	product concentration present at equilibrium
$[E]$	enzyme concentration
$[E_0]$	enzyme concentration present at zero time
$[E_t]$	enzyme concentration remaining after reaction for time t
$[E\text{-}S]$	concentration of enzyme-substrate complex
$[E_{inh}]$	concentration of enzyme required in the presence of inhibitor
kE	total quantity of enzyme present in a reactor
v	initial rate of enzyme catalysed reaction
v'	initial rate of enzymic reaction in the reverse direction
v''	initial rate of reaction in the presence of inhibitor
γ	initial rate of reaction/maximum rate of reaction
σ	substrate concentration/K_m
X	degree of conversion of substrate to product
η	stationary effectiveness factor
η'	operational effectiveness factor
η''	effectiveness factor obtained in the presence of inhibitor
a_m	the surface area per unit volume of cells
K_L	liquid mass transfer coefficient
Da	Damköhler number
D	diffusion coefficient
D_e	effective diffusion coefficient
x	distance
a_m	ext. surface area of cell
V_m	volume of cell
a	external area of permease zone

a'	external area of cell per unit volume
c	concentration
V	velocity
$L\alpha d_p$	Thickness of particle
N_{Pe}	Peclet number
ψ	porosity of particle
τ	toruosity factor
h_s	external transport coefficient
μ_s	external diffusional coefficient
ϕ	Thiele modulus
λ	decay constant for enzyme activity
Q	heat of reaction generated per unit of reactant
V	working volume of reactor
π	residence time
q	volumetric flow rate of substrate into a reactor
H	average residence time of substrate in a reactor
D_c	dispersion coefficient
Bo	Bodenstein number
$\mathcal{P}$	productivity of a reactor
μ	interstitial velocity
$t/2$	half-life of enzyme activity
Mi/Mx	fluctuation in throughput or conversion achieved
l_c	height of column
T_c	time constant
N_r	number of reactors employed
n	number of half-lives an enzyme is used for
Lf	transport efficiency/energy dissipated in a reactor
Δf	proportion of tracer eluted in any time interval
N	number of stages in a multi-stage reactor
EER	excess enzyme requirement
PIRS	proportional increase in reactor size
$\Delta P/L$	pressure-drop per unit length of reactor
α	the proportion of a continuous reactors period of operation that a batch reactor operates for.

3.1 INTRODUCTION

In a jocular fashion, enzymes have been compared with midwives, as both ease delicate biological materials from one metastable state to another! More realistically, enzymes are a large and ubiquitous class of protein molecules that act as biologically catalysts, catalysing all of the reactions that constitute cellular metabolism. Thus they provide the means whereby complex functions such as the synthesis of genetic materials and structural polymers and other cellular materials are carried out as well as providing the basis for large-scale traditional industries such as brewing and baking.

Enzymes consist of L-amino acids covalently linked together in a defined sequence, termed the primary structure, and coiled in a complex fashion, a zwitterionic structure with an active site being formed. The active site consists of the relatively few amino acids that have a direct role in binding the substrate and catalysing the reaction characteristic of each particular form of enzyme molecule. The secondary structure of an enzyme is those sections of the polypeptide chain that assume certain well-defined structures such as the α-helix, while the tertiary structure refers to the overall coiled structure of the polypeptide as maintained by secondary forces, including ionic, hydrogen and hydrophobic bonds. The quaternary structure describes the way in which certain, complex, usually intracellular enzymes consist of a number of polypeptide chains associated by means of secondary forces to form multi-subunit enzymes.

It is interesting to note that it is not necessary to know the amino acid composition of the enzyme or its three dimensional structure in order to use it effectively on a large scale in industrial, analytical, or medical applications. Several thousand enzymes possessing different substrate specificities are known, and it is virtually certain that a very large number still await discovery. However, only a very few have been isolated in a pure form and crystallized, and these enzymes tend not to be those enzymes that have acquired the greatest practical applications.

Enzymes vary considerably in size, for instance the β-subunit of RNA polymerase has a molecular weight of 155 000, whereas brain acyl phosphatase has a molecular weight of only 9000. Enzymes are classified into six different groups depending on the type of reaction they catalyse. These are oxidoreductases, transferases, hydrolases, lyases, isomerases, and ligases, although groups 2, 3 and 6 are all strictly speaking transferases that transfer groups without altering the oxidation state of the reactants.

(1) Oxidoreductases catalyse oxidation-reduction reactions involving oxygenation, such as $C-H \rightarrow C-OH$, or overall removal or addition of hydrogen atom equivalents, for example $CH(OH) \rightarrow C = O$.

(2) Transferases mediate the transfer of various groups, such as aldehyde, ketone, acyl, sugar, phosphoryl, and so on, from one molecule to another,

(3) Hydrolases. The range of hydrolysable groups is very broad. It includes esters, amides, peptides, and other $C-N$-containing functions, anhydrides, glycosides, and several others.

(4) Lyases catalyse additions to, or formation of, double bonds such as $C=C$, $C=O$ and $C=N$.

(5) Isomerases. Various types of isomerizations, including racemization, are catalysed.

(6) Ligases. Such enzymes are often termed synthetases. They mediate the formation of $C-O$, $C-S$, and $C-N$ bonds.

Each class is then further subdivided until individual enzymes are identified by a chemically meaningful six figure code. (For further details see Dixon & Webb (1979).)

Some difficulties do arise occasionally with this definition; for instance some enzymically active protein molecules possess more than one enzyme activity, acting on different substrates at different active sites remote from each other in the primary structure of the protein. This phenomenon occurs when the m-RNA molecules for two enzymes of differing substrate specificity, but which map close to each other in the genome, become fused owing to the loss of the intervening nucleic acid. A single polycistronic m-RNA molecule, and hence a single multi-enzyme protein is formed. An example is the fused histidinol dehydrogenase and aminotransferase of *Salmonella typhimurium* (Yourno *et al.* 1970). Another exception arises when the substrate specificity of an enzyme becomes altered by combination with a lower molecular weight molecule; for instance mammalian lactose synthetase is derived from the combination of a N-acetyl galactotransferase with α-lactalbumin.

Enzymes extracted from different sources and which catalyse the same chemical reaction (that is, homologous enzymes) may not be chemically identical. The tertiary structures in the critical area of the active site may be sufficiently similar to bind the same substrate and to catalyse the same reaction. However, the primary structures in other parts of the enzyme molecule may be sufficiently different to produce a subtle change in the overall tertiary structure and thus alter the optimum conditions under which catalysis will proceed. It is conventional to name enzymes only in relation to the reactions they catalyse. Thus α-amylase catalyses the hydrolysis of α 1,4: linkages between D-glucose units in polyglucoses such as amylose. But α-amylase from the fungal organism *Aspergillus oryzae* has a pH optimum of 4.7 and an optimum temperature of 50°C, while α-amylase from the bacterium *Bacillus lichenformis* has a pH optimum of 7.5 and an optimum temperature of 90°C.

Although enzymes are sufficiently large to be regarded as heterogenous catalysts, because of their solubility they are usually classed as being homogenous catalysts. Obviously, immobilised enzymes in which the particle size of the catalyst is at least an order of magnitude bigger than the enzyme are genuine heterogenous catalysts.

3.2 ASSAY OF ENZYME ACTIVITY (see Clinical examples in Chapter 5)

Enzyme catalysed reactions can be monitored spectrophotometrically, as many of the substrates or products of enzymes absorb visible or non-visible light. This is the commonest and simplest method of assaying enzymes, and it allows complete progress curves to be obtained by direct and continuous monitoring of the reaction. Spectrophotometric assays are especially common with enzymes that use nicotinamide coenzymes that absorb at 340 nm in their reduced forms.

Other assays depend on following the change in pH as a reaction proceeds with a pH meter, monitoring oxygen consumption or evolution during the course of a reaction using an oxygen electrode, or more generally following gas evolution with a manometer. In some cases assays can be made considerably more sensitive when the absorption of fluorescent light is followed. The very early events during the initiation of an enzyme catalysed reaction can be followed using rapid reaction techniques such as the stopped-flow apparatus (Gutfreund 1972).

A standard unit for enzyme activity is the Katal, defined as the amount of enzyme activity required to convert one mole of substrate to product per second. When using enzymes in industrial applications their activity should always be quoted in a form, and estimated under conditions, that relate most closely to that application, whether it is in a clinical assay or in large-scale biochemical processing. Thus in some cases it is more useful to express enzyme activity as the rate of change in solubility or viscosity of a material per unit weight or volume of enzyme added, rather than in more conventional terms such as the nmoles of product formed per mg of enzyme per second. Note that batch assays may give an inexact impression of the potential activity of an enzyme when used in a reactor particularly if a much lower substrate concentration is used in the batch assay or if only an initial rate determination is made. In particular a batch assay will give an underestimate of the activity achieved in columns when either substrate inhibition or product activation occurs, and can overestimate activity when substrate activation or product inhibition occurs.

3.3 COFACTORS

Not all enzymes are capable of acting alone; many require the presence of non-protein cofactors to enable catalysis to be carried out. Such cofactors, which are in reality cosubstrates because they undergo chemical transformation during the reaction include simple metal ions or organic molecules; and may in some cases be covalently bound to the enzyme. They must be continually reconverted back to their original form in order that catalysis can continue, a process usually referred to as regeneration. Regeneration can take place spontaneously in aerobic aqueous conditions by hydrolysis or oxidation reactions such as in the cases of tetrahydrofolate, biotin, oxidized flavins, and pyridoxyl phosphate coenzymes. However, in most cases, including ATP, coenzyme A, folic acid, NAD^+ and $NADP^+$ regeneration can only be achieved by directly coupling with the oxidation of high energy substrate molecules, either by cytochromes or by enzymes. In the case of the nicotinamide cofactors regeneration is especially difficult as they slowly degrade in aqueous media. Coenzymes are usually expensive and available only in relatively small quantities but are invariably reaction specific, for instance pyridoxyl phosphate, thiamine pyrophosphate and flavin mononucleotide and nicotinamide adenine nucleotides participate specifically in reactions involving the transfer of amino and aldehyde groups and hydrogen atoms respectively. Note that an important recent advance is the use of cross-linked NAD for the affinity precipitation of NAD-requiring enzymes.

3.4 THE DISTINCTIVE FEATURES OF ENZYMES AS CATALYSTS

Enzymes are very powerful catalysts in four respects. Firstly, they are very efficient, catalysing reactions often 10^8-10^{11} times more rapidly than the corresponding nonenzymic catalysts, as measured by the (V_{max}/K_m) value, such that as many as 10^6 substrate molecules can be metabolized per enzyme active site per minute, this parameter being termed the molecular center activity of the enzyme. These rates are achieved even though the enzyme catalysed reactions require much less extreme conditions, such as temperature, pH, and pressure, and reactions proceed in the cheapest, safest, and most abundant solvent, water.

Secondly, the range of reactions catalysed is extremely broad such that a much greater variety of reactions can be catalysed than with chemical catalysts. The advent of genetically engineered enzymes and enzyme analogues or 'synzymes', for instance those formed by the *in vitro* polymerization of amino acids in a defined order by solid phase (Merrifield) synthesis catalysts, gives the possibility that reactions other than those catalysed by naturally occurring enzymes can be specifically carried out at rates comparable to natural enzymes and under similarly mild conditions. A good example of such enzyme analogues are crown ethers (Stoddart 1980). Others include a glutamic acid phenylalanine copolymer with high lysozyme activity, the decapeptide Glu-Phe-Ala-Ala-Glu-Glu-Ala-Ala-Ser-Phe which has glycosidase activity, methylenimidazole (catalytic) and dodecyl (binding) groups attached to the polymethylenimine chain which confers arylsulphatase activity and palladium again attached to polyethylenimine which act as a hydrogenation catalyst (Coleman & Royer 1980).

Thirdly, enzymes are usually very specific in terms of the type of reaction catalysed. However, in a very few cases an enzyme has been shown to act on its substrate in two different ways, for instance malate and isocitrate dehydrogenases. EC 1.1.1.38, 40 and 42 can either reduce or decarboxylate their substrates (Dixon & Webb, 1979). Enzymes can be specific for the structures of the substrate and product formed, such that often only a single chemical present in a mixture of very similar chemicals is transformed with complete fidelity to a single product. This can result in higher product yields and fewer potentially polluting side-products. This specificity is very impressive since most substrate molecules possess a number of potentially reactive functionalities. Specificity is due to the ability of the enzyme to bind the substrate and organize reactive groups so that a specific reaction transition state is particularly favoured. Substrate specificity is usually not absolute, as other chemically related but non-physiological substrates are often accepted as substrates, that is bind to the active site of the enzyme, while maintaining full stereospecificity in the conversion. For instance glucose oxidase is a very well known and widely used enzyme, but recently Alberti & Kibanov (1982) have shown that it also catalyses the conversion of benzquinone into hydroquinone. Also, alkaline phosphatase has been used "in reverse" as a phosphoylating enzyme (Martinek *et al.* 1980). A rather better example is the fact that the glucose isomerase used in large quantities in industry to form high fructose syrups from glucose is naturally a xylose isomerase forming xyulose rather than fructose (Bucke 1977). In fact most of the industrially important enzymes are not used on their natural substrate. A luciferase has been successfully used to detect the very low concentrations of trinitrotoluene emanating from land mines such that a successful sensing device could be constructed, and new organisms capable of degrading TNT have also been found (Naumova *et al.* 1982). Obviously TNT is not usually a naturally occurring substrate for enzymes! This reactivity of enzymes to nonphysiological substrates may prove very important in the degradation of pollutants especially as the substrate specificity of enzymes can sometimes be modified by mutation (Betz *et al.* 1974). Specificity us not only regiospecific but also stereospecific such that chiral carbon atoms are discriminated between, a startling exhibition of the asymmetric three-dimensional nature of the active site of the enzyme which can impose specificity on the binding of the substrate, the catalytic step, or on both steps. For instance enzymes invariably distinguish between paired chemically alike substituents such as the two hydrogen atoms on prochiral carbons such as CH_2XY.

Fourthly, enzymes are naturally subject to a number of controls both in terms of the gross control of how fast they are synthesised and degraded, and in the fine control of their activity by the binding of small molecule modifiers which can either increase or decrease the activity of the enzymes. It is important for industrial users of enzymes to appreciate that because of their high affinities for their substrates, enzymes are especially preferred as catalysts when very low substrate concentrations are encountered, but that the high substrate specificity of many enzymes is often not required, as in many processes the substrate is pure and so contains only one molecule that can be used as a substrate.

3.5 ENZYME CATALYSIS

Enzymes are stereospecific biological catalysts and are subject to the same kinetic and thermodynamic constraints as are chemical catalysts, that is, they alter the rate at which a reaction proceeds but not the final position of equilibrium between the substrate(s) and product(s). Catalysis can be effected either by proceeding by an alternative reaction mechanism and thus an alternative transition state with a lower free energy; by lowering the free energy of the conventional transition state intermediate, or by providing an environment that decreases the free energy of the product so that the nett ΔG is decreased.

Three phases in the time-course of an enzyme reaction are generally recognized. In the first, which lasts for only a fraction of a second, the concentration of free enzyme falls drastically, most of it becoming combined with substrate in a dynamic equilibrium that persists so long as fresh substrate molecules are available. Note that the rate of reaction is very sensitive to the energy with which the substrate is bound, such that a difference in free energy of only 2 kcals/mole between two competing substrates will result in a 97:3 ratio between their rates of reaction. In the second phase, all the reactants including the enzyme molecules are in a dynamic equilibrium with the maximum activity being expressed, since the high substrate concentration means that enzyme molecules can reform enzyme—substrate complexes very rapidly immediately after having transformed the previous substrate molecule(s). In the third phase of the reaction the substrate concentration becomes appreciably diminished, and thus the rate of the enzyme catalysed reaction falls asymptotically. This phase of reaction is especially important in many industrial reactions where complete reaction is desired so as to maximize the yield and concentration of product obtained, such that the interactions of parameters such as substrate and enzyme concentrations and reaction conditions are of crucial importance. This phase can often occupy the majority of the reaction time especially when product inhibition takes place, which is especially likely when high substrate concentrations are employed. Given sufficient time a dynamic equilibrium is set up between the substrate(s) and product(s) in which the net free energy change (ΔG) is zero, that is

$$\Delta G \text{ products} - \Delta G \text{ reactants} = 0 \ . \tag{3.1}$$

ΔG is related to the heat of reaction by the relationship

$$\Delta G = \Delta H + T\Delta S \tag{3.2}$$

where ΔS and ΔH are the entropy and enthalpy of the reaction respectively, ΔG is the Gibbs free energy, and T is the absolute temperature. The rate at which the reaction proceeds, that is the rate at which equilibrium is approached, is independent of the magnitude of ΔG. The rate is dependent on the activation energy which must be supplied to the reaction so as to form a high energy transition state complex ($\Delta G^{\#}$) (3.1). That is:

$$k \text{ (observed)} = \frac{RT}{nh} \ e^{-\Delta G^{\#}/RT} \tag{3.3}$$

where k is a rate constant, R the universal gas constant and h Plank's constant.

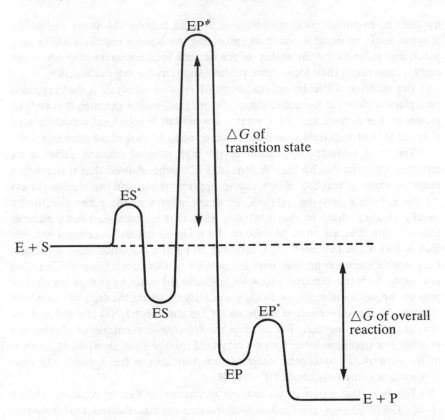

Fig. 3.1 — Changes in the free energy of the reactant molecules as an enzyme catalysed reaction proceeds to the formation of product(s) *via* an enzyme–substrate complex.

Catalysts thus exert their dramatic rate enhancing effects by decreasing $\Delta G^{\#}$, but do not affect ΔG, or the final position of equilibrium. Formation of the E–S complex is associated with a decrease in entropy in many cases such as those of chymotrypsin and carboxypeptidase. However, increases in entropy have also been observed for other enzymes such as pepsin and trypsin, presumably caused by important changes in the conformation of the enzyme or in the amount of water bound by the enzyme, for instance due to polarity changes resulting from combination with the substrate.

Enzyme activity is characterized by the formation of an enzyme–substrate complex and by the polyfunctional nature of the catalysis, since a number of active groups, for example carboxyl, hydroxyl, sulphydryl, amino and imidazole, together with some other liganded metal ions, are used in combination. Each group is individually a rather poor reagent but their action is facilitated firstly by their hydrophobic local environment, created because the active site of the enzyme tends to exclude water so enhancing the organic reactions which take place; and secondly by the ability of the enzyme to concentrate substrate molecules, so increasing their local concentration and thus the rates of reaction.

The inherent difficulty in the study of enzyme catalysis is that reactions take place enclosed in the active site of the enzyme, so that although the catalytic groups of the enzyme and the reactants are present in high local concentrations they are at least transiently covalently linked and so are part of the same molecule.

The most valuable explanation of the high rates of reaction achieved by enzymes was provided by Page & Jencks (1971) who showed that it is not that enzymes exert a 'magical' effect due to approximation and orientation effects on the substrate and the catalytically active groups in the active site, or by usually creating strain in the substrate. Rather, it is reactions taking place in free solution that are slow because of the translational and rotational entropy that is lost when two molecules combine to form a transition state. Chemical bimolecular nucleophilic reactions are subject to this adverse factor. This does not apply to intramolecular processes, because the reacting groups are already part of the same molecule, or to enzyme catalysis, where the enzyme–substrate complex is also effectively a single entity, as the substrate(s) and enzyme are covalently linked together. For instance the effective concentrations of substrate possible due to proximity in enzyme catalysed reactions are about 10 M, whereas in the equivalent bimolecular condensation reactions in free solution effective concentrations are only about $10^7 - 10^9$ M.

The best description of an enzyme mechanism is that of lysozyme (Blake et al. 1967, Vernon 1967). X-ray diffraction studies showed that a hexa–hexose fragment of the substrate fits into a long deep cleft in the enzyme, and on binding to the enzyme the sugar ring is distorted into the half-chair (or sofa) conformation of the carbonium ion. Reaction then proceeds via a carbonium ion intermediate which is stabilized by the ionized carboxylate of Asparagine 52. The expulsion of the alcoholic moiety is facilitated by distortion of the sugar ring into the half-chair conformation.

Enzymes usually act as the solvation shell for charges that are developed on the substrate, and the enzyme is distorted on binding the substrate such that it becomes complementary to the transition state. The transition state, rather than the substrate, is stressed by being bound as tightly as possible, so maximizing the parameter k_{cat}/K_m, the second order rate constant for the reaction of enzyme with free substrate molecules. The substrate is bound only weakly such that the K_m evolves to a value only just above the substrate concentration usually encountered in vivo. If very tight binding of the substrate did occur then the enzyme–substrate complex would be stabilized, so making the enzyme less reactive.

Thus an enzyme which has evolved to maximize its rate of reaction has as high a k_{cat}/K_m as possible and a relatively low K_m value. Then the upper limit on the rate of reaction is the rate of diffusion controlled encounters, that is, $10^8 - 10^9$ sec^{-1} M^{-1}. For instance, carbonic anhydrase has a k_{cat}/K_m of 1.2×10^8 sec^{-1} M^{-1} and a K_m/S of greater than 16, and triose phosphate isomerase has a k_{cat}/K_m of 2.4×10^8 and a K_m/S of 150, so they cannot become more rapid catalysts (Fersht 1980).

Further evolution for greater catalytic efficiency (high efficiency indicated by high k_{cat}/K_m ratio) must therefore proceeed in other directions such as the fidelity or specificity with which the reaction is catalysed. For instance, very high copying accuracy is required in protein synthesis and DNA replication so as to maximize the 'signal-to-noise' ratio during these processes, so that aminoacyl t_{RNA} synthetases have evolved in addition to their synthetic activity, a hydrolytic activity which deacylates mistakenly activated products with a high turnover number compared to the rate with which the required product is attacked (Fersht & Kaethner 1976). Thus a copying accuracy much higher than that permitted by simple equilibrium thermodynamics is permitted, albeit at a high cost in terms of the amounts of energy liberated by the hydrolase activities.

Some 'manmade' progress in improving enzymes has been achieved by chemical modification; for instance flavin has been immobilized close to the active site of papain, forming a stereoselective oxidoreductive catalyst (Levine et al. 1977, Slama et al. 1981). Another example is the attachment of a diphosphine rhodium complex to a protein by Wilson & Whitesides (1978) in order to facilitate assymmetric hydrogenation. Such synthetic enzymes are likely to be most useful for reactions for which no natural enzyme can be discovered but are unlikely to be able to compete successfully with naturally occurring enzymes when they exist in a readily available form (see review by Wiseman (1984)).

3.6 ENZYME KINETICS (see Part A, Sections 3.18, 3.20, 3.21 and 3.24: each on aspects of enzyme kinetics)

The activity of an enzyme is determined by the enzyme concentration, the substrate concentration and its availability, the concentration of cofactors and/or allosteric effectors, the presence, concentration, and type of inhibitors, and ionic strength, pH, and temperature. The way in which these parameters affects enzyme activity is the study of enzyme kinetics. It gives an understanding of the reactions being studied and allows control to be exerted.

For single substrate (S) to single product (P) conversions the following general scheme of reaction applies:

$$S + E \underset{k_{-1}}{\overset{k_1}{\rightleftharpoons}} ES \overset{k_2}{\rightarrow} E + P . \tag{3.4}$$

When the concentration of subtrate is much greater than the concentration of enzyme $[E]$ the rate of reaction is zero order with respect to the reactants, that is, it depends only on the concentration of enzyme present. Thus the velocity of the reaction remains essentially constant until nearly all the substrate has been consumed, when the rate of reaction becomes dependent on the prevailing substrate concentration and is first order with respect to the substrate concentration.

Enzyme kinetics are best described by the well known Michaelis–Menten relationship (Fig. 3.2) (Dixon & Webb 1979) which takes into account the formation of an enzyme–substrate complex.

$$v = \frac{k_2 [E] [S]}{[S] + K_m} \qquad (3.5)$$

where $K_m = \dfrac{k_{-1} + k_2}{k_1}$ and represents the substrate concentration which gives half the maximum rate of reaction. That is, when the substrate concentration is ten times the K_m value the reaction proceeds at 91% of the maximum rate of reaction (V_{max}), and if one hundred times the K_m at 99% of the V_{max} value, provided excess substrate inhibition does not occur and no inhibitors are produced by the enzyme reaction. In this case

$$V_{max} = k_2 [E] . \qquad (3.6)$$

Michaelis–Menten kinetics assume that the substrate and the enzyme are both soluble and are homogeneously mixed, which does not always occur in industrial processes where very concentrated, viscous substrates are often used and solid substrates are sometimes used. In this latter case the surface area presented by the substrate particles is often a more useful term to use than concentration, particularly as some enzymes such as lipases appear to adsorb onto the surface of such solid substrate particles.

Numerical values of V_{max} and K_m can be derived graphically following a transposition of the Michaelis–Menten equation to give:

$$\frac{1}{v} = \frac{1}{V_{max}} + \frac{K_m}{V_{max}} \cdot \frac{1}{[S]} \qquad (3.7)$$

which can be plotted as $1/v$ vs. $1/[S]$ to give values of $-1/K_m$ and $1/V_{max}$ from the intercepts with the $1/S$ and $1/v$ axes respectively; the slope of the line being K_m/V_{max}. This plot is called the Lineweaver–Burk plot.

Alternatively a normalized version of the Michaelis–Menten equation is sometimes easier to use. That is,

$$\gamma = \frac{\sigma}{1 + \sigma} \qquad (3.8)$$

where $\gamma = v/V_{max}$ and $\sigma = S/K_m$.

Then the 'Lineweaver–Burk plot' consists of a graph of $1/\gamma$ vs. $1/\sigma$.

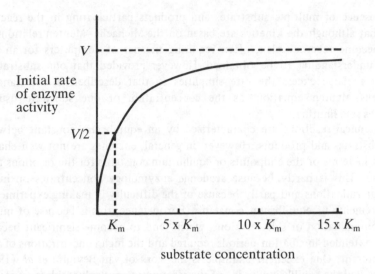

Fig. 3.2 – Michaelis–Menten plot of initial rate of reaction (v) vs. substrate concentration ($[S]$). This reaches half the maximum velocity (V_{max}) when the substrate concentration reaches a concentration equal to the Michaelis–Menten constant (K_m).

In many cases the rate at which the enzyme–substrate complex breaks down into free enzyme and product molecules is relatively slow, and so limits the rate of enzyme catalysed reaction. Then the rate constant k_2 is very small and K_m approximates to k_{-1}/k_1 and so equals k_s, the dissociation constant of the enzyme–substrate complex ($[E]$ $[S]/[ES]$). Since under these conditions the value of k_2 governs the rate of reaction, it is often referred to as k_{cat} and represents the turnover number of the enzyme in moles of substrate/active site/unit time.

Enzyme reactions are best described as involving three activated enzyme–substrate complexes by the following scheme:

$$E + S \rightleftharpoons EP^* \rightleftharpoons ES \rightleftharpoons ES^{\#} \rightleftharpoons EP \rightleftharpoons EP^* \rightleftharpoons ES \qquad (3.9)$$

where ES^* and EP^* are the activated enzyme–substrate and enzyme–product complexes and $ES^{\#}$ the transition state complex respectively (Fig. 3.1). Note that the Michaelis complex is not activated in the physical sense but still requires thermal activation in order to react. The combination of enzyme with substrate may be exergonic or endergonic. The free energy of the transition state represents the energy barrier that must be overcome before the reaction can proceed. The free energy of the transition state in an uncatalysed reaction is very much higher than when a catalyst, such as an enzyme, is employed, so that the un-catalysed reaction proceeds less rapidly.

Further complications in the kinetic description of reactions are caused by the presence of multiple substrates and products participating in the reaction such that although the kinetics are based on the Michaelis—Menten relationship they become very complex, often requiring the use of computers for an adequate understanding to be obtained. However provided that one substrate is present in large excess the rate simplifies to that described in the standard Michaelis—Menten equation, as the concentration of one substrate usually becomes rate limiting.

Chemical reactions are characterized by an equilibrium constant between the substrates and products. However, in general, enzymes are not well characterized in terms of the endpoints or equilibrium constants for the reactions they catalyse. This is partly because academic enzymology concentrates on initial rate determinations and partly because of the difficulty in making experimental measurements of equilibrium constants. For instance, this is because of microbial contamination or side reactions, which tend to become significant because of the extended incubation periods required and the high concentrations of product present. One excellent example is the work of van Beynum *et al.* (1980) who studied the equilibrium point of the reaction of amyloglucosidase on starch.

The most important influence on the position of equilibrium of an enzyme catalysed reaction and thus the yield of product that can be expected is the overall free energy change of the reaction as described in the equation

$$\Delta G = -RT \ln K \ . \tag{3.10}$$

The closeness to which the position of chemical equilibrium can be reached in an enzyme catalysed reaction is also determined by factors such as product inhibition, so that the susceptibility of individual enzymes to product inhibition may vary appreciably. This parameter is an important determinant of the actual position of equilibrium reached, especially when high substrate concentrations, which are preferred in industrial enzymology, are used.

The position of equilibrium can also be influenced by the immobilization of the enzyme whereby the local concentration of product in contact with the enzyme can be decreased by selectively partitioning the substrate away from the support. Thirdly, the position of equilibrium can also be influenced when the enzyme is present in high concentrations relative to the reactants such that an appreciable proportion of these reactants can be present as a complex with the enzyme, such that their effective concentrations are reduced. This factor may be important in industrial enzymology where relatively high concentrations of enzyme are often employed, high degrees of conversion of substrate to product are usually needed and enzymes may be required to work "in reverse".

For a more detailed treatment please refer to 'Symbolism and Terminology in Enzyme Kinetics' published by the Nomenclature committee of the Int. Union of Biochem. (See *Biochem. J.* (1983) **213**, 516—7 and *Eur. J. Biochem.* (1982) **128**, 281—291.

3.7 THE EFFECT OF pH ON ENZYME ACTIVITY

In general, enzymes are active over only a limited pH range. A definite optimum pH for enzyme activity is usually observed because, like other proteins, enzymes possess many ionizable groups so that pH changes may alter the conformation of the enzyme, the binding of the substrate, and the catalytic activity of the groups in the active site of the enzyme. Effects may be due to a change in the maximum reaction rate (V_{max}), a change in the affinity of the enzyme for the substrate (K_m), or an alteration in the stability of the enzyme. Stability depends on the length of time the enzyme has been maintained at the unfavourable pH. Similarly ionizable groups in the substrate can be affected by pH and may be important in forming the enzyme—substrate complex. All these effects usually occur together but may occur in isolation. Thus when broad pH optima are observed, for example invertase, this is usually because the substrate cannot be ionized. In many industrial enzyme reactions the pH is not maintained but changes during the reaction, depending on the buffering capacities of the substrates and products involved. pH also affects enzyme stability such that the optimum operational pH is often a compromise between the effects on enzyme activity and enzyme stability.

3.8 THE EFFECT OF TEMPERATURE ON ENZYME ACTIVITY

Enzyme catalysed reactions like other chemical reactions increased in rate with rises in temperature, this effect being described by an Arrhenius relationship:

$$k = A e^{-E/RT} \qquad (3.11)$$

where k is the reaction rate constant, A the Arrhenius constant, E the energy of activation, R the gas law constant, and T the absolute temperature. Activation energies for enzyme catalysed reactions are normally in the range 4—20 kcal/mole range such that the rate of reaction increases by a factor of between 0 and 1 with each 10°C rise in temperature. Enzyme stability is also influenced by temperature in a similar fashion with enzyme denaturation having activation energies in the range 40—130 kcal/mole such that the optimum temperature for reaction to take place is a compromise between these dual effects of increased temperature. The use of higher temperatures causes an increase in activity and a decrease in stability, taking into account the period of time the enzyme is used, although a few enzymes can be heated to above 100°C and still retain activity, for instance adenylate kinase can retain activity even after having been maintained at pH 1.0 and 100°C (Chiga & Plaut 1960). On the contrary some enzymes have been noted as being less stable when cooled, presumably because hydrophobic forces, which decrease in strength with decreases in temperature, play a large part in maintaining the active conformation.

Temperature also influences the position of the equilibrium in the reaction (K) as described by the following equation

$$\frac{\partial \ln k}{\partial T} = \frac{\Delta H}{RT^2} \cdot \tag{3.12}$$

3.9 ENZYME INHIBITION

Inhibition of enzymes is an important topic because inhibition studies have yielded valuable information concerning the structure and mechanism of enzymes and because obviously inhibition must be minimized in order to maximize the activity of an enzyme catalysed reaction. In fact on some occasions enzymes have been characterized by the pattern of their response to different chemically related inhibitors (Fullbrook & Slocombe 1970).

Four common types of inhibition are recognized. Irreversible inhibition occurs when the inhibitor molecule combines irreversibly with the enzyme, chemically modifying its structure and abolishing or greatly reducing its activity. An example is di-isopropylfluorophosphate (DFP) which inhibits enzymes which have reactive serine groups in their active sites, such as acetylcholinesterase. Thus they act as probes of the active site and as nerve poisons. Another example is heavy metals such as lead or mercury which react with SH groups in the active site of enzymes.

Not all enzyme inhibitors, however, create problems; many have useful applications such as clavulinic acid which is an important antibiotic as it inhibits microbial β-lactamases which inactivate penicillins.

Three types of reversible inhibition occur. Competitive, which can be reduced by increasing the substrate concentration; non-competitive, which cannot be so reduced; and excess substrate inhibition caused by the formation of non-productive complexes. These can be decreased by using a loower concentration of substrate i.e. E $\rightleftharpoons$ ES $\rightarrow$ E + P. Competitive inhibitors compete with the normal substrate molecules to occupy the active site of the enzyme forming a reversible enzyme—inhibitor complex which is characterized by the K_i value, the dissociation constant of this complex. With non-competitive inhibition the inhibitor also forms a reversible complex with the enzyme but at a site other than the active site, so that inhibition is not normally reduced by increasing the substrate concentration.

The effect of low concentrations of these inhibitors on the Michaelis—Menten equation are described in the following equations for competitive (3.13) and non-competitive (3.14) and substrate inhibition (3.15) respectively:

$$v = \frac{V_{max} [S]}{[S] + K_m (1 + [I]/K_i)} \tag{3.13}$$

$$\nu = \frac{V_{max} [S]}{(1 + [I]/K_i) [S] + K_m}$$ (3.14)

$$v = \frac{V_{max} [S]}{[S] + K_m + [S]^2/K_s}$$ (3.15)

where $[I]$ is the inhibitor concentration, K_i the inhibition constant which is defined as the concentration of inhibitor at which the V_{max} of the enzyme is halved (non-competitive) or K_m doubled (competitive), and K_s the dissociation constant of the enzyme—substrate complex.

Anomolous K_m values are obtained when competitive and substrate inhibition occur, and anomolous V_{max} values occur when non-competitive and substrate inhibition occur. For instance in competitive inhibition an apparent K_m is measured which is greater than the true K_m of the enzyme obtained in the absence of the inhibitor but no change in the value of the V_{max} is obtained, while in non-competitive inhibition V_{max} is reduced in the presence of inhibitor but the K_m value is unaffected by the inhibitor.

Inhibition by substrate is less common but occurs at high substrate concentrations even though inhibition is not readily apparent and Michaelis—Menten kinetics are obeyed at low substrate concentrations. Inhibitors are often present in low concentrations in many substrates and inhibition may only become serious when continuous high throughputs of substrate are used, necessitating strict quality control of the feedstocks used. Many more complex forms of inhibition also occur, usually less commonly. Care must also be taken when activators of the enzyme are present in the substrate or are produced during the reaction. For instance the activity of an enzyme determined in a batch assay will give an underestimate of the activity that can be achieved in a column reactor when either substrate inhibition or product activation occur, and can overestimate the activity achieved in a column when substrate activation or product inhibition occur. An important industrial example of enzyme inhibition is by galactose on β-galactosidase (lactase) during the hydrolysis of lactose in cheese whey.

3.10 THE VARIOUS TYPES OF ENZYMIC CATALYST

Enzymes have been the subject of intense academic interest for many decades and are currently poised to become important industrial catalysts. Enzyme technology can be viewed as a struggle to obtain the advantages of enzymic catalysis, such as stereospecificity under mild conditions, while overcoming or circumventing their inherent disadvantageous features such as instability. Thus the application of enzymes in industry is a compromise between what is scientifically and technologically possible and what is commercially desirable. At present this effort to apply our scientific understanding of biochemistry and microbiology is at a very similar stage to that of chemistry towards the end of the last century when its potential industrial applications were being explored, the field of chemical engineering, and a host of new industries, being the end result.

Several main types of enzymic catalyst can be recognized, including whole cells, organelles, and enzymes used in both free and immobilized forms. However, the distinctions between the different types are sometimes blurred, for instance when cell growth takes place in immobilized cell preparations, when immobilized cells lyse during use, or when enzymes are immobilized onto the surface of cells.

The use of fermenting cells, and to a lesser extent soluble enzymes, is of long standing interest, but currently the use of immobilized enzymes and cells in industry is increasingly common, although very recently much interest has been shown in the use of soluble enzymes in two-phase reaction mixtures.

No single form of catalyst can be expected to be universally applicable and successful. The ideal should be a versatile, fully understood range of industrially suitable biological catalysts, one or more of which can be applied easily to any particular task. With a better understanding of immobilization techniques and a greater appreciation of the advantages of using immobilized enzymes and cells, many more useful practical applications of enzymes should be expected.

3.11 A COMPARISON OF ENZYMES WITH CHEMICAL CATALYSTS

When compared with chemical catalysts, enzymes have a number of advantages. These include the variety of reactions catalysed and the mild conditions employed which are especially important when labile reactants are used, as well as minimizing the amounts of energy used. For instance chemically, in the Haber process for fixing nitrogen, pressures of 200–1000 atms. and 500°C are used. Other advantages include the high degrees of conversion that can be obtained and the smaller amounts of pollutants produced, and the specific reactions obtained. In particular many organic chemists are attracted to enzyme based catalysts because of the abilities of the enzyme to distinguish between enantiomers and to distinguish enantiotropic groups and the faces of molecules possessing prochiral centers (Jones 1976a, b), and because such specific reactions can be carried out in aqueous solvents. The speed and specificity of biochemical reactions may also be of use in the preparation of radiochemicals containing shortlived isotopes. However, biological catalysts, such as immobilized cells, often have complex requirements such as cofactors, and are usually more labile than chemical catalysts, so requiring precise control of the reaction conditions. Economic factors also influence the choice of catalyst; for instance, chemical processes utilize large amounts of energy and are often based on petrochemicals, whereas biological catalysts do not require large energy inputs and usually use renewable biological resources. Thus the present-day use of biological catalysts in the food and pharmaceutical industries may be followed by their use in the heavy chemical industries in the future.

3.12 A COMPARISON OF ENZYMES WITH FERMENTATIONS

At present fermenting micro-organisms are the dominant form of industrial biological catalyst, their great advantage being their versatility, as illustrated by the wide range and sophistication of the processes in current use. However, complex and often expensive nutrients are required, and during fermentation much of this nutrient material is used as an energy source, even if the required biochemical is formed in the stationary phase after cell multiplication has been completed. Continuous operation may also be difficult with fermentations, the operational life of many isolated enzymes being longer than that of fermenting cells. Fermentation capital and running costs are high, especially when high rates of aeration and sterile conditions are required. Furthermore, after the fermentation has finished isolation of the desired product from complex dilute media is often difficult and expensive. Also, the cell mass and exhausted fermentation medium must be collected and disposed of, and significant batch to batch variations must be taken into account. Note that serious environmental problems can be caused by the disposal of wastes from large-scale fermentors which have BODs of $200-400$ mg l^{-1} and may contain bioactive molecules, such as antibiotics that can interfere with normal biological waste treatment systems.

By comparison, immobilized enzymes or cells are convenient to handle, appear to be less susceptible to microbial contamination, and permit easy separation of product from the catalyst. The use of immobilized enzymes enables greater control throughout the reaction, which should ultimately improve the yield and quality of the product. Because the stabilities of enzymes are usually increased by immobilization continuous use or re-use in batch operations is made easier. Furthermore, because non-dividing immobilized cells require only maintenance energy, yields of product will be greater than in fermentation methods. Use of immobilized cells enables the conditions for cell growth and product formation to be optimized more fully, since these steps are carried out separately, whereas in fermentations they often occur simultaneously or are at least carried out in the same apparatus such that the conditions of operation are a compromise between the two requirements.

Venkatasubramanian *et al.* (1978) have carried out an economic comparison between the production of the flavouring agent monosodium glutamate by batch fermentation, and continuously by using enzymes associated with collagen-immobilized *Corynebacterium* cells. The total fixed capital investment, operating and overhead costs were lower for the immobilized cell plant, but overall production costs were higher because of the need to replace the immobilized cells at intervals. However, because of the lower investment required by the immobilized cell process, this method resulted in a 50% return on capital invested compared to a 36% return obtained when using the fermentation process.

3.13 IMMOBILIZED BIOCATALYSTS (for recommendations concerning methods for characterising immobilized biocatalysts please refer to *Enzyme and Microbial Technology* 1983 **5** 304–307)

Immobilization has been defined as the process whereby the movement of enzymes, cells, organelles, etc. in space is completely or severely restricted, usually resulting in a water-insoluble form of the enzyme. Immobilized enzymes are also sometimes referred to as bound, insolubilized, supported or matrix-linked enzymes. Note that immobilization of catalysts is not unique to enzymes, the use of immobilized chemical catalysts being advantageous when expensive metal catalysts are used. Also immobilized enzymes are naturally found in soil when enzymes released during the decay of plant, animal or microbial sources become adsorbed onto soil particles and immobilized cells exist in the form of natural mycelial pellets.

Enzymes or cells are often used in an immobilized form in industry because, most importantly, re-use or continuous use of the biocatalyst is made possible. Continuous use is especially important in maintaining a constant environment for the immobilized biocatalyst, a factor which is very important in maintaining enzyme stability and because the enzyme or cell is prevented from contaminating the product or polluting a waste-stream. Thus both the capital and recurrent costs of enzyme catalysed processes can be reduced at a stroke by immobilizing the enzyme (Fig. 3.3).

Free cells or enzymes are very difficult to re-use or to use continuously, as both catalysts are too small to filter, and recovery by centrifugation is too expensive. Water-soluble immobilized enzymes can be formed by linking the enzyme to soluble polymers; for instance Marshall & Humphries (1977) formed a soluble catalase–dextran conjugate, and Charles *et al.* (1974) coupled lysozyme to alginic acid which could be reversibly on order to enhance activity towards macromolecular substrates simply by raising the pH above pH4. Immobilized enzyme/cell processes are easier to automate and enable the advantages of various reactor configurations to be exploited. These include rapid pH and temperature control, good gas transfer in stirred reactors, and the minimization of product inhibition in packed-bed reactors. Also as enzyme is used more efficiently, fewer cells have to be grown, and so less fermentation wastes have to be disposed of. When immobilized cells are used, flowrates through the reactor can be used which are greater than those at which the maximum specific growth rate of growing cells in a fermentation culture would be exceeded, so that washout of the cells and undue increases in the viscosity of the substrate can be avoided.

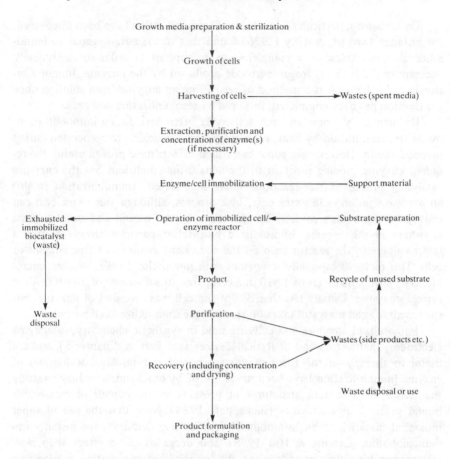

Growth media preparation & sterilization

Growth of cells

Harvesting of cells ⟶ Wastes (spent media)

Extraction, purification and
concentration of enzyme(s)
(if necessary)

Enzyme/cell immobilization ⟵ Support material

Exhausted ⟵ Operation of immobilized cell/ ⟵ Substrate preparation
immobilized   enzyme reactor
biocatalyst
(waste)

Product   Recycle of unused substrate

Waste   Purification
disposal
                                          Wastes (side products etc.)

Recovery (including concentration
and drying)

                                          Waste disposal or use

Product formulation
and packaging

Fig. 3.3 – A generalized representation of a typical sequence of operations
involved in industrial processes which use enzymic catalysts as the chief catalytic
step.

Often the stability of the enzyme or cell is increased by immobilization,
and/or the concentration of the enzyme is increased such that a much smaller
reactor can be used to achieve the same productivity when using immobilized
rather than free biocatalysts, as well as being able to operate the process in a
much more convenient fashion. Immobilization enables the enzyme or cell to
be evenly distributed throughout the reactor, so ensuring an even supply of
substrate to each enzyme or cell. Lastly, an important general advantage of using
immobilized cells and enzymes is that the cells and enzymes are buffered by the
support against changes in the pH, temperature, or ionic strength etc. in the bulk
solvent. Such protection is reflected in enhanced stabilities. Protection from
shear forces particularly at air—liquid interfaces and from microbial contaminants
and proteases derived from them is also effected.

On occasions, particular advantages of immobilization have been discovered, for instance Duvnjak & Lilly (1976) found that it was advantageous to immobilize glucose oxidase to a manganese oxide support in order to catalytically decompose the toxic hydrogen peroxide produced by the enzyme. Immobilization can also be used as a method of recovering an enzyme from solution once the reaction has been completed, followed by resolubilization and reuse.

Unfortunately, some enzyme activity is often lost during immobilization, owing to denaturation by heat, pH changes, free radicals, etc., generated during immobilization. However, in some cases such effects have proved useful, the required enzyme proving resistant to the denaturing conditions but the enzymes catalysing unwanted side-reactions being inactivated. Immobilization is also an extra operation, and extra cost, in a process, although this extra cost can usually be more than covered by reductions in the capital and running costs elsewhere in the process. Immobilized biocatalyst particles always occupy a larger volume in the reactor than do the equivalent amounts of free enzyme or cells. This factor is especially important with immobilized cells, because microbial cells have a diameter of $1-10\ \mu m$, compared to a diameter of $30-100\ \text{Å}$ for typical enzymes. Usually less than 1% of the cell mass consists of enzyme, but the remaining cell mass still takes up space in the immobilized cell particle.

Immobilized enzymes and cells are used in synthetic chemistry, in enzyme electrodes, thermistors and analytical devices (see Part A, Chapter 5), and are useful in therapy in the form of extracorporeal shunts. Also, techniques of enzyme immobilization have been made use of by biochemists wishing to study the tertiary and subunit structures of proteins and as models of membrane-bound proteins and enzymes (Bickerstaff 1984). Note that the use of other biological materials in an immobilized form is also common, for instance immunoglobulins (Genung & Hsu 1978), and drugs so as to effect their slow release over long time periods when the immobilized preparation is used as a clinical implant.

The recovery of immobilized biocatalysts, or their retention in the reactor, is achieved by the use of filters, as in packed-bed reactors, or by settling under gravity, or centrifugation if the biocatalyst particles are very small. Any requirement for ancillary equipment — such as basket centrifuges — is obviously undesirable and so it is preferable to maintain the biocatalyst as an integral part of the reactor (see Fig. 3.3).

The use of immobilized cells rather than fermenting cells or immobilized enzymes has become popular recently. Entrapment in gellified media is the most useful and widely used method of immobilizing cells. Ideally, the method should be mild, and not denaturing any of the required enzyme activity of the cells. Therefore the generation of heat, pH changes, or free radicals during the immobilization procedure should be avoided. The method of immobilization should be permanent, safe, cheap, simple, versatile, and easy to scale-up, and the reagents nontoxic so as to facilitate large-scale applications. The immobilized cell particles should also be small, so as to minimize internal diffusional restrictions, even-sized, spherical, and smooth, so as to produce even flow of substrate solutions through packed-bed columns of immobilized cells. Also, they should be mechanically strong, so as to be resistant to compaction in columns and to abrasion in stirred reactors.

At present, immobilized cells have been chiefly used in two ways. Firstly, in bioconversions catalysed by one enzyme (or only a very few enzymes) associated with the cells. These are often simple hydrolysis and isomerization reactions. Such processes are thus very similar to conventional immobilized enzyme processes. The cells do not need to retain viability or even structural integrity, and cofactor and energy source regeneration usually does not need to take place, for instance gitoxigenin can be bioconverted to 5-β-hydroxygitoxigenin using immobilized *Daucus carota* cells (Jones & Veliky 1981), sucrose isomerized to isomaltulose using immobilized *Erwinia rhapontici* cells (Cheetham *et al.* 1982) and di-menthyl succinate stereo-selectively hydrolyzed by gel entrapped *Rhodotorula menata var. texensis* cells, with the substrate dissolved in organic solvent (Omata *at al.* 1981). However, on some occasions more complex bioconversions take place; sorbosone formation from sorbose by *Gluconobacter melanogenus* requires an intact respiratory chain (Martin & Perlman 1976), and coenzyme A formation (Shimizu *et al.* 1975) is of interest.

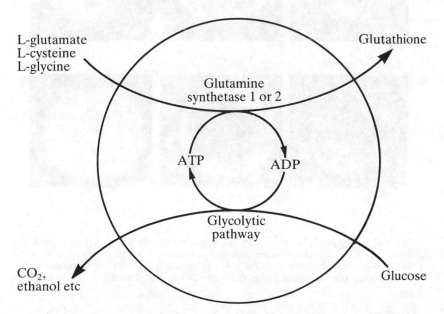

Fig. 3.4 − A schematic representation of a system for production of glutathione. Circular line in the figure shows the boundary of the polyacrylamide gel lattice (Murata *et al.* 1978). Note glutamine synthetase 1 and 2 from air dried *E. coli*, and the glycolytic pathway from acetone dried *S. cerevisae*.

Secondly, immobilized cells are often used in the form of growing immobilized cells where the immobilized cells are supplied with a growth medium containing the substrate to be transformed. Such reactions are more akin to conventional fermentations, particularly solid-state fermentations; primary metabolites usually being produced. Complex multistep conversions can be achieved concomitant with the growth and division of the immobilized cells, and cofactor regeneration and ATP renewal can take place, but the outgrowth of cells from the support and the need to maintain sterility can be major problems. Thus 1-isoleucine has been produced using immobilized *Serratia marcescens* (Wada *et al.* 1980a), glutathione by immobilized *S. cerevisiae* cells (Murata *et al.* 1978, 1981) (Fig. 3.4), bacitracin by a bacillus sp. (Morikawa *et al.* 1979), and ethanol by *Saccharomyces cerevisiae* cells (Wada *et al.* 1980b) (Fig. 3.5).

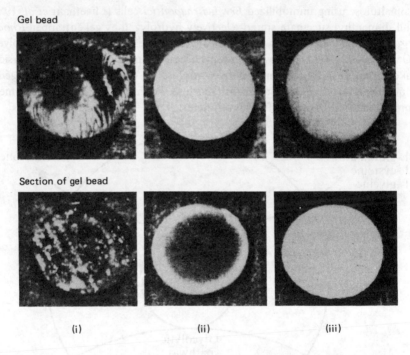

Gel bead

Section of gel bead

(i) (ii) (iii)

Fig. 3.5 — Photographs of sections of immobilized cell pellets, (i) of a small inoculum of cells, before incubation with nutrients, (ii) after incubation with nutrients and the growth of cells inside the gel, and (iii) of a high concentration of cells immobilized homogenously in the pellets (Wada *et al.* 1980b).

Immobilized cells can also be used as a source of enzymes, for instance proteases are secreted by immobilized *Streptomyces fordiae* cells (Kokuba *et al.* 1981) and α-amylase by *B. subtilis* cells (Kokuba *et al.* 1978) (Fig. 3.6).

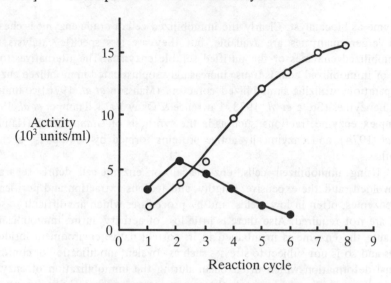

Fig. 3.6 – α-Amylase production by immobilized whole cells.
(o) immobilized whole cells; (•) washed cells. (Kokubu *et al.* 1978).

Subcellular organelles, plant cells, and even animal cells can be immobilized by similar techniques; for instance Aizawa *et al.* (1980) immobilized non-phosphorylating electron transport particles from beef heart mitochondria in agar and used them to oxidize NADH or succinate such that they could be used as an electrochemical measuring device when linked to an oxygen electrode. Carbon dioxide has been shown to be fixed by immobilized chloroplasts. (Kierstan & Bucke, 1977; Karube *et al.* 1979b)..

Rat pancreatic β-cells have been encapsulated in an alginate–polylysine complex where they are accessible to glucose but inaccessible to antibodies, so that they are not rejected when implanted in diabetic rats (Lim & Sun 1980). The pancreatic islet cells continued to secrete insulin, reversing the experimentally-induced diabetic state of the rats. Lastly a very interesting development is the report that whole filaments of the cyanobacter *Anabaena* entrapped in alginate gel sustained nitrogen fixation for at least 130 h. Over 90% of the nitrogen fixed was released from the cells as ammonia. This was achieved by inhibition of glutamine synthetase, the first enzyme involved in ammonia assimilation, although a significant loss in the stability of the cells was incurred by this treatment (Musgrave *et al.* 1982).

3.14 A COMPARISON OF IMMOBILIZED ENZYMES AND CELLS (see extensive review by Cheetham (1980))

The choice between the use of immobilized enzymes or immobilized cells is similar to the more familiar choice between the use of purified or crude soluble enzyme as biocatalyst. Clearly the immobilized cell or crude enzyme is cheaper, and larger quantities are available, but they are less specific catalysts than immobilized enzymes or the purified soluble enzymes. The alternatives to the use of immobilized cells is to use increasingly sophisticated immobilized enzyme preparations utilizing immobilized cofactors (Mansson *et al.* 1976) co-immobilized enzymes (Srere *et al.* 1973, Lawrence & Okay 1973, Tramper *et al.* 1978), complex enzyme fractions, such as in the synthesis of *Gramicidin S* (Hamilton *et al.* 1974), and enzymically active proteins formed by solid-phase chemical synthesis.

Using immobilized cells, enzyme-cell or enzyme-cell debris separation is avoided, and the expensive, lengthy, and tedious extraction and purification of enzymes, often in low yields, and by procedures which are difficult to scale-up, are not required. Also there is little loss of activity during immobilization, because the enzyme is maintained in its correct native environment inside the cells and so is not subject to stresses such as covalent modification or conformational deformation which often occur during the immobilization of enzymes. The cell membrane also protects the enzyme from mechanical denaturants such as shear forces and gas bubbles, and excludes many chemical denaturants such as heavy metals and organic solvents. In most immobilized cell preparations the cells are made inaccessible to potentially invasive microorganisms, and nitrogen-free substrates are used, so that the maintenance of sterile operating conditions is not required. Thus the operational stabilities of immobilized cell enzyme activities are often greater than those of the corresponding immobilized enzyme, particularly if the immobilized cell activities can be regenerated *in situ*.

The use of immobilized cells should be preferred to immobilized enzymes for synthetic endergonic reactions requiring the use of coupled enzymes, expensive co-factors and energy sources such as NADH or ATP; the use of complete metabolic pathways or the complete metabolism of the whole cells. Immobilized cells vary widely in their physiological status, various descriptive terms are used such as living/dead, viable/non-viable, whole/disrupted etc. Some agreement needs to be reached on a rigorous nomenclature for immobilized cells.

There is an absolute need for cell immobilization when an enzyme cannot be extracted in a stable form, is easily inactivated during immobilization, or when the regulating mechanisms associated with the metabolic pathways are required. Cell immobilization cannot be used for processes involving extracellular enzymes unless it is acceptable for the enzymes to continuously leak out of the immobilized cells during use, such as the α-amylase produced by *Bacillus subtilis* entrapped in polyacrylamide (Kokubu *et al.* 1978). In batch use the α-amylase formed was three times that formed by washed cells, with the enzyme produced increasing with re-use until a steady state was reached after 7 cycles. (Fig. 3.6). Immobilized cells are in an artificial stationary or resting phase induced by nutrient deficiency or by lack of space available for new cells, so that cell growth and product formation are not associated. It is clear that the synthetic activity of the cells can only be maintained for a limited period, depending on the stabilities of the relevant enzymes. Decay can be due to reversible or irreversible inactivation of the enzymes, loss of intermediary metabolites, or inhibition by metabolites. Because of the lack of essential nutrients only non-growth associated metabolic functions are carried out by the cell, such as osmotic and pH regulation and the turnover of macromolecules. This maintenance energy requirement varies with the type of cell and the environmental conditions. Prolonged cell viability is associated with a low rate of endogenous metabolism closely matched to maintenance energy requirements, which may explain the great stability of immobilized cell activities. The operational stabilities may often be greatly enhanced by regeneration of the enzyme activities of the immobilized cells. Regeneration may be achieved either by re-induction, or by causing the immobilized cells to divide *in situ*.

Immobilized whole cells are likely to produce less pure products than immobilized enzymes because unwanted side-reactions and further metabolism of the desired biochemical may occur. Therefore, screens of potentially useful microorganisms for cell immobilization should include a survey of side-reactions. These undesired reactions can often be eliminated by selective denaturation or inhibition of the unwanted enzyme activities of the cell, or by use of substrates of greater purity. For instance, formation of the side-product succinic acid during malic acid production can be prevented by the addition of bile salts to the substrate solution (Sato *et al.* 1976). The role of the bile salts may be to selectively denature the enzymes required for succinic acid formation, or more probably to impair cell wall integrity, so allowing loss of the cofactors required for succinic acid formation. This latter hypothesis is especially likely, since after bile salt treatment the activity of the immobilized cells was enhanced, presumably owing to the loss of the diffusion barrier to substrate and product transfer caused by an intact cell membrane. Such permeablisation procedures make cytosolic enzymes functionally equivalent to membrane-bound or periplasmically-located enzymes. Further metabolism of the desired product can also be prevented. For instance, in the production of urocanic acid the enzyme urocanase is selectively heat denatured at 70°C for 30 min. (Yamamoto *et al.* 1974b). Alternatively, selection of cells possessing a large ratio of synthetic to degradative activity can be used, such as in the synthesis of 6-amino penicillic acid where the penicillin amidase activity of immobilized *Escherichia coli* cells was 10 times greater than the penicillinase activity (Chibata *et al.* 1977). Selection of the best microroganism is also important; for instance, in the formation of L-citrulline most organisms converted the citrulline into L-ornithine using ornithine transcarbamylase, but Yamamoto *et al.* (1974a) discovered that *Pseudomonas putida* did not further metabolize the desired product. Using immobilized cells, the yield of product can be lower than that obtained with the corresponding immobilized enzyme because of the substrate consumed to provide maintenance energy for the cells.

Contamination of the final product by cells or materials derived from the immobilized cells may also occur, particularly as cell lysis has been frequently observed when cells have been used for long periods. The cells lost may be either the originally immobilized cells, or more usually cells produced by multiplication of the immobilized cells. Only the originally immobilized cells can be leached out of the column when the reactor is operated in the absence of nutrients and in the presence of antibiotics. However, cell leakage is likely to be especially marked when the complete metabolism of the cell is retained after immobilization so that division of immobilized cells is possible.

The sizes of immobilized cell particles are often an order of magnitude larger then those of immobilized enzymes. This partially reflects the larger size of cells as compared with enzymes. But it may also reflect the higher catalyst densities of the later ($\times 10-100$) and the consequently more severe diffusional limitations on activity which restricting the size of the immobilized enzyme particles used.

Because of the relatively large size of immobilized cells, high molecular weight substrates can only be metabolized by immobilized cells used in ultra-filtration cells, the low molecular weight products passing through the ultra-filtration membrane. Additional disadvantages are that the activity of the immobilized cells per unit reactor volume will always be lower than that of the corresponding immobilized enzyme, and so larger reactors or longer residence times will be required to achieve the same productivity. The catalyst site density of immobilized cell preparations will be lower than in the corresponding immobilized enzyme preparations, so that competition between simultaneous enzymic and nonenzymic reactions will be enhanced. An example of such a nonenzymic reaction is psicose formation during glucose isomerization (Bucke 1977). Diffusional limitations (mass-transfer effects) on reaction rates using immobilized cells will be large because of the additional diffusional barrier presented by the cell wall and membrane, so emphasizing the need for highly porous immobilization supports and permeablization of the cell, for instance by exposure to toluene, dimethyl sulphoxide or use of the antibiotic nystatin which forms small pores reversibly (Mosbach 1982). Finally, the activity and stability of immobilized cell preparations may be adversely affected by the action of intracellular proteases, necessitating the use of protease inhibitors.

3.15 AN ASSESSMENT OF IMMOBILIZATION SUPPORTS AND METHODS
(also see Part A, Chapter 4)

Great ingenuity has been devoted to the choice of suitable immobilization supports and methods of immobilization, for instance enzymes have been immobilized to millipore membranes which were then used packed in a column reactor and to the ceramic monoliths originally developed as catalytic after-burners for internal combustion engines (Benoit & Kohler 1975). It was found that substantial improvements in the activity of immobilized RNA-ase activity could be obtained by supplying the yeast RNA substrate in an aerosol form (Kirwan *et al* 1974) One of the first process uses of immobilized enzymes is illustrated in Fig. 3.7.

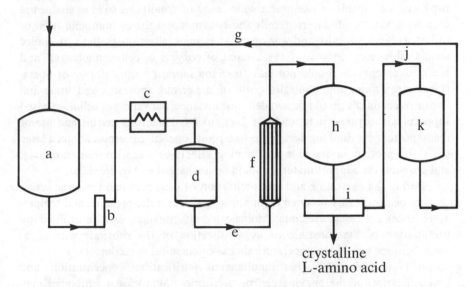

crystalline
L-amino acid

Fig. 3.7 – A schematic description of a continuous process for the formation and separation of L-amino-acids from a racaemic (DL) mixture using immobilized aminoacyclase. Redrawn from Chibata *et al.* (1976). (a) is the acetyl-DL-amino acid reservoir, (b) a filter, (c) a heat exchanger, (d) the enzyme column, (e) the L-amino acid stream, (f) a heat exchanger, (g) the DL-acetyl amino acid stream, (h) the crystallizer, (i) a separator, (j) the acetyl-D-amino acid stream, and (k) the racemization tank.

Immobilization methods can be either 'active' for instance covalently coupled enzymes or 'passive', for instance absorbed enzymes. A great variety of natural or synthetic, organic, or inorganic materials have been used as immobilization supports with these materials, differing in size, shape, density, and porosity, and being used in the form of sheets, tubes, fibres, cylinders, or most popularly spheres. For a review see White & Kennedy (1980). The size of the support particle used is a crucial factor as it helps determine the extent of internal diffusional restrictions on enzyme activity, the pressure drop generated in packed-bed columns, the velocity of fluid flow required to fluidize columns, and the power input needed to suspend particles in stirred reactors. Specialized requirements are often apparent; for instance, when an immobilized enzyme produces hydrogen peroxide, use of co-immobilized catalase or manganous oxide will catalyse the decomposition of the peroxide before it can denature the enzyme.

The immobilization method of choice is simple and reproducible. It should also be mild, cheap, safe, versatile, and easy to use on a large scale. The actual immobilization method used depends on a variety of scientific, engineering, and economic aspects of the process. The method should also be cheap, generate no fines, and be capable of use over a wide range of conditions so as to enable use of a wide variety of different cells and enzymes and the co-immobilization of different types of cells and enzymes. The immobilization method of choice should allow easy control of the amount of enzyme or cells immobilized, and the cells or enzyme should not leak from the support during storage or operational use. During immobilization, use of dangerous apparatus and toxic and corrosive chemicals should be avoided. For instance, activation of cellulose based supports with cyanogen bromide or 2-amino 4,6-dichloro-S-triazine and use of many protein crosslinking agents are hazardous operations, especially on a large scale. To prevent undesirable partitioning effects between support, substrate, and product, the support material should be uncharged and hydrophilic.

During the harvesting and immobilization of cells maximum enzyme activities can be most easily retained by minimising pH, osmotic pressure, and temperature shocks. During use, microbiological contamination can be avoided by prefiltration of the substrate or by sterilization of the substrate with ultraviolet light, or by pH or temperature shocks or chemical bacteriocides.

It is also desirable that simultaneous purification, concentration, and immobilization of the enzyme can be performed. Monoclonal antibodies may eventually be cheap and available enough to use as immobilization reagents in this way. No generally accepted 'best' method for immobilizing enzymes or cells is recognized, although some companies and research groups may have their favourite methods. Thus early on in experimentation it is very important to carry out a screening exercise whereby the available methods are compared. This is because the activity, stability, and ease of use of the immobilized biocatalyst can vary enormously and unpredictably with the immobilization method used. On occasions it has been found that even when a high proportion of activity is retained when assayed immediately after immobilization, it decays very rapidly, presumably having been destabilized during the immobilization reaction.

Many factors must be considered when choosing an immobilization method. These include the chemical nature and kinetic features of the reaction, the costs and the chemical and physical stability of the reactants and biocatalyst, and the effect immobilization may have on the activity and stability of the biocatalyst, the yield and purity of product required, as well as many other more specific effects. Thus a compromise must be reached since the important variables rarely act independently of each other, and several minor often unrelated effects can combine to produce a very undesirable effect such as loss of stability or excessive formation of side-product. Such behaviour can be simplified by compiling an 'interaction diagram' which illustrates the causal relationships between the various parameters semi-quantitatively. The eventual scale and conditions of operation in a production facility and the intended uses of the product must also be borne in mind.

3.15.1 Characterisation of immobilized biocatalysts

The final activity of an immobilized enzyme preparation is determined by the activity of the original cells or enzyme, the concentration of enzyme or cells that can be permanently immobilized. the extent to which the enzyme activity is retained after immobilization, and the effect of partitioning and diffusional effects on the enzyme activity. Note that because of the high substrate concentrations that are often used, increases in the K_m of an enzyme upon immobilization may have little effect on its performance, whereas decreases in V_{max} will of course be deleterious.

In a typical immobilization experiment the objective is to achieve as high an active enzyme loading per unit volume of immobilization support as possible. Therefore the following parameters are normally measured: the volume, enzyme activity, and protein content of the free enzyme used; the weight, particle size distribution, porosity, and the chemical and physical nature of the support used; the enzyme activity of the immobilized enzyme, and the volume, enzyme activity, and protein concentration of any free enzyme remaining after immobilization has been completed. Similar measurements are made when cells are immobilized. Note should also be taken of any swelling, shrinking, aggregation, or fragmentation of the support during immobilization or upon contacting with substrate, of any leakage of the biocatalyst from the support, or any carry over of immobilization reagents or free enzyme in the pores of the support, and also of the ease with which substrate can diffuse into the biocatalyst particles.

Usually, immobilization is irreversible, and so the immobilization process is dominated by the slow diffusion of enzyme in the pores of the support material. The distribution of immobilized enzyme is initially nonuniform with the enzyme concentrated around the outside of the particle, an effect that can sometimes be demonstrated by staining the immobilized enzyme (Carleysmith *et al.* 1980). Eventually the enzyme becomes uniformly distributed throughout the support given sufficient time for the equilibrium to be set up, and provided that the distribution of immobilization sites within the support is uniform and that there is sufficient enzyme available to saturate the support. When enzyme is non-uniformly distributed around the surface it tends to be more active than equivalent preparations in which the enzyme is uniformly distributed. This is because diffusional restrictions are less important owing to the shorter diffusion pathway. However, these preparations often deactivate more rapidly than the former because they tend to be more exposed to denaturants. Thus a compromise can be made between an immobilized enzyme in which the enzyme is concentrated near the surface which is more active, but less stable, and a uniformly distributed immobilized enzyme which although being less active and sometimes less selective if multiple reactions occur, is more stable. This choice parallels the difference between free and immobilized enzymes in which the latter, although usually being less active, are often more stable.

For methods of immobilization see Part A; Chapter 4. In our work we have entrapped cells of *Erwinia rhapontia* in gel pellets of calcium alginate on a pilot plant scale to form isomaltulose (Cheetham *et al.* 1982).

3.16 CO-IMMOBILIZED ENZYMES

Usually, single enzymes are immobilized, but an exciting new area of research has begun on the co-immobilization of two or more enzymes so as to form multifunctional catalysts. These should be capable of more complex conversions than are possible using a single enzyme. For instance, glucose oxidase and catalase have been co-immobilized so that the hydrogen peroxide produced by the glucose oxidase is degraded by the catalase before it can inactivate the glucose oxidase. Similarly, xanthine oxidase has been stabilized by co-immobilization with catalase and superoxide dismutase, to rapidly remove the denaturant molecules produced by the oxidation reaction (Tramper *et al.* 1978). Adenylate kinase and acetate kinase have also been co-immobilized in poly-acrylamide gel and used to regenerate ATP and/or ADP using acetylphosphate as the phosphorylating agent. Another example is the use of co-immobilized enzymes still associated with their parent cells to convert sorbose into 2-keto-1-gulonic acid, a precursor of vitamin C (Martin and Perlman 1976). Other potentially useful examples include: co-immobilized glucoamylase and glucose isomerase; and glucoamylase and pullulanase. A further development is the co-immobilization of enzymes to form 'metabolic pathways' which do not occur naturally in order to carry out novel syntheses.

Co-immobilized systems are likely to be more efficient than the use of separately immobilized enzymes since intermediates can diffuse between the different enzymes much more easily. Co-immobilized enzymes usually exhibit shorter lag periods prior to the establishment of a steady-state compared with separately immobilized enzymes. The main difficulty with this technique appears to be the establishment of satisfactory operating conditions, such as pH and temperature, appropriate to the activity and stability characteristics of all the enzymes being used, for example glucoamylase (amyloglucosidase) and glucose isomerase have similar stabilities but different pH optima (pH 4.5 and 8.0 respectively); and β-galactosidase and glucose isomerase not only have different pH optima but the β-galactosidase is less heat stable.

Further scope for improved systems probably lies in the formation of composite biocatalysts containing co-immobilized cells, enzymes, and organelles from different sources, for instance D'Souza & Nadkarni (1980) used glucose oxidase bound to yeast cells induced for maximum invertase and catalase activities to convert sucrose into fructose and gluconic acid. Another approach is to immobilize coenzymes with enzymes. By this means the coenzyme can be covalently linked such that an allosteric site on an enzyme is filled, thus increasing the activity of the enzyme. Alternatively an enzyme which can recycle the cofactors, and the cofactor, can be separately bound to the cofactor requiring enzyme so that the cofactor may move into the active site of each enzyme in turn, being alternately oxidized and reduced (Mosbach, 1982).

3.17 TWO-PHASE REACTIONS

Rather than use immobilized enzymes or cells, another approach which is gaining in popularity is to conduct reactions in the presence of a second aqueous-phase such as an organic solvent (Lilly 1983). Thus high concentrations of substrates which are normally water-insoluble can be maintained; for instance steroids. Alternatively the second phase can compromise a water-immicible substrate, for instance a hydrocarbon. Water-miscible solvents are not normally useful because they are potent denaturants of enzymes, although there are some exceptional enzymes such as the isoprenoid alcohol kinase from *S. aureus* which is soluble in butanol and other solvents but insoluble in water (Sandermann 1974). (Fig. 3.8). On the contrary when a water immiscible solvent is used the enzyme is exposed to a very much lower concentration of solvent molecules and can be much more stable. Conventionally the reaction is carried out in a stirred reactor in order to maintain a fine emulsion with the enzyme present in the aqueous phase, which may be either the continuous or the dispersed phase.

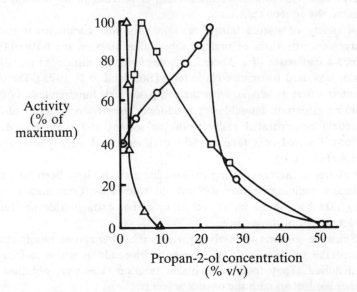

Fig. 3.8 – The effect of the concentration of the water-miscible organic solvent propan-2-ol on cholesterol oxidase activity, o, Δ, and Ψ are forms of the enzyme extracted using Triton X-100, trypsin and phosphate buffer respectively (Cheetham *et al.* 1982).

Since microorganisms and enzymes are generally believed to act on dissolved substrates, the low water solubility may greatly restrict product yields and production rates. Substrates with low water solubility can be made more available to cells and enzymes through the use of surfactants, water-miscible solvents, and by grinding to small particle sizes, although the mechanisms of mass-transfer in such systems are rather unclear. Water-insoluble organic solvents serve as a vehicle for the rapid transport of substrates and products between the organic and aqueous phases. These solvents, by serving as a reservoir for substrates and products, can minimize the effects of substrate or product inhibition of the reaction. In addition, these solvents can substantially reduce the volume of the reaction mixture needed to produce a given amount of product because large amounts of the water-insoluble substrate can be dissolved in the organic phase. Another potential advantage is that oxygen solubility is usually much greater in organic solvents than in water. This property may be advantageous for conducting reactions that have an oxygen requirement.

Schwartz & McCoy (1977) found that the yields of epoxide derivatives of octadiene were increased fivefold by adding 20–50% cyclohexane to cultures of *Pseudomonas oleovorans*. Up to 90% of the octadiene added was converted to epoxide products. The cyclohexane was not metabolized, and the yield increases were not due to increases in the dissolved oxygen concentration. The mono-epoxide product was quite inhibitory to the microorganism, and the organic solvent minimized this inhibition by reducing the amount of epoxide present in the aqueous phase

A variety of steroid transformations have been conducted in the presence of large concentrations of organic solvent; for a review see Butler (1979). For instance a cell paste of a *Nocardia rhodochous* sp. suspended in carbon tetrachloride was used to oxidize cholesterol (Buckland *et al.* 1975). The only water present was that associated with the cell paste. One hundred grams of cell paste in 200 ml of carbon tetrachloride produced cholestenone at a rate of 7 g/h. The cells could be separated easily from the solvent and then re-used for more reactions. Re-used cells retained 50% of their initial activity after seven batch reactions (69 h use).

Cell-free cofactor-requiring oxido-reductases also have been used in reaction mixtures containing organic solvents. In studies by Cremonesi *et al.* (1973, 1975), 50% by volume butyl acetate or carbon tetrachloride was found to be the best solvent of those tested.

Since the solvents themselves partially inhibit enzymes, Bhasin *et al.* (1976) replaced the solvents with silicone polymer beads in which steroid substrate was dissolved. They found that higher reaction rates were obtained with the polymer beads than with the organic solvents.

A somewhat different use of a water-immiscible solvent is to lower the water activity so as to reverse hydrolytic reactions, Klibanov *et al.* (1977) and Semenov *et al.* (1981) found that they could immobilize chymotrypsin on glass beads and then suspend the beads in a chloroform solution containing ethanol and N-acetyl tryptophan. The immobilized enzyme catalysed the reverse of the hydrolysis reaction and produced esterified amino acid in 100% yield, whereas in water the yield of peptide is only 0.1%. The structured water layer that surrounds the beads (that is, the Nernst layer) apparently protected the enzyme from denaturation by the solvent. Enzymatic hydrolysis reactions have also been reversed using water-miscible compounds such as glycerol and ethanol to lower water activity (Ingalls et al. 1975).

Enzymes are likely to be stabilized against denaturation by the organic solvent phase or organic solvent molecules dissolved in the aqueous phase by immobilizing them in a porous support in which an aqueous microenvironment can be maintained. In this case use of rather hydrophobic support matrixes may be important so as to encourage partitioning of the substrate molecules to the enzyme. For instance β-hydroxysteroid dehydrogenase was markedly stabilized by immobilization to cyanogen bromide-activated Sepharose, and it retained 60% of its original activity after 2 months' continuous usage in a water-ethyl-acetate biphasic system. The organic solvent acted as a weak inhibitor of the enzyme (Carrea et al. 1979).

An interesting new development is the use of two or more mutually immiscible aquous phases containing for instance dextran and polyethylene glycol (Pollack & Whitesides 1976). Such two-phase aqueous emulsions are characterized by a much lower interfacial tension ($0.1-100$ $\mu N/m$) than is common for conventional oil–water or water-immiscible solvent-water emulsions ($1000-2000$ $\mu N/m$). Therefore the power input required to produce and maintain the aqueous emulsions is smaller than is needed for the former type of emulsion. Use of two-phase aqueous mixtures also often allows the recovery of enzymes or cells in one phase and the products of reaction in the other; for instance Kuhn (1980) has carried out alcoholic fermentations in aqueous two-phase systems based on dextran, polyethylene glycol, and water.

3.18 INDUSTRIAL ENZYME KINETICS (also see Sections 3.6, 3.20, 3.21 and 3.24)

The quantification and prediction of the performance of complex systems such as immobilized enzymes and cells can be achieved by using mathematical models. Modelling may enable the prediction of the behaviour of large-scale reactors, give an understanding of the way in which processes act as an integrated whole, identify specific areas in which improvements can be made, and facilitate the transition from laboratory to industrial practice. Equations should be simple and general, that is, they should apply to a number of enzyme or cell substrate systems even though the enzymes and cells in different systems will not be genetically, morphologically, and biochemically identical.

Initial rate determinations of enzyme activities are made in classical enzymology by measuring substrate consumption or product formation over short time periods. This is done in aqueous solutions and employs dilute enzyme and substrate solutions, such that the rate of reaction equals V_{max}. Industrially, however, we often want the complete reaction of concentrated substrate solutions using concentrated enzyme preparations. This may be required to occur continuously for long time periods, often in nonhomogenous solutions containing immobilized enzymes or cells, and sometimes in organic solvents. It is well understood that mass transfer limitations may have great influence on the intrinsic kinetics. Transient (rather than steady-state behaviour) is only of interest in order to predict response times and recovery times between start-up and shut-down periods caused by cleaning and maintenance etc.

Because there is often a low average substrate concentration during industrial reactions, first, or mixed first and zero-order kinetics, rather than zero-order Michaelis–Menten kinetics apply (Fig. 3.9) such that the rate of reaction depends directly on the prevailing substrate concentration, that is,

$$v = \frac{V_{max}\,[S]}{K_m} = \frac{k_2\,[E]\,[S]}{K_m} = \frac{-\mathrm{d}[S]}{\mathrm{d}t} = \frac{\mathrm{d}[P]}{\mathrm{d}t}.\qquad(3.16)$$

Therefore the rate of reaction is usually controlled by the amount of substrate and enzyme present such that the time required to achieve a given product concentration can be controlled by the concentration of enzyme present. If a batch reactor is used, the maximum rate of reaction occurs very soon after starting the reaction. The catalytic efficiency of the enzyme decreases, however, during the course of the reaction such that increasingly large concentrations of enzyme are needed to maintain the initial rate of reaction. This is especially important when progressively more highly converted substrates are required. Here the increased contact times of reactants with the enzyme (which are required to

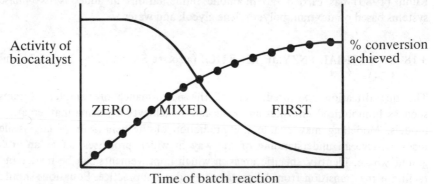

Time of batch reaction

Fig. 3.9 – An illustration of the effect of the time for which a batch enzyme catalysed reaction is allowed to proceed on the activity of the enzyme (–) and the extent of conversion of substrate to product (●——●). Reproduced with permission from *Principles of Biotechnology* (Ed. Wiseman, A.) Surrey University Press, Glasgow and London 1983.

achieve these high degrees of conversion) allow the possibility of chemical or biological side-product formation or further metabolism of the desired product. Such problems are increased if the reaction is carried out at elevated temperatures. An interesting side-reaction is the formation of colour and psicose during the production of high-fructose syrups by immobilized glucose isomerase, and another is the hydrolysis of the β-lactam ring of benzyl penicillin during enzymic deacylation to 6-aminopenicillinic acid. A useful parameter to measure is the half-time or life of a reaction, that is the time required for half of the substrate originally present to be converted to products by the enzyme. This term should be distinguished from the half-life of the enzyme, a measure of its stability.

The reaction velocity in the Michaelis–Menten equation is the velocity of the enzyme catalysed reaction when the concentration of products tends to zero. This is a very useful parameter when studying the properties of enzymes, but a form of the equation integrated with respect to time is of greater value in industrial operations where the reaction is usually required to proceed to complete conversion or equilibrium. That is:

$$V_{max} = K_m \ln \frac{[S]}{[S_t]} + ([S] - [S_t])$$ (3.17)

where

t is the time of reaction,

$[S]$ is the initial substrate concentration,

and $[S_t]$ is the substrate concentration after time t.

$([S] - [S_t])$ is the product concentration after time t.

This equation illustrates the relationship between the time required to achieve a particular product concentration (that is, $[S] - [S_t]$) and the enzyme concentration (since $v = k + 2 [E]$), assuming that the enzyme is stable. This relationship is important in industrial applications since the rate of reaction is usually controlled by varying the enzyme concentration used and/or the period of reaction rather than the substrate concentration, and the degree of conversion achieved rather than the rate of reaction is the most important parameter to measure.

The time of reaction (t) required for a certain concentration of enzyme to change the concentration of substrate from $[S]$ to $[S_t]$ is given by the relationship:

$$t = \frac{SX + K_m \ln [1/(1-X)]}{Ek_2}$$ (3.18)

where X is degree of conversion and E is total enzyme activity expressed, which is useful in estimating and minimizing the concentration of enzyme and incubation time required. The half-life of the reaction ($t/2$) is the reaction time required for the substrate concentration to fall to half of its original value. That is:

$$t/2 = \ln 2 \, \frac{K_m}{V_{max}} + \frac{S}{2V_{max}} \, . \tag{3.19}$$

Thus the lower the half-life of the reaction then the more active the enzyme that is used, and/or the lower the concentration of substrate that is being used. (Fulbrooke, 1982).

In equation 3.18 the term E is the product of the weight of enzyme present in the reactor and its activity (V). Thus the size of the reactor required to achieve a given productivity can be calculated by substituting the term $2\pi r^2 l$. $\rho . V$. for E where r is the radius of the reactor, l its height and ρ the bulk density of the enzyme preparation.

3.18.1 Effects on equilibria

A difficulty arises because enzymes do not affect the position of equilibrium in a chemical reaction only when present at very low concentrations, whereas in many commercially important reactions the concentrations of enzymes used are very high, and so much of the substrate can be present as an enzyme—substrate complex. If the equilibrium constant is very high the reaction can be treated as going to completion. If, however, the equilibrium constant is relatively low, such as that between glucose and fructose in the reaction catalyzed by glucose isomerase, it is necessary to separate the desired product from the reactor output stream followed by recycling of the remaining substrate to achieve a high enough degree of conversion. If two or more substrates are involved they can both only be completely converted in a single pass through the reactor if they are supplied in the correct stoichiometric proportions. When one reactant is very much cheaper than the other(s) it may be advantageous to supply it in excess so as to maximize utilisation of the more expensive reactant(s).

When a reversible enzyme catalysed reaction is taking place, which is especially likely when high degrees of conversion of substrate to product are being achieved, the mechanism is:

$$[E] + [S] \, \frac{k_1}{k_{-1}} \, [ES] \, \frac{k_2}{k_{-2}} \, [E] + [P] \, . \tag{3.20}$$

The rate equation for the forward reaction is expressed by (3.21) using the Michaelis—Menten constant for the forward and reverse reactions:

$$v = \frac{V_{max}{}' \, (S - P/K_{eq})}{K_m \, (1 + P/K_m{}') + S} \tag{3.21}$$

where K_m and $K_m{}'$ are the Michaelis constants of the forward and reverse reactions respectively. K_{eq} is the equilibrium constant and is expressed as the Haldane relationship:

$$K_{eq} = P_{eq}/S_{eq} = V_m \cdot K_m{}'/V_m{}' \cdot K_m \, , \tag{3.22}$$

where P_{eq} and S_{eq} are the concentrations of product and substrate at equilibrium, respectively. V_m' is the maximum reaction rate of the reverse reaction, and K_m' the Michaelis–Menten constant for the reverse reaction.

In a plug flow type reaction, the time-conversion equation is derived by integrating (3.21) and is given by:

$$V'_{max} \cdot t = \frac{K_{eq} (K_m' - K_m)}{K_m' (K_{eq} + 1)} S \cdot X$$

$$- \frac{K_{eq} [K_{eq} \cdot K_m' (K_m + S) + K_m (K_m' + S)]}{K_m' (K_{eq} + 1)^2} \tag{3.23}$$

$$X \ln (1 - X - X/K_{eq}) \text{ (Yamamoto } et\ al.\ 1977).$$

Note that when the breakdown of the enzyme–substrate complex is reversible it is possible, depending on the relative magnitudes of the K_m and V_{max} values of the forward and reverse reactions, for transitorily high concentrations of 'product' to be obtained when the enzyme is used in reverse, even when no 'product' can normally be detected at equilibrium (Kashe & Galunsky 1982).

3.19 EFFECTIVENESS FACTORS

The activities of immobilized enzymes are usually expressed in terms of the number of grams of product formed per gram of immobilized enzyme/cells used per hour, or as g of product formed per ml of reactor volume per unit time.

The combined effects of the factors which affect the intrinsic properties of the enzyme are expressed as effectiveness factors. These are sometimes referred to as coupling efficiencies or activity yields. The effectiveness factor is the activity of the immobilized enzymes (or cells) divided by the activity of an equivalent quantity of free enzyme (or cells). Assay must be under the same conditions. Stationary effectiveness factors (η) are measured under initial rate conditions and so compare the activities of the free and immobilized enzymes under zero order kinetic conditions with respect to the substrate concentration (that is, excess substrate). Operational effectiveness factors (η_o) are measured under the conditions required or expected in a production facility, and so are usually measured under first-order kinetic conditions such that the activity of the enzyme is limited by the amount of substrate available (the limitation is due to the need to achieve complete conversion of the substrate into product, and substrate concentration must decrease therefore towards zero). The most useful guide to the practical usefulness of an immobilized enzyme preparation is the operational effectiveness factor. Where an effectiveness value of 1.0 is obtained, this indicates that the activity of the enzymes or cells has not been reduced appreciably by immobilization or diffusional restrictions. If values less than 1.0

are obtained, this indicates overall the amount of activity lost. Such loss is the sum of loss during immobilization and the extent of diffusion limitations and/or adverse partitioning effects on the activity of the enzyme. It is important to note therefore that when using an expensive or scarce enzyme it is desirable to obtain a high effectiveness value. Alternatively, using a cheap abundant enzyme, high turnover of substrate is obtainable by use of a high enzyme loading. One can obtain effectiveness factors higher than unity, in several situations: (i) when immobilized cells lyse, so making intracellular enzymes more available to their substrate, (ii) when cell division occurs inside immobilized cell preparations, (iii) when inhibitors are selectively excluded from the support matrix and thus from the environment of the enzyme, (iv) when heat transfer limitations in the particle combined with an exothermic reaction cause the temperature of the support particles to rise so increasing the activity of the enzyme, (v) simply because the immobilized enzyme activities are often more stable than the corresponding free enzyme such that when assayed after a period of use the immobilized enzyme can often be much more active.

The Damköhler and Thiele modules are dimensionless numbers that denote the ratio between the maximum rate of enzyme reaction (V_{max}) and the maximum rate of substrate transfer in the vicinity of the immobilized enzyme or cell. Values demonstrate what factor(s) limits the effective activity of an immobilized biocatalyst and give an indication of how far that activity falls short of the maximum possible activity. Under steady-state conditions the rate of reaction of substrate must equal the rate of substrate transfer to the enzyme/cell such that the Damköhler number (Da), which can be defined as:

$$Da = \frac{[S_b] - [S_s]}{[S_s]} \qquad (3.24)$$

equals 1.0 (Kashe 1983). When Da is >1.0 there is a steep concentration gradient through the immobilized biocatalyst particle because the enzyme is very active, because the diffusion of substrate is restricted, or both, such that there is a large difference between $[S]_b$ the bulk substrate concentration and $[S_s]$ the substrate concentration at the surface of the cell or enzyme. That is, the reaction is diffusion or mass-transfer controlled.

By contrast, when Da is <1.0 the reaction rate depends solely on the activity of the immobilized biocatalyst, and there is only a negligible difference between $[S_b]$ and $[S'']$, that is the system is reaction controlled.

In graphical terms, if the observed rate of reaction is plotted against the observed activity expressed as a function of the maximum possible rate of reaction (V_{max}), then a graph results in which the observed rate of reaction increases where $Da = 1.0$ and then continues as a plateau at the same value over the region where $Da > 1.0$. Note also that from the above definition of Da, $[S_s]$ can be estimated when $[S_b]$ is known (Kashe 1983).

3.20 STEADY-STATE KINETICS (see Part A, Sections 3.6, 3.18 and 3.24)

For a system where the substrate diffusion coefficient is independent of the substrate concentration, where the enzyme is evenly distributed on the surface of the support and all the enzyme molecules are equally active, the rate of substrate transfer equals the rate of reaction at steady state. That is

$$a_m K_L (S - S_s) = \frac{V_{max}^s S_s}{S_s + K_m} \,, \tag{3.25}$$

where V_{max}^s is the maximum reaction rate per unit external surface area, K_L is the liquid mass transfer coefficient (cm s^{-1}), S is the bulk substrate concentration, S_s is the substrate concentration at the surface of the particle, and a_m is the surface per unit volume (cm^{-1}).

When an enzyme is immobilized evenly inside a spherical support and no partitioning of substrate occurs and no temperature or pressure gradients are set up within the immobilized enzyme particle, then the rate of enzymic reaction is governed by the rate of diffusion of substrate molecules which is best described by Ficks Law:

$$\frac{dS}{dx} = D \,, \tag{3.26}$$

where D is the measured diffusion coefficient of the substrate and x is the distance from the centre of the support. Thus when a steady state has been set up where the rate of consumption of substrate is equal to and limited by the flux of substrate through each successive transport zone, the kinetics of the immobilized enzyme is described by

$$\frac{d^2S}{dx^2} + \frac{2}{x} \cdot \frac{dS}{dx} = \frac{V_{max}^v S}{D(K_m + S)} \tag{3.27}$$

where V_{max}^v is the maximum enzyme activity per unit volume of support, and x the distance the molecule must travel.

The kinetics of immobilized cells can usually be treated in a similar manner to that described above. However, the kinetic behaviour of immobilized whole cells is complicated by the possibility of cell division after immobilization, by the presence of an additional barrier to substrate and product diffusion – the cell wall and membrane, and by the presence of active transport systems called permeases which transport some large molecules, such as sugars, into the cell. The movement of most molecules, including oxygen, is by diffusion and so is governed by Fick's Law, but the active transport of nutrients is described by a Michaelis–Menten type relationship. Very often it is necessary to permeabilize the immobilized cells so as to facilitate substrate transfer. Permeabilization is somethimes effected by immobilization, particularly when the less mild methods are used, for instance E. coli cells possessing aspartase activity are permeabilized by immobilizing in polyacrylamide gel. More usually, permeabilization has to be carried out deliberately in a separate step, for instance in the formation of glucose-6-phosphate by *Achromobacter butyrii* cells which are treated with organic solvents (Murata, *et al.* 1979), and *Brevibacterium ammoniagenes* cells which are treated with bile salts when producing malic acid (Yamamoto *et al.* 1977). Simple, accurate kinetic descriptions of immobilized cell systems are, however, more difficult when the reaction is carried out by growing immobilized cells supplied with nutrients in addition to the substrate for the desired reaction. This latter situation is close to that encountered in conventional fermentations.

A model for a single metabolizing cell or enzyme consists of it being surrounded by outer concentric transport zones. If substrate transfer occurs entirely by diffusion, then assuming unidirectional flow and an absence of partitioning effects the substrate flux due to diffusion is described by Fick's Law. As the cell or enzyme can be regarded as a substrate sink, activity can be reduced to a form of Michaelis–Menten kinetics where the rate coefficients for the individual steps are condensed into the coefficients. This is because the kinetics of the complete metabolic pathway are chiefly determined by the rate limiting reactions. Thus when immobilized biocatalysts are used continuously, an equilibrium is set up.

The model holds for smaller molecules such as oxygen. For larger substrates, particularly those which act as carbon sources and are transported by permeases, the microorganism is surrounded by two concentric transport zones, transport through the outer zone being diffusion mediated, and transport through the inner zone being mediated by the permease (which is powered by cellular energy reserves). Permease mediated transport is not affected by the bulk substrate concentration but is described by a Michaelis–Menten or Langmuir isotherm type relationship (Cohen & Monod 1957). A steady state is achieved in which the rate of consumption of substrate by the cell is equal to, and limited by, the flux of substrate through each successive transport zone. Thus:

$$a_m \left(-D \frac{ds}{dx} a_m \right) = a_p \left(\frac{\alpha_p [S']}{\beta_p + [S']} \right) = V_R \left(\frac{\alpha R [S'']}{\beta R + [S'']} \right) \qquad (3.28)$$

where $[S']$ is the substrate concentration external to the permease region, α_p and β_p are coefficients, a is the external area of the permease zone, $[S'']$ is the substrate concentration in the metabolic zone, D is the diffusion coefficient of the substrate, and ds/dx represents the substrate area and volume of single micro-organisms respectively.

Provided that the cell mass functions and biological properties remain constant with time, a differential mass balance for the immobilized cell preparation can be established, in which the rate of transport of substrate is equal to its removal by the cells and the appearance of product. That is:

$$D_e \left(\frac{d^2 S}{dx^2} \right) - \frac{a' \alpha S}{\beta + S} = 0 \ , \tag{3.29}$$

where a' is the external surface area of viable microorganisms per unit volume of immobilized cell pellet; α represents either α_p or $\dfrac{\alpha_m V_m}{\alpha_m}$, and β is a rate equation coefficient. That is, when $x = L$ (the thickness of the immobilized cell preparation) $dc/dx = 0$, and when $x = 0, S = S'$ (Atkinson 1974). D_e represents the effective molecular diffusion coefficient within the tortuous intrapellet pores, and reflects the hydrogen bond mediated interactions of substrate with the support materials and steric hindrance between substrate and the support.

However, division of the entrapped cells causes complications. Often the mechanical strength of the immobilized cell preparation is weakened, and in extreme cases the granules may be broken. Although the concentration of immobilized cells has been increased, the new cells are not evenly dispersed within the support material, which leads to increased diffusional restrictions.

The problems of getting product out of the immobilized cell preparations are very similar to those concerned with substrate movement into the cell. However, in many cases leakage of intermediary metabolites out of the cell may be greatly enhanced when the cell has lost its chemi-osmotic integrity.

As has been shown for immobilized enzymes where diffusional limitations exist in heterogeneous systems, inhibition of the enzyme activity of the immobilized cells affects both the inherent activity of the enzyme and the rate of depletion of substrate. Thus:

$$v'' = \eta'' \frac{V_{max} S}{K_m + S} \tag{3.30}$$

where η'' is the effectiveness factor in the presence of inhibitor and v'' the initial rate of reaction in the presence of inhibitor. Thus the presence of chemical inhibitors, either supplied in low concentrations in the substrate or produced during the enzyme catalysed reaction, can be masked by the presence of diffusional restrictions so that often diffusional and chemical inhibition are operating at the same time. Their combined effect is less than the sum of their separate effects, and so the effect of inhibitors may be masked under severe diffusional limitations. The effects may later become apparent, and can cause a slow but progressive loss in the activity of the immobilized cells and when a porous support material is being used the activity is first lost on the outside of the pellet and then the cells located further inside the pellet are gradually inactivated. The presence of chemical inhibitors, either supplied in low concentrations in the substrate or produced during the enzyme catalysed reaction, was masked by the presence of diffusional restrictions because the inhibited enzyme, provided it is not too great a proportion of the total activity, would not have been expressed owing to the diffusional restrictions.

3.21 INTRINSIC ACTIVITY OF ENZYMES – MODIFYING FACTORS

3.21.1 Introduction

Six main factors have been identified which modify the intrinsic catalytic properties of enzymes either during immobilization or after they have been immobilized. In heterogenous catalysis such as is carried out by immobilized enzymes the rate of reaction is determined not simply by the prevailing pH, temperature, and substrate solution, but on the rates of proton, heat and substrate transport, through the support matrix to the immobilized cells or enzymes. Diffusion of large molecules inside the immobilized enzyme support will obviously be limited by steric interactions within the matrix, and it is generally found that the activity of immobilized enzymes to high molecular weight substrates is lower than when low molecular weight substrates are used. For instance, immobilized proteases are highly reactive towards low molecular weight artificial substrates, but are much less active towards high molecular weight natural substrates such as bovine serum albumin.

The intrinsic kinetics of an enzyme are defined as being those of the soluble enzyme, and the kinetics of the immobilized biocatalyst in the absence of, and in the presence of, modifying factors are called the inherent and effective kinetics respectively.

The main factors are: (a) Conformational effects, due to chemical modification of the enzyme protein during immobilization. These can have especially serious effects on the enzyme activity when amino acids which form part of the active site or which are important in maintaining the tertiary structure of the enzyme are involved. (b) Steric effects, which occur because often a proportion of the enzyme molecules are immobilized in a position relative to the support surface such that the active site is relatively inaccessible to substrate molecules. (c) Microenvironmental effects, which occur when the immobilized enzyme is maintained in a very different physical environment from that encountered in the bulk solvent. Partitioning effects are common causes of microenvironmental effects; for instance, hydrophilic substrates will be selectively attracted to the surface and pores of hydrophilic supports, thus increasing their local concentration; whereas hydrophobic substrates will be repelled from hydrophilic supports but selectively attracted into hydrophobic supports with similar effects on the local concentration of substrate present in the immediate vicinity of the support (3.10). Similarly, positively charged substrates and protons will be attracted into negatively charged supports, giving a local high substrate concentration and low pH inside the support; also, the interiors of many supports are relatively anaerobic, gases having been 'salted-out' by charged groups on the support matrix.

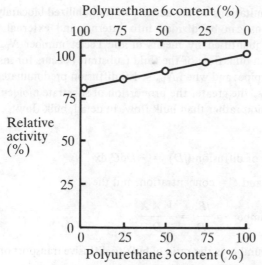

Fig. 3.10 — The effect of gel hydrophobicity on the relative hydrolytic activity of the polyurethane-entrapped cells (PU-3 gives a hydrophobic gel and PU-6 a hydrophilic gel). The reaction was carried out in water-saturated *n*-heptane containing 39 mM dl-menthyl succinate (Omata *et al.* 1981).

Such partitioning effects can cause the pH optimum and/or K_m of the immobilized enzyme to differ from the values obtained for the free enzyme. In the case of the positively charged substrate and negatively charged matrix, quoted above, the K_m of the immobilized enzyme will be decreased, and the pH optimum shifted to a more alkaline pH. Opposite effects of a similar nature will be observed when using positively charged molecules and positively charged supports; that is, the pH optimum is shifted to a more acidic pH and the K_m is increased. Partitioning effects often occur because cells usually have an overall negative charge which may cause shifts in the pH and substrate concentration optima of the enzyme activities of immobilized cells. For instance the optimum pH of *Escherichia coli* and *Azotobacter agilis* cells immobilized to anion exchange resins were higher than for the same cells in suspension (Hattori 1973). Such shifts may be disadvantageous, but can often be usefully exploited to allow operation in previously impractical conditions. As the enzymes of immobilized cells do not directly interact with the support material microenvironment, conformational and steric effects of this type are unlikely.

Diffusional restrictions occur because before reaction can take place the substrate must diffuse to the immobilized enzyme. Here, simultaneous diffusion, (Fick's Law describes this), and enzymic reaction, occur, such that concentration gradients of substrate and product are set up inside (and around) the pellets. This is important when a highly active enzyme is used. Initially, while equilibrium is being set up, the gradients may be anomolous; however, once a steady-state is established, gradients are invariably observed.

3.21.2 Diffusional limitations on the activity of immobilized biocatalysts

Diffusional limitations can be divided into internal and external diffusional effects, and can be quantified by means of the Peclet number N_{Pe} such that when $N_{Pe} > 1.0$, then bulk flow of the fluid (substrate) occurs for instance as in the centre of a large pipe; and when $N_{Pe} < 1.0$ diffusion predominates. Thus the lower the value of N_{Pe} the greater the proportion of substrate molecules that are transported by diffusion rather than bulk flow. In detail, bulk flow:

$$(jB) = V \times C \qquad\qquad (3.31)$$

and the rate of diffusion $(jD) = -DdC/dx$ $\qquad\qquad (3.32)$

where $V =$ flow rate and $C =$ concentration, and the

$$\text{Peclet number} = \frac{jB}{jD} = \frac{V \times X}{D} .$$

In short, the Peclet number is the ratio of bulk to diffusive transport of molecules.

3.21.3 External diffusional limitations

External diffusional limitations are caused by the rate of diffusion of substrate being restricted in the thin film of poorly mixed fluid surrounding each support particle (the Nernst–Plank layer). External diffusional restrictions can be decreased but not prevented by increasing the degree of agitation in well-mixed reactors. Another method is to increase the flow-rate in tubular reactors; or to use a more concentrated, or less viscous, substrate (Horvath & Engasser, 1974). Diffusional effects often give rise to extended lag periods before a steady state is set up, because the substrate molecules must diffuse through the bulk solvent, then through the film of badly mixed solvent around the support particle, the pores in the support, and then into the cell in the case of immobilized cells, before reaction can take place. Then the same extended diffusion pathway must be negotiated by product molecules in reverse before they can be detected and monitored in the bulk solvent.

3.21.4 Internal diffusional restrictions

Internal diffusional restrictions are caused by the small size (and tortuosity) of the pores in the support. This prevents forced flow of fluid inside the pores of the pellets under normal operating pressures. A reduction (localized) in the measured diffusion coefficient compared to the value obtained in free solution results, and thus

$$D_e = \frac{D\psi}{\tau} \qquad\qquad (3.33)$$

where D is the diffusion coefficient measured in free solution, D_e is the effective diffusion coefficient measured inside the support particles, ψ is the porosity of the particles, and τ is a tortuosity factor which represents the path-length which must be traversed by molecules diffusing between two points within a particle. The effective diffusion coefficient is proportional to the water content of the support and inversely proportional to the molecular weight of the substrate, as a general rule.

Internal diffusional restrictions of this kind have been part circumvented by the use of pellicular immobilized enzymes. Here, a thin layer of active enzyme overlays a catalytically inactive core, which was designed to impart mechanical strength to the particle. It is unfortunate that only relatively low immobilized enzyme activities can be obtained by this type of immobilized enzyme. There is, however, an additional advantage in that the pH in the vicinity of the enzyme can be controlled to some content.

High loadings of immobilized enzyme can also cause internal diffusional restrictions such that all of the potential immobilized enzyme activity is not all expressed; that is, some of the activity is latent. This is because the enzyme molecules located in the outer portions of the support particle consume most or all of the substrate which enters the particles, and the enzyme molecules located deeper in the particle have little opportunity to attack the substrate. Even when low internal diffusional restrictions occur in the presence of a low concentration of immobilized biocatalyst there can be appreciable restrictions when higher concentrations of immobilized biocatalyst are used. Therefore an optimum effective biocatalyst loading can be determined experimentally.

The significant variables which affect the extent of internal diffusional restrictions are quantitated by means of the Thiele modulus (ϕ). This has the form:

$$\phi = L\,\lambda = L\sqrt{\frac{V''_{max}\,[E]}{K''_m\,D_e}} \tag{3.34}$$

where L is half the thickness of the particle, $[E]$ the enzyme content of the particle, and V''_{max} and K''_m the kinetic parameters of the immobilized enzyme. For a slab $\phi = L\,\lambda$, for a cylinder $\phi = r/2\,\lambda$, and for a sphere $\phi = r/3\,\lambda$, where r is the radius of the particle. Thus the Thiele modulus is proportional to the molecular centre activity of the enzyme in question, that is, the number of molecules of substrate transformed per molecule per second, to the diffusion path presented to the substrate molecules by the support and to the affinity of the enzyme for the substrate, provided of course that the enzyme is evenly distributed within the support particles. When ϕ is large the effectiveness factor for the immobilized biocatalyst is low, and when ϕ approaches zero the effectiveness factor tends to unity (Wang et al. 1979).

Internal diffusional restrictions are preset if the activity of an immobilized enzyme preparation increases as the size of the support particle is reduced, particularly when it is crushed to a very small particle size, that is, when the length of the diffusion pathway followed by the substrate molecules is greatly reduced. Internal diffusional restrictions can also be detected by a lowering in the apparent activation energy for the reaction, and may be expected to be a major influence on immobilized cell activities because immobilized cell particles are usually much larger than the corresponding immobilized enzyme preparations. Thirdly, internal diffusional restrictions can also be made apparent if changes in the degreee of conversion of substrate to product occur when the height of a column of immobilized enzyme is changed while keeping the residence time of substrate in the reactor constant. Another method of estimating pore diffusion effects is to compare the effects of temperature on the activities of the free and immobilized enzymes. If the Arrhenius plot for the immobilized enzyme is a straight line with a slope equal to that of the soluble enzyme, then no diffusional restrictions occur, as they normally cause a decrease in the slope at higher temperatures.

Internal diffusional restrictions can be minimized by using a low molecular weight substrate, a high substrate concentration, a low biocatalyst concentration, and highly porous support particles in which the pores are as large, nontortuous, and interconnected as possible, by immobilizing the enzyme only to the outside surface of the support, or very importantly by using the smallest size of particles available (Fig. 3.11). Ideally, a particle size of zero, corresponding to the free enzyme or cell, should be used! Sometimes it may be advantageous to increase the surface area to volume ratio and thus allow substrate to diffuse into the particles more easily, but of course small and irregularly shaped particles would be more prone to abrasion damage in stirred reactors, and may cause uneven flow and compression in packed-bed reactors especially if there is a large particle size variation.

Thus a compromise can be reached between forming an extremely active immobilized preparation with a low effectiveness factor and a less active preparation with a higher effectiveness factor. Indeed, the effectiveness factor will vary throughout the support particles, being high near the surface and lower nearer the centre of the particle and vary along the length of a packed-bed reactor, being relatively high near the inlet of the column and relatively low near the outlet.

Diffusional restrictions, it has been observed, will underemphasize the inhibitory effects of enzyme inhibitors, such that the combined effects of diffusional restrictions and chemical inhibition are less than the sum of their separate effects. In addition, an inhibitor present in the substrate itself will inhibit the first enzyme molecules encountered. These are usually the enzyme molecules located on the outside of the support and at the input end of a packed-bed column. Then there is progressive inhibition of enzymes located deeper and deeper into the support particle (and further along the column too).

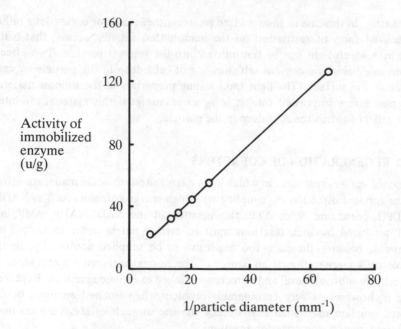

Fig. 3.11 — Immobilized esterolytic activity as a function of particle diameter for chymotrypsin immobilized to several size fractions of rock magnetite (Monroe *et al.* 1977).

In addition to diffusional restrictions on mass-transfer to immobilized biocatalysts, heat and gas transfer limitations can also be important on occasions. For instance, when exothermic enzymic reactions take place in large reactors with only a slow rate of mass transfer the heat generated may not be dissipated efficiently, leading to appreciable increases in temperature and consequently losses in enzyme activity due to thermal denaturation.

Because of the low solubility of oxygen in aqueous solutions, oxygen transfer limitations can often seriously limit the rate of oxygen requiring reactions involving cells, or oxygen requiring enzymes, especially when a high catalyst density is used and when the cells or enzymes are used in an immobilized form. Several approaches have been made to overcome this problem. These include the generation of oxygen *in situ*, for instance by the addition of hydrogen peroxide and catalase (Holst *et al.* 1982) or the co-immobilization of oxygen producing algae. Alternatively, solvents which have a greater solubility for oxygen than has water have been used, such as carbon tetrachloride (Buckland *et al.* 1975) or perfluorochemicals (freons) (Adlercreutz & Mattiasson 1982), although the former present safety problems and the latter are very expensive.

Lastly, in the case of immobilized photosynthetic cells or organelles a rather specialized form of restriction on the immobilized activity occurs, that is the ease with which light can be transmitted into the support particle. Two effects occur, one being the obvious self-shielding of cells deep in the particle by cells closer to the surface. The light transmitting properties of the support material also play a very important role, it being an advantage if this material can internally reflect light to the cells deep in the particle.

3.22 REGENERATION OF COFACTORS

As stated earlier, reactions in which a net energy input must be made, are driven in the forward direction by coupling with high energy cofactors such as NADH, NADPH, coenzyme A, or ATP. Regeneration of the NAD, NADP, AMP, and ADP produced by such reactions must be carried out in order to sustain the reactions, because they are too expensive to be supplied continually. In the whole cells regeneration is an aspect of the normal integrated metabolism of the cell, via substrate level and cytochrome linked exergonic reactions. Regeneration is, however, a very considerable problem when isolated enzymes or disrupted cells are used such that usually the use of such catalysts are confined to degradative or isomerization reactions.

ATP appears to be the simpler coenzyme to regenerate. Banghn *et al.* (1978) have used a cyclic ATP regeneration system using three different enzymes immobilized by a novel condensation copolymerization reaction, and have studied various organic syntheses coupled with ATP regeneration (Wong *et al.* 1981). Murata *et al.* (1981) have produced glutathione, involving simultaneous ATP regeneration using immobilized *S. cerevisiae* cells (Fig. 3.4). Marshall (1973) used carbamyl phosphokinase from *S. facalis* immobilized on glutaraldehyde treated alkylamine glass to regenerate ATP from ADP and carbamyl phosphate derived from potassium cyanate and potassium phosphate, and other workers have carried out continuous ATP formation from ADP or adenosine using creatinine kinase, phosphofructokinase, and bakers' yeast co-immobilized in polyamphoteric gels.

Reduced pyridine nucleotide regeneration is a very difficult problem since very high fidelity is required to make the process cheap and efficient enough to attempt to use in industrially oriented syntheses. Wykes *et al.* (1975) performed cofactor recycling in enzyme reactors using free and immobilized alcohol and lactate dehydrogenases, but were limited by the relative instability of the coenzyme. Similarly, NADPH has been regenerated using glucose-6-phosphate and glucose-6-phosphate dehydrogenases (Wong *et al.* 1981a, b) or immobilized chloroplasts (Karube *et al.* 1980); NAD^+ and FMN have also been regenerated using *Alcaligenes entrophus* cells and hydrogen gas (Klibanov & Puglisi 1980), and NADH covalently bound to polyethylene glycol using L-leucine dehydrogenase and formate dehydrogenase in a continuous reactor (Wichmann *et al.*

1981). Another approach is to use water-soluble polyacrylamide derivatives of NAD and ATP (m. wts 5–10,000) which can enter permeabilized cells in flow-through systems and be regenerated (Mosbach 1978). Similar dextran-linked derivatives of NAD/NADH have been used by Chang et al. (1982), in this case encapsulated within artificial microcapsules. A more specific approach is to co-immobilize an enzyme, alcohol dehydrogenase, and a spacer-extended NAD^+ analogue to the same support, which allows continuous regeneration while converting ethanol into acetaldehyde, and lactaldehyde into propandiol (Lejoy et al. 1980). Similarly, alcohol dehydrogenase and myo-inositol-1-phospho-synthetase have been co-immobilized with NAD^+ which is again continuously regenerated by entering the active sites of adjacent enzymes in alternative fashion. Thus the NAD^+ in effect behaves more like a prosthetic group than a coenzyme (Mansson et al. 1978, 1979).

Enzymes can also be used to carry out other reactions using cofactors, for instance NAD can be transformed into NADP by the NADP kinase of whole cells of *Brevibacterium ammoniagenes* (Hayashi et al. 1979, Murata et al. 1981). It proved advantageous to permeabilize the cells so as to allow phosphates into the cell; the cells had a NADP synthesizing activity with a half-life of 8 days.

3.23 BIOCHEMICAL REACTORS

3.23.1 Introduction
An enzyme reactor is, basically, the container in which a reaction catalysed by free or immobilized enzymes or cells occurs, (plus the associated mixing, sampling, and monitoring devices). The reactor's role can be defined as to make a specified product at a given rate from the defined reactants — at a minimum cost! Enzyme reactors differ from chemical reactors because they function at low temperatures and pressures, and comparatively little energy is consumed or generated during reaction. Enzyme reactors differ from fermentations in not behaving in an auto-catalytic fashion, for there the growing cells continually regenerate themselves. As with chemical reactors, enzyme reactors are judged by their productivity and specificity.

Reactors are classified as homogeneous or heterogeneous. This depends of course on whether their contents are homogeneous, in which only one phase is present, or heterogeneous where more than one phase is present. In hetero-geneous reactors there is often a continuous liquid substrate phase and a discon-tinuous solid immobilized biocatalyst phase. Reactors can also be classified in other ways, such as whether reaction takes place batchwise or continuously, or whether the reactor is open or closed, that is, whether the catalyst passes out of the reactor during use or not. The most important classification is according to the extent of mixing which takes place within the reactor, the idealized extreme types being perfectly mixed and plug-flow reactors. The concentration profile of reactants within reactors of different configurations may vary appreciably. In well-mixed reactors the reactant molecules are maintained in a constant state of agitation, and in the later plug-flow reactors fluid elements move through the reactor as a plug without mixing with the previous or subsequent fluid elements applied to the reactor (Fig. 3.12). Thus in batch well-mixed reactors the composition of the reactants varies during the course of the reaction (although remaining constant throughout the reactor). In a continuous stirred reactor in steady-state the composition of the reactor contents is uniform and should not vary with time (in the absence of enzyme inactivation). In plug-flow reactors the composition of the reactants is time invariant but does vary along the length of the reactor. These concentration profiles may affect both the activity and stability of the immobilized enzymes, for instance, endogenously produced inhibitors are constantly swept out of continuous reactors so that critically high concentrations are not usually reached. Thus in plug-flow reactors the effectiveness factor will be high near the input end and lower near the exit end of the reactor. This is because the substrate concentration decreases as reactants pass through the reactor. In batch reactors the effectiveness factor is uniform throughout the reactor. However, in practice, despite careful design and operation, the ideal types of mixing can only be approached (fairly closely in many cases). In particular, non-ideal flow in packed-bed columns is a fact of life, especially when used on a large scale.

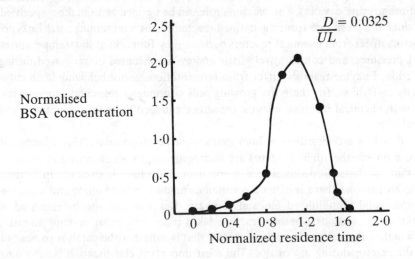

Fig. 3.12 – An illustration of the residence time distribution typical of a packed-bed column. In this case bovine serum albumin was applied to the column as the tracer molecule (Ghose & Chand 1978).

3.23.2 The various types of biochemical reactor (see Part A, Chapter 4 also)
In addition to the basic reactors configurations depicted in Fig. 3.13, other more complicated types have recycle loops to facilitate complete conversion of substrate when high volumetric throughputs must be maintained, and specialized fittings depending on the particular reaction that is being catalysed. Choice of the most appropriate type of reactor (as well as their efficient design and operation) is very important in industry.

It is important to consider temperature and pH control, the requirement to supply or remove gaseous reactants, the presence of desired or undesired solid particles in the feedstock. Another consideration is the chemical and biological stabilities of the substrates and products. Also, it is necessary to allow for the frequency of catalyst replacement required. Also, there is the possibility of substrate and/or product inhibition, the intended scale of operation, in addition to the uses intended for the product.

Batch, semi-batch, or continuous operation in stirred, thin-membrane, multichambered, packed-bed, expanded-bed, or fluidized bed reactors are possible, with constant flow packed-bed and batch stirred reactors being the most popular. Thin-membrane reactors include hollow fibre and ultrafiltration reactors. Continuous operation of the reactor has economic advantages, especially where a large output is required, as capital and labour costs are decreased, and constant operation is most easily controlled and automated. However, catalyst replacement and regeneration are more difficult. The activity per unit reactor volume is very important, as this parameter chiefly determines the size of reactor required to achieve the desired productivity; it also influences operational factors such as the amount of bed compression encountered. When activity is calculated on the basis of reactor volume, packed-bed reactors have an advantage over stirred reactors, as in the latter the particulate catalyst may occupy only 10% of the total reactor volume.

The enzyme activity of the original cells or enzymes, the weight of cells or enzyme immobilized per litre of support material, the retention of activity after immobilization, and the effect of diffusional effects on the operational activity of the immobilized enzyme or cells, the stabilities of the enzyme activities of the immobilized enzyme on cells, the stability of the reactants and the desired yield and purity of product, safety, and other operational requirements, must be taken into account. For example, a fluidized bed or stirred reactor is especially advantageous when good gas mixing, or efficient pH or temperature control, or use of high flow rates is required, or when substrates containing undissolved solids are to be used; but very friable support materials cannot be used, and untapered fluidized beds can only be used over a narrow range of flow rates. The final choice of temperature, pH, substrate concentrations, and the conversion at which the reactor will operate, depends on an economic assessment taking into account the operating costs of substrates, labour, overheads, etc. and the capital costs of the equipment involved.

Type of reactor	Diagrammatic representation	Equation describing performance
Batch	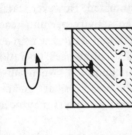 $S \longrightarrow S_t$	$XS - K_m \cdot \ln(1-X) = \dfrac{kE.t}{V}$
Tubular (plug-flow)	Substrate — qS → ▯ → Product qS_t	$XS - K_m \cdot \ln(1-X) = \dfrac{kE}{q} = \dfrac{kE\pi}{V}$
Continuous stirred	Substrate — qS → ▯ → Product qS_t	$XS + K_m \cdot \dfrac{X}{1-X} = \dfrac{kE}{q}$

Fig. 3.13 – A diagrammatic and algebraic description of the common reactor types. Equations are derived from an enzyme acting on a single substrate in the absence of inhibition. X is the proportion of substrate converted, kE is the total enzyme activity in the reactor, S and S_t are the initial and final substrate concentrations, q the flow rate of substrate into the reactor, t the time of operation, K_m the Michaelis–Menten constant and V is the working volume of the reactor. Reproduced with permission from *Principles of Biotechnology* (Ed. Wiseman, A.), Surrey University Press, Glasgow and London 1983.

3.23.3 Assessment of the performance of biochemical reactors (see Fig. 3.13)

The performance of a reactor should be assessed under simulated full-scale production conditions in which good manufacturing practises are observed. Performance is measured by its activity, stability, and selectivity; and by the yield of product obtained, the degree of conversion of substrate to product which can be achieved, and the concentration of reactants that can be used.

The aim is to achieve a particular productivity as rapidly as possible and by using as small a reactor as possible. Thus the running and capital costs of the process are minimized. A high activity is essential when a substrate or product is unstable (or if side-products are formed). This problem will be minimized if the residence times of reactants in the reactor are as low as possible. Activity is usually expressed volumetrically, as the g of product formed/L of reactor volume used/h of operation, or a similar term. Volumetric activities are determined by the concentration of the immobilized biocatalyst in the reactor and by the activity of the immobilized biocatalyst. This activity is due to the activity of the free biocatalyst, the amount of biocatalyst loaded on to the support material, the proportion of this activity which remains enzymically active after immobilization, and the extent of diffusional and partitioning effects. Note that because of the high degrees of conversion of substrate to product that are usually required enzymes with low K_m values, that is, a high affinity for their substrates, are preferred because of their greater activities when only low substrate concentrations remain as the last few critical degrees of conversion are being achieved.

In enzyme reactors a high enzyme concentration is desirable in order to reduce the reactor volume and the residence time of reactants in the reactor. The half-life (operational stability) of an immobilized cell/enzyme represents the time required for the activity of the reactor to decay to half of its original value. Stability may depend on initial activity (effectiveness factor) due to the action of intra- and interparticle diffusional restrictions. Biocatalyst stability (half-life value) should be high, so that enzyme production and immobilization need only be carried out at infrequent intervals.

The productivity of a reactor depends on the activity and stability of the enzymes and the yield of products obtained. During the use of immobilized cells or enzymes, multiple reactions often take place, generating undesired side-products. These may be caused chemically, by further metabolism of the desired product, by another enzyme or group of enzymes metabolizing part of the substrate, or by a more complex mechanism. It may be that the byproduct(s) are useful and can be easily purified and sold, but usually the aim is to stop or minimize their formation by a judicious choice of reactor configuration and reaction conditions as they can act as enzyme inhibitors or denaturants. Note that one consequence of the use of high concentrations of substrates is that substances which are not normally regarded as being substrates for a particular enzyme become so, because the high K_m of the enzyme for them is approached, whereas normally the enzyme does not act on them because binding to the enzyme is very weak.

The yield of the reaction can be expressed either as a relative or operational yield. Relative yields represent the percentage of substrate molecules reacted which have formed the desired product. This parameter falls sharply with increasing conversion. Alternatively, the operational yield is the percentage of substrate molecules fed into the reactor which have reacted to form the desired product.

The selectivity of the reaction is defined as the number of molecules of desired product formed, divided by the number of molecules of substrate transformed into unwanted product. Reactions in which back-mixing occur, such as in stirred tank reactors, generally have reduced selectivity because products are mixed with reactants, and the reactant concentration available for reaction is reduced. Yield and selectivity will therefore be greatest in tubular reactors approaching ideal plug-flow behaviour. Very selective reactions are valuable because feedstocks can be exploited most efficiently and economically and excellent yields of product can be obtained particularly when maximum conversion of substrate to product can be maintained. Purification and isolation of the desired product is facilitated by the relative absence of impurities. Selective reactions are encouraged by the use of pure enzymes and substrates, where possible, and by minimizing non-ideal flow in the reactor. Isolation of the product is also favoured if high concentrations of reactant can be used, particularly when the biocatalyst is very selective and irreversible. Quantities of solvent that must be removed plrioer to recovery are greatly reduced.

The rate of formation of product by an immobilized cell or enzyme is expressed volumetrically, for instance as g product/l reactor volume/h reaction, by

$$\frac{\text{concentration of product in the eluate}^\dagger \times \text{flow rate}}{\text{volume of reactor}} \tag{3.36}$$

or by integrating the area under a graph of volumetric activity vs time of operation. and can be related to the flow rate of substrate (q) though a reactor over a time period (t) by Productivity $= \int_0^t qdt$ or more specifically when activity decays exponentially and constant conversion is maintained, by:

$$\text{Productivity} = \int_0^t q \exp -\ln 2 \, \frac{t}{t/2} \, dt \tag{3.37}$$

where q is the initial feedrate of substrate and $t/2$ is the half-life of the enzyme activity (Pitcher 1978).

† Or the difference in concentration between the inflow and outflow streams.

Monitoring of the material or mass balance of the reactor is also important. Generally, for reactors operating in a steady state the rate of accumulation of material within the reactor is zero. Although some losses may be incurred when the substrates and/or products are volatile or unstable. Transiently, that is, during the start-up of a reactor before equilibrium is reached:

The rate of flow reactants into the reactor − the rate of flow of reactants out of the reactor − the rate of removal of reactants by reaction = the rate of accumulation of material within the reactor   (3.38)

For a batch reactor the first two terms are zero, and the rate of accumulation of products or the rate of disappearance of the reactants equals the reaction rate. For a continuous tubular reactor or a continuous stirred reactor no material should accumulate inside the reactor, and so the rate of removal of reactants is balanced by the difference between the inflow and outflow rates.

The heat balance in the reactor can also be monitored, although because enzyme reactions are not usually associated with large energy uptake or outputs this parameter is not as important as in many chemical reactions. At a steady state for any given reactant:

Rate of inflow of heat = Rate of inflow of component × Q (the heat of reaction generated per unit component reacted)   (3.39)

For a practical study of such effects using amyloglucosidase is packed-bed reactors see Marsh & Tsao (1976). Note that heat transfer in enzyme reactions can be important since enzymes are comparatively labile to heat; for instance immobilization supports which are poor conductors of heat can maintain the enzyme at temperatures substantially above that of the bulk solvent.

3.23.4 Batch reactors
Batch reactors are the traditional form of reactor (for example as used in brewing) which fit in well with use of soluble enzymes (Fig. 3.14) (or when small outputs or infrequent operation is required). They usually consists of a tank in which substrate or enzyme is mixed, the contents of the reactor being discharged once the required degree of reaction has been achieved. Disadvantages of the batch reactor are due to changing conditions during the course of the reaction, thus necessitating more complex control equipment. Also, there is the difficulty in maintaining good mixing when the reactor is scaled-up, because of the impossibility of ensuring proportional power input as the reactor increases in size. Other problems are the batch-to-batch variations in the quality of the product; the 'downtime' lost in between batch reactions; the uneven demands on services and ancillary equipment and labour made by batch rather than continuous operation; and also an inability to re-use the enzyme or a tendency to lose enzyme during its recovery between batch reactions (although semi-continuous use is possible). These disadvantages can be outweighed especially when

good mass or gas transfer is required. For instance, production of 6-amino-penicillanic acid is done in a batch stirred reactor containing immobilized penicillin acylase. Here there is the need to neutralize the acid formed during the reaction by the continuous addition of alkalis (Carleysmith & Lilly 1979). The plug-flow column is now most commonly employed in industry. In several batchwise applications of enzymes such as the production of malt extract or brewers' wort it is necessary to have a high enzyme activity during reaction but for no residual active enzyme to be left in the product. This is usually achieved by increasing the temperature in the reactor such that the last few molecules of substrate are consumed just before the last enzyme molecules are denatured.

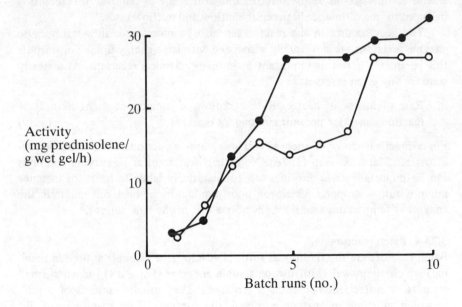

Fig. 3.14 – The repeated batchwise transformation of cortisol to prednisolone by immobilized *A. simplex* 2 g (wet weight). *A. simplex* gel were incubated in 285 ml 0.5% (w/v) peptone; pH 7.0 (•), or 285 ml 0.1% (w/v) peptone + 0.2% glucose, pH 7.0 (○). Reactions was initiated by adding 15 ml 20 mM cortisol (dissolved in methanol). After each run, i.e. when all cortisol was converted, the gel was washed and reincubated with fresh cortisol-containing nutrient medium (Ohlson *et al.* 1978).

3.23.5 Continuous stirred tank reactors

CSTRs (back-mix reactors), are versatile and cheap, and are particularly useful when liquid-phase reactions are being carried out. The reactor contents are well mixed such that the concentration of reactants is virtually uniform throughout the vessel. Note that the product stream leaving the reactor has the same composition as the mixture within the tank, (substrate enters at same rate as contents of the reactor are removed).

The supply of gas (also, pH and temperature control) is easily carried out in such well-mixed reactors. Note that spatial variations in pO_2, pH, and temperature will occur within a stirred tank depending on its size and on the mixing efficiency. Fresh catalyst can be easily added to the reactor, and substrates containing particulate materials can be tolerated without causing fouling. One disadvantage is that a relatively high power input is required to give efficient agitation. Therefore it is possible that unwanted abrasion of the immobilized biocatalyst particles may occur. Another problem is that only a low biocatalyst concentration can be maintained in a stirred reactor compared to the high concentrations possible in plug-flow columns. therefore, a particular rate of product formation can be achieved by using smaller reactors in the case of plug-flow columns. However, in practice a stirred reactor may still be preferred because large stirred tanks are easily constructed (and are cheap), and can be easily switched to different tasks. An example of a CSTR in industrial practice is in the removal of unwanted sugars during beet sugar refining using immobilized α-galactosidase in U-shaped vessels, each of which is fitted with agitators and screens to keep the immobilized enzyme in the reactor as the syrup is pumped through the reactor.

The maximum conversion in a CSTR is always less than in the equivalent batch reactor for a given residence time. The reaction is likely to be less selective in the stirred reactor than in a plug-flow reactor because of the good mixing. One problem however, is that the enzyme or cell suspension is less selective and more susceptible to product inhibition in the stirred reaction, and thus high degrees of conversion are more difficult to obtain. However, on the other hand a well mixed reactor will tend to exploit the advantages of substrate-inhibited and product-activated reactions more than a plug-flow reactor. For instance, in the former case substrate can be added in batches at intervals during the reaction so that its concentration never becomes so high that substantial inhibition occurs (see Lilly & Sharp, 1968).

Approximate plug-flow behaviour together with efficient mixing can be achieved by using a number of stirred tanks connected in series, although this is a comparatively complicated and expensive alternative, and has not been exploited on a large scale (Carley-Smith & Lilly, 1979).

3.23.6 Plug-flow column reactors (or tubular reactors)
These consist of columns packed with immobilized biocatalyst particles. They are kinetically more efficient than stirred reactors when compared on a volumetric activity basis and are also simple and easy to operate and automate. Large-scale and continuous operation minimize labour costs and overheads, and facilitate control and produce products of a more reliable quality than the

traditional batch processes. Packed columns allow a high concentration of biocatalyst to be used. The product molecules or other inhibitors produced during the reaction are constantly swept out of the reactor so that the concentration of products are minimized – so less product inhibition occurs. However, plug-flow reactors are more prone to substrate inhibition than CSTRs. For a study of pulse responses in packed-bed immobilized enzyme reactors carrying out reversible and consecutive reactions see Adachi *et al.* (1981).

Upflow operation offers the advantages of good fluid particle contact and minimum pressure drop. The downflow mode allows operation under essentially atmospheric pressure where the fluid flow through the bed can be controlled readily by the hydraulic head above the bed. Particle attribution and fluid channeling problems are also minimized in downflow operation, and so is the problem of fluid distribution at the reactor entrance. Downflow reactors can also be operated under pressure, which allows for greater bed depth. The preferred reactor design in industrial use appears to be a number of fixed bed columns operating in the parallel mode in which fluid flow occurs in the downflow fashion.

Packed-bed columns can be prone to self-compression (also to fouling by particulate materials in the substrate stream and are more difficult to operate hygienically and to add fresh biocatalyst than one batch reactors. The other occasions when a plug-flow reactor is kinetically less advantageous then a well-mixed reactor are when substrate inhibition and/or product activation occur. Despite the difficulty of keeping temperature and pH constant, tubular reactors are very useful. Another problem is the supply of gaseous reactants (use recycle loops therefore).

Columns can be operated in an expanded bed form at higher flow rates, so preventing fouling and compression and improving mass transfer. Alternatively packed biocatalyst particles can be used in a trickle-flow mode as in some waste-water processes employing microbial flocs.

Note that when immobilizing cells or enzymes for use in packed beds a compromise usually has to be made between the use of small particles which tend to have a low diffusional restrictions but high compressibilities and larger particles with higher diffusional restrictions but lower compressibilities.

3.23.7 Fluidized bed reactors
In fluidized beds biocatalyst particles are suspended and mixed by the rapid upwards flow of substrate or gas. Thus substrates containing solid particles can be used. pH control, temperature control, and the supply of gases are easy when using fluidized beds. Owing to the large void volume of fluidized beds high biocatalyst concentrations cannot be achieved. Another problem is that the high flow rates of substrate that are required to fluidize the biocatalyst particles may cause washout of the catalyst and allow only a low degree of reaction to take place (Fig. 3.15). Recycling of substrate is a method of avoiding wash-out and

a way of achieving complete conversion of substrate to product. Another way is the use of several fluidized beds in series or fluidized beds that are tapered in shape.

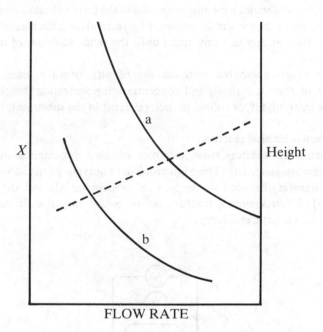

Fig. 3.15 — The influence of the flow-rate of substrate through a fluidized bed on the degree of conversion achieved (X), and the height of the fluidized-bed. The use of (a) relatively concentrated and (b) dilute substrate concentrations are depicted. Reproduced with permission from *Principles of Biotechnology* (Ed. Wiseman, A.), Surrey University Press, Glasgow and London 1983.

Alternatively, if the enzyme is immobilized to magnetic particles it can be maintained static by the application of an electromagnetic field (Sada *et al.* 1981), or prevented from leaving the reactor by installing a magnetic collar at the exit of the reactor (Gelft & Boudrant 1974) even when very high flow rates through the reactors are used. However, despite their attractive features fluidized beds have not been achieved widespread or large-scale use because they are costly to run and difficult to scale-up.

3.23.7 Ultrafiltration reactors
Ultrafiltration reactors are well-mixed reactors. Because of the selective nature of the ultrafiltration membrane this type of reactor can be used to separate low and high molecular weight reactants. Therefore they are most useful for carrying out depolymerization reactions. This application can best be done with soluble

enzymes so as to ensure good contacting with macromolecular substrates. See successful use of cephalosporin acetylesterase in an ultrafiltration reactor (Abbott *et al.* 1976).

Selective membranes have also been used in the form of hollow-fibre reactors. A plug-flow mode of use can be achieved by recirculating the reactants through the hollow fibre apparatus many times until the required degree of reaction has been achieved.

Disadvantages associated with the use of ultrafiltration reactors are the small sizes of reactor available and concentration polarization (blockage of the membrane by solid, fat, or colloid particles present in the substrate).

3.23.8 Electrochemical reactors

In electrochemical reactors redox enzymes are used in combination with electrodes to supply electricity. There is a problem of obtaining satisfactory coupling (electron transfer) between the various redox components and the electrodes (Fig. 3.16). Electrochemical reactors can be useful as fuel cells, as analytical probes, or in synthetic chemistry.

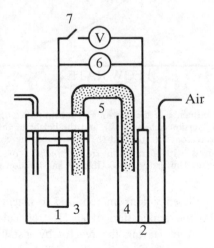

Fig. 3.16 – A schematic diagram of a biochemical fuel cell (1) immobilized whole cell-platinum black electrode, (2) carbon electrode, (3) catholyte, (4) anolyte, (5) salt bridge (1.5 cm diam), (6) recorder, (7) switch (Karube *et al.* 1977, 1977a).

3.24 ENZYME KINETICS IN REACTORS (see also Part A: Sections 3.6, 3.18, 3.20, 3.21 and 3.24)

The kinetic expressions used here start with reactions which involve a single product without substrate or product inhibition occurring. Subsequently the effects of non-ideal flow of reactants and diffusional restrictions are introduced. The final choice of temperature, pH, substrate concentration, and the extent of conversion at which the reactor will operate depends on an economic assessment taking into account the operating costs of substrates, labour, overheads, etc. and the capital costs of the equipment required. The overall effect of this approach, therefore, is to try to minimize the size of reactor or the concentration of enzyme required to achieve a given productivity.

The Michaelis–Menten kinetic approach is adapted to provide the kinetic description of batch reactors, but the equation is best used integrated with respect to time. For an uninhibited, irreversible reaction employing a single enzyme in a batch reactor under isothermal conditions:

$$XS - K_m'' \cdot \ln(1 - X) = \frac{KE\pi}{V} \tag{3.40}$$

where kE is the maximum activity of the total enzyme in the reactor in mols reacted per sec, V is the working volume of the reactor, π is the time period of operation, and X is the degree of conversion $\dfrac{S - S_t}{S}$, where S is the starting substrate concentration and S_t the concentration remaining after a period, t, of reaction.

In the same way, for a plug-flow reactor:

$$XS - K_m'' \cdot \ln(1 - X) = \frac{kE}{q} = \frac{kE\pi}{V} \tag{3.41}$$

where q is the volumetric rate of supply of substrate in sec^{-1} and π is the average residence time of substrate solution in the reactor in seconds such that if a graph of XS vs. $\ln(1 - X)$ is plotted, the slope of the resulting points equals 2.303 times the apparent K_m of the enzyme.

Thus account is taken of the degree of conversion of substrate to product, flow rate of substrate through the reactor. Another factor is the degree of mixing of fluid in the reactor. Effects of inhibitors can also be allowed for. Note that the enzyme located at the inlet end of the reactor is exposed to a high concentration of substrate irrespective of the final degree of conversion achieved by the reactor, and so operates under effectively zero-order kinetic conditions with substrates present in excess. The enzyme located towards the outlet end of the reactor will, however, only be exposed to a low substrate concentration if a high degree of conversion is being achieved and so will operate under first-order kinetic conditions.

For a well-mixed continuous flow reactor:

$$XS + K_m'' \cdot \frac{X}{(1 - X)} = \frac{kE}{q} = \frac{kE\pi}{V}. \tag{3.42}$$

Where S/K_m is low, that is, where a high degree of conversion has been obtained (or a low substrate concentration used) the reaction rate is first-order, with the time of reaction required to give a particular degree of conversion being directly proportional to S/K_m. The reaction is essentially zero-order at high values of S/K_m, and in that case the time taken to reach a given degree of conversion is virtually independent of S/K_m. In industrially practicable processes high degrees of conversion of substrate to product are usually required. Where the immobilized enzyme usually operates under first-order kinetic conditions, especially considering the very high densities of enzyme that are often used in commercial immobilized enzyme preparations. In that situation the time required for a relatively small increase in conversion is critically dependent on the value of S/K_m achieved in operation at the time, so this parameter is of great experimental importance.

From a consideration of the equations describing plug-flow and continuous stirred reactors the performances of both types of reactor are virtually identical when high substrate concentrations are used with $S \gg K_m$, while at low substrate concentrations $S \ll K_m$ the plug-flow reactor may be as much as 20–30 times as efficient as a CSTR if a high degree of conversion is required, as well as having the additional advantage of possessing a much higher density of enzyme molecules.

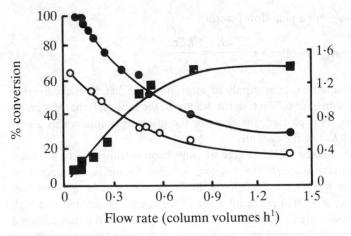

Fig. 3.17 – The effect of the residence time of substrate (sucrose) on the isomaltulose synthesizing activity of *Erwinia rhapontici* cells immobilized in alginate gels and used in a column reactor, (■) activity of cells, (●) degree of conversion, (○) molar isomaltulose concentration in the column eluate (Cheetham *et al.* 1982). Reproduced by permission from *Nature* Copyright © 1982, Macmillan Journals Ltd.

In continuous flow packed columns (or stirred reactors) an important relationship has been observed between the flow rate through the reactor (q)

and the degree of conversion (Fig. 3.17). The precise shape of the curve varies
with factors such as the extent of product inhibition but should be constant
for any size of reactor provided the hydrodynamic properties remain unaltered.
When a high S/K_m is used and low degrees of conversion are being achieved the
behaviour of both reactors are almost identical. However at low values of S/K_m,
(when the order of the reaction is greater than zero) the performance of tubular
reactors becomes favourable to an even greater extent as compared to that of
well-mixed reactors. In this situation, high degrees of conversion can be obtained
at much higher flow rates in plug-flow than in well-mixed reactors which are of
comparable value in other ways.

In each reactor it is the amount of enzyme in active form that determines
the extent of the reaction. Reactors are therefore required to be compared on
the basis of the amounts of enzyme required to carry out a particular reaction.
Figure 3.18 shows the amounts of enzyme required to give a range of conversions in a continuous stirred reactor compared with the amount of enzyme
required in a packed column reactor. The amounts of required enzyme are
plotted for various substrate concentrations. The extreme values are identical
to the sizes of plug flow and CST reactors required to produce the same amount
of product for zero- and first-order reactions where no change in volume and no

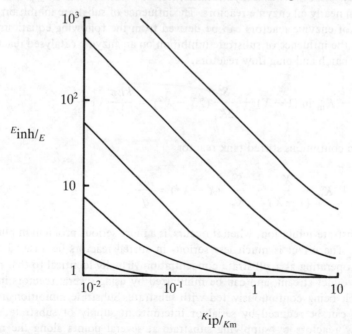

Fig. 3.18 — The effect of noncompetitive inhibition by product (K_{ip}) in a plug-
flow reactor on the amount of enzyme required to obtain 90% conversion at
five given flow rates. E_{inh} and E are the amounts of enzyme required in the
presence and absence of inhibition (Lilly & Dunnill 1972).

inhibition occurs. In the case of batch reactors the amount of enzyme required will be higher than for a continuous reactor because of the 'downtime' lost between batches. This if α is the proportion of the continuous reactor operation time that the batch reactor is in operation, then the amount of enzyme and thus the size of the reactor required to give an overall productivity equal to that of the plug-flow (P) will be:

$$\frac{P \times 100}{\alpha}.$$
(3.43)

3.24.1 Inhibition in enzyme reactors

Note that in most cases where enzymes are used as industrial catalysts, concentrated substrate solutions are used and high, or close to equilibrium, degrees of conversion of these substrates are desired. Since these conditions are very different from those encountered in academic enzymology, in particular the high concentrations of product produced, some products which are not normally thought of as inhibitors because of their relatively high K_i values do exert significantly inhibitory effects. Thus enzyme inhibition of some kind and to some degree occurs in nearly all enzyme reactors. The influence of substrate inhibition on the kinetics of enzyme reactors can be derived from the following equations which describe the influence of substrate inhibition on an enzyme catalysed reaction.

For batch and plug flow reactors:

$$XS - K_m \ln(1 - X) + \frac{S^2}{2K_s}(2X - X^2) = \frac{kE\pi}{V} \text{ or } \frac{kE}{q},$$
(3.44)

and for a continuous stirred tank reactor

$$XS + K_m \frac{X}{(1 - X)} + \frac{S^2}{K_s} \cdot (X - X^2) = \frac{kE}{q}.$$
(3.45)

Thus substrate inhibition, when it occurs, is a more serious problem in plug-flow reactors. The effect is much less serious in stirred reactors because all the enzyme is operating at a substrate concentration virtually identical to that present in the product stream, and can be minimized by using several reactors in series with each being continuously fed with substrate. Substrate inhibition in batch reactors can be reduced by serial or intermittent supply of substrate, and in plug-flow reactors by supplying substrate at several points along the reactor, although none of these methods appears to be industrially practicable.

The behaviour of batch and plug-flow reactors when subject to competitive inhibition by product is described by:

$$\left(\frac{1 - K_m}{K_{ip}}\right) XS - \left(\frac{1 - S}{K_{ip}}\right) K_m \ln(1 - X) = \frac{kE\pi}{V} \text{ or } \frac{kE}{q}, \tag{3.46}$$

and for continuous stirred tank reactors by:

$$XS + K_m \cdot \frac{X}{1 - X} + \frac{K_m}{K_{ip}} \cdot \frac{X^2 S}{1 - X} = \frac{kE}{q} \tag{3.47}$$

when S/K_{ip} is small product inhibition is not a serious problem, but at high values of S/K_{ip} and when high degrees of conversion are being obtained competitive product inhibition becomes a severe problem. This effect occurs in both plug-flow and stirred reactors, but is more severe in the latter because all the contents of the reactor are at the same high product concentration.

When a batch reactor is subject to noncompetitive inhibition its kinetic behaviour can be described by:

$$XS \left(1 + \frac{S}{K_{ip}} - \frac{K_m}{K_{ip}}\right) - S^2 \frac{(2 - X)X}{2K_{ip}} - \left(1 + \frac{S}{K_{ip}}\right) K_m \ln(1 - X) =$$

$$\frac{kE, \pi}{V} \text{ or } \frac{kE}{q}, \tag{3.48}$$

and for continuous stirred reactors by:

$$XS + K_m \cdot \frac{X}{1 - X} + \frac{X^2 S^2}{K_{ip}} \cdot 1 + \frac{K_m}{S(1 - X)} = \frac{kE}{q}. \tag{3.49}$$

A greater extent of inhibition is obtained in the presence of a noncompetitative rather than a competitive inhibitor at the same concentrations and with the same substrate concentration and degree of conversion. Again, inhibition is more severe in a CSTR than in a plug-flow reactor.

These effects can be conveniently expressed as the ratio of enzyme required when inhibitor is present to the amount of enzyme required to achieve the same conversion in the absence of inhibitor. Figure 3.18 shows this effect for noncompetitive product inhibition in a plug-flow reactor achieving 90% conversion.

Although the above equations describing the performance of biochemical reactors may appear to be rather esoteric at first sight, they can prove very useful in practice. For example, given comparatively simple information such as K_m and K_i values they can give a good idea of what the maximum 'ceiling' performance could be when operation is fully optimized. Thus even at a comparatively early stage in a development programme once can predict what the 'best-case' productivity could be, thus allowing costings to be done, and estimates made for the capacities likely to be required for equipment upstream and downstream from the enzyme reactor.

3.25 THE EFFECT OF NON-IDEAL FLOW ON BIOCHEMICAL REACTOR PERFORMANCE

Non-ideal flow occurs when different elements of substrate move through the reactor at varying rates such that they are in contact with the enzyme for variable periods owing to their differing residence times in the reactor giving an opportunity for different extents of reaction to take place and for further reaction of the desired product and side-reactions to take place. Non-ideal flow can be caused by factors such as back-mixing, channelling, or the presence of stagnant flow regions within the reactors. The extent of non-ideal flow can be estimated by measuring the residence time distribution of a slug of tracer, for instance a dye injected into the inlet of the reactor, as it leaves the reactor (Fig. 3.12); although sometimes allowance must be made for the hold-up of tracer molecules in porous particles.

Channelling is caused by excessive pressure drops, irregular packing of the support, and the uneven application of the substrate to the column. Thus the fluid tends to find a preferential path through the column, usually down one side when friction of substrate with packing material is greater than that with the walls, or down the centre of the column when friction of substrate with the walls slows fluid flow. Therefore some of the particles in the column do not come into contact with the substrate, and contact with the rest of the particles is variable.

Eddy diffusion also causes inhomogeneous distribution of fluid. This effect occurs when there are variations in the size and shape of the support particles and because of the tendency of the larger particles to become packed close to the walls of the column such that the flow rate varies over the cross-section of the column.

Back-mixing in columns is quantified by means of a dispersion number which is defined as:

$$\frac{D_c}{\mu l_c} \tag{3.50}$$

where D_c is the dispersion coefficient, μ is the interstitial velocity, and l_c is the depth of the bed. Thus the dispersion number increases roughly with the Reynolds number. The reciprocal of this value $\mu l_c/D_c$ is the Bodenstein number. The Bodenstein number (Bo) equals ∞ when plug-flow is operating and zero when perfect mixing is taking place. Usually in immobilized enzyme/cell columns there will be significant deviations from ideal flow, often because the void spaces between the support particles act as mixing chambers, such that Bo values less than 1.0 are obtained (see Fig. 3.19). However, the published data suggests that divergences from plug-flow are very low; for instance, close approximation to plug-flow was obtained using immobilized Streptomyces cells in polyester sacs (Ghose & Chand 1978), and for glucoamylase immobilized to porous glass particles by Marsh & Tsao (1976). Calculations support this conclusion, so that it is safe to assume that most properly operated reactors achieve approximate plug-flow behaviour. This is an important factor in reactor operation since it has been demonstrated that the size of reactor required to achieve a given productivity rises dramatically as deviations from plug-flow become appreciable (Kobayashi & Moo-Young 1971, Fig. 3.19).

The dispersion number can also be expressed in terms of the Peclet number ($N_{pe} = \mu d_p/D_c$), that is

$$\frac{D_c}{\mu l_c} = \frac{d_p}{l_c N_{pe}} \quad . \tag{3.51}$$

In columns, mixing can be either transverse or longitudinal or both, the latter being the greatest problem. Longitudinal mixing is thought to be caused by the interstitial spaces between the support particles acting as mixing chambers. Whether longitudinal mixing is important or not depends on the ratio of the length of the column to the diameter of the support particles, becoming appreceable, when this ratio exceeds 100. Thus even comparatively large support particles can be tolerated in large-scale columns. The experimental study of fluid flow in immobilized enzyme systems is difficult. One very interesting approach to this problem in the use of flow-through microfluorimetry which can help to elucidate the interplay between the intrinsic activity of the immobilized enzyme and its fluid dynamic properties (Hofmann & Sernatz 1983).

In stirred reactors more effective mixing can only be achieved by increasing the stirrer speed, or by decreasing the viscosity of the substrate. Other ways of achieving better mixing are to decrease the concentration of the biocatalyst particles or by baffling the reactor more effectively. In packed-bed reactors poor packing, imperfect support particles, and temperature gradients lead to non-ideal flow, which can be expressed in terms of the increase in the reactor size required to achieve the required productivity (Fig. 3.19). When S/K_m is large, the reaction is zero-order and non-ideal flow has no effect on the output of the reactor. However when S/K_m is very small, that is, when a high percentage conversion has been attained or when a low concentration of substrate is used or when the enzyme(s) have a low affinity for the substrate, then back-mixing becomes very important and the volume of the reactor needed to achieve a given productivity increases, particularly in stirred reactors. To achieve plug-flow one should employ even-sized, smooth, spherical, evenly packed support particles and arrange the absence of accumulated solids or gases in the column.

The yield of a reactor is affected by back-mixing in several ways. Some substrate becomes mixed with the product stream leaving the reactors, and so is no longer available to form product (Fig. 3.19). Also back-mixing decreases the substrate concentration, and so lowers the rate of reaction. Another effect is to increase the product concentration at which reaction takes place. Non-ideal flow also causes a marked decrease in the selectivity and thus the yield of the reaction because the contact time of some reactant molecules will be much greater or less than average so giving a greater chance for formation of side-products and further metabolism of the desired product, or non-reaction of substrate respectively.

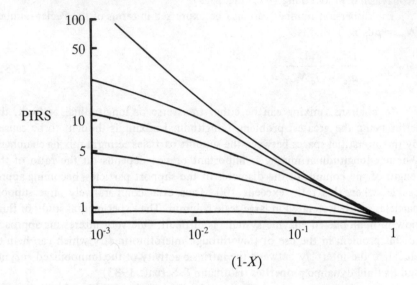

Fig. 3.19 – An illustration of the proportional increase in reactor size (PIRS) required to compensate for deviations from ideal plug-flow behaviour over a range of degrees of conversion of substrate $(1-X)$ and a S/K_m of 0.5. The degree of back-mixing decreases with the Bodenstein number (Bo), a measure of the extent of back-mixing which equals infinity under plug-flow conditions and zero in perfectly mixed conditions. Redrawn from Kobayashi & Moo-Young (1971).

3.26 PHYSICAL PROBLEMS IN BIOCHEMICAL REACTORS USING IMMOBILIZED BIOCATALYSTS

The main mechanical problems when using enzyme reactors are abrasion in stirred reactors, and compression and fouling in packed columns. Such mechanical factors will only really apply when immobilized enzymes are used on a large scale, and not in small-scale analytical and clinical use. Another possible problem is overheating of the reactor when an exothermic reaction is taking place, leading to denaturation of the enzyme(s), although this is only a serious problem when large reactors are employed since heat transfer is some two orders of magnitude more rapid than rates of chemical diffusion.

3.26.1 Abrasion

Abrasion of biocatalyst particles in CSTRs or fluidized beds increases with the shear rate and the fraction of the reactor volume occupied by the particles, and decreases as the viscosity of the suspending fluid and the strength of the support particles are increased; for instance, see Regan *et al.* (1974), who studied the attrition of β-galactosidase immobilized on AE-cellulose in a stirred reactor. A novel approach to this problem is to agitate enzyme immobilized to a magnetic support by means of an electromagnetic field, so eliminating particle–impeller collisions (Sada *et al.* 1980); although particle–particle interactions remain as a potential cause of abrasion.

3.26.2 Compression

Compression of biocatalyst particles in packed-bed reactors occurs when excessive pressure drops through the column are generated by friction between the fluid and the support particles or by partial blockage of the column bed by particulate materials. The pressure drop is described by the Carmen–Kozeny and other similar equations (Coulsen & Richardson 1970), the length of the column and the flow rate of fluid through it being the chief determinants. Compaction of the package bed is especially likely when small or irregularly shaped or packed particles are used in large, tall columns for long periods. In noncompressible columns under conditions which give low Reynolds numbers the pressure drop is a linear function of flow rate, the resistance to flow being constant. With deformable or compressible columns pressure drops increase exponentially as flow rates are increased. Compression usually increases rapidly at first, the rate of compression falling exponentially with time, this being a 'creep' effect well known in materials science: creep being the slow but continuous deformation of materials that are subject to relatively small stresses for long periods such as the deformation of fluids of extremely high viscosity, such as glass, under their own weight. The usual result is that the support particles are not fractured unless very high or sudden forces are exerted, but that they deform viscoelastically such that the intra-pellet spaces in the column are lost as the length of the packed bed decreases resulting in a reduction in the flow rate through the column and an increased pressure drop across the column. This pressure drop tends to cause further compression, in a vicious circle, eventually terminating in complete blockage of the column (Fig. 3.20). Compaction can be minimized by the use of relatively large, incompressible, smooth, spherical and evenly packed support particles, by the use of sectionalized columns, by fluidizing the column at intervals, or by decreasing the height to diameter ratio of the column, but there are practical difficulties in packing columns. It is very difficult to compare the compressive modulii of different immobilized preparations since they are usually formed and used under very difficult conditions.

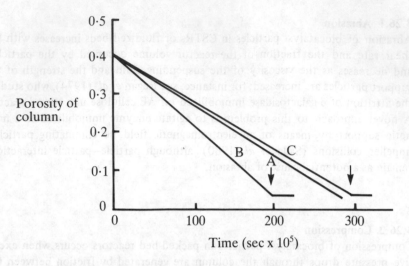

Fig. 3.20 — Measurement of the rate of compression of alginate gel pellets in packed columns over long periods. Columns were run in a downflow mode using, (A) distilled water and (B) 1.6 M sucrose as solvent. (C) is a similar column to (B) except that the sucrose was pumped up the column. ↓ represents the point at which blockage of the column occurred (Cheetham 1979). Reprinted by permission of the publishers, Butterworth & Co. Ltd ©.

3.26.3 Fouling

Fouling causes a loss in the activity of immobilized biocatalysts. This is due to the deposition of solid or colloidal materials (originally suspended in the substrate, around, or in, the pores of the support) preventing access of substrate to the enzyme. In stirred reactors fouling of exit or inlet filters is the main problem. In packed-bed reactors, however, fouling of the void spaces between particles (and of the pores of the particles) can be observed. The degree of this fouling will vary with the amount, size distribution, and chemical and physical nature of the fouling materials. Another important factor is the location of the biocatalyst either in (or on) the support material. Fouling is also accentuated when concentrated viscous substrates are used, and can occur even when optically clear substrates are used, the solid material being generated by a change in conditions, for example pH on the column causing precipitation. Even extremely low concentrations of undissolved solids can cause dramatic blockages especially if the solids are all deposited as a thin layer very close to the inlet of the column and/or when very large throughputs of substrate are encountered. Fouling can be decreased by clarification of the substrate by pretreatment with heat or chemicals or by filtration or centrifugation which, however, are expensive and inconvenient procedures and are impossible with some natural material suitable for sale in that form, such as milk.

Fouling is the major factor preventing the application of immobilized biological catalysts in many areas of the food, beverage, and pharmaceutical industries where only clear substrate streams can be tolerated, so necessitating the removal of any undissolved solids in a separate unit operation before biological catalyst can be exposed to the substrate. The ability to process unrefined substrates would also be advantageous if cheap, relatively unrefined substrates derived from agricultural surpluses are to be cheaply processed as chemical feedstocks.

Lipids, colloids, proteins, and suspended solid particles are the most common fouling materials; for example yeast homogenates or whole whey will very rapidly clog packed columns . Fouling can be separated into two main components. External fouling is the blockage of the interstitial spaces between biocatalyst particles, resulting initially in channelling and eventually a complete blockage of the column such that fluid cannot be forced through the column even under extreme pressures. Internal fouling of the pores in each individial biocatalyst particle, whether it be in the form of a thin layer as in ultrafiltration membranes or as porous pellets can occur such that even though fluid can freely pass through a packed column, little substrate can penetrate into the support particles and thus come into contact with the immobilized enzyme or cell. This type of fouling can be detected if activity is increased, and if the fouling material is released, for example when the immobilized biocatalyst is ground to a small size. Both external and internal fouling often occur together in the same system, although either may occur alone. Fouling of columns may be prevented, however, by intermittent repacking, fluidization, or back-flushing of the column bed.

Problems in the use of fouling substrate streams are often compounded because these substrates are often very viscous, perhaps because of the physical nature of the fouling solids, but often because it is commercially desirable to use a concentrated substrate solution so as to decrease the volume of fluid which has to be processed. Use of such viscous fluids can even make the prefiltration or precentrifugation of the substrate stream more difficult.

Fouling is conventionally prevented by clarification of the substrate stream by centrifugation or filtration, but these steps are expensive especially where large quantities of low priced product are being made, and cannot be used when the fouling solids are an essential part of the product, for instance colloids cannot be removed from milk prior to processing, as the milk cannot be subsequently reconstituted to its original state. External fouling could be prevented by using stirred or fluidized bed reactors provided that the fouling particles are appreciably smaller than the catalyst particles. However, these reactor configurations require a greater power input; many immobilized catalyst support particles are easily abraided in stirred reactors, and there may be kinetic disadvantages also. In practice, external fouling of packed bed reactors is normally temporarily corrected by repacking the column, by back-flushing, or by providing an intermittent upflow of air, all of which necessitate departures from continuous operation. Alternatively, recycling substrate at high flow rates may help to prevent fouling. Recently the use of nonporous magnetic immobilized enzymes in stirred or fluidized reactors has been proposed as a method of overcoming fouling by suspended solids when the reactant is a small molecule (for a review see Halling & Dunnill (1980); for instance, chymotrypsin immobilized to magnetic supports has been used to hydrolyze casein (Monroe *et al.* 1981). An even more novel approach is to use a soluble enzyme to carry out the reaction and then to completely and specifically recover it by a bioaffinity procedure using a ligand which binds to the enzyme of interest, and which is itself attached to a magnetic support (Halling & Dunnill 1979).

The problems associated with the presence of undissolved particles in the substrate solution are obviously at their most extreme when the solid particles are the desired substrate. Examples are the hydrolysis of casein and the enzymic digestion of microbial cell walls in order to release the contents of the cytosol. Rather than use a soluble enzyme, use of a soluble enzyme where the enzyme is linked to a soluble polymer such as dextran is a possible solution.

3.26.4 Microbial contamination
When manufacturing food or pharmaceutical products it is usually necessary to maintain sterile or, more commonly, hygienic conditions. Thus contamination by microorganisms must be stringently monitored and prevented. Contamination is especially likely when the substrate provides nutrients for the growth of the microorganisms, when residence times are long and when there are areas in the reactor, such as stagnant regions or rough surfaces, which the contaminants can easily colonize. Prevention of contamination is important not only because the presence of cells in the medium is undesirable in itself, but because they can foul the column and produce enzymes and metabolites which can degrade the desired product, produce undesirable side-products, and degrade the immobilized enzyme activity support. Contamination is minimized when the product produced discourages the growth of microorganisms, for instance antibiotics, ethanol, and organic acids. Inclusion of bacteriosides, bacteriostats, organic solvents, or other materials in the substrate, or by treating the reactors with them intermittently has been used with some degree of success although such procedures are hindered by the lack of knowledge of the mode of action of common antibacterial agents such as sorbic acid. For instance, Baret (1980) used a substituted diethylenetriamine to disinfect a column of immobilized *A. niger* lactase used for hydrolysis of the lactose in whey permeate. Daily treatment greatly reduced the microbial count in the effluent from the column and maintained its activity at a higher level than in columns which had not been disinfected.

A novel approach which has been used on a laboratory scale is the use of enzymes as antibacterial agents, for instance immobilized peroxidase (Henry *et al.* 1974), immobilized lysozyme acting on *B. subtilis* contaminants in an immobilized β-galactosidase column (Mattiasson 1973) (Fig. 3.21), and bacterio-lytic enzymes from *Achromobacter lunata* immobilized in collagen (Karube *et al.* 1977).

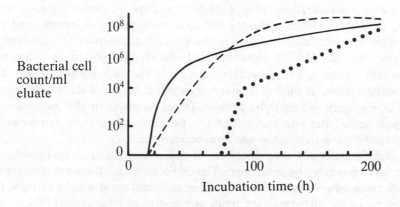

Fig. 3.21 – The time-dependent appearance of bacterial cells in the eluate from enzyme columns containing β-galactosidase (——) lysozyme immobilized in the upper section of the column with β-galactosidase below (. . . .), and co-immobilized lysozyme-β-galactosidase (- - - - -) bound to Sepharose 4B. Substrate solution used was 10 mM lactose in 0.1 M potassium phosphate, pH 7.0 containing 10^6 cells of *Bacillus subtilis* per ml. Experiments were carried out at room temperature (Mattiasson 1977).

3.27 THE STABILITY OF IMMOBILIZED BIOCATALYSTS (see Part A, Section 4.4)

3.27.1 Introduction

The stability of an immobilized enzyme depends on the intrinsic nature of the enzyme and the conditions under which it is immobilized and used, that is, the type of reactor and the reaction conditions used. The factors which stabilize and inactivate enzymes are not systematically understood, and individual enzymes vary much in stability; for instance enzymes derived from thermophilic organisms can be exceptionally stable not only to heat but to other denaturants. Loss of activity, for instance due to an inhibitor, should be distinguished from the loss of activity when the enzyme is denatured, for instance by heat. In practice a number of different effects occur together, the relative importance of each individual cause, such as oxidation or proteolytic degradation, depending on the system under study. It is well known that immobilization of an enzyme can influence its operational stability, both increases and decreases in stability having been observed experimentally. For instance proteases are more resistant to autodigestion when used in an immobilized from, covalent linkage to a support can stabilize the tertiary structure of enzymes because of covalent modification, and enzymes are often stabilized when immobilized simply because they are also concentrated by immobilization — most enzymes being more stable when maintained in a high local concentration of protein(s).

Enzyme denaturation can be considered as a special case of enzyme inhibition, diffusional effects increasing the apparent activities of immobilized cell activities by a similar mechanism to the effect first described in immobilized enzymes by Ollis (1972). Denaturation is usually thought to be a two-step procedure consisting of a reversible uncoiling of the molecule followed by the irreversible chemical modification or aggregation of the molecule leading to loss of solubility and catalytic activity. Direct estimates of the proportion of enzyme active sites left unmodified and therefore catalytically active can be obtained by fluorescent active site titration methods.

Some reports of the enhancement of operational stabilities by immobilization may, however, be artifactual. This is because of diffusional restrictions, which cause only a fraction of the total immobilized enzyme to be active at the beginning of use, the remainder acting as a reserve of fresh activity which only comes into action as some of the activity expressed initially is denatured (Fig. 3.22). Thus a loss of activity with time is only observed after a period of time sufficient for this 'buffer' enzyme activity to be lost. In the absence of diffusional restrictions, activity decays exponentially with time, whereas when diffusional limitations are present, activity decays linearly with time (Figs. 3.23, 3.24). A good example of this phenomenon is the glucose isomerase activity of pelleted cells, where loss of activity was linear when diffusion limited activity, but at the end of the operation when little activity remained exponen-

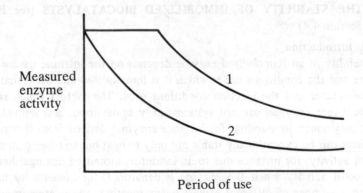

Fig. 3.22 — A diagrammatic representation of the operational stabilities of (1) immobilized and (2) free (soluble) enzyme preparations versus time. Note that the immobilized preparation is often more stable then the soluble enzyme and displays a period during which no enzyme activity appears to be lost. Reproduced with permission from *Principles of Biotechnology* (Ed. Wiseman, A.), Surrey University Press, Blackie & Son Ltd., Glasgow and London 1983.

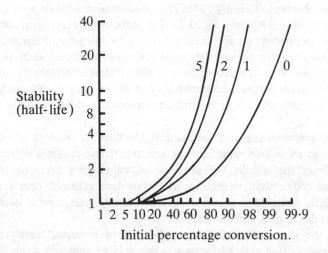

Fig. 3.23 – The influence of intraparticle diffusional restrictions on the stability of an immobilized enzyme, defined as the time required for the fractional conversion of substrate into products to become half of its original value (Kobayashi et al. 1980). The figures next to the lines are values for the Thiele modulus, that is, the line labelled zero refers to conditions in which there are no internal diffusional restrictions.

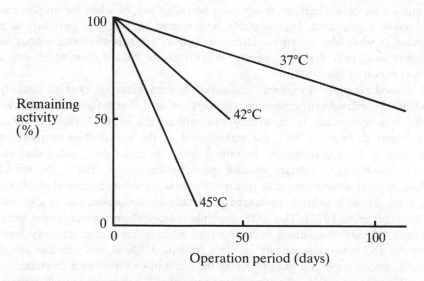

Fig. 3.24 – The operational stability of the aspartase activity of immobilized E. coli cells operated at three different temperatures. The enzyme is more active, but less stable as the operating temperature is increased (Chibata et al. 1976a).

tial decay was observed (Regan 1977). This phenomenon is dealt with in detail by Kobayashi *et al.* (1980), see Fig. 3.23. For multi-enzyme systems, continued activity is also dependent on the preservation of the integrity of the cell membrane. Even when a complex of enzymes are used in concert such as in the formation of ethanol from glucose, activity still declines exponentilly, just like a single immobilized enzyme, presumably because one enzyme is usually more labile than the other and so limits and determines the stability of the overall system.

Microbial contamination or mechanical blockage of reactors often occur suddenly and so result in a rapid, abrupt loss in activity; thus it is important to actually measure the stability of an immobilized enzyme over the required period of use rather than to extrapolate from data gathered over a shorter period. Similarly, stabilities obtained on a laboratory scale cannot necessarily be easily reproduced on a larger scale.

Usually the successful industrial application of biological catalysts such as free and immobilized cells and enzymes can only be achieved when they are used in ways consistent with existing chemical engineering principles and practices, taking into account the special properties of enzymic catalysts compared with chemical catalysts, such as their stereospecific reaction mechanisms and their comparatively labile nature. Both the initial activity of the immobilized enzyme and its stability directly influence its usefulness in industrial applications. Highly active preparations of comparatively low stability can be tolerated when the support materials are cheap and abundant and so can be easily discarded after only a comparatively short period of use, or when the support can be easily regenerated. Highly stable immobilized enzymes are obviously at a premium when the enzyme and/or the support are expensive and cannot be regenerated, such that the operator is reluctant to discard them after only a short period of operation.

Good examples of enzyme stabilization by immobilization exist, particularly when immobilized cells and/or gel supports are used. For instance the preservation of a notoriously labile hydrogenase within cells of *Alcaligenes eutrophus* (Klibanov & Puglisi 1980), the stabilization of the isomaltulose synthesizing enzyme of *Erwinia rhapontici* by immobilizing the intact cells, rather than extracted enzyme, in calcium alginate gel (Cheetham *et al.* 1982), the use of *Pseudomonas dentrificans* cells to denitrify drinking water (Nilsson *et al.* 1980), and the use of alginate immobilized plant cells for the production of alkaloids (Brodelius *et al.* 1979). This latter example is especially interesting as the metabolism of the immobilized cells appeared to have been diverted away from growth and towards secondary product formation. Metal ions can also act as stabilizers, an important example being the effect of Ca^{2+} ions on α-amylases.

When some viable cells remain, the activity of the immobilized cell preparation can often be increased, often to levels greater than originally present, by the periodic supply of nutrients so as to allow the growth of *de novo* cells with fresh enzyme activity (Ohlson *et al.* 1979). Enzymes immobilized by multiple points of attachment to a support can be stabilized; for instance ribonuclease was about 10°C more thermostable, and malate dehydrogenase could be made 'halophilic' by the same approach, enzyme activity being retained even in 4 M NaCl solutions (Koch-Schmidt & Mosbach 1977, Koch-Schmidt *et al.* 1979).

Several factors can be implicated as affecting the stability of immobilized cells and enzymes. Obvious denaturants include excessive heat, heavy metals, and materials which disrupt the tertiary structure of enzymes. These factors may act directly or indirectly. Direct effects include chemical modification of the enzyme, for instance immobilized proteases may be stabilized by acylation (Maneepum & Klibanov 1982). Spontaneous oxidation of amino acids important in maintaining the correct confirmation of the enzyme is important, and especially those amino acids which are catalytically active in the active site, such that the relatively anaerobic environment provided by many immobilization matrices may be beneficial. A second effect is attack by proteases; these may be secreted from contaminant microorganisms, or in the case of immobilized cells they may be endogenous in origin. Thus it is important to minimize microbial contamination by operating at elevated temperatures, to include bacteriostatic agents in the substrate, to cool the cell broth prior to and during harvesting so as to prevent microbial contamination, and to minimize the protease content of the cells by selecting for strains of microorganisms with minimal protease activity. Another ploy is to use growth media which yield cells with a minimum protease content, for instance by omitting proteins and by harvesting the cells at a phase in growth in which the protease content of the cells is at a minimum. Deliberate inactivation of contaminating protease activities from enzyme preparations may also be useful, as was found for glucose isomerase (Adler *et al.* 1979). Other factors which have been implicated include the maintenance of a low water activity environment, for instance by maintaining high reactant concentrations, and the maintenance of the integrity of immobilized cells, presumably because the nature environment of the cell enhances stability. There are also several well documented instances where enzyme stability has been improved by the attachment of an enzyme to support by multiple bonds, the degree to which the enzyme structure is stabilized increasing with the number of enzyme—support bonds.

Indirect effects include fouling of immobilized preparations such that the enzyme activity is lost because the substrate molecules are physically prevented from having access to the enzyme, by, for instance, the pores in the immobilized enzyme matrix being blocked by particulate material. Lastly, certain reactants appear to stabilize the enzymes, the phenomenon of substrate stabilization of enzymes is well known, the rationalization being that the enzyme—substrate complex is more stable than the simple enzyme; similar effects have been described for product stabilization especially where the product is a chemical which discourages microbial contamination. Immobilization can also serve to mechanically stabilize fragile cells, for instance surface-independent hybridoma or lymphoblastoid cells, have been successfully used in an immobilized form for the continuous formation of monoclonal immunoglobulins and lymphokines. The cells could thus be used in high densities, and were mechanically stabilized by immobilization, so the need for methods to separate the cells and their products was obviated (Nilsson *et al.* 1983).

A major problem in experimentation is the time and resources required to carry out stability tests, particulerly when the immobilized enzyme or cell already has a good operational stability and when the tests must be carried out on a large scale of operation. Sometimes 'accelerated stability tests' are resorted to where the reactor is deliberately operated at a temperature well in excess of that expected to be used in practice such that factors affecting the rate of decay of the immobilized preparation are revealed more rapidly. Another difficulty is that the maximum possible or 'ceiling' stability possible for any particular enzyme remains unknown during experimentation so that at any time one does not know whether one has achieved 1% or 99% of the maximum possible longevity.

Lastly, it is important to recognize that in some cases it is the exceptional stability of enzymes that is a problem in industry; for instance bacterial α-amylase which has a number of important large-scale applications in the food industry is deactivated only relatively slowly even when boiled, and so active enzyme may contaminate the product with very unfortunate consequences. Similar problems arise with microbial rennets, which have in fact been made more heat labile by chemical modification.

3.27.2 The stability of biochemical reactors employing immobilized enzymes or immobilized cells

The stability of an immobilized biocatalyst in actual operation is affected by a number of factors, including irreversible inhibition by substances present in the substrate. Important factors are denaturation by temperature pH or ionic strength or shear forces. In addition there is attack by microorganisms or enzymes, particularly proteases, derived from microorganisms fouling the column. Other problems are the leakage of cells or enzyme from the support, dissolution or fragmentation of the support, and poor enzyme–substrate contacting. Self-digestion occurs in the case of proteases.

Activity decays more slowly, when the reaction is diffusion controlled as often it is, and this is true whether the diffusional restriction is caused by a higher enzyme concentration or the use of larger particles, or by less rapid diffusion of substrate within the particle, or by the use of lower substrate concentration. Sometimes the decay of activity is linear with time until the biocatalyst activity has been reduced to a point where the reaction is diffusion controlled. Concave decay of activity can be caused by leakage of enzyme from the reactor during use, and a convex decay by cumulative inhibition or denaturation. For first-order decay, $-\mathrm{d}\,E_o/\mathrm{d}_t = \lambda\,E_o$ that is,

$$\ln \frac{E_o}{E_t} = \lambda t \; , \tag{3.52}$$

where E_o is the original activity of the reactor, E_t is the activity remaining after time t, and λ is the decay constant. In practice the stability of the reactor is usually described in terms of a half-life, which is the time required for half of the original activity to be lost $(t/2)$, although it is more correct to quote a first-order decay constant.

$$\text{half-life } (t/2) = \ln 2/\lambda \quad \text{that is } (0.693/\lambda) \; , \tag{3.53}$$

or

$$\text{half-life} = \frac{t \ln 2}{\ln (E_o/E_t)} \; . \tag{3.54}$$

The use of the term half-life must be distinguished from the usage sometimes encountered when describing batch reactions where the half-life represents the time taken for the reaction to be completed.

After any given period of operation (t) the activity remaining is given by the expression:

$$\text{activity at time } t = \text{initial activity, } \exp{(-t/T_c)} \; , \tag{3.55}$$

where T_c is a time constant representing the time required for the immobilized activity to decay to 36.8% of its original activity. In practice, reactors are operated for a period after which it is calculated that it becomes more effective to replace or regenerate the catalyst rather than to continue to use the original enzyme preparation.

The quest to find general methods of stabilizing the activity of immobilized cells carrying out multi-step synthetic reactions constitutes one of the major challenges in this field of research. In addition to the operational stability the storage stability of an immobilized biocatalyst must be known because this property dictates the user's ability to transport the immobilized cells or enzyme and to store them, especially during hold-ups in operations. Thus amylases are often stabilized with starch hydrolysates, sugars, or their derivatives, while proteases are stabilized by peptides produced from proteins. Very good storage can often be obtained, for example glycerol dehydratase, adenosidase and 3'5' cyclic AMP phosphodiesterase entrapped in Sephadex G-50 stored dried at around 0°C remained active after several years (Schneider & Friedman 1981).

3.27.3 Regeneration of biocatalyst activities
A few enzymes can undergo reversible denaturation and so can be regener-

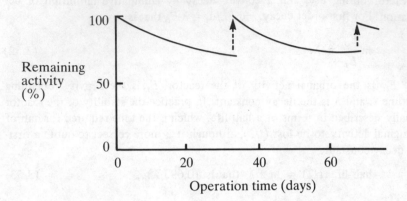

Fig. 3.25 – The stability and regeneration of aminoacylase immobilized on DEAE-Sephadex and used in a packed-bed column. Regeneration (denoted by ↑) was achieved by desorbing the exhausted enzyme and adsorbing fresh active enzyme onto the original ion-exchange resin. Redrawn from Chibata *et al.* (1976).

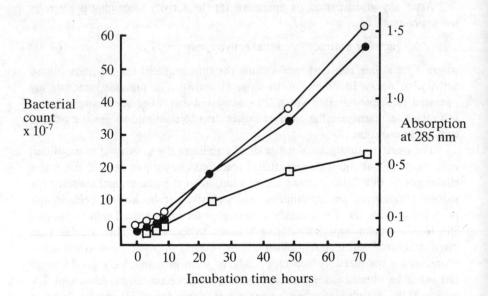

Fig. 3.26 – The relation between the activity (absorbance at 285 nm) of *A. simplex* cells and the number of cells at various stages of activation by nutrients. □ = Δ¹-dehydrogenase activity of immobilized cells, ● = Δ¹-dehydrogenase activity of free cells after solubilization of the alginate support; ○ = number of living cells (bacterial weight). (Ohlson *et al.* 1979).

ated. Otherwise regeneration must be confined to desorbing the potentially inactivated enzyme followed by immobilization of fresh, unreactive enzyme (Fig. 3.25). This procedure is not easy to carry out *in situ*, unless the enzyme is immobilized by adsorption and the sites for enzyme production, support regeneration and the factory using the immobilized enzyme are close together, because the costs of transporting new and exhausted immobilized enzyme can be prohibitive. Thus the ideal support material is cheap and abundant enough to be discarded after a single use and replaced with fresh immobilized enzymes or cells. Regeneration of the activity of immobilized viable cells is achieved either by re-inducing enzyme(s), or by supplying growth medium and allowing cell growth to take place *in situ*, although it may not be possible to use the growth medium originally used for the cell, a recovery type medium sometimes proving more effective (Fig. 3.26). Cell growth however, may lead to leakage of cells from the support particles and increased diffusional restrictions inside the pellets because the new cells either block up some of the pores in the support or a problem may arise due to the fresh cells not being evenly distributed throughout the support. Cell growth *in situ* can also physically disrupt pellets because of the internal pressures which are generated.

3.27.4 Constant productivity with biocatalyst reactors

Constant productivity (by adding fresh biocatalyst during operation) is not always operationally convenient or even possible when reactors of the packed-bed type are used. Another useful way is to control the flow-rate through the reactor. The flow rate can be continuously or intermittently decreased so as to maintain the degree of conversion constant. During the period of operation, however, the quantity of product produced per unit time will fall. Here one must express the results as the flow rate required to give the desired product quality in a given time, rather than the activity or percentage conversion achieved in a given time. Latent periods caused by diffusional restrictions in which the expressed activity does not fall for some time after operation of the reactor is commenced can contribute to the achievement of constant degrees of conversion. Constant product quality can be arranged by increasing the temperature of operation. Thus, as the reaction proceeds, the loss in enzyme activity with time is offset by the increased enzyme activity at the higher temperatures; then the rate of change of the temperature of operation with time should be quoted. This procedure as used for glucose isomerase is described by Park *et al.* (1981). Even comparatively complex changes in temperature can be easily and accurately controlled by microprocessor devices. The activities of biocatalysts are invariably much less stable when higher temperatures are used. An example of manipulating the operating temperature is seen in the industrial scale production of malic acid using the fumarase activity of immobilized *Brevibacterium ammoniagenes*. cells.

Use of a number of columns in series, but of different ages, out of phase with each other, often in combination with one of the above procedures, is commonplace. Verhoff & Schlager (1981) found that this procedure enabled better enzyme utilization but required a larger volume of reactor than when conversion was maintained by continuously decreasing the flow rate. In this case the amount of immobilized biocatalyst used does not change with time despite the constantly declining productivity obtained from any individual column because exhausted columns are replaced continually with fresh columns. The greater the number of reactors that are employed, then the smaller the variation in the aggregate product concentration that will ensue. Also, fewer problems with compression of the columns are encountered when using a number of smaller columns.

By employing an adequate number of reactors, the variation in throughput level, while maintaining constant conversion, can be maintained within any desired limits. Typically, a ±5% flow rate fluctuation is an acceptable limit. The minimum number of reactors required for a given production rate can be calculated from eq. (3.56). By staggering the start-up and reloading times, it is possible to achieve smooth reactor operation with minimal fluctuations in throughput or conversion level (Venkatrasubramanian 1980).

$$Mi/Mx = \exp(n \ln 2/N_r) \tag{3.56}$$

where N_r = number of reactor,
      n = number of half-life utilizations of the catalyst,
      Mi/Mx = ratio of minimum to maximum production rate.

Most commercial plants have at least six columns in the reactor battery controlled by microprocessor feedback devices. While increasing the number of reactors provides greater operational flexibility and minimal flow rate variation, it is nevertheless important to accept the resulting higher costs for reactor piping, valves, and instrumentation. There is also a marginal increase in operational costs due to more frequent catalyst changes.

Both series and parallel modes of operation of the reactor system are possible. Series operation provides fewer streams to control and better catalyst exhaustion capability. However, the potential for pressure drop and compaction problems are greatly enhanced in series operation. Parallel operation offers the greatest operational flexibility and each reactor can be operated essentially independently of the others, because each column can be brought into and taken out of service quite readily. The actual mode of operation in a production facility is determined by the resultant of a number of interrelated operational parameters, some of the most important of which are the cost of the immobilization support, the flow rate of substrate through the reactor, and the activity and stability of the immobilized cells or enzyme (also see Fig. 3.26).

3.28 SCALE-UP

During its development a process will be scrutinized for economic potential by, for instance, calculating the likely cash flow if commercialization is proceeded with, and sensitivity analysis whereby the variation in the potential profitability of the process and/or product with factors such as the cost of raw materials, conversion efficiency, and scale of operation are calculated.

The unit cost of virtually any product falls with about the two-thirds power of the volume of that product produced. This relationship dictates that in order to obtain these economics of scale there is a strong tendency to build production facilities as large as possible, provided that the diseconomies of operating a large plant at less than maximum capacity can be avoided. Two advantages of large-scale operation are that because of the comparatively large financial investments that must neccessarily be made there is a strong disincentive for competitors to enter the field unless they are very strongly motivated; secondly, in large-scale operations even very modest improvements can result in large financial rewards.

Scale-up is a very important factor in progressing a process towards commercialization via the steps of process invention, process development, process design, design engineering, plant construction, and plant commissioning.

When scaling-up an immobilization step it is advisable to define the rates and extents of mixing of enzyme, support, and reagents and the time needed for each step, so that these can be transferred to the larger scale operation. If the immobilization process is an exo- or endo-thermic process a large-scale immobilization vessel may need to be cooled or heated. In general, methods of immobilization need to be developed for large-scale enzyme and cell immobilization which minimize the labour required, use cheap abundant support materials, and which can be carried out reproducibly under conditions consistent with good manufacturing practice.

When scaling-up a process, geometric and dynamic similarly should be maintained if possible, but the presence of existing readily available equipment which could be used should not be neglected. On a larger scale, effects which were of minimal importance on a smaller scale can become vitally important, such as channelling and compression in packed-bed reactors, the growth of contaminant microorganisms in 'dead spots' in reactors, and in piping, or the positioning of monitoring devices in order to obtain the most accurate and representative information about reactant concentrations, pH, temperature, etc. The most obvious change on scale-up is that laboratory equipment is predominantly made of glass, whereas plant scale equipment is constructed in stainless steel. On a large scale many procedures which were easy to carry out can become difficult or impossible to perform. For instance, high speed mixing becomes increasingly difficult as the scale of operation is increased, also buffers can only rarely be used to control pH because of their expense and because they may contaminate the final product. However, other procedures, such as the batchwise use of ion-exchange resins in basket centrifuges, are actually easier to perform on a larger scale than on a smaller scale. Note also that the financial advantages of using very large reactors must be balanced against the risk of losing the contents of a reactor through accident or operator error.

Unlike fermentations, enzyme and cell reactors do not have to be operated under microbiologically sterile conditions, chiefly because rich sources of nutrients and good conditions for microbial growth are not deliberately supplied. Nevertheless, hygienic operation should be aimed at, particularly when manufacturing products intended for the human or animal food, beverage, or pharmaceutical markets, because substantial microbial growth can still occur even if only trace amounts of nitrogen source and other nutrients are present. One should aim for rapid throughputs of reactants, use of elevated operating temperatures, incorporation of bacteriostatic agents in the substrate and efficient reactor design and construction (such that dead spaces where microorganisms can lodge and divide are eliminated, the number of openings, such as sampling points, which provide potential access to microorganisms, is minimized, and direct connection of hygienic and non-hygienic areas is avoided). Use of high concentrations of substrate such that the resulting high osmotic pressure and low water reactivity discourage microbial growth is also useful, and of course hygiene is facilitated if the process makes antibiotics, organic acids, or ethanol.

It is also important that the reactor is sanitized adequately between uses. Back-flushing with acidified water or with water containing hydrogen peroxide or quaternary ammonium compounds has been used. In general, simplicity of process is desirable, such that the number of steps and the use of expensive and/or complicated equipment are minimized.

Good operating procedures are essential both to protect the worker and the consumer of the product under manufacture. Important factors include enclosure of process equipment, use of protective equipment, regular cleaning of equipment, and regular monitoring for chemical and microbiological contamination. Microprocessors are admirably suited to control automatically the start-up, shut-down, and cleaning of equipment, and will help to reduce human access to relatively unsafe areas of plant. Several useful texts deal with this subject; these include *Patty's Industrial Hygiene and Toxicology* (Wiley Interscience), the *CRC Handbook of Laboratory Safety,* and the *Hygienic design and operation of food plant* (Ellis Horwood Ltd.).

3.29 DISCUSSION

This chapter has attempted to survey the principles of enzyme engineering, relating to the industrial use of enzymes and similar biocatalysts. The field still appears to be in its infancy, and many major problems need to be overcome. These include: overcoming of permeability barriers and the supply of oxygen to immobilized cells or enzymes which require it as a substrate; the use of water-insoluble and water-immiscible substrates such as steroids and hydrocarbons; the discovery of new enzymes with novel enzyme activities, the use of less easily available classes of enzymes such as membrane-bound enzymes, the capability to use substrates containing organic solvents or undissolved solid particles which tend to foul conventional immobilized biocatalysts; the ability to regenerate cofactors *in situ* so that synthetic endergonic reactions can be performed and the problems caused by the high cost and limited availability of cofactor overcome; and the use of enzymes acting 'in reverse', and co-immobilized cells, enzymes organelles etc., so that complex multistep reactions can be carried out. Efficient methods for stabilizing enzymes and for harvesting cells, and versatile methods of manipulating the pH and temperature optima, stabilities, positions of equilibrium, and substrate specificities of enzymes, need to be developed. In particular, details of the biosynthesis, destruction, and control of enzymes in intact cells need to be elucidated both so that enzymes can be extracted most efficiently and so they can be used as catalysts in their own right. Very often immobilized biocatalysts need to be characterized more fully so as to facilitate scale-up studies. Only then can the diverse catalytic potential of enzymes be realized on an industrial scale, not only to produce high value, low bulk products, such as fuels and solvents, from renewable natural resources, which are at present produced by chemical catalysts from non-renewable petrochemical feedstocks.

A vital objective is the elucidation of general principles and concepts concerning topics such as enzyme stabilization, scale-up of immobilization, use of water-insoluble substrates etc., but understandably, most of the work carried out in such areas is empirical in nature and oriented towards the formation of satisfactory laboratory-scale preparations rather than an improved understanding of basic principles.

Provided that some of the above problems can be resolved in a satisfactory manner and a genuine working relationship can be established between scientists, technologists, and marketing people who are aware of the processes and products for which there is a real need, then enzyme technology can avoid the trap of being a technology in search of a product, and the tremendous amount of research work which has gone into this area in the recent past can be used and justified.

ACKNOWLEDGEMENTS

The author wishes to gratefully acknoweldge the cooperation of the following publishers in granting permission for the reproduction of figures and tables in this chapter:

Academic Press, The Biochemical Society, Butterworth & Company, Ellis Horwood Ltd., Macmillan Journals, Plenum Publishing Corp., Springer-Verlag, Surrey University Press (Blackie & Son Ltd.), Wheatland Journals, and John Wiley & Sons.

NOTE IN PROOF

Recently, several promising techniques for manipulating the substrate specificity and other properties of enzymes have been developed. These include renaturation techniques, the redesign of enzyme active sites by site-directed mutagensis, the selective use of non-specific antibodies and the coupling of enzymes with chemical catalysts.

Sarawathi & Keyes (1984) have shown that renaturation of a protein in the presence of competitive inhibitor to the desired enzyme activity, followed by removal of the inhibitor by dialysis and cross-linking of the protein to maintain its new conformation, can generate *de novo* enzyme activities. For instance, an 'acid-esterase' has been formed from RNase by treatment with indole proprionic acid.

Winter *et al.* (1982) and Wilkinson *et al.* (1984) have demonstrated that a single point mutation leading to a selective change in one amino acid of *B. stearothermophilus* tyrosyl t_{RNA} synthetase can decrease the K_m of the enzyme for ATP 100-fold. This approach requires a detailed knowledge of the 3-dimensional structure of the enzyme and *r*DNA methodology for the parent organism, and also presumes a good idea of the catalytic function of each amino acid.

Soloman *et al.* (1984) found that monoclonal antibodies raised to carboxypeptidase could be separated into fractions that would selectively inhibit either its peptidase or its esterase activities.

P. S. J. Cheetham
12.11.84

REFERENCES

Abbott, B. J., Cerimela, B. & Fukuda, D. S. (1976) *Biotechnol. Bioeng.* **18** 1033–1042.
Adachi, S., Hashimoto, K., Miyai, H., Kurome, R., Matsuno, R. & Kamikubo, T. (1981) *Biotechnol. Bioeng.* **23** 1961–1976.
Adler, D., Lim, H., Cottle, D. & Emery, A. (1979) *Biotechnol. Bioeng.* **21** 1345–1359.
Adlercreutz, P. & Matthiasson, B. (1982) *Eur. J. Appl. Microbiol. Biotechnol.* **16** 165–170.
Aizawa, M., Wada, M., Kato, S. & Suzuki, S. (1980) *Biotechnol. Bioeng.* **22** 1769–1783.
Alberti, B. N. & Klibanov, A. M. (1982) *Enzyme & Microbial. Technol.* **4** 47–49.
Archer, M. C., Ragnarsson, J. O., Tannenbaum, S. R. & Wang, D. I. C. (1973) *Biotechnol. Bioeng.* **15** 181–196.
Atkinson, B. (1974) in: *Biochemical Reactors,* Pion Press Ltd. pp. 191–214.
Banghn, R. L. & Whitesides, G. M. (1981) *J. Amer. Chem. Soc.* **103** 4890–4899.
Baret, J. L. A. G. (1980). Patent Applic. GB 2065137A assigned to Corning Glass Works, Corning, N.Y.
Benoit, M. R. & Kohler, J. T. (1975) *Biotechnol. Bioeng.* **17** 1617–1626.
Betz, T. L., Brown, P. R., Smyth, M. J. & Clarke, P. H. (1974) *Nature* **247** 261–264.
Bhasin, D. P., Gryte, C. G. & Studebaker, J. F. (1976) *Biotechnol. Bioeng.* **18** 1777–1792.
Bickerstaff, G. F. (1984) *Topics in Enzyme and Fermentation Biotechnology* Vol. 9, Chapter 4, pp. 162–201, Ellis Horwood Limited.
Blake, C. C. F., Johnson, L. N., Main, G. A., North, A. C. T., Phillips, D. C. & Sharma, V. R. (1967) *Proc. Royal Soc. London* **B167** 378–388.
Bucke, C. (1977) in *Topics in Enzyme and Fermentation Biotechnology* **1**, (Ed. Wiseman, A.). Ellis Horwood Limited, pp. 147–168.

Buckland, B. C., Dunnill, P. & Lilly, M. D. (1975) *Biotechnol. Bioeng.* **17** 815−826.
Bunting, P. S. & Laidler, K. J. (1974) *Biotechnol. Bioeng.* **16** 119−134.
Butler, L. C. (1979) *Enzyme & Microbial. Technol.* **1** 253−259.
Brodelius, P., Deus, B., Mosbach, K. & Zenk, M. H. (1979) *FEBS Lett.* **103** 93−97.
Carleysmith, S. W. & Lilly, M. D. (1979) *Biotechnol. Bioeng.* **21** 1057−1073.
Carleysmith, S. W., Eames, M. B. L. & Lilly M. D. (1980) *Biotechnol. Bioeng.* **22** 957−967.
Carrea, G., Colombi, F., Mazzola, G., Cremonesi, P. & Antonini, E. (1979) *Biotechnol. Bioeng.* **21** 39.
Chang, T. M. S., Yu, Y. Y. & Grunwald (1982) in: *Enzyme Engineering.* (Chibata, I., Saburo, F. & Wingard jr, L. B., eds.). Plenum Press, Vol. 6 451−456.
Chang, T. M. S. (ed.) (1977) *Biomedical Applications of Immobilized Enzymes and Proteins.* Vols. 1 & 2: Plenum Press, New York.
Charles, M., Coughlen, R. W. & Hasselberger, F. X. (1974) *Biotechnol. Bioeng.* **11** 1553−1556.
Cheetham, P. S. J. (1980 in *Topics in Enzyme and Fermentation Biotechnology,* Vol. 4 (Ed. Wiseman, A.) Ellis Horwood Limited, pp. 189−238.
Cheetham, P. S. J., Blunt, K. W. & Bucke, C. (1979) *Biotechnol. Bioeng.* **22** 2155−2168.
Cheetham, P. S. J., Dunnill, P. & Lilly M.D. (1982) *Biochem. J.,* **201** 515−521.
Cheetham, P. S. J. *Enz. and Microb. Technol.* **1** 183−188.
Cheetham, P. S. J., Imber, C. E. & Isherwood, J. (1982) *Nature* **299** 628−631.
Cheyan, M., Vay Wyk, P. J., Ohlson, N. F. & Richardson, T. (1975) *Biotechnol. Bioeng.* **17** 585−598.
Chibata, I., Tosa, T. & Sato, T. (1967) in: *Methods in Enzymology* Vol. 44 (Ed. Mosbach, K.) Academic Press, pp. 739−746.
Chibata, I., Tosa, T., Sato, T. & Mori, T. (1976) *Methods in Enzymology,* Vol. 44 (Ed. Mosbach, K.: Academic Press, pp. 746−759.
Chiga, M. & Plaut, G. W. E. (1960) *J. Biol. Chem.* **235** 3260.
Cohen, G. N. & Monod, J. (1957) *Bacteriol. Rev.* **21** 169−174.
Coleman, D. R. & Royer, G. P. (1980) **45** 2268.
Coulsen, J. M. & Richardson, J. F. (1970) *Chemical Engineering* 2nd ed. Pergamon Press, Oxford, pp. 1−19.
Cousineau, J. & Chang, T. M. S. (1977) *Biochem. Biophys. Res. Commun.* **79** 24−31.
Cremonesi, P., Carrea, G., Sportoletti, G. & Antonini (1973) *Arch. Biochem. Biophys.* **159** 7−10.
Cremonesi, P., Carrea, G;, Ferrara, L. & Antonini, E. (1975) *Biotechnol. Bioeng.* **17** 1101−1108.
D'Souza, S. F. & Nadkarni, G. B. (1980) *Biotechnol. Bioeng.* 2179−2190.
Dixon, M. & Webb, E. C. (1979) The Enzymes. 3rd ed. Longmans Group, London.
Duvnjak, Z. & Lilly, M. D. (1976) *Biotechnol. Bioeng.* **18** 737−740.
Engasser, J. M. & Horvath, C. (1975) *Ind. Eng. Chem. Fundam.* **14** 107−110.
Fersht, A. R. (1977) *Enzyme structure and mechanism* W. H. Freeman, San Francisco.
Fersht, A. R. (1980) in: *Enzymic and Non-enzumic Catalysis* (ed. Dunnil, P., Wiseman, A. & Blakeborough, N.) Ellis Horwood Limited/Society of Chemical Industry, pp. 13−27.
Fersht, A. R. & Kaethner, M. M. (1976) *Biochemistry* **15** 3342−3346.
Freeman, A. & Aharonowitz (1981) *Biotechnol. Bioeng.* **23** 2747−2759.
Fullbrook, P. & Slocombe, B. (1970) *Nature* **226** 1054.
Fullbrook, P. (1983) in *Ind. Enzymol.* (eds. Godfrey, A. & Reichelt, J.); Macmillan 8−40.
Gelft, G. & Boudrant, J. (1974) *Biochim. Biophys. Acta* **334** 467−.
Genung, R. K. & Hsu, H. W. (1978) *Biotechnol. Bioeng.* **20** 1129−1142.
Ghose, T. K. & Chand, S. (1978) *J. Ferm. Technol.* **56** 315−322.
Giard, D. J., Loeb, P. H., Tilly, W. G., Wang, D. I. C. & Levine, D. W. (1979) *Biotechnol. Bioeng.* **21** 433−442.
Guilbault, G. G. (1970) *Enzymatic Methods of Analysis,* Pergamon Press, Long Island City, New York.
Guilbault, G. (1980) *Enz. and Microbiol. Technol.* **2** 258−264.
Gupta, K. G., Jain, A. K. & Dhawan, S. (1979) *Biotechnol. Bioeng.* **21** 649−658.
Gutfreund, H. (1972) *Enzymes: Physical Principals,* Wiley−Interscience.
Hahn-Hagerdal, B. & Mattiasson, B. (1982) *Eur. J. Appl. Microbiol. Biotechnol.* **14** 140−143.
Halling, P. J. & Dunnill, P. (1979) *Eur. J. Appl. Microbiol and Biotechnol.* **6** 195−205.
Halling, P. J. & Dunnill, P. (1980) *Enz. & Microbiol. Technol.* **2** 2−10.
Hattori, R. (1973) *J. Gen. Appl. Microbiol* **18** 319−325.
Hayashi, T., Tanaka, Y. & Kavashima, K. (1979) *Biotechnol. Bioeng.* **21** 1019−1030.
Henry, S., Koczan, J. & Richardson, T. (1974) *Biotechnol. Bioeng.* **16** 289−291.

Hofmann, J. & Sernatz, M. (1983) *Trends in Anal. Chem.* **2** 172–175.
Holst, O., Enfors, S. O. & Mattiasson, B. (1982) *Eur. J. Appl. Microbiol. Biotechnol.* **14** 64–68.
Horvath, C. & Engasser, J. M. (1974) *Biotechnol. Bioeng.* **16** 909–923.
Ingalls, R. G., Squires, R. G. & Butler, L. B. (1975) *Biotechnol. Bioeng.* **17** 1627–1637.
Jones, A. & Veliky, I. A. (1981) *Eur. J. Appl. Microbiol. and Biotech.* **13** 84–89.
Jones, J. B. (1976a) in: *Applications of Biochemical Systems in Organic Chemistry* (Jones, J. B., Sih, C. J. & Perlman, D. eds.) John Wiley, Chapters 1 and 6.
Jones, J. B. (1976b) in: *Methods in Enzymology* **44** (Ed. Mosbach, K.) Academic Press 831–844.
Jowitt, R. (Ed.) (1980) *Hygienic design & operation of food plant.* Ellis Horwood Limited.
Karube, I., Suganama, T. & Suzuki, S. (1977a) *Biotechnol. Bioeng.* **19** 301–309.
Karube, I., Suganama, T., Tsuru, S. & Suzuki, S. (1977b) *Biotechnol. Bioeng.* **19** 1727–1734.
Karube, I., Mitsuda, S. & Suzuki, S. (1979a) *Eur. J. Microbiol. and Biotechnol.* **7** 343–350.
Karube, I., Aizawa, K., Ieda, S. & Suzuki, S. (1979b) *Biotechnol. Bioeng.* **21** 253–260.
Karube, I., Otsuka, T., Kayano, H., Matsunaga, T. and Suzuki, S. (1980) *Biotechnol. Bioeng.* **22** 2655–2665.
Karube, I., Matsunaga, T., Otomine, Y. & Suzuki, S. (19810 *Enz. & Microbiol. Technol.* **3** 309–312.
Kashe, V. (1983) *Enz. Microb. Technol.* **5** 2–13.
Kashe, V. & Galunsky, B. (1982) *Biochem. Biophys. Res Communs.* **104** 1215–1222.
Kierstan, M. & Bucke, C. (1977) *Biotechnol. Bioeng.* **19,** 387.
Kirwan, D. J.., Enright, J. T. & Gainer, J. L. (1974) *Biotechnol. Bioeng.* **16** 551–553.
Klibanov, A. M. & Puglisi, A. W. (1980) *Biotech. lett.* **2** 445–450.
Klibanov, A. M., Samokkin, G. P., Martinek, K. & Berezin, I. V. (1977) *Biotechnol. Bioeng.* **19** 211–218.
Kluepfel, D., Biron, L. & Ishaque, M. (1980) *Biotechnol. Lett.* **2** 309–314.
Kobayashi, T. & Moo-Young, M. (1971) *Biotechnol. Bioeng.* **13** 893–910.
Kobayashi, T., Katagiri, K., Ohmiya, K. & Shiniau, S. (1980) *J. Ferm. Technol.* **58** 23–31.
Koch-Schmidt, A-C. & Mosbach, K. (1977) *Biochem.* **16** 2101.
Koch-Schmidt, A.C., Mosbach, K. & Werber, M. M. (1979) *Eur. J. Biochem.* **100** 213.
Kokubu, T;. Karube, I. & Suzuki, S. (1978) *Eur. J. Appl. Microbiol.* **5,** 233–240.
Kokubu, T., Karube, I. & Suzuki, S. (1981) *Biotechnol. Bioeng.* **23** 29–41.
Kuhn, I. (1980) *Biotechnol. Bioeng.* **22** 2393–2398.
Lawrence, R. L. & O'Kay, V. (1973) *Biotechnol. Bioeng.* **15** 217–221.
Legoy, M. D., Larreta Garde, V., Le Moullerc J. M., Ergan, F. & Thomas, D. (1980) *Biochemie* **62** 341.
Levine, H. L., Jakagawa, Y. & Kaiser, E. T. (1977). *Biochem. Biophys. Res. Comm.* **76** 64–70.
Lilly, M. D. (1983) *Phil. Trans. Royal Soc.* **B300** 391–397.
Lilly, M. D. and Sharp, A. K. (1968) *Chem. Eng. (Lond.)* **215** CE12–19.
Lilly, M. D. & Dunnill, P. (1972) *Proc. Biochem. Biotechnol. Bioeng. Symp. No. 3* 221–227.
Lim, F. & Sun, A. M. (1980) *Science* **210** 908–910.
Maneepum, S. & Klibanov, A. M. (1982) *Biotechnol. Bioeng.* **24** 483–486.
Marsh, D. R. & Tsao, G. T. (1976) *Biotechnol. Bioeng.* **18** 349–362.
Mansson, M. O., Larsson, P-O. & Mosbach, K. (1978, 1979) *Eur. J. Biochem,* **86,** 455 and *FEBS Lett.* **98,** 309 resp.
Marshall, D. L. (1973) *Biotechnol. Bioeng.* **15** 447–453.
Marshall, V. J. & Humphries, J. D. (1977) *Biotechnol. Bioeng.* **19** 1739–1760.
Martin, C. K. A. & Perlman, D. (1976) *Biotechnol. Bioeng.* **18** 217–237.
Martinek, K., Moxhaev, V. V., Smirnov, M. D. & Berezin, I. U. (1980) *Biotechnol. Bioeng.* **22** 249–251 (also *Bioorg Khum* (1980) **6** 600).
Mattiasson, B. (1977) *Biotechnol. Bioeng.* **19** 777–780.
Merrill, A. M. & McCormick, D. B. (1979) *Biotechnol. Bioeng.* **21** 1629–1638.
Monroe, P. A., Dunnill, P. & Lilly, M. D. (1977) *Biotechnol. Bioeng.* **19** 101–124.
Monroe, P. A., Dunnill, P. & Lilly, M. D. (1981) *Biotechnol. Bioeng.* **23** 677–689.
Monti, J. C. & Jost, R. (1978) *Biotechnol. Bioeng.* **20** 1173–1185.
Morikawa, Y., Karube, I. & Suzuki, S. (1979) *Biotechnol. Bioeng.* **22** 1015–1023.
Mosbach. K. (1978) *Advs. Enzymol.* **46,** 205–278.
Murata, K., Tani, K., Kato, J. & Chibata, I. (1978) *Eur. J. Applied Microbiol.* **6** 23–27.
Murata, K., Kato, J. & Chibata, I. (1979) *Biotechnol. Bioeng.* **21** 887–895.
Murata, K., Tani, K., Kato, J. & Chibata, I. (1981) *Eur. J. Appl. Microbiol. Biotechnol.* **11** 72–77.

Musgrave, S. C., Kerby, N. W., Codd, G. A. & Stewart, W. O. P. (1982) *Biotech. Letts.* 4 647–652.
Naumova, R. P., Belousova, T. O. & Gilyazova, R. M. (1982) *Appl. Biochem. & Microbiol.* 18, 73–77.
Nilsson, I., Ohlson, S., Haggstrom, L., Molin, N. & Mosbach, K. (1980) *Eur. J. Appl. Microbiol. & Biotechnol.* 10 261–274.
Ohlson, S., Larsson, P. O. & Mosbach, K. (1978) *Biotechnol. Bioeng.* 20 1267–1284.
Ohlson, S., Larsson, P. O. & Mosbach, K. (1979) *Eur. J. Appl. Microbiol.* 7 103–110.
Omata, T., Iwamoto, N., Kimura, T., Tanaka, A. & Fukui, S. (1981) *Eur. J. Appl. Microbiol. Biotechnol.* 11 199–204.
Page, M. I. & Jencks, W. P. (1971) *Proc. Natl. Acad. Sci. US* 68 1678.
Park, S. H., Lee, S. B. & Ryu, D. D. Y. (1981) *Biotechnol. Bioeng.* 23 1237–1254.
Pastore, M. & Morisi, F. (1976) in: *Methods in Enzymol.* 44, (Ed. Mosbach, K.) Academic Press, pp. 822–830.
Patty's Industrial Hygiene and Toxicology (3rd edn.) 3 Vols. 1978–80. Wiley–Intersicence.
Pitcher, W. H. (1978) in *Advances in Biochemical Engineering* 10 (Ed. Ghose, T. K., Fiechter, A. & Blakeborough, N.) Springer Verlag, p. 1–26.
Pitcher, W. H. (1978) in: *Enzyme Engineering* 4 (Ed. Broun, G. B., Manecke, G. & Wingard, L. B.) Penum Press, pp. 67–76.
Poulsen, G. & Zittan, L. (1976) in *Methods in Enzymology* 44 (Ed. Mosbach, K.). Academic Press, 809–821.
Regan, D. L., Dunnill, P. & Lilly, M. D. (1974) *Biotechnol. Bioeng.* 16 333–343.
Regan, R. (1977) Stone & Webster Biochem. Symp. Toronto, paper 2.
Roland, J. F. (1981) *Enz. and Microbiol. Technol.* 3 105–110.
Ryu, D. D. Y. & Mandels, M. (1980) *Enz. and Microbiol. Technol.* 2 91–102.
Sada, E., Katoh, S. & Terashima, M. (1980) *Biotechnol. Bioeng.* 22 243–246.
Sada, E., Katoh, S., Shiozawa, M. & Fukui, T. (1981) *Biotechnol. Bioeng.* 23 2561–2567.
Sandermann, H. (1974) *Eur. J. Biochem.* 43 415.
Sarawathi, S. & Keyes, M. H. (1984) *Enzyme & Microbial Technol.* 6 98–100.
Sato, T., Mori, T., Tosa, T., Chibata, I., Karai, M., Yamashita, K. & Sumi, A. (1975) *Biotechnol. Bioeng.* 17 1797–1804.
Sato, T., Tosa, T. & Chibata, J. (1976) *Eur. J. Appl. Microbiol.* 2 153–160.
Satoh, I., Karube, I. & Suzuki, S. (1976) *Biotechnol. Bioeng.* 18 269–272.
Schneider, Z. & Friedman, H. C. (1981) *J. Appl. Biochem.* 3 135–146.
Schwartz, R. D. & McCoy, C. J. (1977) *J. Appl. Envir. Microbiol.* 34 47.
Seliger, H., Teufel, E. H. & Philipp, M. (1980) *Biotechnol. Bioeng.* 22 55–64.
Semenov, A. N., Berezin, I. V. & Martinek, K. (1981) *Biotechnol. Bioeng.* 23 355–360.
Shimizu, S., Morioko, H., Tani, Y. & Ogita, K. (1975) *J. Ferm. Technol.* 53 77–83.
Slima, J. T., Oruganti, S. R. & Kaiser, E. T. (1981) *J. Amer. Chem. Soc.* 103 6211–6213.
Soloman, B., Moaw, N., Pines, G. & Karchalski-Katzir, E. (1984) *Mol. Immunol.* 21, 1–11.
Srere, P. A., Mattiasson, B. & Mosbach, K. (1973) *Proc. Natl. Acad. Sci. (US)* 70 1534–1538.
Stoddart, J. F. (1980) in: *Enzymic and Non-enzymic catalysis* (Ed. Dunnill, P., Wiseman, A. & Blakeborough, N.) Ellis Horwood Limited, pp. 84–109.
Szalay, A. A., Mackay, C. J. & Langridge, W. H. R. (1979) *Enz. and Microbiol. Technol.* 1 153–224.
Tramper, J., Muller, F. & Van der Plas, H. C. (1978) *Biotechnol. Bioeng.* 20 1507–1522.
Van Beynum, G. M. A., Roels, J. A. & Van Tilburg, R. (1980) *Biotechnol. Bioeng.* 22 643–649.
Van Ginkel, C. G., Tramper, J., Layket, K. Ch. A. M. & Klapwizk, A. (1983) *Enzyme & Microbial Technol.* 5 297–303.
Venkatasubramanian, K., Constantinides, A. & Vieth, W. R. (1978) in: *Enzyme Engineering* 3 (Ed. Pye, E. K. & Weetall, H. H.) Plenum Press, p. 29–43.
Venkatasubramanian, K. (1980) in *Food Process Engineering* 2 (Ed. Linko, P. & Arinkari, J.) Applied Science.
Verhoff, F. H. & Schlager, S. T. (1981) *Biotechnol. Bioeng.* 23 41–60.
Vernon, C. A. (1967) *Proc. Royal Soc. London* B167 448.
Vieth, W. R. & Venkatasubramanian, K. (1976) in: *Methods in Enzymology* 44 (Ed. Mosbach, K.) Academic Press, pp. 768–776.
Wada, M., Uchida, T., Kato, J. & Chibata, I. (1980a) *Biotechnol. Bioeng.* 22 1175–1188.
Wada, M., Kato, J. & Chibata, I. (1980b) *Eur. J. Appl. Microbiol. Biotechnol.* 275–257.
Wang, D. I. C., Cooney, C. L., Demain, A. L., Dunnill, P., Humphrey, A. E. & Lilly, M. D. (1979) *Ferm. & Enz. Tech.* J. Wiley Ltd.
White, F. H. & Portno, A. D. (1978) *J. Inst. Brew.* 84 228–230.

White, C. A. & Kennedy, J. F. (1980) *Enz. & Microbial. Technol.* 2 82–90.
Wichmann, R., Wandrey, C. Buckmann, A. F. & Kula, M. R. (1981) *Biotechnol. Bioeng.* 23 2789–2802.
Wilkinson, A. J., Fersht, A. R., Blow, D. M., Carter, P. & Winter, G. (1984) *Nature* 307, 187–8.
Wilson, M. E. & Whitesides, G. M. (1978) *J. Am. Chem. Soc.* 100 306–307.
Winter, G. Fersht, A. R., Wilkinson, A. J., Zoller, M. & Smith, M. (1982) *Nature* 299, 756–758.
Wiseman, A. (1984) in: *Topics In Enzyme and Fermentation Biotechnology* Vol. 9. (Ed. Wiseman, A.) Ellis Horwood Limited, pp. 202–212.
Wong, C., Daniels, L., Orme-Johnson, W. H. & Whitesides, G. M. (1981a) *J. Amer. Chem. Soc.* 103 6227–6228.
Wong, C. H., Gordon, H., Cooney, C. L. & Whitesides, G. M. (1981b) *J. Org. Chem.* 46 4676–4679.
Wykes, J. R., Dunnill, P. & Lilly, M. D. (1975) *Biotechnol. Bioeng.* 17 51–68.
Yamamoto, K., Sato, T., Tosa, T. & Chibata, I. (1974a) *Biotechnol. Bioeng.* 16 1589–1599.
Yamamoto, K., Sato, T., Tosa, T. & Chibata, I. (1974b) *Biotechnol. Bioeng.* 16 1601–1610.
Yamamoto, K., Tosa, T., Yamashita, K. & Chibata, I. (1976) *Eur. J. Appl. Microbiol* 3 169–183.
Yamamoto, K., Tosa, T., Yamashita, K. & Chibata, I. (1977) *Biotechnol. Bioeng.* 19 1101–1114.
Yourno, J., Koho, T. & Roth, J. R. (1970) *Nature* 228 820–824.

CHAPTER 4

Principles of immobilization of enzymes

Professor J. F. KENNEDY, Research Laboratory for the Chemistry of Bioactive Carbohydrate and Proteins, Department of Chemistry, University of Birmingham B15 2TT, England and The North East Wales Institute, Deeside, Clwyd CH5 4 BR, Wales and
Dr. C. A. WHITE, Vincent Kennedy Ltd, 47 Conchar Road, Sutton Coldfield, B72 1LL, England

4.1 CLASSIFICATION OF IMMOBILIZED ENZYMES

There are several ways of classifying the various types of immobilized enzymes. They depend to some extent on the definition used to describe an immobilized enzyme. The term immobilized enzyme includes:

(a) enzymes modified to a water-insoluble form by suitable techniques;
(b) soluble enzymes used in reactors equipped with a semipermeable ultrafiltration membrane, allowing the passage of reaction products resulting from the hydrolysis of high molecular weight substrates, but retaining the enzyme molecules inside the reactor;
(c) enzymes, the mobility of which has been restricted by their attachment to another macromolecular, but the overall composite molecule which is formed remains water soluble.

The classification must therefore be based on a combination of the nature of the interaction responsible for immobilization and the nature of support. Figure 4.1 describes the classification system used herein and the individual methods which are discussed in this chapter (with more details in Part B).

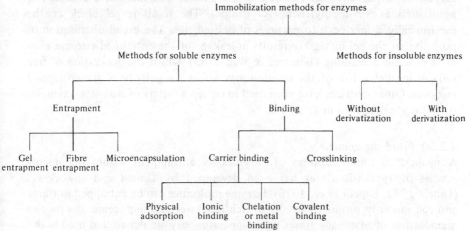

Fig. 4.1 — Classification of immobilization methods for enzymes.

4.2 TECHNIQUES OF ENZYME IMMOBILIZATION (see Part B for Chemistry)

Over the last few years the number of methods available for enzyme immobilization has increased dramatically. In the following sections the general methods are described within the classification system described above.

4.2.1 Entrapment

The entrapment method of immobilization is based on the localization of an enzyme within the lattice of a polymer matrix or membrane in such a way as to prevent the released protein whilst allowing penetration of substrate. This immediately restricts the application of the product to those reactions which involve relatively small substrate and product molecules. Since the enzyme does not bind in any way to the gel matrix or membrane, the method can be applied more generally than any other method for the entrapment of any kind of enzymes, microbial cells, and organelles of different sizes and properties with little regard for the properties of the enzyme, etc., with little destruction of biological activity. Entrapment can be subclassified into gel entrapment, fibre entrapment, and microencapsulation.

4.2.1.1 *Gel entrapment*

The gel entrapment method involves entrapping of the enzymes within the interstitial spaces of crosslinked water insoluble polymer gels. Occlusion within crosslinked polyacrylamide gels was the first employed technique of entrapment and was the one with which Bernfeld & Wan (1963) immobilized trypsin (EC 3.4.21.1), papain (EC 3.4.22.2), β-amylase (EC 3.2.1.2), and D-fructose biophosphate aldolase (EC 4.1.2.13). The usual method of gel formation is to polymerize acrylamide in an aqueous solution of the soluble enzyme and a crosslinking agent such as N,N-methylenebis(acrylamide). The resulting gel block can be mechanically dispersed into particles of desired size. The broad distribution of pore size of the gel inevitably results in leakage of the entrapped enzyme even after prolonged washing (Bernfeld & Wan 1963), whilst the generation of free radicals in the course of the reaction may affect the activity of the entrapped enzymes. Other matrices have been used to entrap a variety of enzymes, examples of which are described in Table 4.1.

4.2.1.2 *Fibre entrapment*

A method of immobilization of enzymes by entrapment within the microcavities of synthetic fibres has been developed by Dinelli and coworkers: (Dinelli 1972, Dinelli *et al.* 1978). Enzyme molecules can be entrapped in fibres and continuously produced by the conventional wet spinning techniques for the manufacture of manmade fibres using apparatus very similar to that used in the textile industry.

This method has several advantages over gel entrapment. High surface area for enzyme binding can be obtained by using very fine fibres. The fibres are resistant to weak acids and alkalis, high ionic strengths, and some organic solvents, and, depending on the polymer used, can show good resistance to microbial attack. Whilst more than one enzyme can be immobilized to produce immobilized multi-enzyme systems, the choice of enzyme is restricted to those which require low molecular weight substrates, and those which are not deactivated by the water immiscible polymer solvents. The most common polymer used is cellulose acetate, owing to its low cost and good biological and chemical resistance, and a number of enzymes have been successfully immobilized (see Table 4.2).

4.2.1.3 *Microencapsulation*

Microencapsulated enzymes are formed by enclosing enzymes within spherical semipermeable polymer membranes having diameters in the $1-100 \, \mu m$ range. Enzymes immobilized in this manner are physically contained within the membrane, whilst substrate and product molecules are free to diffuse across the membrane provided that their molecular sizes are sufficiently small enough to allow this.

Whilst the encapsulation of dyes, drugs, and other chemicals has been known for some time, it was not until the mid 1960s that such a method was first applied to enzymes (Chang 1964). Since that first report a number of other enzymes (see Table 4.3) have been successfully immobilized via microencapsulation, using a number of different materials and methods to prepare the microcapsules (see Part B).

Apart from the major disadvantages of this method described above, other disadvantages include: the occasional inactivation of the enzyme during microencapsulation; the high enzyme concentration required; and the possibility of enzyme incorporation into the membrane wall and subsequent leakage with some of the microencapsulation methods used. The major advantages of this method of immobilization are: the extremely large surface area for the contact of substrate and enzyme within a relatively small volume; and the real possibility of simultaneous immobilization of many enzymes in a single step.

4.2.2 Carrier binding

Attachment of an enzyme to an insoluble carrier is the oldest (Nelson & Griffin 1916) and most prevalent method for enzyme immobilization. Whilst this is the method which normally comes to mind when considering immobilization of enzymes, and which is probably the most investigated, one must not be misled into believing that this is a discovery of the 20th century. Indeed, as is so often discovered, 'Nature was there first', and it may be said that in the human body the greater proportion of biologically active molecules of the body exist at some time in an immobilized form.

The artificial immobilization of enzymes can be subclassified according to the binding mode of the enzyme into: physical adsorption, ionic binding, chela-

Table 4.1
Examples of gel entrapped enzymes

Material	Enzyme	Reference
Polyacrylamide	Acetylcholinesterase (EC 3.1.1.7)	Ngo & Laidler (1978)
	Chymotrypsin (EC 3.4.21.1)	Martinek et al. (1977 & 1980), Kuan et al. (1980), Halwachs et al. (1978)
	Asparginase (EC 3.5.1.1)	Mori et al. (1976)
	D-Glucose dehydrogenase (EC 1.1.1.47)	Chen et al. (1979)
	Alcohol dehydrogenase (EC 1.1.1.1)	Chen et al. (1979)
	β-D-Galactosidase (EC 3.2.1.23)	Danielson et al. (1979)
	Phenol-2-monooxygenase (EC 1.14.13.7)	Kjellén & Neujahr (1979 & 1980)
	Urease (EC 3.5.1.5)	Sada et al. (1980)
	β-D-Fructofuranosidase (EC 3.2.1.26)	Adachi et al. (1980)
	Penicillin amidase (EC 3.5.1.11)	Szewczuk et al. (1979)
	Catalase (EC 1.11.1.6)	Buchholz & Gödelmann (1978)
	D-Glucose oxidase (EC 1.1.3.4)	Buchholz & Gödelmann (1978),
	Glucoamylase (EC 3.2.1.3)	Sada et al. (1981)
	AMP-diaminase (EC 3.5.4.6)	Moriyama et al. (1980)
		Karube et al. (1977a)
Acrylamide/ glycidylmethacrylate	β-D-Fructofuranosidase (EC 3.2.1.26)	Fukui et al. (1978)
Ethyleneglycol/2-hydroxy ethylmethacrylate	D-Glucose isomerase (EC 5.3.1.5)	Fukui et al. (1978)
2-Hydroxyethylmethacrylate	α-Amylase (EC 3.2.1.1)	Kumakura et al. (1977), Kaetsu et al. (1979)

Material	Enzyme	Reference
κ-Carrageenan	Glucoamylase (EC 3.2.1.3)	Kaetsu et al. (1979)
	Cellulase (EC 3.2.1.4)	Kaetsu et al. (1979)
	α-D-Glucosidase (EC 3.2.1.20)	Kaetsu et al. (1979)
	D-Glucose oxidase (EC 1.1.3.4)	Kaetsu et al. (1979)
	Aminoacylase (EC 3.5.1.14)	Tosa et al. (1979)
	Aspartate ammonia-lyase (EC 4.3.1.1)	Tosa et al. (1979)
	Fumarate hydratase (EC 4.2.1.2)	Tosa et al. (1979)

Table 4.2

Examples of fibre entrapped enzyme

Material	Enzyme	Reference
Cellulose acetate	Aminoacylase (EC 3.5.1.14)	Bartoli et al. (1978)
	Fumarate hydratase (EC 4.2.1.2)	Marconi et al. (1975)
	Glucoamylase (EC 3.2.1.3)	Corno et al. (1972)
	D-Glucose isomerase (EC 5.3.1.5)	Giovenco et al. (1973)
	Dihydropyrimidinase (EC 3.5.2.2)	Snamprogetti (1976)
	β-D-Fructofuranosidase (EC 3.2.1.26)	Marconi et al. (1974a)
	β-D-Galactosidase (EC 3.2.1.23)	Morisi et al. (1973)
	Penicillin amidase (EC 3.5.1.11)	Marconi et al. (1973)
	Tryptophan synthetase	Marconi et al. (1974b), Zaffaroni et al. (1975)
Poly(vinyl alcohol) and cellulose	Dipeptidyl peptidase (EC 3.4.14.1/2)	Pardin et al. (1977)
	Urate oxidase (EC 1.7.3.3)	Kitano et al. (1980)
	Urease (EC 3.5.1.5)	Kitano et al. (1980)

Table 4.3

Examples of microencapsulated enzymes

Material	Enzyme	Reference
Phase separation method		
Nitro-cellulose	β-D-Galactosidase (EC 3.2.1.23)	Wadiak & Carbonell (1975)
	Asparaginase (EC 3.5.1.1)	Chang (1973)
Collodion	Alcohol dehydrogenase (EC 1.1.9.1)	Campbell & Chang (1976)
	Malate dehydrogenase (EC 1.1.1.37)	Campbell & Chang (1976)
	Catalase (EC 1.11.1.6)	Mogensen & Vieth (1973)
	Pyruvate kinase (EC 2.7.1.40)	Campbell & Chang (1975)
	Hexokinase (EC 2.7.1.1)	Campbell & Chang (1975)
	β-D-Galactosidase (EC 3.2.1.23)	Paine & Carbonell (1975)
	Urease (EC 3.5.1.5)	Mogensen & Vieth (1973)
Interfacial polymerization method		
Nylon	β-D-Galactosidase (EC 3.2.1.23)	Ostergaard & Martiny (1973)
	Asparaginase (EC 3.5.1.1)	Mori *et al.* (1973)
Polyurea	Asparaginase (EC 3.5.1.1)	Mori *et al.* (1973)
Liquid membrane method		
	Glucoamylase (EC 3.2.1.3)	Gregoriadis *et al.* (1971)
	Nitrate reductase (EC 1.7.99.4)	Mohan & Li (1974)
Liquid drying method		
Ethyl-cellulose	Triacylglycerol lipase (EC 3.1.1.3)	Kitajema *et al.* (1969)
Polystyrene	Catalase (EC 1.11.1.6)	Kitajema *et al.* (1969)
	Triacylglycerol lipase (EC 3.1.1.3)	Kitajema *et al.* (1969)
	Urease (EC 3.5.1.5)	Kitajema *et al.* (1969)

tion or metal binding, and covalent binding. In any of these methods the selection of the insoluble carrier as well as the binding method is of paramount importance. The ideal support for a given application is one which would increase substrate binding, decrease product inhibition, shift the apparent pH optimum to the desired value, discourage microbial growth, and be readily recovered for re-use. For use in a reactor the carrier should be stable in solution and should not deteriorate under the reaction conditions. It should be mechanically rigid and show little compaction in high flow rate continuous operations using fixed bed reactors. If the immobilized enzyme is to be used in a medical application it should not evoke immune responses or clotting reactions. Messing (1975) indicates a method for the selection of the appropriate carriers for industrial uses, based on their properties.

Of the many organic or inorganic, natural or synthetic, carriers that have been suggested for enzyme immobilization it appears that methacrylate or silicone rubber are most appropriate for medical uses, whilst inorganic supports can best fulfil the requirements for industrial applications. The major interest in organic supports stems from the ease with which enzymes can be attached, but they are limited in many applications owing to their poor dimensional stability and, in most cases, their inability to allow regeneration of the catalyst by simple means. However, if physical adsorption or ionic binding techniques are used then organic carriers can be used, as in the case of the industrial production of L-aminoacids using aminoacylase (EC 3.5.1.14) immobilized on DEAE-Sephadex® (Tosa *et al.* 1966). However, in special instances reversible covalent chemistry has been designed to allow reversible covalent attachment of the enzyme and hence regeneration, for example via disulphide linkages (Kennedy & Zamir 1975).

As we have indicated, carriers can be classified into organics and inorganics, with organics being subclassified into natural and synthetic polymers (Table 4.4 lists some of the more commonly used carriers), but this classification is not wholly adequate for a full description of the carrier. Parameters such as surface area and pore diameter will affect the loading of the enzyme, and therefore a further classification based on morphology can be considered, as nonporous and porous carriers. It is more appropriate to consider any morphology within the classification used for Table 4.4.

4.2.2.1 *Physical adsorption*

Physical adsorption is historically the earliest artificial method of enzyme immobilization with β-D-fructofuranosidase (invertase, EC 3.2.1.26) being adsorbed onto aluminium hydroxide (Nelson & Griffin, 1916). This method is the easiest method of preparing immobilized enzymes, being based on the physical adsorption of the enzyme molecules onto the surface of solid matrices. The method consists of allowing an aqueous solution of the enzyme to come into contact with the carrier, and this is brought about by one of the four distinct methods available:

Table 4.4
Examples of insoluble carriers

Classification	Carrier
Organic carriers — natural	
	Activated carbon
	Agar
	Agarose
	Albumin
	Cellulose
	Chitin
	Chitosan
	Collagen
	Dextran
	Gelatin
	Silk
	Starch
Organic carriers — synthetic	
	Acrylamide-based polymers
	Acrylic/methacrylic acid copolymers
	Maleic anhydride copolymers
	Nylon
	Polystyrene
	Poly(vinyl alcohol)
Inorganic carriers	
	Alumina
	Bentonite
	Calcium phosphate
	Celite
	Hydrous metal oxides
	Hydroxyapatite
	Kaolinite
	Magnetite
	Pumice
	Sand
	Silica gel
	Titania
	Zirconia

(a) static procedure;
(b) electrodeposition;
(c) reactor loading process; and
(d) mixing or shaking bath loading.

Of the four techniques the most frequently employed for laboratory preparation is the mixing or shaking bath loading, in which the carrier is placed in the enzyme solution and mixed with a stirrer or continuously agitated in a shaking water bath, resulting in a uniform loading of enzyme. The preferred method for commercial purposes is the reactor loading process, in which the support is loaded into the reactor that will be used for the processing system, and the enzyme is then added to the reactor and recirculated or agitated through the reactor. In the electrodeposition procedure the carrier is placed next to one of the electrodes in an enzyme bath, the current is turned on, the enzyme migrates to the carrier and is deposited upon the surface. The static procedure is the most inefficient of the four techniques and requires more time. In this method no agitation or stirring is used resulting in low, nonuniform enzyme loading unless the carrier is exposed to the enzyme for many days.

Because no reactive species are involved there is little or no conformational change in the enzyme on immobilization, and therefore a derivative with a specific activity similar to that of the soluble counterpart may be obtained. The adsorption is dependent on the experimental variables such as pH, nature of the solvent, ionic strength, quantity of enzyme and adsorbent, time, and temperature. A close control of these variables is required owing to the relatively weak binding forces between protein and adsorbent (hydrogen bonds, Van der Waals forces, hydrophobic interactions, etc.).

A major factor influencing the quantity of enzyme adsorbed to a solid support is enzyme concentration exposed to the unit surface of carrier during the immobilization process. The activity increases with increasing enzyme concentration, approaching a saturation value asymptotically at higher enzyme concentrations. Both time and temperature are important parameters in immobilization by adsorption, particularly with porous carriers, since diffusion is an important factor for immobilizing enzymes in such carriers.

In spite of the advantages already described, desorption of the protein from the carrier occurs during use owing to the weakness of the involved binding forces, with subsequent loss of catalytic activity and contamination of products, is a strong disadvantage. High concentrations of salt or substrate have been shown to enhance the rate of desorption of the enzyme. This disadvantage can render some enzymes active for only a short period, and thus adsorption techniques are of limited reliability when absolute immobilization of an enzyme is desired.

Some immobilized enzymes which have been prepared by physical adsorption are listed in Table 4.5.

Table 4.5

Examples of immobilized enzymes prepared by physical adsorption

Carrier	Enzyme	Reference
Organic supports		
Phenoxyacetyl-cellulose	Alkaline phosphatase (EC 3.1.3.1)	Dixon et al. (1979)
	Glyceraldehyde-3-phosphate dehydrogenase (EC 1.2.1.12)	Dixon et al. (1979)
	β-D-Xylosidase (EC 3.2.1.27)	Ogumtimein & Reilly (1980)
Palmitoyl-cellulose	Triacylglycerol lipase (EC 3.1.1.3)	Horiuti & Imamura (1978)
Tannin-aminohexyl-cellulose	Aminoacylase (EC 3.5.1.14)	Watanabe et al. (1979)
	Naringinase	Ono et al. (1978)
Tannin-TEAE-cellulose	Pullulanase (EC 3.2.1.14)	Ohba & Ueda (1980)
Concanavalin A-agarose	Phosphodiesterase 1 (EC 3.1.4.1)	Schiger et al. (1980)
	β-D-Fructofuranosidase (EC 3.2.1.26)	Woodward & Wiseman (1978)
	Acid phosphatase (EC 3.1.3.2)	Torchilin et al. (1977a), Van Etten & Saini (1977)
Lecithin – agarose	D-Glucose oxidase (EC 1.1.3.4)	Mattiasson & Borrebaeck (1978)
	Peroxidase (EC 1.11.1.7)	Mattiasson & Borrebaeck (1978)
	Catalase (EC 1.11.1.6)	Mattiasson & Borrebaeck (1978)
	α-D-Glucosidase (EC 3.2.1.20)	Mattiasson & Borrebaeck (1978)

Heptyl-agarose	Chlorophyllase (EC 3.1.1.14)	Sud'ina et al. (1979)
Hexyl-agarose	Guanylate cyclase (EC 4.6.1.2)	Garbers (1978)
Octyl-agarose	Guanylate cyclase (EC 4.6.1.2)	Garbers (1978)
	Xanthine dehydrogenase (EC 1.2.1.37)	Tramper et al. (1979)
Decyl-agarose	Guanylate cyclase (EC 4.6.1.2)	Garbers (1978)
Amino hexadecyl-agarose	Chlorophyllase (EC 3.1.1.14)	Sud'ina et al. (1979)
Amino hexaundecyl-agarose	Chlorophyllase (EC 3.1.1.14)	Sud'ina et al. (1979)
Activated carbon	Glucoamylase (EC 3.2.1.3)	Cho & Bailey (1979)
Glassy carbon	D-Glucose oxidase (EC 1.1.3.4)	Boudillon et al. (1979)
Poly(tetrafluoroethylene)	Lactate dehydrogenase (EC 1.1.1.27)	Danielson & Siergiej (1981)
Chitin or chitosan	Trypsin (EC 3.4.21.4)	Goodman & Peanasky (1982)
Inorganic supports		
Molecular sieve 4A	Trypsin (EC 3.4.21.4)	Mukherjea et al. (1980)
Zeolite	D-Glucose oxidase (EC 1.1.3.4)	Pifferi et al. (1980)
Silica gel	Trypsin (EC 3.4.21.4)	Buchholz et al. (1979)
	Alcohol dehydrogenase (EC 1.1.1.1)	Mickel'sone et al. (1979)
Octadecyl silica	Adenosine deaminase (EC 3.5.4.4)	Melander & Horvath (1978)
Methyl aerosil	Chlorophyllase (EC 3.1.1.14)	Sud'ina & Golod (1979)
Aluminium hydroxide	Urease (EC 3.5.1.5)	Grunwald et al. (1979)

4.2.2.2 *Ionic binding method*
Immobilization via ionic binding is based, mainly, on ionic binding of the enzyme molecules to solid supports containing ion-exchange residues. In some cases physical adsorption may, however, also take place. The main difference between ionic binding and physical adsorption is the strength of the enzyme to carrier linkages which are much stronger for ionic binding although less strong than in covalent binding. The preparation of immobilized enzymes using ionic binding uses the same simple procedures described for physical adsorption.

Owing to the ionic nature of the binding forces, leakage to enzyme from the carrier may occur in some situations, as with physical adsorption, when high substrate concentrations, high ionic strength solutions, or pH variations are used. However, the mild conditions used for immobilization do mean that confromational changes only occur to a small extent, if at all, resulting in immobilized enzymes with high enzymic activities.

The first enzyme reported to be immobilized by this method was catalase (EC 1.11.1.6) using DEAE-cellulose (Mitz 1956), whilst the first industrial application of immobilized enzymes used aminoacylase (EC 3.5.1.14) immobilized on DEAE-Sephadex® (Tosa *et al.* 1966) in order to produce L-amino acids from racemic mixtures of *N*-acetyl-DL-amino acids. Some of the immobilized enzymes that have been prepared by ionic binding are listed in Table 4.6.

The supports used for ionic bonding are ion-exchangers prepared most frequently from organic supports, although inorganic forms (especially silica) with the same or similar ion-exchange residues have been used. The organic polymers are derivatives of polysaccharides, mainly dextran and cellulose, or synthetic polymers based mainly on polystyrene derivatives. The type of ion-exchange residue on the carrier results in the carrier being classified as anion or cation exchangers depending on their ability to exchange anions (chloride or hydroxyl) or cations (hydrogen and sodium ions) of the carrier with anionic or cationic residues of the enzymes. The most common anion exchanges used include DEAE-, TEAE- and ECTEOLA—derivatives, whilst CM-derivatives are the most common cation exchangers used.

4.2.2.3 *Chelation or metal binding*
A relatively new technique involves the use of transition metal compounds as a means of activating the surface of the support and allowing direct coupling of the enzyme etc. without prior derivatization of the activated support, through formation of chelates. Supports which have been used include glass, chitin, celite, alginic acid, gelatin, poly-(4- and 5-acrylaminosalicyclic acids) and cellulose (Cardoso *et al.* 1978, Kennedy & Doyle 1973, Kennedy & Epton 1973, Kennedy & Watts 1974, Kennedy *et al.* 1974, 1980a), and they have been used to immobilize enzymes and antibiotics. The methods have been recently reviewed in depth (Kennedy 1979). An extension to this process has been to

derivatize the activated support with, for example, 5-amino-
increase the degree and strength of binding of the enzyme etc. w
to this support by diazotization (Cardoso *et al.* 1978, Kennedy &

It is also possible to form the hydroxide or hydrous oxide
metals in the presence of, or immediately prior to contact with,
to be immobilized, and such supports have been used for the immc   ....ion of
amino acids, peptides, proteins (including enzymes), carbohydrates, and antibiotics (Kennedy & Humphreys 1976, Kennedy & Kay 1976, Kennedy & Pike
1978, and Kennedy *et al.* 1976, 1977a, b, 1980a). The metals used include
titanium(III), titanium(IV), and zirconium(IV) which form polymeric oxides
and hydroxides. Examples of immobilized enzymes which have been prepared
by this method are listed in Table 4.7. An improvement in the process uses
alkylamine derivatives of the titanium(IV)-activated supports to increase the
binding forces (Cabrel *et al.* 1982a, b).

Although this method is described as chelation or by some workers as
partially covalent, desorption can occur, as with many immobilizations, even
covalent binding, usually under operation or long-term storage. This has led
some authors to classify the method under adsorption, but the more recent
developments of chelation show good success and high degrees of operational
stability can be achieved (Cabral *et al.* 1982b).

4.2.2.4 *Covalent binding*

The covalent binding method is based on the covalent attachment of enzymes
to water insoluble matrices, and is the most widespread and one of the most
investigated approaches to enzyme immobilization. The selection of conditions
for immobilization by covalent binding is more difficult than in the cases of the
other carrier binding methods described, and frequently involves more complicated and less mild conditions than the other methods, but, since covalent bonds
are being formed, stable immobilized enzyme preparations which do not leach
enzyme into solution in the presence of substrate or high ionic concentration
solutions are formed in almost all cases.

The immobilization of an enzyme by covalent attachment to a support
matrix must involve only functional groups of the enzyme that are not essential
for catalytic action, and therefore no reagents must be used which could affect
the binding and active sites of the enzyme. To achieve higher activities in the
resulting immobilized enzyme preparation by prevention of inactivation reactions with the essential amino acid residues of the active site, a number of
methods have been devised (Zaborsky (1973):

(i) covalent attachment of the enzyme in the presence of a competitive
   inhibitor or substrate;
(ii) a reversible covalently linked enzyme−inhibitor complex;

Table 4.6

Examples of immobilized enzymes prepared by ionic binding

Carrier	Enzyme	Reference
Anion exchangers		
DEAE – cellulose	Glucoamylase (EC 3.2.1.3)	Maeda et al. (1979)
	Phosphodiesterase 1 (EC 3.1.4.1)	Aukati et al. (1978)
	D-Glucose isomerase (EC 5.3.1.18)	Huitron & Limon-Lason (1978)
	Inulinase (EC 3.2.1.7)	Guiraud et al. (1981)
	Methanol oxidase	Baratti et al. (1978)
	Dextransucrase (EC 2.4.1.5)	Kaboli & Reilly (1980)
	Aminoacylase (EC 3.5.1.14)	Szwajcer et al. (1981)
	Choridazon dihydrodial dehydrogenase A and B	Keller et al. (1979)
AE – cellulose	D-Glucose isomerase (EC 5.3.1.18)	Huitron & Limon-Lason (1978)
TEAE – cellulose	Pullulanase (EC 3.2.1.41)	Ohba et al. (1978)
DEAE – Sephadex®	D-Glucose isomerase (EC 5.3.1.18)	Huitron & Limon-Lason (1978)
	Phenol 2- monooxygenase (EC 1.14.13.7)	Kjellén & Neujahr (1979 & 1980)
	Dextransucrase (EC 2.4.1.5)	Kaboli & Reilly (1980)
	β-D-Fructofuranosidase (EC 3.2.1.26)	Woodward & Wiseman (1978)
QAE – Sephadex®	Hexokinase (EC 2.7.1.1)	Miura et al. (1979)
	Creatine kinase (EC 2.7.3.2)	Miura et al. (1979)
	Carbamoyl phosphate synthetase (EC 6.3.4.16/6.3.5.5)	Miura et al. (1979)

DEAE – Bio-Gel® A	Phenol 2-monooxygenase (EC 1.14.13.7)	Kjellén & Neujahr (1979 & 1980)
Amberlite® IRA 93	D-Glucose oxidase (EC 1.1.3.4)	Klei et al. (1978)
Amberlite® IRA 94	β-D-Fructofuranosidase (EC 3.2.1.26)	Ooshima et al. (1980)
Amberlite® IRA 910	D-Glucose oxidase (EC 1.1.3.4)	Klei et al. (1978)
Amberlite® IRA 938	D-Glucose oxidase (EC 1.1.3.4)	Klei et al. (1978)
Cation exchanges		
CM – cellulose	Penicillin amidase (EC 3.5.1.11)	Carleysmith et al. (1980)
Cellulose phosphate	Aminoacyl-tRNA synthetases	Yamada (1978)
CM-Sephadex®	β-D-Fructofuranosidase (EC 3.2.1.26)	Woodward & Wiseman (1978)
SP-Sephadex®	Dextransucrase (EC 2.4.1.5)	Kaboli & Reilly (1980)
	Glucoamylase (EC 3.2.1.3)	Adachi et al. (1978)
Dextran sulphate	Lactate dehydrogenase (EC 1.1.1.27)	Klinov et al. (1979)
Amberlite® IRC-50	Cholesterol oxidase (EC 1.1.3.6)	Cheetham (1979)
	β-D-Fructofuranosidase (EC 3.2.1.26)	Ooshima et al. (1980)
Amberlite® IRC-200	Rennet	Gouges et al. (1979)

Table 4.7

Examples of immobilized enzymes prepared by chelation or metal binding

Carrier	Enzyme	Reference
Organic supports		
Cellulose	Nuclease P$_1$	Rokugawa *et al.* (1979)
	β-D-Fructofuranosidase (EC 3.2.1.26)	Woodward & Wiseman (1978)
	Pectin lyase (EC 4.2.2.10)	Hanish *et al.* (1978)
DEAE − cellulose	Papain (EC 3.4.22.2)	Kennedy & Pike (1979 & 1980)
Glass fibre paper	Papain (EC 3.4.22.2)	Kennedy & Pike (1979)
Polypropylene	D-Glucose dehydrogenase (EC 1.1.1.47)	Bisse & Vonderschmitt (1978)
Nylon	Triacylglycerol lipase (EC 3.1.1.3)	Kobayashi *et al.* (1980)
Duolite A-7	D-Glucose isomerase (EC 5.3.1.5)	Bhatt *et al.* (1979)
Polystyrene sulphonate		
Inorganic supports		
Alumina	Glucoamylase (EC 3.2.1.3)	Allen *et al.* (1979)
	β-D-Xylosidase (EC 3.2.1.37)	Ogumtimein & Reilly (1980)
Stainless steel	β-D-Xylosidase (EC 3.2.1.37)	Ogumtimein & Reilly (1980)
Horneblende	Glucoamylase (EC 3.2.1.3)	Flynn & Johnson (1977)

Support	Enzyme	References
Magnetic iron oxide	α-Amylase (EC 3.2.1.1)	Kennedy et al. (1977b), Kennedy & White (1979)
Controlled pore glass	Pectin lyase (EC 4.2.2.10)	Hanish et al. (1978)
	Glucoamylase (EC 3.2.1.3)	Cabral et al. (1981)
Controlled pore silica	Glucoamylase (EC 3.2.1.3)	Emery & Cardoso (1978)
Soda glass	β-D-Glucosidase (EC 3.2.1.21)	Kennedy & Watts (1974)
Lead glass	Glucoamylase (EC 3.2.1.3)	Cardoso et al. (1978), Emery & Cardoso (1978), Kennedy & Chaplin (1979)
Silica gel	Trypsin (EC 3.4.21.4)	Volkova et al. (1979)
Silochrome	Pronase (see EC 3.4.24.4)	Bogatskii et al. (1979)
Titanium(IV) oxide	Dextranase (EC 3.2.1.11)	Kennedy & Kay (1977a)
Hydrous metal oxides prepared from Ti(IV), Zr(IV), Fe(III), Sn(II) or V(III)	Chymotrypsin (EC 3.4.21.1)	Kennedy et al. (1976)
	Dextranase (EC 3.2.1.11)	Kennedy et al. (1976 & 1981)
	Glycoamylase (EC 3.2.1.3)	Kennedy et al. (1981)
	D-Glucose oxidase (EC 1.1.3.4)	Kennedy et al. (1976)
	β-D-Glucosidase (EC 3.2.1.21)	Kennedy et al. (1976)
	Papain (EC 3.4.22.2)	Kennedy & Pike (1978), Kennedy et al. (1980a)
	Peroxidase (EC 1.11.1.7)	Kennedy et al. (1981)
	Trypsin (EC 3.4.21.4)	Kennedy et al. (1976)

(iii) a chemically modified soluble enzyme whose covalent linkage to the matrix is achieved by newly incorporated residues;

(iv) a zymogen precursor.

The wide variety of binding reactions and of carriers with functional groups capable of covalent coupling, or susceptible to being activated to give such groups, makes this a generally applicable method of immobilization. Nevertheless, the compositional and structural complexity of proteins has not allowed, except in a very limited number of cases, the application of general rules by means of which the method best suited for a specific task could be predicted. Clearly, if the primary structure, the tertiary structure, and the structure of the active site of the enzyme were able to be determined with ease, potential linkage positions and methods could be selected which are least likely to cause deactivation.

Enzymes are of course heteropolymers built up of more than twenty different amino acids, many of which have functional groups suitable to be used for linking to a carrier matrix (see Table 4.8). Absent from this list are the

Table 4.8

Reactive residues of proteins[†]

Residue	Originating amino acid
$-NH_2$	ϵ-Amino of L-lysine and N-terminus amino group
$-SH$	Thiol of L-cysteine
$-COOH$	Carboxyl of L-aspartate and L-glutamate and C-terminus carboxyl group
	Phenolic of L-tyrosine
	Guanidino of L-arginine
	Imidazole of L-histidine
$-S-S-$	Disulphide of L-cystine
	Indole of L-tryptophan
CH_3-S-	Thioether of L-methionine
$-CH_2OH$	Hydroxyl of L-serine and L-threonine

†see Srere & Uyeda (1976).

amino acids with amide groups (L-glutamine and Lasparagine), those with hydrocarbon side chains (L-alanine, L-leucine, L-isoleucine, L-valine, L-phenylanine, and L-proline) and glycine, owing to the relatively low abundance in protein molecules and, more importantly, to their hydrophobic natures which increases the chances of a residue being buried inside the protein. Even if

Table 4.9

Average percent composition of proteins (reactive residues only)[†]

Residue	Percent
L-Serine	7.8
L-Lysine	7.0
L-Threonine	6.5
L-Aspartic acid	4.8
L-Glutamic acid	4.8
L-Arginine	3.8
L-Tyrosine	3.4
L-Cysteine	3.4
L-Histidine	2.2
L-Methionine	1.6
L-Tryptophan	1.2

[†] see Dayhoff & Hunt (1972).

Table 4.10

The number of reactions in which amino acids can partake

Amino acid	Number of reactions
L-Cysteine	31
L-Lysine	27
L-Tyrosine	16
L-Histidine	13
L-Methionine	7
L-Tryptophan	7
L-Arginine	6
L-Glutamic acid	4
L-Aspartic acid	4
L-Serine	0
L-Threonine	0

immobilization via one of these residues could be achieved, severe disruption to the protein structure would occur, resulting in the loss of enzyme action. Table 4.9 shows the average amino acid composition of a number of proteins, from the point of view of the relative reactivity of the residues. What is more important is the comparison with the number of reactions in which each amino acid can partake (Means & Feeney 1971) (see Table 4.10).

Most of the coupling reactions involving the active side chains which are described in detail in Part B are classified as carbonyl-type reactions with the nucleophilic groups of the protein (amino, thiol, or hydroxyl). In terms of nucleophilic reactivity that involving thiol groups is one or two orders of magnitude greater than that involving amino or hydroxyl groups, but the thioesters are much less stable than esters which are in turn less stable then the substituted amines which are formed. A consequence of all these factors is that the most convenient residues for involvement in immobilization are, in descending order, L-lysine, L-cysteine, L-tyrosine, L-histidine, L-aspartic acid, L-glutamic acid, L-arginine, L-tryptophan, L-serine, L-threonine, and L-methionine.

The functional group of the carriers which can react directly with enzymes are listed in Table 4.11, together with the group with which it reacts in the enzyme. However, only a few supports contain these reactive groups for direct coupling of enzymes, including maleic anhydride-based copolymers, methacrylic acid anhydride-based copolymers, nitrated fluoroacryl methacrylic copolymers, and idoalkylmethacrylates. The majority of support materials used do not possess these reactive groups, but have hydroxyl, amino, amide, or carboxyl groups which require activation before they can be used for immobilization. The reactions which have been developed to date can be classified into the following:

    (a) diazotization;
    (b) amide bond formation;
    (c) alkylation and arylation;
    (d) Schiffs base formation;
    (e) Ugi reaction;
    (f) amidination reactions;
    (g) thiol-disulphide interchange;
    (h) mercury-enzyme interactions;
    (i) γ-irradiation induced coupling.

Examples of each of these classes are described in detail in Part B. Principal examples of enzymes immobilized by those techniques are listed in Tables 4.12–4.15.

Table 4.11
Immobilization methods of enzymes by covalent binding, showing
the matching between carrier and enzyme functional groups

Reactive groups of carrier	Reacting groups of enzyme	Coupling reactions
(diazonium salt)	$-NH_2$ $-SH$ —OH	Diazo linkage
(acid anhydride)	$-NH_2$	Amide bond formation
$-CH_2CON_3$ (acyl azide)	$-NH_2$ $-SH$ —OH	Amide bond formation
(imidocarbonate)	$-NH_2$	Amide bond formation
$-R-NCS$ (isothiocyanate)	$-NH_2$	Amide bond formation
—NCO (isocyanate)	$-NH_2$	Amide bond formation
$-CH_2COCl$ (acyl chloride)	$-NH_2$	Amide bond formation
(cyclic carbonate)	$-NH_2$	Amide bond formation

Table 4.11 — continued

Reactive groups of carrier	Reacting groups of enzyme	Coupling reactions

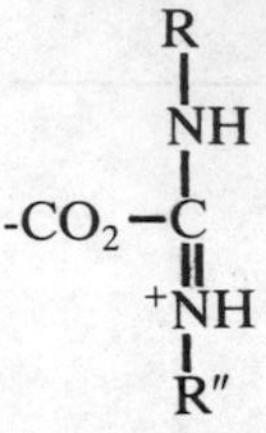

$$\begin{array}{c} R \\ | \\ NH \\ | \\ -CO_2-C \\ \| \\ {}^+NH \\ | \\ R'' \end{array}$$

$-NH_2$

Amide bond formation

(O-acylisourea)

$-CO_2-C=CH-CO-NH-C_2H_5$

$-NH_2$ Amide bond formation

SO_3-

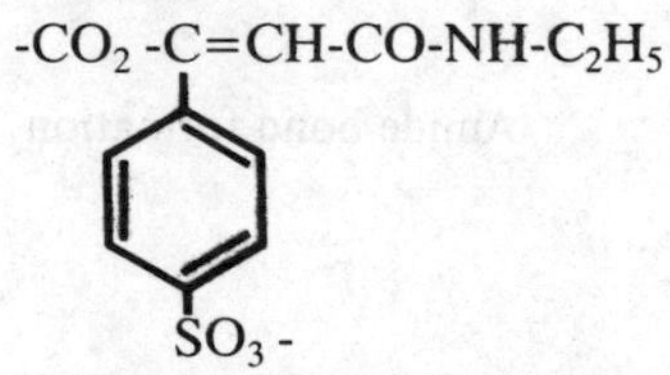

(Woodward's reagent k derivative)

F

$-NO_2$

$-NH_2$ Arylation

NO_2

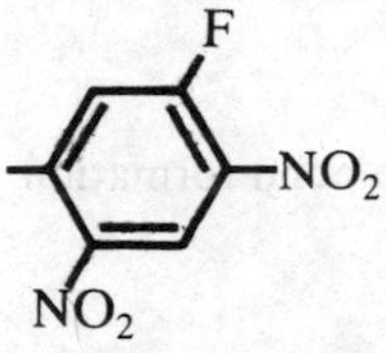

(5-fluoro-2,4-dinitroanilide)

Cl

$-O-$ triazine $-NH_2$ Arylation

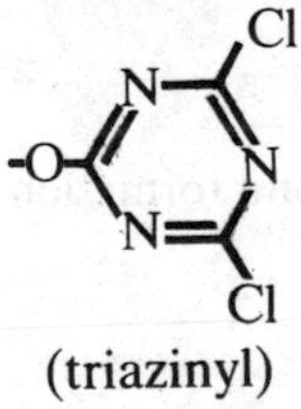

Cl

(triazinyl)

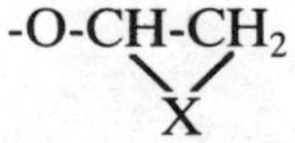

$-O-CH-CH_2$
$\quad\quad \diagdown\,\diagup$
$\quad\quad\quad X$

$-NH_2$

$-OH$ Alkylation

$X = {>}NH, {>}O, {>}S$ $-SH$
(e.g. oxirane)

Table 4.11 – continued

Reactive groups of carrier	Reacting groups of enzyme	Coupling reactions
-O-CH$_2$-CH$_2$-SO$_2$-CH = CH$_2$ (vinylsulphonyl)	-NH$_2$ -SH -OH	Alkylation
(vinyl keto)	-NH$_2$ -SH -OH	Arylation
-CHO (aldehyde)	-NH$_2$	Schiffs base formation
	-CO$_2$H -NH$_2$	Ugi reaction
NH ‖ -C-OC$_2$H$_5$ (imido ester)	-NH$_2$	Amidination reaction
-CN (cyanide)	-NH$_2$	Amidination reaction
(disulphide residue)	-SH	Thiol-disulphide interchange

Table 4.11 — continued

Reactive groups of carrier	Reacting groups of enzyme	Coupling reactions
![phenyl-HgCl structure] —HgCl (mercury derivative)	-SH	Mercury-enzyme interaction
$M^\bullet$ (matrix radical)	$E^\bullet$ (Enzyme radical)	γ-Irradiation induced coupling
$-NH_2$ (amine)	$-NH_2$ $-CO_2H$	Amide bond formation (in presence of condensing reagents
$-CONHNH_2$ (acyl hydrazide)	$-NH_2$ $-CO_2 H$	Amide bond formation (in presence of condensing reagents)

Table 4.12

Examples of immobilized enzymes prepared by diazotization coupling (see Part B, p. 386–387)

Carrier	Enzyme	Reference
Organic supports		
Cellulose	β-D-Galactosidase (EC 3.2.1.23)	Beddows *et al.* (1980)
	D-Glucose oxidase (EC 1.1.3.4)	Beddows *et al.* (1980)
	Trypsin (EC 3.4.21.4)	Beddows *et al.* (1980)
	Papain (EC 3.4.22.2)	Beddows *et al.* (1980)
	Pepsin (EC 3.4.23.1)	Beddows *et al.* (1980)
	β-Amylase (EC 3.2.1.2)	Ohba & Ueda (1980)
Enzacryl® AA	β-D-Xylosidase (EC 3.2.1.37)	Ogumtimein & Reilly (1980)
	Dextransucrase (EC 2.4.1.5)	Kaboli & Reilly (1980)
	Tyrosinase (EC 1.10.3.1/1.14.18.1)	Iborra *et al.* (1977)
Agar	Protein kinase (EC 2.7.1.37)	Kozlova *et al.* (1978)
Styrene/maleic anhydride	Chymotrypsin (EC 3.4.21.1)	Lai & Cheng (1978)
Glycidyl methacrylate/ ethylene dimethacrylate	Penicillin amidase (EC 3.5.1.11)	Drobnik *et al.* (1979)
Polyacrylamide/nylon	Urease (EC 3.5.1.5)	Shemer *et al.* (1979)
Polyethylene terephthalate	Trypsin (EC 3.4.21.4)	Blassberger *et al.* (1978)
Inorganic supports		
Controlled pore glass	Aldehyde dehydrogenase (EC 1.1.1.3)	Lee (1978)
	Phospholipase A₂ (EC 3.1.1.4)	Adamich *et al.* (1978)
Controlled pore silica	Hydrogenase (EC 1.18.3.1)	Hatchikian & Monsan (1980)

Table 4.13

Examples of immobilized enzymes prepared by amide bond formation

Carrier	Enzyme	Reference
Acid anhydride derivatives		
Ethylene/maleic anhydride	Alkaline phosphatase (EC 3.1.3.1)	Zingaro & Uziel (1970)
	Naringinase	Goldstein et al. (1971)
Butanediol divinylether/ anhydride	Lactate dehydrogenase (EC 1.1.1.27)	Brümmer et al. (1972)
	Trypsin (EC 3.4.21.4)	Brümmer et al. (1972)
	Chymotrypsin (EC 3.4.21.1)	Brümmer et al. (1972)
	Papain (EC 3.4.22.2)	Brümmer et al. (1972)
	Ficin (EC 3.4.22.3)	Brümmer et al. (1972)
	Bromelain (EC 3.4.22.4)	Brümmer et al. (1972)
	Subtilisin (EC 3.4.21.14)	Brümmer et al. (1972)
	Subtilopeptidase B	Brümmer et al. (1972)
	Pronase (see EC 3.4.24.4)	Brümmer et al. (1972)
Methylvinylether/ maleic anhydride	Alkaline phosphatase (EC 3.1.3.1)	Zingaro & Uziel (1970)
	Naringinase	Goldstein et al. (1971)
Isobutyl vinylether/ maleic anhydride	Naringinase	Goldstein et al. (1971)
Acyl azid derivatives		
Enzacryl® AH	β-D-Xylosidase (EC 3.2.1.37)	Ogumtimein & Reilly (1980)
	Tyrosinase (EC 1.10.3.1/1.14.18.1)	Iborra et al. (1977)
	Dextransucrase (EC 2.4.1.5)	Keller et al. (1979)
	Penicillin amidase (EC 3.5.1.11)	Drobnik et al. (1979)
Glycidyl methacrylate/ ethylene dimethacrylate		
Poly(acryloylmorpholine)	Carbonate dehydratase (EC 4.2.1.1)	Epton et al. (1979)

Support	Enzyme	Reference
Acrylic acid/isothiocyanate-styrene	Papain (EC 3.4.22.2)	Manecke et al. (1978)
Polyethylene terephthalate	Trypsin (EC 3.4.21.4)	Blassberger et al. (1978)
Poly(vinylalcohol)	β-D-Glucosidase (EC 3.2.1.21)	Manecke et al. (1978)
Nylon/polyacrylamide	β-D-Galactosidase (EC 3.2.1.23)	Beddows et al. (1979)
	Papain (EC 3.4.22.2)	Beddows et al. (1979)
Collagen	L-Iditol dehydrogenase (EC 1.1.1.14)	Paul et al. (1978)
	D-Glucose oxidase (EC 1.1.3.4)	Coulet et al. (1980)
	Asparate aminotransferase (EC 2.6.1.1)	Arrio-Dupont & Coulet (1975)
Controlled pore glass	Parathion hydrolase	Munnecke (1979)
Controlled pore silica	Parathion hydrolase	Munnecke (1979)

Cyclic imidocarbonate derivatives

Support	Enzyme	Reference
Agarose	Citrate synthase (EC 4.1.3.7/28)	Mukherjee & Srere (1978)
	Tyrosinase (EC 1.10.3.1/1.14.18.1)	Iborra et al. (1977)
	β-D-Galactosidase (EC 3.2.1.23)	Danielson et al. (1979)
	Xanthine dehydrogenase (EC 1.2.1.37)	Tramper et al. (1979)
	Phenol 2-monooxygenase (EC 1.14.13.7)	Kjellén & Neujahr (1979)
	Hydroxysteroid dehydrogenase (EC 1.1.1.50 etc)	Carrea et al. (1979)
	Xanthine oxidase (EC 1.2.3.2)	Johnson & Coughlan (1978), Tramper et al. (1978)
	Superoxide dismutase (EC 1.15.1.1)	Tramper et al. (1978)
	Catalase (EC 1.11.1.6)	Tramper et al. (1978)
	D-Fructose bisphosphate aldolase (EC 4.1.2.13)	Janasik et al. (1978)
	Fumarate hydratase (EC 4.2.1.2)	Erekin & Friedmann (1979)
	L-Malate dehydrogenase (EC 1.1.1.37)	Erekin & Friedmann (1979)
	Trypsin (EC 3.4.21.4)	Mozhaev et al. (1979), Martinek et al. (1980)
	Colipase	Patton et al. (1978)
	Triacylglycerol lipase (EC 3.1.1.3)	Patton et al. (1978)

Table 4.13 – *continued*

Carrier	Enzyme	Reference
Cellulose	Xanthine oxidase (EC 1.2.3.2)	Johnson & Coughlan (1978)
	Nuclease P₁	Rokugawa et al. (1979)
	L-Asparaginase (EC 3.5.1.1)	Jackson et al. (1979)
	Trypsin (EC 3.4.21.4)	Mozhaev et al. (1979)
Dextran	Poly-D-galacturonase (EC 3.2.1.15)	Rexova–Benkova et al. (1980)
2-Hydroxyethylmethacrylate	Trypsin (EC 3.4.21.4)	Pittner et al. (1980)
Polythiol/4-vinylpyridine	Xanthine oxidase (EC 1.2.3.2)	Johnson & Coughlan (1978)
Hornblende		
Glass beads	Chymotrypsin (EC 3.4.21.1)	Janasik et al. (1978)
Isocyanate and		
isothiocyanate derivatives		
Cellulose	β-Amylase (EC 3.2.1.2)	Maeda et al. (1978)
Enzacryl® AA	Tyrosinase (EC 1.10.3.1/1.14.18.1)	Iborra et al. (1979)
Polyurethane	β-D-Fructofuranosidase (EC 3.2.1.26)	Fukushima et al. (1978)
Polypropylene glycol	Trypsin (EC 3.4.21.4)	Lipatova et al. (1979)
Polyethylene terephthalate	Trypsin (EC 3.4.21.4)	Blassberger et al. (1978)
Glycidyl methacrylate/	Penicillin amidase (EC 3.5.1.11)	Drobnik et al. (1979)
ethylene dimethacrylate		
Activated carbon	D-Glucose oxidase (EC 1.1.3.4)	Boudillon et al. (1979)
Sylochrome	Pronase (see EC 3.4.24.4)	Bogatskii et al. (1979)
Acyl chloride derivatives		
Amberlite® IRC-50	Catalase (EC 1.11.1.6)	Schreiner (1966)
Cyclic carbonate derivatives		
Cellulose	Dextranase (EC 3.2.1.11)	Cheetham & Richards (1973)
	β-D-Glucosidase (EC 3.2.1.21)	Barker *et al.* (1971), Kennedy &

Support	Enzyme	Reference
	Trypsin (EC 3.4.21.4)	Kennedy & Zamir (1973)
	Chymotrypsin (EC 3.4.21.1)	Kennedy et al. (1973)

Carbodiimide mediated deravatives

Support	Enzyme	Reference
Agarose	Riboflavin kinase (EC 2.7.1.26)	Merrill & McCormick (1979)
	Phenol 2-monooxygenase (EC 1.14.13.7)	Kjellén & Neujahr (1979 & 1980)
	β-D-Xylosidase (EC 3.2.1.37)	Ogumtimein & Reilly (1980)
	Cholesterol oxidase (EC 1.1.3.6)	Cheetham (1979)
Cellulose	Papain (EC 3.4.22.2)	Kucera & Kuminkova (1980)
	Chymotrypsin (EC 3.4.21.1)	Kucera & Luminkova (1980)
	α-Amylase (EC 3.2.1.1)	Kucera & Kuminkova (1980)
	Glucoamylase (EC 3.2.1.3)	Kucera & Kuminkova (1980)
	Poly-D-galacturonase (EC 3.2.1.15)	Kucera & Kuminkova (1980)
Nylon	Lactate dehydrogenase (EC 1.1.1.27)	Daka & Laidler (1980)
	Phenol 2-monooxygenase (EC 1.14.13.7)	Kjellén & Neujahr (1980)
Nylon/acrylonitrile	β-D-Fructofuranosidase (EC 3.2.1.26)	Abdel-Hay et al. (1980)
	Pepsin A (EC 3.4.23.1)	Abdel-Hay et al. (1980)
	Acid phsophatase (EC 3.1.3.2)	Abdel-Hay et al. (1980)
	Alkaline phosphatase (EC 3.1.3.1)	Abdel-Hay et al. (1980)
Nylon/acrylic acid	Alkaline phosphatase (EC 3.1.3.1)	Beddows et al. (1981)
	β-D-Galactosidase (EC 3.2.1.23)	Beddows et al. (1981)
Acrylamide/acrylic acid	Chymotrypsin (EC 3.4.21.1)	Torchilin et al. (1977b)
Glycidyl methacrylate ethylene dimethacrylate	Penicillin amidase (EC 3.5.1.11)	Drobnik et al. (1979)
Amberlite® IRC-50	β-D-Fructofuranosidase (EC 3.2.1.26)	Ooshima et al. (1980)
Glassy carbon	D-Glucose oxidase (EC 1.1.3.4)	Boudillon et al. (1979)
Activated carbon	D-Glucose oxidase (EC 1.1.3.4)	Cho & Bailey (1979)
	Glucoamylase (EC 3.2.1.3)	Cho & Bailey (1978)
	D-Gluconolactonase (EC 3.1.1.17)	Cho & Bailey (1978)
Glass	Trypsin (EC 3.4.21.4)	Borchet & Buchholz (1979)
Controlled pore glass	Aldehyde dehydrogenase (EC 1.2.1.3)	Lee (1978)

Table 4.14

Examples of immobilized enzymes prepared by alkylation and arylation

Carrier	Enzyme	Reference
Halageno acetyl derivatives		
Chloroacetyl-cellulose	Aminoacylase (EC 3.5.1.14)	Sato *et al.* (1972)
Bromoacetyl-cellulose	Aminoacylase (EC 3.5.1.14)	Sato *et al.* (1972)
	Glucoamylase (EC 3.2.1.3)	Maeda & Suzuki (1972)
Iodoacetyl-cellulose	Aminoacylase (EC 3.5.1.14)	Sato *et al.* (1971)
	Glucoamylase (EC 3.2.1.3)	Maeda & Suzuki (1972)
Triazinyl derivatives		
Cellulose	Phospho-D-glucomutase (EC 2.7.5.1)	Shimizu & Lenhoff (1979)
	D-Glucose-6-phosphate dehydrogenase (EC 1.1.1.49)	Shimizu & Lenhoff (1979)
Filter paper	Dextransucrase (EC 2.4.1.5)	Kaboli & Reilly (1980)
Agarose	Chymotrypsin (EC 3.4.21.1)	Finlay *et al.* (1978)
	Trypsin (EC 3.4.21.4)	Finlay *et al.* (1978)
	Lactate dehydrogenase (EC 1.1.1.27)	Finlay *et al.* (1978)
Polystyrene	α-Amylase (EC 3.2.1.1)	Fischer *et al.* (1978)
Oxirane derivatives		
Glycidyl methacrylate	Glucoamylase (EC 3.2.1.3)	Švec *et al.* (1978)
Polyacrylamide	β-D-Galactosidase (EC 3.2.1.23)	Friedrich *et al.* (1980)
Vinylketo derivatives		
Polyhydroxyalkyl methacrylate	Chymotrypsin (EC 3.4.21.1)	Stambolieva & Turkova (1980)
	Trypsin (EC 3.4.21.4)	Stambolieva & Turkova (1980)

Table 4.15

Examples of immobilized enzymes prepared by Schiff base formation

Carrier	Enzyme	Reference
Cellulose	Penicillinase (EC 3.5.2.6)	Klemes & Citri (1979)
	β-D-Galactosidase (EC 3.2.1.23)	Beddows et al. (1980)
	D-Glucose oxidase (EC 1.1.3.4)	Beddows et al. (1980)
	Trypsin (EC 3.4.21.4)	Beddows et al. (1980)
	Papain (EC 3.4.22.2)	Beddows et al. (1980)
	Pepsin A (EC 3.4.23.1)	Beddows et al. (1980)
	L-Glutamate dehydrogenase (EC 1.4.1.2)	Sundaram & Joy (1978)
	Urease (EC 3.5.1.5)	Sundaram & Joy (1978)
Agarose	Acid phosphatase (EC 3.1.3.2)	Torchilin et al. (1977a)
Nylon	Urate oxidase (EC 1.7.3.3)	Sundaram et al. (1978)
	D-Glucose dehydrogenase (EC 1.1.1.47)	Bisse & Vonderschmidtt (1977)
	L-Arginase (EC 3.5.3.1)	Carvajal et al. (1978)
	Lactate dehydrogenase (EC 1.1.1.27)	Daka & Laidler (1980)
	D-Glucose dehydrogenase (EC 1.1.1.47)	Sundaram et al. (1979)
Nylon/polyethylene	β-D-Fructofuranosidase (EC 3.2.1.26)	Abdel-Hay et al. (1980)
Nylon/polyacrylonitrile	Pepsin A (EC 3.4.23.1)	Abdel-Hay et al. (1980)
	Acid phosphatase (EC 3.1.3.2)	Abdel-Hay et al. (1980)
	Alkaline phosphatase (EC 3.1.3.1)	Abdel-Hay et al. (1980)
Polyacrylonitrile	Choline oxidase (EC 1.1.3.17)	Matsumoto et al. (1980)
	Glucoamylase (EC 3.2.1.3)	Carleysmith & Lilly (1979)
Chitosan	Malate dehydrogenase (EC 1.1.1.37)	Spettoli et al. (1980)
	β-D-Galactosidase (EC 3.2.1.23)	Leuba & Widmer (1979)
	Pepsin A (EC 3.4.23.1)	Hirano & Miura (1979)

Table 4.15 – *continued*

Carrier	Enzyme	Reference
	Alkaline phosphatase (EC 3.1.3.1)	Hirano & Miura (1979)
	Trypsin (EC 3.4.21.4)	Stanley et al. (1978)
Chitin	Urease (EC 3.5.1.5)	Iyengar & Rao (1979)
Polyacrylamide	β-D-Xylosidase (EC 3.2.1.37)	Ogumtimein & Reilly (1980)
	Glucoamylase (EC 3.2.1.3)	Klyosov & Gerasimar (1979)
Amberlite® XAD-7	Penicillin amidase (EC 3.5.1.11)	Carleysmith et al. (1980)
	Cholesterol oxidase (EC 1.1.3.6)	Cheetham (1979)
Glycidyl methacrylate	Glucoamylase (EC 3.2.1.3)	Švec et al. (1978)
	Penicillin amidase (EC 3.5.1.11)	Drobnik et al. (1979)
Styrene/maleic anhydride	Chymotrypsin (EC 3.4.21.1)	Lai & Cheng (1978)
	Pyruvate decarboxylase (EC 4.1.1.1)	Beitz et al. (1980)
	Formate dehydrogenase (EC 1.2.1.2)	Rodinov et al. (1977)
Polyaminostyrene	Urease (EC 3.5.1.5)	Mattiasson et al. (1978)
	Glucoamylase (EC 3.2.1.3)	Lee et al. (1980)
Controlled pore glass	Alcohol dehydrogenase (EC 1.1.1.1)	Johnson (1978)
	Xanthine oxidase (EC 1.2.3.2)	Johnson & Coughlan (1978)
	Acetylesterase (EC 3.1.1.6)	Konecny & Sieber (1980)

Support	Enzyme	Reference
Controlled pore silica	Trypsin (EC 3.4.21.4)	Monsan (1978)
	Glucoamylase (EC 3.2.1.3)	Bohnekamp & Reilly (1980)
	β-Amylase (EC 3.2.1.2)	Bohnekamp & Reilly (1980)
	Hydrogenase (EC 1.18.3.1)	Hatchikian & Monsan (1980)
	Dextransucrase (EC 2.4.1.5)	Kaboli & Reilly (1980)
	β-D-Xylosidase (EC 3.2.1.37)	Ogumtimein & Reilly (1980)
	Catalase (EC 1.11.1.6)	Chang & Reilly (1978)
	D-Glucose oxidase (EC 1.1.3.4)	Chang & Reilly (1978)
	D-Glucose isomerase (EC 5.3.1.5)	Chang & Reilly (1978)
Silica gel	Pepsin A (EC 3.4.23.1)	Voivodov et al. (1979)
Sand	Trypsin (EC 3.4.21.4)	Pavanakrishnan & Bose (1980)
Brick	Acetylesterase (EC 3.1.1.6)	Konecny & Sieber (1980)
Alumina	Glucoamylase (EC 3.2.1.3)	Allen et al. (1979)
Hornblende	Alcohol dehydrogenase (EC 1.1.1.1)	Johnson (1978)
	Xanthine oxidase (EC 1.2.3.2)	Johnson & Coughlan (1978)
	Urate oxidase (EC 1.7.3.3)	Johnson & Coughlan (1978)
	Glucoamylase (EC 3.2.1.3)	Flynn & Johnson (1978)
Titania	Ribonuclease (EC 3.1.27.2 etc)	Dale & White (1979)
Ferrite	Chymotrypsin (EC 3.4.21.1)	Halling et al. (1979)
	Trypsin (EC 3.4.21.4)	Halling et al. (1979)
	Ribonuclease (EC 3.1.27.2 etc.)	Halling et al. (1979)
	Lysozyme (EC 3.2.1.17)	Halling et al. (1979)
Attapulgite	D-Amino acid oxidase (EC 1.4.3.3)	Parkin & Hultrin (1979)
Glass beads	D-Glucose oxidase (EC 1.1.3.4)	Wasserman et al. (1980)
	Catalase (EC 1.11.1.6)	Wasserman et al. (1980)
	Hydrogenase (EC 1.18.3.1)	Epton et al. (1972)

Table 4.16

Examples of immobilized enzymes prepared by intermolecular crosslinking

Crosslinking agent	Enzyme	Reference
Glutaraldehyde	Alcohol dehydrogenase (EC 1.1.1.1)	Sodini *et al.* (1974)
	Glutamate dehydrogenase (EC 1.4.1.2)	Ahn *et al.* (1975)
	Penicillin amidase (EC 3.5.1.11)	Carleysmith *et al.* (1980)
	Catechol 1,2-dioxygenase (EC 1.13.11.1)	Neujahr (1980)
	Phenol 2-monooxygenase (EC 1.14.13.7)	Kjellén & Neujahr (1979)
	Phosphatase (EC 3.1.3.1/2)	Tashiro & Matsuda (1978)
	Carboxypeptidase A (EC 3.4.17.1)	Quiocho & Richards (1964 & 1966)
	Ribonuclease (pancreatic) (EC 3.1.27.5)	Avrameas & Ternynck (1969)
1,5-Difluoro-2,4-dinitro-benzene	Carboxypeptidase A (EC 3.4.17.1)	Quicho & Richards (1966)
	Ribonuclease (pancreatic) (EC 3.1.27.5)	Marfey & King (1965)
Diazobenzidine	Carboxypeptidase A (EC 3.4.17.1)	Quicho & Richards (1966)
Tannic acid	Pullulanase (EC 3.2.1.41)	Ohba *et al.* (1978)
	Invertase (EC 3.2.1.26)	Negoro (1972)
	β-D-Galactosidase (EC 3.2.1.23)	Olson & Stanley (1979)

4.2.3 Crosslinking

This method is based on the formation of covalent bonds between enzyme molecules, by means of bi- or multi-functional reagents, leading to three-dimensional crosslinked aggregates which are completely insoluble in water but which do not require the use of water insoluble carriers. This method involves the addition of the appropriate amount of crosslinking agent to an enzyme solution under conditions which give rise to the formation of multiple covalent bonds. Optimum conditions for obtaining maximum insolubility whilst retaining high enzymic activity must unfortunately be determined for each system by trial and error until sufficient is known about the enzyme's primary, secondary, and tertiary structures to allow prediction of the best conditions which will exhibit minimal distortion of the enzyme upon its reaction to give a crosslinked immobilized enzyme.

The reagents required for crosslinking possess two identical functional groups (homobifunctional reagents) or two or more different functional groups (heterobi- or heteromulti-functional reagents), the latter being more common in binding enzymes to insoluble carriers than in intermolecular crosslinking reactions. The main reagents used contain:

(a) carboxyl functional groups which react with L-lysine residues by Schiff's base formation;
(b) diazo groups which react with L-lysine, L-histidine, L-tyrosine, L-arginine, or L-cysteine residues by diazo coupling reactions;
(c) isocyanate groups which react by amide (peptide) bond formation;
(d) alkyl iodides which react with nucleophilic residues by alkylation reactions;
(e) iodoacetamides which react with L-cysteine residues by alkylation.

The first report of an immobilization reaction by this method involved the crosslinking of carboxypeptidase A (EC 3.4.17.1) with glutaraldehyde (Quiocho & Richards 1964) which provides an immobilized enzyme with intermolecular linkages which are irreversible and can survive extremes of pH and temperature. However, the major disadvantages of the method, which severely limit the application of the method, are: the difficulties in controlling the reaction; the need for large quantities of enzyme, much of which looses its activity by involvement of the active site in bond formation or by being at the centre of the crosslinked aggregate and out of contact with the substrate; and the gelatinous nature of the final product. Some other immobilized enzymes which have been prepared by this method are listed in Table 4.16.

4.2.4 Immobilized soluble enzymes

All the methods for immobilization of enzymes described thus far have involved modification of the enzyme or its microenvironment with resulting alterations in pH and temperature profiles and kinetics, which frequently result in a reduced activity relative to the corresponding free enzyme. In order to use an enzyme in its native (soluble) state continuously over a long period, methods of physically confining the enzyme have been devised which utilize semipermeable membranes, hollow bore fibres, or ultrafiltration membranes. The methods used can involve chemical derivatization, although this is not essential.

4.2.4.1 *Immobilization without enzyme derivitization*
The membrane etc. used is impermeable to enzyme molecules but permeable to product, and in some cases to substrate, molecules; and, because no chemical modification occurs, the method allows the study of soluble enzymes and their operational stability in continuous reactors.

The method is especially suited for conversion of high molecular weight water-soluble or insoluble substrates, as it allows the intimate contact of the soluble enzyme with substrate, achieving an efficient conversion of these types of substrate, unlike the insoluble immobilized enzymes which usually have lower catalytic efficiencies towards the same substrates.

Other advantages of this method are: the simplicity of the method required to immobilize the enzymes by placing the enzyme in solution, on one side of a semipermeable membrane; simultaneous immobilization of many enzymes; selectivity control of substrates and products through membrane selectivity; large ratio of surface area to volume (hollow fibres); protection of enzyme from access by microorganisms; absence of enzyme leakage when properly constructed membranes are chosen; and the favourable ease with which the membrane reactors can be loaded with enzymes, operated, cleaned, sterilized, and regenerated compared with other methods of immobilization.

However, the disadvantages inherent in the method include: the possible reduction of reaction velocity as a result of the permeability resistance of the membrane; the difficulty of working with very low substrate concentrations due to substrate adsorption by membranes; the possibility of enzyme inactivation due to high shear forces or vigorous agitation (ultrafiltration membrane cells); and, among others, the need for a careful control of the residence time of low molecular substrates in order to achieve high conversions.

Table 4.17 lists some examples of enzymes immobilized by this method.

Table 4.17
Examples of immobilized soluble enzymes

Procedure	Enzyme	Reference
Ultrafiltration	β-D-Fructofuranosidase (EC 3.2.1.26)	Cantarella *et al.* (1977)
Membrane method	Amidase (EC 3.5.1.4)	Wandrey *et al.* (1979)
	Acid phosphatase (EC 3.1.3.2)	Greco *et al.* (1980)
	β-D-Galactosidase (EC 3.2.1.23)	Roger *et al.* (1976)
Hollow fibre devices	α-D-Galactosidase (EC 3.2.1.22)	Silman *et al.* (1980)
	β-D-Fructofuranosidase (EC 3.2.1.26)	Silman *et al.* (1980)
	D-Glucose oxidase (EC 1.1.3.4)	Besserdich *et al.* (1980)

4.2.4.2 Immobilization with enzyme derivitization

Recently several reports (Marshall & Rabinowitz 1976, Ugarova et al. 1977, Vegarud & Christensen 1977) have described chemical modification of enzymes without insolubilization, using low or high molecular weight compounds. Although the modification of enzymes with low molecular weight compounds is often of limited utility, there are some situations in which it can serve a specific and useful purpose; for example, the acylation of enzymes with low molecular weight reagents can have a stabilizing effect (Ugarova et al. 1977).

In the preparation of water-soluble, enzyme—polymer conjugates, reactions similar to those employed in the chemical coupling of enzymes on to insoluble polymers, are used. The bonding of enzymes to soluble polymers may be achieved by one of the following procedures: reactions of the enzyme with an activated soluble polymer, reaction of the enzyme with an activated insoluble polymer followed by solubilization of the enzyme—polymer conjugate, or copolymerization of monomers with enzyme.

The first reports on water-soluble derivatized enzymes were by Katchalski & Sela 1958, Glazer et al. 1962), who had prepared them in order to elucidate the interrelationship between the electrostatic potential of the polymer chain and the displacement of the optimal pH of the enzyme. In other reports (Marshall & Rabinowitz 1976, Vergarud & Christensen 1977) the water-soluble enzyme derivatives have been prepared in order to increase the effective molecular size of the enzyme to prevent its release from membrane dependent devices, and to improve the mechanical properties and operational stability of the enzyme.

In fact, when using a native soluble enzyme in a membrane device, the instability of the enzyme over long periods and the (associated) need to limit the porosity of the membrane to prevent loss of enzyme in some applications can be disadvantageous. These disadvantages can be overcome by using a water-soluble, enzyme—polymer conjugate, which allows a choice of ultrafiltration membranes of higher porosity with consequential faster diffusion away of products and therefore reduction of end product inhibition. At the same time a stabilization of the enzyme can be enforced by attachment of the enzyme to a polysaccharide (Solomon & Levin 1974), or by forming a stable environment of definite electrostatic nature around the enzyme (Wykes et al. 1971).

Another advantage of using soluble derivatized enzymes is in the hydrolysis of macromolecular or insoluble particulate substrates, such as cellulose, as the treatment of such substrates with conventional immobilized enzyme catalysts is accompanied by severe diffusional resistances.

The main disadvantage of preparing these water-soluble conjugates is the more laborious purification needed after polymer activation and reaction with the enzyme. The excess reagent and unreacted enzyme have to be separated by precipitation, gel filtration, ultrafiltration, or dialysis.

Some examples are derivatized enzymes are listed in Table 4.18.

Table 4.18

Examples of soluble derivatized enzymes

Soluble polymer	Enzyme	Reference
Dextran	α-Amylase (EC 3.2.1.1)	Charles et al. (1974)
	Chymotrypsin (EC 3.4.21.1)	Vergarud & Christensen (1977)
	β-D-Glucosidase (EC 3.2.1.21)	Vergarud & Christensen (1977)
	Lysozyme (EC 3.2.1.17)	Vergarud & Christensen (1977)
	Trypsin (EC 3.4.21.4)	Marshall & Rabinowitz (1976)
Ethylene-maleic anhydride copolymer	Glucomylase (EC 3.2.1.3)	Soloman & Levin (1974)
Styrene-maleic anhydride copolymer	Glucoamylase (EC 3.2.1.3)	Soloman & Levin (1974)
Alginic acid	Lysozyme (EC 3.2.1.17)	Charles et al. (1974)

4.2.5 Miscellaneous methods

Although enzymes have been immobilized by one of the earlier described methods, sometimes immobilization of enzymes with greater efficiency can be achieved by combination of more than one specific method. This is largely used for crosslinking of enzymes previously immobilized, for instance, by adsorption, owing to the low operational stability of the immobilized enzyme preparations obtained by the adsorption method only. This double method of immobilization eliminates the disadvantage of lack of mechanical properties of the preparations obtained with only crosslinking with multifunctional reagents. With the crosslinking of adsorbed enzymes a monolayer of immobilized enzymes can be formed; however, the experimental conditions must ensure good adsorption of the enzyme on the support — and there is also a necessity that no aggregation of individual colloidal particles occurs.

The efficiency of the entrapment method of immobilization can be improved by use of intramolecular crosslinking reactions to attach enzyme molecules to the walls of the membrane or to each other. The product of the entrapment method is treated with a bifunctional reagent, usually glutaraldehyde, although chelation or metal binding methods have been used (Kennedy & Kalogerakis 1980) to effect crosslinking and in so doing increase the mechanical stability of the final product to withstand operational conditions.

To eliminate the loss of enzyme activity due to diffusion effects caused by crosslinking reactions enzymes, at low concentrations, are crosslinked via glutaraldehyde to a nonenzymic protein which is rich in L-lysine residues (such as bovine serum albumin), and the resultant immobilized enzyme can be insolubilized by gel entrapment methods. This method known as co-crosslinking, devised by Broun (1976), is particularly useful when, owing to its chemical nature, insolubilization of an enzyme cannot be achieved by glutaraldehyde crosslinking alone.

The ability of inorganic supports to convalently bind enzymes can be brought about by the methods described above (subsection 4.2.2.4) or alternatively by coating the inorganic carrier with organic materials which possess the required functional groups to bind proteins. Several attempts have been made, including the use of polymeric 1,3-diaminobenzene (Kennedy & Kay 1977b, Kennedy *et al.* 1977b, 1977c, 1980b), aminobenzoic acid-formaldehyde resin (Chaplin & Kennedy 1976) and alkylamine derivatives (Royer & Uy, 1973).

Formation of soluble derivatives of enzymes (as described in subsection 4.2.4.2) can be used as an intermediate step in immobilization by making the enzyme more susceptible to covalent bonding via, for example, carbohydrate constituents attached to the protein (Solomon & Levin 1974) or to ionic binding via highly charged residues attached to the protein (Wykes *et al.* 1971). In the latter case the enzyme is attached to a water-soluble copolymer of acrylic acid and maleic hydride or ethylene and maleic acid, and the resulting derivatives bind strongly to ion exchange materials in a practically irreversible manner.

Table 4.19

Examples of miscellaneous methods of immobilization

Procedure	Enzyme	Reference
Entrapment & crosslinking (collagen)	Lipase (EC 3.1.1.3)	Sato et al. (1977)
	Urokinase (EC 3.4.21.31)	Karube et al. (1977b)
	Glucoamylase (EC 3.2.1.3)	Gondo & Koya (1978)
	Asparaginase (EC 3.5.1.1)	Morikawa et al. (1978a)
	D-Glucose oxidase (EC 1.1.3.4)	Gondo et al. (1980)
	D-Glucose isomerase (EC 5.3.1.5)	Gondo et al. (1980)
	Alcohol dehydrogenase (EC 1.1.1.1)	Morikawa et al. (1978b)
	Lactate dehydrogenase (EC 3.5.1.5)	Bollmeier & Middleman (1979)
Entrapment and crosslinking (gelatin)	Urease (EC 3.5.1.5)	Bollmeier & Middleman (1979)
	Glucoamylase (EC 3.2.1.3)	Kennedy & Kalogerakis (1980)
	Urate oxidase (EC 1.7.3.3)	Remy et al. (1978)
Co-crosslinking (albumin & glutaraldehyde)	β-D-Fructofuranosidase (EC 3.2.1.26)	D'Souza & Nadkarni (1981)
	L-Glutamate dehydrogenase (EC 1.4.1.2)	Barbotin & Breuil (1978), Barbotin & Thomasset (1979)
	2-Acetamido-2-deoxy-β-D-hexosidase (EC 3.2.1.52)	Yeung et al. (1979)
	Catechol-1,2-dioxygenase	Neujahr (1980)
	Urease (EC 3.5.1.5)	Vallin & Tran-Minh (1979)
	α-Steroid dehydrogenase	Legoy et al. (1980)

Method	Enzyme	Reference
Adsorption & crosslinking (collagen + glutaraldehyde)	β-D-Fructofuranosidase (EC 3.2.1.26)	Ludolph et al. (1979)
Adsorption & crosslinking	Chymotrypsin (EC 3.4.21.1)	Halling & Dunnill (1979)
Adsorption & crosslinking (NiO. Fe O_2 Mn−Zn Ferrite)	D-Glucose oxidase (EC 1.1.3.4)	Krishnaswamy & Kittrel (1978)
Adsorption & crosslinking (Kieselguhr + glutaraldehyde)	D-Glucose oxidase (EC 1.1.3.4)	Wassermann et al. (1980)
	Catalase (EC 1.11.1.6)	Wassermann et al. (1980)
Adsorption & crosslinking (Glass + glutaraldehyde)	β-D-Glucosidase (EC 3.2.1.21)	Gray et al. (1974), Chaplin & Kennedy (1976)
	Dextranase (EC 3.2.1.11)	Kennedy & Kay (1977b), Kennedy et al. (1980b)
	Papain (EC 3.4.22.2)	Gray et al. (1974), Kennedy et al. (1977c)
	Urease (EC 3.5.1.5)	Kennedy et al. (1977c)
	Cholinesterase (EC 3.1.1.8)	Kennedy et al. (1977c)
	Peroxidase (EC 1.11.1.7)	Gray et al. (1974)
	Catalase (EC 1.11.1.6)	Gray et al. (1974)
	Uricase (EC 1.7.3.2)	Gray et al. (1974)
	Glucoamylase (EC 3.2.1.3)	Gray et al. (1974)
	α-Amylase (EC 3.2.1.1)	Kennedy et al. (1977b)
Organic coating of inorganic support	Chymotrypsin (EC 3.4.21.1)	Bessmertnaya & Antonov (1973)
Ionic binding and derivatization	Trypsin (EC 3.4.214)	Yarovaya et al. (1975)
	Glucoamylase (EC 3.2.1.3)	Soloman & Levin (1974)

A number of examples of immobilized enzymes prepared by these miscellaneous methods are listed in Table 4.19.

4.3 CHOICE OF IMMOBILIZATION METHOD

Although many methods of immobilization techniques have been developed and applied to many enzymes, it is now well recognized that no one method can be regarded as the universal method for all applications or all enzymes. This is because of the widely different chemical characteristics and composition of enzymes; different properties of substrates and products, and the different uses to which the product can be applied. Therefore, for each application of an immobilized enzyme it is necessary to find a procedure which is simple and inexpensive to perform and which gives a product with good retention of activity and high operational stability.

However, from the vast amount of information which has been generated on the characteristics of the support matrices and the effects of the methods which have been used, it is possible to make generalizations which can be used to form the basis for selection of a method which may be applicable for a specific case, although a guarantee of success cannot be given, and more than one method may have to be attempted before a suitable process is obtained. Table 4.20 gives a brief summary of the relative advantages and disadvantages of the difficult methods of enzyme immobilization which are available.

When immobilization is accompanied by a chemical reaction, as in the crosslinking and covalent binding methods, conformational changes in the protein molecule must be kept to a minimum to avoid partial deactivation due to involvement of the active site in the immobilization reaction. This requires the use of the mildest conditions possible to effect immobilization. However, once an enzyme is successfully immobilized by chemical means, the operational stability of the product is high, owing to the strength of the bonds between enzyme molecules (in the case of crosslinking) or between enzyme and carrier (in the case of covalent binding), and the reluctance of these bonds to disruption by substrate or salt solutions. Crosslinking is generally not a suitable method for large-scale industrial applications, because of the lack of mechanical stability of the final product, whilst covalently bound enzymes which utilize organic matrices can rarely be regenerated and are again unattractive for large-scale industrial use on this account.

Physical adsorption, ionic binding, and chelation or metal binding are attractive methods for enzyme immobilization owing to mild conditions involved in the binding reaction. However, because the binding forces are generally weaker than for chemical binding methods, operational stabilities are lower through loss of enzyme from the matrix as a result of changes in ionic concentra-

Table 4.20

Comparison of the attributes etc. of different classes immobilization techniques

Characteristic	Crosslinking	Physical adsorption	Ionic binding	Chelation or metal binding	Covalent binding	Entrapment
Preparation	Intermediate	Simple	Simple	Simple	Difficult	Difficult
Binding force	Strong	Weak	Intermediate	Intermediate	Strong	Intermediate
Enzyme activity	Low	Intermediate	High	High	High	Low
Regeneration of carrier	Impossible	Possible	Possible	Possible	Rare	Impossible
Cost of immobilization	Intermediate	Low	Low	Intermediate	High	Intermediate
Stability	High	Low	Low	Intermediate	High	High
General applicability	No	Yes	Yes	Yes	No	Yes
Protection of enzyme from microbial attack	Possible	No	No	No	No	Yes

tion, pH, substrate concentration, or temperature of the reaction medium. To offset this disadvantage the possibility of regeneration does mean that industrial applications using these techniques are possible.

With the entrapment methods of immobilization, high retention of activity is possible as a result of there being no binding between enzyme and carrier, but limitation of enzyme activity can occur owing to diffusion effects of large molecular weight substrates and products. Therefore the use of entrapment methods must be limited to reactions involving small molecular weight substrate and product molecules.

4.4 OUTLINE OF PROPERTIES OF IMMOBILIZED ENZYMES

It is essential to understand the changes in physical and chemical properties which an enzyme could be expected to undergo upon immobilization, if the best use is to be made of the various techniques available. Changes have been observed in the stability of enzymes, and in their kinetic properties because of the micro-environment imposed upon them by the supporting matrix and by the products of their own action.

4.4.1 Stability (see also Part A, Chapter 3.27)

The stability of enzymes might be expected to either increase owing to the stabilizing effects of the microenvironment or decrease owing to the micro-environment having a denaturing effect on the enzyme. The hydrophobic microenvironment of some supports has been considered to be the cause of the loss of activity upon lyophilization of enzymes which are stable when stored in aqueous suspension. In some cases this destabilization can be overcome by in-cluding a material into the solution, prior to lyophilization, which can overcome the hydrophobic effects and provide hydrophilic microenvironment. One such material is sorbitol (Kennedy & Kay 1976). Immobilization of proteolytic enzymes should reduce the ability of the enzyme to cause its own inactivation through autodigestion by isolating enzyme molecules from mutual attack. A number of systems have been reported when the immobilized protease could be stored at 4°C for several months without significant loss of activity.

The stability of many immobilized enzymes can be improved by using inorganic supports such as glass or ceramic materials rather than organic poly-mers, owing to the greater dimensional stabilities of the inorganic supports (Messing 1974), whilst the use of sulphonamide linkages to covalently bind enzymes to inorganic supports generally gave less stable products than those coupled by azo linkages (Weetall 1970).

Stability towards denaturing agents may also change on immobilization, with a number of cases being reported of increased stabilities (although it is probable that decreases in stability are frequently not reported). For entrapped enzymes, product inhibition can be increased owing to the diffusion effects of the membrane preventing rapid removal of product. Thermal stability of immobilized enzymes can be increased or decreased, and very little has been reported on the effects of the carrier toward thermal stability.

4.4.2 Kinetic properties (see also Part A, Chapter 3)

On immobilization of an enzyme its specific activity frequently decreases owing to denaturation of the enzymic protein molecule caused by the immobilization procedure. Once immobilized, however, the enzyme comes under the influence of the microenvironment support which may be drastically different from that existing in free solution. The microenvironment may be a result of the physical and chemical character of the support matrix, or it may be due to interactions of the matrix with substrate or product molecules involved in the enzymic reaction.

Shifts in pH optima on immobilization have been found for many enzymes. Comparison of the activity of an enzyme bound to a matrix with the activity of a freely dissolved enzyme at various pH levels has shown that if the enzyme is attached to a negatively charged matrix, the pH optimum is shifted toward the alkaline side: the immobilized enzyme reaches maximum activity at apparently higher alkalinity. This effect is due to the negatively charged groups of the matrix attracting a thin 'film' of positive hydrogen ions, thereby creating a microenvironment for the bound enzyme that has a higher hydrogen ion concentration (lower pH) than the concentration in the surrounding solution where the pH is actually measured (Katchalski *et al.* 1971). Similarly, for a positively charged matrix the apparent shift to optimum pH is to the alkaline side (Fig. 4.2). At high ionic strengths this effect disappears. The apparent Michaelis constant has been found to decrease by more than one order of magnitude when substrate of opposite charge to the support matrix is used. Again, this only happens at low ionic strengths. Both effects can be treated mathematically by including a Maxwell—Boltmann distribution of charge into the treatment of electrostatic potential and Michaelis—Menten equations respectively.

The diffusion of substrate from the bulk solution to the microenvironment of an immobilized enzyme can be a major factor in the rate of the enzyme reaction. By postulation of a diffusion film (Lilly *et al.* 1968) which covers the surface of the immobilized enzyme, and within which the substrate concentration is lower than the bulk solution.The rate at which substrate passes over the immobilized enzyme effects the thickness of the diffusion film, which in turn determines the concentration of substrate in the vicinity of the enzyme and hence the rate of reaction. The effect of this diffusion control is that the rate of reaction varies with the rate of stirring of flow rate of substrate solution in the case of a packed bed. The effect of molecular weight of the substrate can be very pronounced since large molecules diffuse at lower rates and are more subject to steric interactions with the matrix. The relative activity of immobilized enzymes towards high molecular weight substrates has been generally found to

be lower than towards low molecular weight substrates. In some cases, this diffusion control can be advantageous in that large inhibitor molecules present in a reaction mixture are prevented from attacking the immobilized enzyme. Diffusion effects can also affect the microenvironment when charged species are produced by a reaction, producing a change in pH in the microenvironment and altered pH-activity profiles.

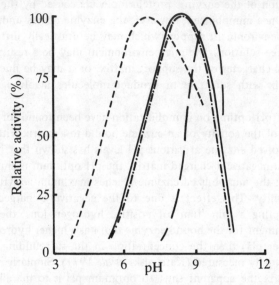

Fig. 4.2 – The effect of microenvironment on the pH activity of immobilized enzyme ——— soluble enzyme cation support, ——— immobilized enzyme; —•—•— immobilized enzyme; rentral support; — — — — immobilized enzyme; amonic support. (From Kennedy, J. F. & White, C.A. (1983) *Bioactive Carbohydrates in Chemistry: in Chemistry, Biochemistry and Biology,* Ellis Horwood Chichester).

4.5 OUTLINE OF ENZYME REACTORS (also for detailed account see Part A, Chapter 3)

Among the applications of immobilized enzymes (see section 4.6), the industrial utilization is, perhaps, the most important field and certainly the most actively reported. A number of reactors have been developed to make the best use of the various properties of the immobilized enzyme and characteristics of the substrate and product. There have been several attempts to provide a classification of enzyme reactors (Wingard *et al.* 1976, Wiseman 1978).

4.5.1 Batch reactors
Batch reactors are the most commonly used type of reactors when soluble enzymes are employed in a process. After the reaction is completed the soluble

enzyme is generally not recovered from the reaction mixture and consequently cannot be re-used; Since one of the major objectives of immobilization of an enzyme is to allow its recovery for re-use, the application of immobilized enzymes in batch reactors can frequently involve an additional process to separate the enzyme. During this recovery process appreciable loss of immobilized enzyme can occur, and some loss of enzyme activity can also occur as a result of, for example, drying out of material. For this reason the use of traditional stirred tank reactors (see Fig. 4.3) has been limited to the production of small quantities of fine chemicals.

a) Stirred tank for soluble enzymes.

b) Stirred tank for immobilized enzymes

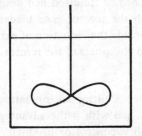

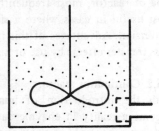

c) Stirred tank with immobilized enzyme basket paddles.

d) Stirred tank with immobilized enzyme basket baffles.

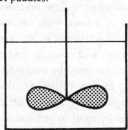

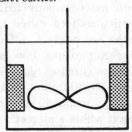

e) Total recycle packed bed reactor.

f) Total recycle fluidized bed reactor.

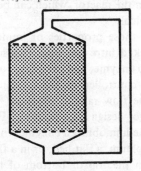

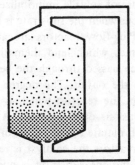

Fig. 4.3 – Types of batch reactors.

The traditional stirred tank reactors consist of a vessel, a stirrer, and frequently baffles attached to the walls of the vessel to improve the mixing of the reactants. Since many immobilized enzymes, particularly those attached to inorganic supports, are broken up by the mechanical stresses in a stirred tank, modifications to the design have been made in an attempt to allow the use of immobilized enzymes whilst overcoming these problems. Such modifications include enclosing the immobilized enzyme in a 'basket' which is part of the impeller blades or baffles of the tank reactor (see Fig. 4.3).

Another alternative is obtained by using a continuous reactor (see subsection 4.5.2) modified to give total or batch recirculation (see Fig. 4.3). This type of reactor, most frequently of the packed bed or fluidized bed design, is most useful in cases where a single pass through the reactor gives inadequate conversion. Advantages of this type of reactor include the reduction of external mass transfer effects by use of high flow rates, and cheapness of the reactor.

4.5.2 Continuous reactors

With the introduction of immobilized enzymes, continuous operation has become a reality in enzyme catalyzed reactions, and with it the advantages of automatic control, ease of operation, and quality control of products. Continuous reactors can be divided into two basic types, depending on the relative flow patterns; continuous feed stirred tank reactors and plug-flow reactors.

Continuous feed stirred tank reactors consist of a tank with separate substrate inlet and product outlet, and the degree of conversion can be controlled by the reactor volume, flow rate through the reactor, and amount and activity of the immobilized enzyme. The immobilized enzyme may be retained within the reactor by filtration of the product stream, incorporating a subsequent settling stage, immobilizing the enzyme onto a magnetically active particle, and retaining it within a magnetic field (which may also be used to stir the particles) or by immobilizing the enzyme to the paddles or baffles of the reactor (see Fig. 4.4). The incorporation of an ultrafiltration process into the reactor will allow the use of soluble immobilized enzymes within the reactor, which can be advantageous when the substrate is insoluble or colloidal.

Plug-flow reactors are a direct result of those properties of immobilized enzymes which lend themselves to being packed into columns. The substrate is then passed through the bed of immobilized enzyme, and product is obtained from the outlet. The degree of conversion is controlled by the residence time within the reactor, which is controlled by the flow rate of substrate and the dimensions of the reactor. A number of reactor designs are available (Fig. 4.5) which dictate the physical form required for the immobilized enzyme. In packed bed reactors the catalyst is retained in a tall column, a flat bed, or in a filtration bed, and the substrate is pumped either from the top or bottom of the bed, whereas in fluidized bed reactors the catalyst is loosely packed into a column and the substrate feed enters the bottom of the column. The flow rate through

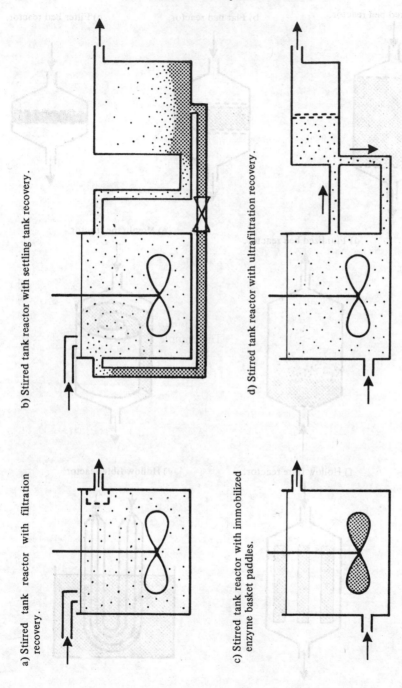

a) Stirred tank reactor with filtration recovery.

b) Stirred tank reactor with settling tank recovery.

c) Stirred tank reactor with immobilized enzyme basket paddles.

d) Stirred tank reactor with ultrafiltration recovery.

Fig. 4.4 – Types of continuous flow stirred tank reactors.

a) Packed bed reactor.

b) Flat bed reactor.

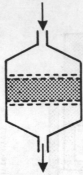

c) Filter bed reactor.

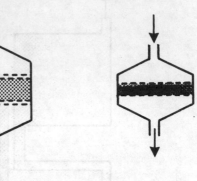

d) Fluidized bed reactor.

e) Membrane reactor.

f) Hollow fibre reactor.

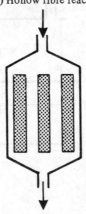

g) Hollow fibre reactor.

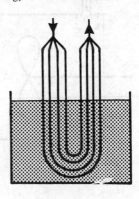

Fig. 4.5 – Types of continuous flow reactors.

the system is balanced to be sufficiently fast to mix the immobilized enzyme within the column but not so fast as to flush the catalyst out of the reactor. This system can be improved by immobilizing the enzyme on a support which has a greater density than the reaction mixture, and is ideal for the treatment of insoluble or colloidal substrates.

An extension of the packed bed reactor are the membrane and hollow fibre reactors in which the walls of the membrane or fibre are impermeable to the enzyme molecules (Fig. 4.5 for the configuration of the reactors). With either type of reactor the substrate can permeate the membrane, etc. to react with enzyme, and the product permeate back across the membrane into the bulk solution; or the substrate can flow across to the side of the membrane to which the enzyme is attached, and the product diffuse through the membrane into the bulk section.

4.6 APPLICATIONS AND FUTURE TRENDS

There have been many proposed uses for immobilized enzymes in the analytical, medical, and industrial fields, but many of these have not been fully developed to the extent of replacing alternative (older) techniques. In the following subsections the description of the various applications is used to illustrate the extent of their development. Further details can be found in Part B, Chapter 3.

4.6.1 Analytical applications
A number of recent advances in the field of immobilization technology in conjunction with potential analytical applications have led to several commercially available systems being made available. Carr & Bowers (1980) have reviewed the area in great detail.

4.6.1.1 *Enzyme electrodes*
Many enzyme electrodes have been proposed, but relatively few have been developed to the extent of being in common use or commericially available. The principal of enzyme electrodes can best be described in terms of the D-glucose electrode which was the first to be prepared (Updike & Hicks 1967). D-Glucose oxidase (EC 1.1.3.4) is immobilized in a polyacrylamide gel and held in place around an oxygen electrode by a piece of cellulose acetate. The immobilized enzyme catalyses the reaction

$$\beta\text{-D-Glucose} + O_2 \longrightarrow \text{D-Glucono-1,5-lactone} + H_2O_2 \ ,$$

and the oxygen electrode measures the depletion of oxygen from the solution at a rate which is dependent on the concentration of D-glucose present. Provided that an oxygen electrode is available, the enzyme electrode is very cheap and easy to prepare and provides a simple method for the specific determination of D-glucose in solution, particularly from opaque or particulate solutions. Develop-

Table 4.21
Enzyme electrodes

Determination	Enzyme	Sensing electrode
D-Glucose	D-Glucose oxidase (EC 1.13.4)	Oxygen
Sucrose	β-D-Fructofuranosidase (EC 3.2.1.26) and D-glucose oxidase (EC 1.1.3.4)	Oxygen
Urea	Urease (EC 3.5.1.5)	Cation or carbon dioxide
Lactic acid	Lactate dehydrogenase (EC 1.1.1.27)	Platinum
Ethanol	Alcohol dehydrogenase (EC 1.1.1.1)	Platinum
D-Amino acids	D-Amino acid oxidase (EC 1.4.3.3)	Cation
Cholesterol	Cholesterol oxidase (EC 1.13.6)	Oxygen
Penicillin	Penicillinase (EC 3.5.2.6)	pH
Amygdalin	β-D-Glucosidase (EC 3.2.1.21)	Cyanide
L-Phenylalanine	L-Amino acid oxidase (EC 3.2.1.21)	Ammonium
	L-Amino acid oxidase (EC 3.2.1.21) and peroxidase (EC 1.11.1.7)	Iodine
Monoamines	Amine oxidase (copper containing) (EC 1.4.3.4)	Oxygen
Glutamine	Glutaminase (EC 3.5.1.2)	Cation

ment of such systems has led to the production of an industrial analyser by the Yellow Springs Instrument Co. A number of other enzyme electrodes which can be prepared are listed in Table 4.21.

4.6.1.2 *Automated analysis*

There are two areas of automated analysis in which immobilized enzymes find general application, namely in the routine analysis of a large number of small samples (for example, blood samples) and in the continuous monitoring of streams of larger volumes of samples. The alternative to adding an aliquot of soluble enzyme to each sample is to immobilize the enzyme and fix it in the sample stream. Whilst the use of columns of immobilized enzyme is not a practical proposition, owing to the resistance to flow it would create, immobilization onto the walls of narrow bore tubes through which the sample passes prior to detection as in, for example, AutoAnalyser® systems, provides a simple means of routine continuous flow analysis. By linking together a number of such immobilized enzyme tubes, containing different enzymes, multiple analyses can be performed on a single sample. A simpler system, suitable for the analysis of smaller numbers of samples, has been devised by Sundaram (1979). An immobilized enzyme pipette, in which the enzyme is immobilized on the inner surface of a nylon tube attached to the disposable tip of an adjustable volume pipette, is used to analyse, for example, urea in sera by holding the sample in the pipette (known as an Impette) for a fixed time prior to expulsion and determination of the products of the reaction (in this case, ammonia).

Where there is a large volume of material to be analysed continuously, it is possible to adopt a different method as described by Mosbach & Danielsson (1974). The principle involved is the measurement of heat generated by the enzyme reaction which is detected and quantified by a thermistor held in the middle of a column of immobilized enzyme as a substrate solution is passed through the column. A major advantage of this system is that the detection system does not rely on optical measurements, and the system can therefore be applied to liquid streams which are not optically clear.

5.6.2 Therapeutic applications

All applications in the biomedical field are still very much in the state of basic studies rather than definite applications, owing to the absence of the necessary information on toxicology, haemolysis, allergenicity, immunological reactions, and chemical stability of the system *in vivo*. Potential applications fall into two major areas, those of enzyme replacement and enzyme therapy.

4.6.2.1 *Enzyme replacement*

There are many diseases that are characterized by the absence of a particular enzyme as a result of genetic malfunction or because of tissue malfunction which lead to the build-up of some products with disastrous results. Such diseases can be controlled, in theory, by injecting the required soluble enzyme from, for example, microbial sources, but in an immobilized form to prevent immunological reactions. Alternatively, an extracorporeal shunt can be used to purify the blood by removal of the accumulating materials.

4.6.2.2 Enzyme therapy

Enzyme therapy differs from enzyme replacement in that the enzyme to be added to the body is either not normally found in the body or is not pathologically diminished in quantity. Such enzymes are added to the body to alter the normal environmental conditions in the body in order to control a diseased state, for example, the use of L-asparaginase (EC 3.5.1.1) in the treatment of certain leukaemias by removal of L-aspartic acid from the blood. The enzyme is administered to the body in an immobilized but biodegradable form such as in the form of polylactic acid capsules.

4.6.3 Industrial applications

Most industrial applications proposed to date have involved the use of hydrolases because they are generally stable and do not require coenzymes. The main areas of application are in the food and pharmaceutical industries, but other potential areas include waste treatment and fine chemical production. To date only three processes are operative on an industrial scale:

(a) production of L-amino acids by the optical resolution of acetyl-DL-amino acids using immobilized aminocyclase (EC 3.5.1.14) in Japan.

(b) production of high fructose syrups using D-glucose isomerase (EC 5.3.1.5) in Europe, Japan, and the USA;

(c) production of 6-aminopenicillanic acid using immobilized penicillin amidase (EC 3.5.1.11) in Europe, Japan, and the USA.

This limited number of industrial applications, in contrast to the large numbers of published reports, is due to several factors, mainly economic ones; carriers or reagent costs for the immobilization procedure are expensive. Other factors are linked with low efficiency of immobilization, poor operational stability, relatively complicated equipment for continuous operation, and low demand for the products being insufficient to justify large-scale working. Further expansion of immobilized enzyme technology must take into account, on one hand, different and more complex processes and, on the other hand, must try to solve the economic and technical limitations.

4.6.4 Future trends

Although the first artificial immobilized enzyme was reported almost 65 years ago, it was only in the late 1960s that industry and academic institutions began to make great strides towards the widespread use of immobilized enzyme technology; but, as described above, the number of applications of this technology is at present very limited.

One of the future areas for the application of immobilized enzyme technology will probably be in the development and evaluation of new reactors using immobilized multienzyme systems requiring multiphase environment and/or cofactor regeneration, which have not yet been developed for industrial use. These applications include the use of immobilized enzymes and cofactors for the production of materials now obtained by fermentation, and the synthesis of new useful products. Treatment of waste water and effluent streams from the food manufacturing industry are expected to be subjected to the application of immobilization technology, whilst rising energy costs and the need for alternative energy sources will see the development of biochemical fuel cells which will be able to produce electricity from sunlight via chemical energy transformations.

One of the major advances in immobilization technology of recent years, which will provide the answer to a number of economic problems inherent to immobilized enzymes, namely the isolation, extraction, and purification of the enzyme in sufficient quantities, and the minimization of enzyme loss, has been the development of immobilization of microbial cells with retention of life (Kennedy 1978) so that they can reproduce and thereby act as an automatically self-renewing form of immobilized enzyme. The immobilization of cells means that enzymes can be used without prior purification in environments which resemble their natural environments and which include all the necessary cofactors etc. For a review of immobilized microbial cells see Kennedy & Cabral (1983).

REFERENCES

Abdel-Hay, F. I., Beddows, C. G. & Guthrie, J. T. (1980) *Polymer Bull. (Berlin)* 2 607–612.

Adachi, S., Kawamura, Y., Nakamishi, K., Matsuno, R. & Kamikubo, T. (1978) *Agric. Biol. Chem.* 42 1707–1714.

Adachi, S., Hashimoto, K., Matsuno, R., Nakanishi, K. & Kamikubo, T. (1980) *Biotechnol. Bioeng.* 22 779–797.

Adamich, M., Voss, H. F. & Dennis, E. A. (1978) *Arch. Biochem. Biophys.* 189 417–423.

Ahn, B. K., Wolfson, S. K. Jr. & Yao, S. J. (1975) *Bioelectrochem. Bioenergetics* 2 142.

Allen, B. R., Charles, M. & Coughlin, R. W. (1979) *Biotechnol. Bioeng.* 21 689–706.

Arrio-Dupont, N. & Coulet, P. R. (1975) *Biochem. Biophys. Res. Comm.* 89 345–352.

Aukati, M. F., Kalashnivaka, T. I., Bubenshchikova, S. N., Kagramanova, V. K. & Baratova, L. A. (1978) *Vestn. Mosk. Univ. Ser.* 2 19 350–352.

Avrameas, S. & Ternynck, T. (1969) *Immunochem.* 6 53–66.

Baratti, J., Couderc, R., Cooney, C. L. & Wang, D. I. C. (1978) *Biotechnol. Bioeng.* 20 333–348.

Barbotin, J.-N. & Breuil, M. (1978) *Biochim. Biophys. Acta* 525 18–27.

Barbotin, J.-N. & Thomasset, B. (1979) *Biochim. Biophys. Acta* 570 11–21.

Barker, S. A., Doss, S. H., Gray, C. J., Kennedy, J. F., Stacey, M. & Yeo, T. H. (1971) *Carbohydr. Res.* 20 1–7.

Bartoli, F., Bianchi, G. E. & Zaccardelli, D. (1978) In: *Enzyme Engineering*, Vol. 4, p. 279–280. Ed. by Broun, G. B., Manecke, G. & Wingard, L. B. Jr. Plenum Press, New York.

Beddows, C. G., Mirauer, R. A., Guthrie, J. T., Abdel-Hay, F. I. & Morrish, C. E. J. (1979)
    Polym. Bull. (Berlin) 1 749−753.
Beddows, C. G., Mirauer, R. A. & Guthrie, J. T. (1980) Biotechnol. Bioeng. 22 311−321.
Beddows, C. G., Guthrie, J. T. & Abdel-Hay, F. I. (1981) Biotechnol. Bioeng. 23 2885−
    2889.
Beitz, J., Schellenberger, A., Lasch, J. & Fischer, J. (1980) Biochim. Biophys. Acta 612
    451−454.
Bernfeld, P. & Wan, J. (1963) Science 142 678−679.
Besserdich, H., Kinstein, D. & Kahrig, E. (1980) Chem. Tech. (Leipzig) 32 243−247.
Bessmertnaya, L. Ya. & Antonov, V. K. (1973) Khim. Proteoliticheskikh Fermentov,
    Mater. Vses. Simp. 43−44.
Bhatt, B. R., Joshi, S. & Kothari, R. M. (1979) Enzyme Microb. Technol. 1 113−116.
Bisse, E. & Vonderschmidtt, D. J. (1977) FEBS Letts. 81 326−330.
Bisse, E. & Vonderschmidtt, D. J. (1978) FEBS Lett. 93 102−104.
Blassberger, D., Freeman, A. & Goldstein, L. (1978) Biotechnol. Bioeng. 20 309−315.
Bogatskii, A. V., Davidenko, T. I., Chuenko, A. V., Vanishpol'skii, V. V., Tertykh, V. A.
    & Chuiko, A. A. Ukr. Biochem. Zh. 51 315−317.
Bohnenkamp, C. G. & Reilly, P. J. (1980) Biotechnol. Bioeng. 22 1753−1758.
Bollmeier, J. P. & Middleman, S. (1979) Biotechnol. Bioeng. 21 2303−2321.
Boudillon, C., Bourgeois, J.-P. & Thomas, D. (1979) Biotechnol. Bioeng. 21 1877−1879.
Brümmer, W., Hennrich, N., Klockow, M., Lang, H. & Orth, H. D. (1972) Eur. J. Biochem.
    25 129−135.
Broun, G. B. (1976). In: Methods in Enzymology Vol. 44, p. 263−280. Ed. by Mosbach, K.
    Academic Press, New York.
Buchholz, K. & Gödelmann, B. (1978) Biotechnol. Bioeng. 20 1201−1220.
Buchholz, K., Duggal, S. K. & Borchert, A. (1979) Dechema Monogr. 84 169−181.
Cabral, J. M. S., Cardoso, J. P. & Novais, J. M. (1981) Enzyme Microb. Technol. 3 41−45.
Cabral, J. M. S., Kennedy, J. F. & Novais, J. M. (1982a) Enzyme Microb. Technol. 4 337−
    342.
Cabral, J. M. S., Kennedy, J. F. & Novais, J. M. (1982b) Enzyme Microb. Technol. 4 343−
    348.
Campbell, J. & Chang, T. M. S. (1975) Biochim. Biophys. Acta 397 101−109.
Campbell, J. & Chang, T. M. S. (1976) Biochem. Biophys. Res. Comm. 69 562−569.
Cantarella, M., Gianfreda, L., Palescondolo, R., Scardi, V., Greco, G., Alfani, F. & Iori, G.
    (1977) J. Solid-Phase Biochem. 2 163−174.
Cardoso, J. P., Chaplin, M. F., Emery, A. N., Kennedy, J. F. & Revel-Chion, L. P. (1978)
    J. Appl. Chem. Biotechnol. 28 775−785.
Carleysmith, S. W. & Lilly, M. D. (1979) Biotechnol. Bioeng. 21 1057−1073.
Carleysmith, S. W., Dunnill, P. & Lilly, M. D. (1980) Biotechnol Bioeng. 22 735−756.
Carr, P. W. & Bowers, L. D. (1980) Immobilized Enzymes in Analytical and Clinical Chem-
    istry, Wiley, New York.
Carrea, G., Colombi, F., Mazzola, G., Cremonesi, P. & Antonini, E. (1979) Biotechnol.
    Bioeng. 21 39−48.
Carvajal, N., Martinez, J., de Oca, F. M., Rodriguez, J. & Fernández, M. (1978) Biochim.
    Biophys. Acta 527 1−7.
Chang, H. N. & Reilly, P. J. (1978) Biotechnol. Bioeng. 20 243−253.
Chang, T. M. S. (1964) Science 146 524−525.
Chang, T. M. S. (1973) Enzyme 14 95−104.
Chaplin, M. F. & Kennedy, J. F. (1976) Carbohydr. Res. 50 267−274.
Charles, M., Coughlin, R. W. & Hasselberger, F. X. (1974), Biotechnol. Bioeng. 16 1553−
    1556.
Cheetham, N. W. H. & Richards, G. N. (1973) Carbohydr. Res. 30 99−107.
Cheetham, P. S. J. (1979) J. Appl. Biochem. 1 51−59.
Chen, A. K., Liu, C. C. & Schiller, J. G. (1979) Biotechnol. Bioeng. 21 1905−1915.
Cho, Y. K. & Bailey, J. E. (1978) Biotechnol. Bioeng. 20 1651−1665.
Cho, Y. K. & Bailey, J. E. (1979) Biotechnol. Bioeng. 21 461−476.
Corno, C., Galli, G., Morisi, F., Bettonte, M. & Stopponi, A. (1972) Stärke 24 420−424.
Coulet, P. R., Sternberg, R. & Thévenot, D. R. (1980) Biochim. Biphys. Acta 612 317−327.
Daka, N. J. & Laidler, K. J. (1980) Biochem. Biophys. Acta 612 305−316.
Dale, B. E. & White, D. H. (1979) Biotechnol. Bioeng. 21 1639−1648.
Danielson, N. D. & Siergiej, R. W. (1981) Biotechnol. Bioeng. 23 1913−1917.
Danielsson, B., Mattiasson, B., Karlsson, R. & Winqvist, F. (1979) Biotechnol. Bioeng.
    21 1749−1766.
Dayhoff, M. O. & Hunt, L. T. (1972). In Atlas of Protein Sequence and Structure Vol. 5,

National Biomedical Res. Fdn., Washington, D.C.
Dinelli, D. (1972) *Process Biochem.* **7(8)** 9–12.
Dinelli, D., Marconi, W., Cecere, F., Galli, G. & Morisi, F. (1978). In: *Enzyme Engineering,* Vol. 3, p. 477–481. Ed. by Pye, E. K. & Weetall, H. H., Plenum Press, New York.
Dixon, J., Andrew, P. & Butler, L. G. (1979) *Biotechnol. Bioeng.* **21** 2113–2123.
Drobnnik, J. Sundek, V., Švec, F., Kálal, J., Vojtišek, V. & Bárta, M. (1979) *Biotechnol. Bioeng.* **21** 1317–1332.
D'Souza, S. F. & Nadkarni, G. B. (1981) *Biotechnol. Bioeng.* **23** 431–436.
Emery, A. N. & Cardoso, J. P. (1978) *Biotechnol. Bioeng.* **20** 1903–1929.
Epton, R., McLaren, J. V. & Thomas, T. H. (1972) *Carbohydr. Res.* **22** 301–306.
Epton, R., Hobson, M. E. & Marr, G. (1979) *Enzyme Microb. Technol.* **1** 37–40.
Erekin, N. & Friedmann, M. E. (1979) *J. Solid-Phase Biochem.* **4** 123–130.
Finlay, T. H., Troll, V., Levy, M., Johnson, A. J. & Hodgkins, L. T. (1978) *Analyt. Biochem.* **87** 77–90.
Fischer, J., Ulbrich, R. & Schellenberg, A. (1978) *Acta Biol. Med. Ger.* **37** 1413–1424.
Flynn, A. & Johnson, D. B. (1977) *Int. J. Biochem.* **8** 243–247.
Flynn, A. & Johnson, D. B. (1978) *Biotechnol. Bioeng.* **20** 1445–1454.
Friedrich, O. H., Chun, M. & Sernetz, M. (1980) *Biotechnol. Bioeng.* **22** 157–175.
Fukui, S., Tanaka, A. & Gelff, G. (1978). In *Enzyme Engineering,* Vol. 4, p. 299–306. Ed. by Broun, G. B., Manecke, G. and Wingard, L. B. Jr., Plenum Press, New York.
Fukushima, S., Nagai, T., Fujita, K., Tanaka, A. & Fukui, S. (1978) *Biotechnol. Bioeng.* **20** 1465–1469.
Garbers, D. L. (1978) *J. Cyclic Nucleotide Res.* **4** 271–279.
Giovenco, S., Norisi, F. & Pansolli, P. (1973) *FEBS Lett.* **36** 57–60.
Glazer, A. N., Bar-Eli, E. & Katchalski, E. (1962) *J. Biol. Chem.* **237** 1832–1838.
Goldstein, L., Lifschitz, A. & Sokolovsky, M. (1971) *Int. J. Biochem.* **2** 440–456.
Gondo, S. & Koya, H. (1978) *Biotechnol. Bioeng.* **20** 2007–2010.
Gondo, S., Morishita, M. & Osaki, T. (1980) *Biotechnol. Bioeng.* **22** 1287–1291.
Goodman, R. B. & Peanasky, R. J. (1982) *Analyt. Biochem.* **120** 387–393.
Gouges, Y., Amen, J. & Sebesi, S. (1979). French Patent, 2, 420, 542.
Gray, C. J., Livingstone, C. M., Jones, C. M. & Barker, S. A. (1974) *Biochim. Biphys. Acta* **341** 457–464.
Greco, G., Albanesi, D., Cantarella, M. & Scardi, V. (1980) *Biotechnol. Bioeng.* **22** 215–219.
Gregoriadis, G., Leathwood, P. D. & Ryman, B. E. (1971) *FEBS Lett.* **14** 95–99.
Grunwald, P., Gunssen, W., Heiker, F. R. & Roy, W. (1979) *Analyt. Biochem.* **100** 54–57.
Guiraud, J. P., Demeulle, S. & Galzy, P. (1981) *Biotechnol. Lett.* **3** 683–688.
Halling, P. J. & Dunnill, P. (1979) *Biotechnol. Bioeng.* **21** 393–416.
Halling, P. J., Asenjo, J. A. & Dunnill, P. (1979) *Biotechnol. Bioeng.* **21** 2359–2363.
Halwachs, W., Wandrey, C. & Schügerl, K. (1978) *Biotechnol. Bioeng.* **20** 541–554.
Hanish, W. H., Rickard, P. A. D. & Nyo, S. (1978) *Biotechnol. Bioeng.* **20** 95–106.
Hatchikian, E. C. & Monsan, P. (1980) *Biochem. Biophys. Res. Comm.* **92** 1091–1096.
Hiraro, S. & Miura, O. (1979) *Biotechnol. Bioeng.* **21** 711–714.
Horiuti, Y. & Imamura, S. (1978) *J. Biochem.* **83** 1381–1385.
Huitron, C. & Limon-Lason, J. (1978) *Biotechnol. Bioeng.* **20** 1377–1391.
Iborra, J. L., Manjón, A. & Lozano, J. A. (1977) *J. Solid-Phase Biochem.* **2** 85–96.
Iborra, J. L., Manjón, A. Tari, M. & Lozano, J. A. (1979) *Gen. Pharmacol.* **10** 143–145.
Iyengar, L. & Rao, A. V. S. P. (1979) *Biotechnol. Bioeng.* **21** 1333–1343.
Jackson, J. A., Halvorson, H. R., Furlong, J. W., Lucast, K. D. & Shore, J. D. (1979) *J. Pharmacol. Exp. Ther.* **209** 271–274.
Janasik, V., Bartha, F., Krettschmer, K. & Lash, J. (1978) *J. Solid-Phase Biochem.* **3** 291–299.
Johnson, D. B. (1978) *Biotechnol. Bioeng.* **20** 1117–1123.
Johnson, D. B. & Coughlan, M. P. (1978) *Biotechnol. Bioeng.* **20** 1085–1095.
Kaboli, H. & Reilly, P. J. (1980) *Biotechnol. Bioeng.* **22** 1055–1069.
Kaetsu, I., Kumakura, M & Yoshida, M. (1979) *Biotechnol. Bioeng.* **21** 847–861.
Karube, I., Hirano, K. I. & Suzuki, S. (1977a) *J. Solid-Phase Biochem.* **2** 41–44.
Karube, I., Suzuki, S., Kusano, T. & Sato, I. (1977b) *J. Solid-Phase Biochem.* **2** 273–278.
Katchalski, E. & Sela, M. (1958), *Advances Protein Chem.* **13** 243–492.
Katchalski, E., Silman, I. & Goldman, R. (1971) *Advances Enzymol.* **34** 445–536.
Keller, E., Eberspächer, J. & Lingens, F. (1979) *Z. Physiol. Chem.* **360** 19–25.
Kennedy, J. F. (1978) In *Enzyme Engineering,* Vol. 4, p. 323–328. Ed. by Broun, G. B., Manecke, G. & Wingard, L. B. Jr. Plenum, New York.
Kennedy, J. F. (1979) *Chem. Soc. Rev.* **8** 221–257.
Kennedy, J. F. & Cabral, J. M. S. (1983), In *Applied Biochemistry and Bioengineering,*

Vol. 4, 189–280. Ed. by Wingard, L. B. & Chibata, I. Academic Press, New York.
Kennedy, J. F. & Chaplin, M. F. (1979) *Enzyme Microb. Technol.*, 1 197–200.
Kennedy, J. F. & Doyle, C. E. (1973) *Carbohydr. Res.* 28 89–92.
Kennedy, J. F. & Epton, J. (1973) *Carbohydr. Res.* 27 11–20.
Kennedy, J. F. & Humphreys, J. D. (1976) *Antimicrob. Agents Chemother.* 9 766–770.
Kennedy, J. F. & Kalogerakis, B. (1980) *Biochimie* 62 549–561.
Kennedy, J. F. & Kay, I. M. (1976) *J. Chem. Soc., Perkin Trans.* 1 329–335.
Kennedy, J. F. & Kay, I. M. (1977a) *Carbohydr. Res., 56* 211–218.
Kennedy, J. F. & Kay, I. M. (1977b) *Carbohydr. Res.* 59 553–561.
Kennedy, J. F. & Pike, V. W. (1978) *J. Chem. Soc. Perkin Trans.* 1 1058–1066.
Kennedy, J. F. & Pike V. W. (1979) *Enzyme Microb. Technol.* 1 31–36.
Kennedy, J. F. & Pike, V. W. (1980) *Enzyme Microb. Technol.* 2 288–294.
Kennedy, J. F. & Watts, P. M. (1974) *Carbohydr. Res.* 32 155–160.
Kennedy, J. F. & White, C. A. (1979) *Stärke* 31 375–381.
Kennedy, J. F. & Zamir, A. (1973) *Carbohydr. Res.* 29 497–501.
Kennedy, J. F. & Zamir, A. (1975) *Carbohydr. Res.* 41 227–233.
Kennedy, J. F., Barker, S. A. & Rosevear, A. (1973) *J. Chem. Soc. Perkin Trans.* 1 2293–2299.
Kennedy, J. F., Barker, S. A. & Zamir, A. (1974), *Antimicrob. Agents Chemother.* 6 777–782.
Kennedy, J. F., Barker, S. A. & Humphreys, J. D. (1976) *J. Chem. Soc. Perkin Trans.* 1 962–967.
Kennedy, J. F., Barker, S. A. & White, C. A. (1977a), *Carbohydr. Res.* 54 1–12.
Kennedy, J. F., Barker, S. A. & White, C. A. (1977b) *Stärke* 29 240–243.
Kennedy, J. F., Barker, S. A. & Pike, V. W. (1977c) *Biochim. Biophys. Acta* 484 115–126.
Kennedy, J. F., Pike, V. W. and Barker, S. A. (1980a), *Enzyme Microb. Technol.*, 2, 126–132.
Kennedy, J. F., Barker, S. A. & Kay, I. M. (1980b), *Carbohydr. Res.* 80 25–36.
Kennedy, J. F., Humphreys, J. D. & Barker, S. A. (1981) *Enzyme Microb. Technol.* 3 129–136.
Kitajema, M., Miyano, S. & Kondo, A. (1969) *Kogyo Kageku Zasshi* 72 493–499.
Kitano, H., Yoshijima, S. & Ise, N. (1980) *Biotechnol. Bioeng.* 22 2643–2653.
Kjellén, K. G. & Neujahr, H. Y. (1979) *Biotechnol. Bioeng.* 21 715–719.
Kjellén, K. G. & Neujahr, H. Y. (1980), *Biotechnol. Bioeng.* 22 299–310.
Klei, H. E., Sundstrom, D. W. & Gargano, R. (1978) *Biotechnol. Bioeng.* 20 611–617.
Klemes, J. & Citri, N. (1979) *Biotechnol. Bioeng.* 21 897–905.
Klinov, S. V., Sugrabova, N. P. & Kurganov, B. T. (1979) *Molek. Biol.* 13 559–566.
Klyosov, A. A. & Gerasimas, V. B. (1979) *Biochim. Biophys. Acta* 571 162–165.
Kobayashi, T., Kato, I., Ohmiya, K. & Shimizu, S. (1980) *Agric. Biol. Chem.* 44 413–418.
Konecny, J. & Sieber, M. (1980) *Biotechnol. Bioeng.* 22 2013–2029.
Kozlova, N. B., Roze, L. V. & Vul'fson, P. L. (1978) *Biochemistry (USSR)* 43 403–411.
Krishnaswamy, S. & Kittrel, J. R. (1978) *Biotechnol. Bioeng.* 20 821–835.
Kuan, K. N., Lee, Y. Y. & Melius, P. (1980) *Biotechnol. Bioeng.* 22 1725–1734.
Kucera, J. & Kuminkova, M. (1980) *Collect. Czech. Comm.* 45 298–306.
Kumakura, M., Yoshida, M., Asano, M. & Kaetsu, I. (1977) *J. Solid-Phase Biochem.* 2 279–288.
Lai, T.-S. & Cheng, P.-S. (1978) *Biotechnol. Bioeng.* 20 773–779.
Lee, C. Y. (1978) *J. Solid-Phase Biochem.* 3 71–83.
Lee, D. D., Lee, G. K., Reilly, P. J. & Lee, Y. Y. (1980) *Biotechnol. Bioeng.* 22 1–17.
Legoy, M. D., Garde, V. L., Le Moullec, J. M., Ergan, F. & Thomas, D. (1980) *Biochemie* 62 341–345.
Leuba, J. L. & Widmer, F. (1979) *J. Solid-Phase Biochem.* 2 257–271.
Lilly, M. D., Kay, G., Sharp, A. K. & Wilson, R. J. H. (1968) *Biochem. J.* 107 5P.
Lipatova, T. E., Konoplitskaya, O. L., Chupina, L. N. & Vasyl'chenko, D. V. (1979) *Ukr. Biochem. Zh.* 51 319–323.
Ludolph, R. A., Vieth, W. R., Venkatsubramanian, K. & Constantinides, A. (1979) *J. Molec. Cat.* 5 197–223.
Maeda, H. & Suzuki, H. (1972) *Agric. Biol. Chem.* 36 1581–1593.
Maeda, H., Tsao, G. T. & Chen, L. T. (1978) *Biotechnol. Bioeng.* 20 383–402.
Maeda, H., Chen., L. F. & Tsao, G. T. (1979) *J. Ferment. Technol.* 57 238–243.
Manecke, G., Pohl, R., Schluensen, J. & Vogt, H. G. (1978). In: *Enzyme Engineering,* Vol. 4, p 409–412. Ed. by Broun, G., Manecke, G. & Wingard, L. B. Jr., Plenum Press, New York.
Marconi, W., Cecere, F., Morisi, F., Della Penna, G. & Rappuolli, B. (1973) *J. Antibiot.*

26 226–232.
Marconi, W., Gulinelli, S. & Morisi, F. (1974a) *Biotechnol. Bioeng.* **16** 501–511.
Marconi, W., Bartoli, F., Cecere, F. & Morisi, F. *(1974b) Agric. Biol. Chem.* **38** 1393–1399.
Marconi, W., Morisi, F. & Mosti, R. (1975) *Agric. Biol. Chem.* **39** 1323–1324.
Marfey, P. S. & King, M. V. (1965) *Biochim. Biophys. Acta* **105** 178–183.
Marshall, J. J. & Rabinowitz, M. L. (1976) *J. Biol. Chem.* **251** 1081–1087.
Martinek, K., Klibanov, A. M., Goldmacher, V. S. & Berezin, I. V. (1977) *Biochim. Biophys. Acta* **485** 1–12.
Martinek, K., Mozhaev, V. V. & Berezin, I. V. (1980) *Biochim. Biophys. Acta* **615** 426–435.
Matsumoto, K., Seijo, H., Karube, I. & Suzuki, S. (1980) *Biotechnol. Bioeng.* **22** 1071–1086.
Mattiasson, B. & Borrebaeck, C. (1978) *FEBS Lett.* **85** 119–123.
Mattiasson, B., Danielsson, B., Hermansson, C. & Mosbach, K. (1978) *FEBS Lett.* **85** 203–206.
Means, G. & Feeney, R. E. (1971) *Chemical Modification of Proteins,* Vol. 1, Holden Day, San Francisco.
Melander, W. & Horvath, C. (1978). In: *Enzyme Engineering,* Vol. 4, p 355–363. Ed. by Broun, G. B., Manecke, G. & Wingard, L. B. Jr., Plenum Press, New York.
Merrill, A. H. Jr. & McCormick, D. B. (1979) *Biotechnol. Bioeng.* **21** 1629–1638.
Messing, R. A. (1974) *Process Biochem.* **9 (II)** 26–28.
Messing, R. A. (1975). In *Immobilized Enzymes for Industrial Reactors* p. 63–78. Ed. by Messing, R. A. Academic Press, New York.
Mikelsone, Z., Mitrofanova, A. N., Poltorak, O. M. & Arens, A. (1979) *Vestn. Mosk. Univ. Ser. 2* **20** 109–113.
Mitz, M. A. (1956) *Science* **123** 1076–1077.
Miura, Y., Miyamoto, K., Urabe, H., Tanaka, H. & Yasuda, T. (1979) *J. Ferment. Technol.* **57** 440–444.
Mogensen, A. D. & Vieth, W. R. (1973) *Biotechnol. Bioeng.* **15** 467–482.
Mohan, R. R. & Li, N. N. (1974) *Biotechnol. Bioeng.* **16** 513–523.
Monsan, P. (1978) *Eur. J. Appl. Microb. Biotechnol.* **5** 1–11.
Mori, T., Tosa, T. & Chibata, T (1973) *Biochim. Biophys. Acta* **321** 653–661.
Mori, T., Sano. R., Iwasawa, Y., Tosa, T. & Chibata, T. (1976) *J. Solid-Phase Biochem.* **1** 15–26.
Morikawa, Y., Karube, I., Suzuki, S., Nakano, Y. & Taguchi, T. (1978a) *Biotechnol. Bioeng.* **20** 1143–1152.
Morikawa, Y., Karube, I. & Suzuki, S. (1978b) *Biochim. Biophys. Acta* **523** 263–267.
Morisi, F., Pastone, M. & Viglia, A. (1973) *J. Dairy Sci.* **56** 1123–1127.
Morriyama, S., Kataoka, S., Nakanishi, K., Matsuno, R. & Kamikubo, T. (1980) *Agric. Biol. Chem.* **44** 2737–2739.
Mosbach, K. & Danielsson, B. (1974) *Biochim. Biophys. Acta* **364** 140–145.
Mozhaev, V. V., Martinek, K. & Berezin, I. V. (1979) *Mol. Biol.* **13** 73–80.
Mukherjea, R. N., Bhattacharya, P., Gangopadhyary, T. & Ghosh, B. K. (1980) *Biotechnol. Bioeng.* **22** 543–553.
Mukherjee, A. & Srere, P. A. (1978) *J. Solid-Phase Biochem.* **3** 85–94.
Munnecke, D. M. (1979) *Biotechnol. Bioeng.* **21** 2247–2261.
Negoro, H. (1972) *J. Ferment. Technol.* **50** 136–142.
Nelson, J. M. & Griffin, E. G. (1916) *J. Amer. Chem. Soc.* **38** 1109–1115.
Neujahr, H. Y. (1980) *Biotechnol. Bioeng.* **22** 913–918.
Ngo, T. T. & Laidler, K. J. (1978) *Biochim. Biophys. Acta* **525** 93–102.
Ogumtimein, G. B. & Reilly, P. J. (1980) *Biotechnol. Bioeng.* **22** 1127–1142.
Ohba, R. & Ueda, S. (1980) *Biotechnol. Bioeng.* **22** 2137–2154.
Ohba, R., Chaen, H., Hayashi, S. & Ueda, S. (1978) *Biotechnol. Bioeng.* **20** 665–676.
Olson, A. C. & Stanley, W. L. (1979). In: *Immobilized Enzymes in Food and Microbial Processes,* p. 51–62. Ed. by Olson, A. C. & Cooney, C. L., Plenum Press, New York.
Ono, M. Tosa, T. & Chibata, I. (1978) *Agric. Biol. Chem.* **42** 1847–1853.
Ooshima, H., Sakimoto, M. & Harano, Y. (1980) *Biotechnol. Bioeng.* **22** 2155–2167.
Ostergaard, J. C. W. & Martiny, S. C. (1973) *Biotechnol. Bioeng.* **15** 561–564.
Paine, M. A. & Carbonell, R. G. (1975) *Biotechnol. Bioeng.* **17** 617–619.
Pardin, M., Bello, C. D., Marani, A., Bartoli, F. & Morisi, F. (1977) *J. Solid-Phase Biochem.* **2** 251–255.
Parkin, K. & Hultrin, H. O. (1979) *Biotechnol. Bioeng.* **21** 939–953.
Patton, J. S., Albertson, P.-A., Erlanson, C. & Borgstrom, B. (1978) *J. Biol. Chem.* **253** 4195–4202.

Paul, F., Coulet, P. R., Gautheron, D. G. & Engasser, J.-M. (1978) *Biotechnol. Bioeng.* **20** 1785–1796.
Pavanakrishran, R. & Bose, S. M. (1980) *Biotechnol. Bioeng.* **22** 919–928.
Pifferi, P. G., Pasquali, C., Tocco, M. G. & Domini Pellerano, I. M. (1980) *Technol. Aliment* **3** 9–14.
Pittner, F., Miron, T., Pittner, G. & Wilchek, M. (1980) *J. Amer. Chem. Soc.* **102** 2451–2452.
Quiocho, F. A. & Richards, F. M. (1964) *Proc. Nat. Acad. Sci. USA* **52** 833–839.
Quiocho, F. A. & Richards, F. M. (1966) *Biochemistry* **5** 4062–4076.
Remy, M.-H., David, A. & Thomas, D. (1978) *FEBS Lett.* **88** 332–336.
Rexova-Benkova, L., Mrackova, M. & Babor, K. (1980) *Collect. Czec. Chem. Comm.* **45** 163–168.
Rodinov, Y. V., Avilova, T. V. & Popov, V. O. (1977) *Biochemistry (USSR)* **42** 1594–1598.
Roger, L. Thapon, J. L., Maubois, J. C. & Brule, G. (1976) *Le Lait* **551** 56–75.
Rokugawa, K., Fujishima, T., Kuminaka, A. & Yoshino, H. (1979) *J. Ferment. Technol.* **57** 570–573.
Royer, G. P. & Uy, R. (1973) *J. Biol. Chem.* **248** 2627–2629.
Sada, E., Katoh, S. & Terashima, M. (1980) *Biotechnol. Bioeng.* **22** 243–246.
Sada, E., Katoh, S. & Terashima, M. (1981) *Biotechnol. Bioeng.* **23** 1037–1044.
Sato, T., Mori, T., Tosa, T. & Chibata, I. (1972) *Arch. Biochem. Biophys.* **147** 788–796.
Sato, I., Karube, I. & Suzuki, S. (1977) *J. Solid-Phase Biochem.* **2** 1–7.
Schiger, H., Teufel, E. H. & Philipp, M. (1980) *Biotechnol. Bioeng.* **22** 55–64.
Schreiner, H. R. (1966), U.S. Patent, 3, 282, 702.
Shemer, L., Granot, R., Freeman, A., Sokolovsky, M. & Goldstein, L. (1979) *Biotechnol. Bioeng.* **21** 1607–1627.
Shimizu, S. Y. & Lenhoff, H. M. (1979) *J. Solid-Phase Biochem.* **4** 75–94.
Silman, R. W., Black, L. T., McGhee, J. E. & Bagley, E. B. (1980) *Biotechnol. Bioeng.* **22** 533–541.
Snamprogetti, S. P. A. (1976), U.S. Patent, 3, 969, 990.
Sodini, G., Baroncelli, V., Canella, M. & Renzi, P. (1974) *Italian J. Biochem.* **23** 121–135.
Solomon, B. & Levin, Y. (1974) *Biotechnol. Bioeng.* **16** 1161–1177.
Spettoli, P., Botlacin, A. & Zamorani, A. (1980) *Technol. Aliment.* **3** 31–34.
Srere, P. A. & Uyeda, K. (1976). In: *Methods in Enzymology,* Vol. 44, p. 11–19. Ed. by Mosbach, K., Academic Press, New York.
Stambolieva, N. & Turkova, J. (1980) *Collect. Czech. Chem. Comm.* **45** 1137–1143.
Stanley, W. L., Watters, G. G., Kelly, S. H. & Olson, A. C. (1978) *Biotechnol. Bioeng.* **20** 135–140.
Sud'ina, E. G. & Golod, M. G. (1979) *Ukr. Biokhim. Zh.* **51** 400–403.
Sud'ina, E. G., Smartsev, M. A., Golod, M. G. & Dovbysh, E. V. (1979) *Ukr. Biokhim. Zh.* **51** 404–408.
Sundaram, P. V. (1979) *Biochem. J.* **179** 445–447.
Sundaram, P. V. & Joy, K. (1978) *J. Solid-Phase Biochem.* **3** 223–240.
Sundaram, P. V., Igloi, M. P., Wassermann, R. & Hinsch, W. (1978) *Clin. Chem.* **24** 1813–1817.
Sundaram, P. V., Blumenberg, B. & Himsh, W. (1979) *Clin. Chem.* **25** 1436–1439.
Švec, F., Kálal, J., Menyailova, I. I. & Nakhapteyan, L. A. (1978) *Biotechnol. Bioeng.* **20** 1319–1328.
Szewczuk, A., Ziomek, E., Mordarski, M., Siewiński, M. & Wieczorek, J. (1979) *Biotechnol. Bioeng.* **21** 1543–1552.
Szwajcer, E., Szewzuk, A. & Mordarski, M. (1981) *Biotechnol. Bioeng.* **23** 1675–1681.
Tashiro, Y. & Matsuda, T. (1978) Japan, Pat. Kokai 78–32191.
Torchilin, V. P., Galka, M. & Ostraowski, W. (1977a) *Biochim. Biophys. Acta* **483** 331–336.
Torchilin, V. P., Tischenko, E. G. & Smirnov, V. N. (1977b) *J. Solid-Phase Biochem.* **2** 19–29.
Tosa, T., Mori, T., Fuse, N. & Chibata, I. (1966) *Enzymologia* **31** 214–225.
Tosa, T., Sato, T., Mori, T., Yamamoto, K., Takata, I., Nishida, Y. & Chibata, I. (1979) *Biotechnol. Bioeng.* **21** 1697–1709.
Tramper, J., Miller, F. Henk, C. & van der Plas, H. C. (1978) *Biotechnol. Bioeng.* **20** 1507–1522.
Tramper, J., Angelino, S. A. G. F., Müller, F. & van der Plas, H. C. (1979) *Biotechnol. Bioeng.* **21** 1767–1786.
Ugarova, N. N., Brovko, L. Y., Rozhkova, G. D. & Berezin, I. V. (1977) *Biochemistry (USSR),* **42** 943–949.
Updike, S. J. & Hicks, G. P. (1967) *Nature* **214** 986–988.

Vallin, D. & Tran-Minh, C. (1979) *Biochim. Biophys. Acta* **571** 321–332.
Van Etten, R. L. & Saini, M. S. (1977) *Biochim. Biophys. Acta* **484** 487–492.
Vegarud, G. & Christensen, T. B. (1977) *Biotechnol. Bioeng.* **17** 1391–1397.
Voivodov, K. I., Galunski, B. R. & Dyankov, S. S. (1979) *J. Appl. Biochem.* **1** 442–445.
Volkova, A. N., Ivanova, L. V., Matveenko, A. P., Yakovlev, V. I. & Kol'tsov, S. I. (1979) *Khim. Khim. Tekhnol.* **22** 844–847.
Wadiak, D. T. & Carbonell, R. G. (1975) *Biotechnol. Bioeng.* **17** 1157–1181.
Wandrey, C., Flaschel, E. & Schügerl, K. (1979) *Biotechnol. Bioeng.* **21** 1649–1670.
Wassermann, B. D., Hultrin, H. O. & Jacobson, B. S. (1980) *Biotechnol. Bioeng.* **22** 271–287.
Watanabe, T., Mori, T., Tosa, T. & Chibata, I. (1979) *Biotechnol. Bioeng.* **21** 477–486.
Weetall, H. H. (1970) *Biochim. Biophys. Acta* **212** 1–7.
Wingard, L. B. Jr., Katchalski-Katzir, E. & Goldstein, L. (1976) *Applied Biochemistey and Bioengineering,* Vol. 1, Academic Press, New York.
Wiseman, A. (1978) *Topics in Enzyme and Fermentation Biotechnology* Vol. 2, Ellis Horwood, Chichester.
Woodward, J. & Wiseman, A. (1978) *Biochim. Biophys. Acta* **527** 8–16.
Wykes, I. R., Dunnill, P. & Lilly, M. D. (1971) *Biochim. Biophys. Acta* **250** 522–529.
Yamada, H. (1978) *J. Biochem. (Tokyo)* **83** 1577–1581.
Yarovaya, G. A., Gulynskaya, T. N., Dotsenko, V. L., Bessermertmaya, L. Ya., Kozlov, L. V. & Antonov, V. K. (1975) *Biorg. Khim.* **1** 646–651.
Yeung, K.-K., Owen, A. J. & Dain, J. A. (1979) *Carbohydr. Res.* **75** 295–304.
Zaborsky, O. R. (1973) *Immobilized Enzymes,* CRC Press, Cleveland.
Zaffaroni, P., Oddo, N., Olivieri, R. & Formiconi, L. (1975) *Agric. Biol. Chem* **39** 1875–1877.
Zingaro, R. A. & Uziel, M. (1970) *Biochim. Biophys. Acta* **213** 371–379.

CHAPTER 5

Enzymes in clinical analysis – principles

Dr. B. J. GOULD, Department of Biochemistry, University of Surrey, Guildford, England and **Dr. B. F. ROCKS,** Department of Pathology, The Royal Sussex County Hospital, Brighton, England

5.1 INTRODUCTION

The topic of this chapter was mentioned only briefly in two chapters of the first edition of *Handbook of Enzyme Biotechnology* (Gould 1975, Barker & Kay 1975). The most frequent use of purified enzymes in clinical analysis is in the measurement of the concentration of substrates (Section 5.2) in plasma or serum. The main development, since the last edition, has been the steady growth of kinetic assays alongside equilibrium assays. Many enzymes are difficult to assay, but in certain cases the addition of other enzymes simplifies their measurement. These analyses (Section 5.3) require the addition of relatively large amounts of pure enzymes. The use of purified enzymes for large numbers of analyses can be expensive, and this is one of the principal reasons for the increasing use of re-usable immobilized enzymes in clinical analyses (Section 5.4). The other important application of enzymes in clinical analysis is in enzyme immunoassay (Section 5.5). There has been a rapid growth in the number of commercially available assays of this type in the last five years. Originally, their main use was in the determination of the blood or urine concentrations of a variety of drugs, but it is now realized that they have many other potential uses.

The rationale for using enzymes in clinical analysis comes from their two best known properties, catalytic efficiency and specificity. As they are extremely good catalysts, the reaction should proceed rapidly even under mild conditions of temperature and pH, and they should not affect the equilibrium position of the reaction. Their specificity ensures that the rate of a particular reaction, or of a limited range of related reactions, is enhanced. This enables the rapid analysis of particular compounds that occur in complex media, such as serum, plasma, and urine, to be performed without prior purification steps. In the case of enzyme immunoassays the essential feature of specificity comes from the specific binding between antigen and antibody, and the enzyme provides an easily detected label which generally amplifies the signal.

The progress curve for a typical enzyme catalysed reaction is shown in Fig. 5.1. This indicates the two parts of the curve from which data are obtained. The *equilibrium method*, which can only be used for the determination of substrates, makes use of data collected when the concentrations of product or substrate are time-independent. Those methods where the data are obtained when the concentration is time-dependent are *kinetic methods*, in which case the data are generally collected from the early linear part of the curve, which is called the initial velocity (rate).

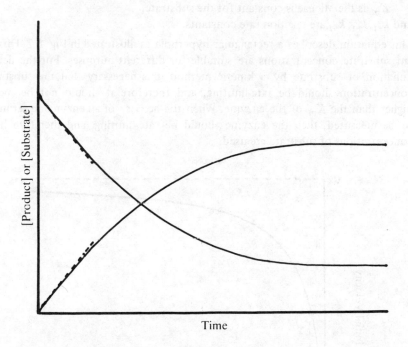

Fig. 5.1 – Progress curves for a typical enzyme catalysed reaction. The reaction can be followed by the loss of substrate or the formation of product. The initial velocity is determined from the tangent(s) to the curves (-----) at zero time. The progress curves have a steady value when the reaction reaches equilibrium.

The simplest enzyme catalysed reaction involves one substrate and one product.

$$E \; + \; S \; \underset{k_{-1}}{\overset{k_{+1}}{\rightleftharpoons}} \; ES \; \overset{k_{+2}}{\longrightarrow} \; E \; + \; P$$

enzyme   substrate   enzyme substrate   enzyme   product

intermediate

The way in which the rate of this reaction is dependent on the substrate concentration is usefully described by the Michaelis–Menten equation:

$$v = \frac{V\,[S]}{K_{\mathrm{m}} + [S]} \tag{5.1}$$

where

v is the velocity (rate) of the reaction,

$V = (k_{+2}\,[E_{\mathrm{o}}])$ is the maximal velocity (rate) of the reaction,

$[S]$ and $[E_{\mathrm{o}}]$ are initial substrate and enzyme concentrations,

K_{m} is the Michaelis constant for the substrate,

and k_{+1}, k_{-1}, k_{+2} are reaction rate constants.

This equation describes a rectangular hyperbola as illustrated in Fig. 5.2. Different substrate concentrations are suitable for different purposes. For the determination of substrate by a kinetic method it is necessary that the substrate concentration should be rate-limiting, and therefore it should not be much higher than the K_{m} of the enzyme. When the activity of an enzyme in serum is to be measured, then the enzyme should be rate-limiting, and therefore high concentrations of substrate are used.

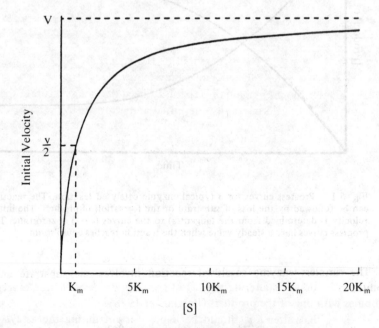

Fig. 5.2 – Theoretical Michaelis plot.

The rate of enzyme catalysed reactions are very dependent on pH and temperature, and it is essential that these two factors are rigorously controlled in all kinetic methods. Many compounds can either reduce, *inhibitors,* or increase, *activators,* the rate of an enzyme catalysed reaction. The presence of these compounds will affect the accuracy of kinetic methods.

5.2 MEASUREMENT OF SUBSTRATE CONCENTRATION WITH ENZYMES

Many substrates, or metabolites, can be measured with the aid of enzymes. Some coenzymes, such as adenosine triphosphate (ATP) and nicontinamide adenine dinucleotide in its oxidized (NAD^+) or reduced from ($NADH + H^+$), can be considered as substrates in this context. The substrates most frequently measured are urea and glucose, but several others are determined in clinical laboratories. For a comprehensive guide to these analyses the four-volume work by Bergmeyer (1974) should be consulted or its ten-volume successor which started to be published in 1983.

5.2.1 Principles of equilibrium methods

These assays are simple to perform, but we need to consider the position of equilibrium and the time needed to reach this point. Consider the equation

$$S + C \xrightleftharpoons[K_{eq}]{enzyme} P + Q$$

where S is the substrate to be measured, C is a coenzyme or coreactant, and P and Q are the products of the enzyme catalysed reactions.

The greater the proportion of S that is utilized, the more accurate the assay. This proportion is determined by the initial amount of $C(C_o)$ relative to $S(S_o)$ and the equilibrium constant (K_{eq}) for the reaction. Table 5.1 shows the calculated ratios of $C_o:S_o$ required to give 99%, 99.5%, and 99.9% conversion. As the equilibrium constant increases, so the relative excess of coreactant needed is reduced. The very high ratios required at low K_{eq} are likely to pose practical problems, for example insolubility of coreactant, possible coreactant inhibition of enzyme, or high initial background absorbance. If the reaction does not approach completion a calibration curve is required covering the anticipated range of substrate concentrations. This also overcomes the nonlinearity of calibration curves that is most pronounced with low K_{eq} (Carr & Bowers 1980). Another solution to this problem is to change the equilibrium position. This can sometimes be done by altering the pH of the reaction or by using a reagent that combines with one of the products (Bergmeyer 1978).

Factors that alter K_{eq} should be kept constant during the assays; this is particularly important with a low K_{eq}. The K_{eq} should not be affected by the presence of enzyme, so long as the molar concentration of enzyme is very low compared to S_o. K_{eq} is affected by temperature and frequently also by pH. Therefore these two parameters should be controlled during the assay.

Table 5.1
Calculated ratio of initial [Coreactant] : Initial [Substrate] required for stated percentage conversion.

| | Values of $C_o:S_o$ | | | |
| | K_{eq} | | | |
Conversion	1	5	10	50
99%	100	20	10	3
99.5%	200	40	20	5
99.9%	1000	200	100	20

The time needed for an anzyme catalysed reaction to reach equilibrium can be calculated from the integrated form of the Michaelis–Menten equation.

$$t = \frac{2.3\, K_m}{V} \log_{10} \frac{S_o}{S_t} + \frac{(S_o - S_t)}{V} \qquad (5.2)$$

where,

V and K_m have the meanings given above,

t is the time of reaction,

S_o is the initial substrate concentration,

S_t is the substrate concentration after time t.

and $(S_o - S_t)$ is the product concentration after time t.

If we accept 99% conversion of substrate to product as adequate for estimation, $\log_{10} \dfrac{S_o}{S_t} \cong 2$ and S_t can be ignored relative S_o. Then equation 5.2 becomes:

$$t = 2.3 \times 2 \frac{K_m}{V} + \frac{S_o}{V}.$$

If $S_o \gg K_m$, then the maximum value of S_o/V becomes K_m/V.

$$t \cong 5.6 \frac{K_m}{V}.$$

For the reaction to be completed within 5 min, K_m/V should be slightly greater than 1 min ml^{-1}. This is achieved if $K_m = 10^{-3}$ mol l^{-1} (which is 1 μmol ml^{-1}) and V is 1 U of enzyme ml^{-1}. The international unit, U, is the amount of enzyme required to convert 1 μmol of substrate to product per minute. Then

$$\frac{K_m}{V} = \frac{1\ \mu\text{mol ml}^{-1}}{1\ \mu\text{mol min}^{-1}}$$

$$= 1\ \text{min mol}^{-1}.$$

This calculation allows the amount of enzyme required per ml of reaction mixture to be calculated when K_m and V, which is proportional to E_o, are known.

Certain factors, which may alter the time required for the assay, are ignored in this calculation. If the product formed in the reaction causes product inhibition or a reverse reaction with a significant rate, V would be reduced. However, these possibilities should usually be unimportant if $S_o \gg K_m$ as assumed above. If $S_o > K_m$ the time will be increased considerably as will the potential problems caused by product formation. The presence of activators in the biological specimen will reduce K_m/V and hence the time required for completion. Alternatively, the presence of inhibitors will increase K_m/V, and the time required to reach equilibrium will be increased.

Typical progress curves are represented in Fig. 5.1. However, it is possible in practice to obtain curves where there is a linear drift after the rapid consumption of substrate. This is shown in Fig. 5.3(a). The most likely cause is an impure nonspecific enzyme which is continuing to catalyse a slow reaction. If there is a 'linear drift' a correction can be made by extrapolation to the zero time. The problem is avoided if pure, specific enzymes are use. Fig. 5.3(b) shows a similar curve except that there is a reaction before the addition of enzyme. The cause is generally the presence of endogeneous enzymes and coreactants. This can usually be overcome by use of a long (for example, 20 min) preincubation before the addition of enzymes. An alternative curve is shown in Fig. 5.3(c) where there is a steady side reaction, such as the oxidation of reduced NAD^+ by dissolved oxygen in the presence of the respiratory chain enzymes. In these cases an extrapolation to zero time is necessary unless the reaction can be prevented. All

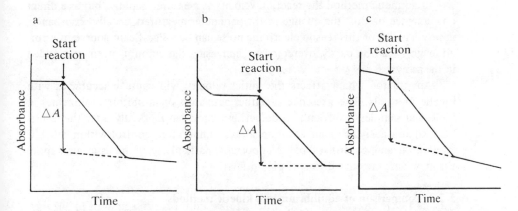

Fig. 5.3 – Atypical progress curves and methods for correction. (a) Shows a slow reaction continuing after the main reaction is completed. This can be corrected for by extrapolation to the time when the reaction was started by the addition of specific enzyme (↓). (b) Preincubation is often necessary to allow endogenous reactions to go to completion before starting the reaction of interest. (c) A continuous side reaction can be corrected for by extrapolation to the time when the reaction was started.

the situations shown in Fig. 5.3 can be overcome, but they are dependent on data collected at several time points, which increases the length of operator time required per assay.

5.2.2 Principles of kinetic methods
The still increasing workload of clinical biochemistry laboratories in the early 1960s was partly solved by the introduction of continuous flow analysers. Part of this increase in the number of analyses involved the determination of the activity of several enzymes. Single time point assays were originally used, but as the limitations of this type of assay were appreciated it became normal to follow the linear early part of the progress curve to determine the initial velocity (Fig. 5.1). This was a slow process which has been overcome since the early 1970s by the introduction of a variety of discrete kinetic analysers (Purdue 1977). These machines follow the progress of the reaction, generally at a number of fixed time intervals, check the linearity of the curve, and print out the initial velocity. The availability of free time on these sophisticated kinetic analysers encouraged the development of kinetic analyses for substrates.

The Michaelis plot (Fig. 5.2) indicates that the initial velocity is a nonlinear function of substrate concentration. The sensitivity to increasing substrate concentrations decreases markedly when $S_o > K_m$, so that $S_o = K_m$ is about the upper limit for the use of kinetic methods. A calibration curve over a range of substrate concentrations is essential to account for the nonlinearity, although if substrate concentrations below $K_m/20$ are used the deviation from linearity is theoretically $< 3.3\%$.

In a kinetic method the reaction velocity is measured rapidly; but as a direct consequence of this, the change in the parameter measured, usually absorbance, is very small. For this reason electronic noise can be a significant source of error, although it can be partly overcome by increasing the amount of enzyme added in the assay.

Any factor which affects the initial velocity will cause inaccuracies with kinetic methods. The presence of either activators or inhibitors of enzymes in biological samples is difficult to detect, but they can markedly alter the results of a kinetic assay. For an accurate assay, temperature control within $\pm 0.1°C$ is essential and control of pH is important, particularly if the assay uses more than one enzyme with different pH optima.

5.2.3 Comparison of equilibrium and kinetic methods
The characteristics of equilibrium and kinetic methods are presented in Table 5.2. Equilibrium methods are generally to be preferred, but they have the major drawbacks of relatively high cost of enzyme and cofactors, and the length of time required per assay. These are the very points where the sophisticated kinetic analysers have advantages. However, kinetic methods are prone to large errors if enzyme activators or inhibitors are present, but undetected, in the bio-

Table 5.2

Comparison of equilibrium and kinetic methods for the assay of substrates

	Equilibrium method		Kinetic method	
	Rating	Comments	Rating	Comments
Change followed, e.g. A	+++	Relatively large and accurate measurement. More sensitive	++	Very small but within ability of sophisticated kinetic analysers. Electronic noise can be a problem
Change in pH	++	May alter K_{eq} but unlikely to cause error unless K_{eq} is small	–	Any change from optimum pH may cause low values – especially if more than one enzyme used in assay
Change in T	+++	Will alter K_{eq} but no serious error unless K_{eq} is small and T change is several degrees	+	Good temperature control ($\pm 0.1\%$) is essential
Presence of inhibitors	++	Lengthen assay time check reaction terminated	–	Low values
Presence of activators	+++	Equilibrium reached more quickly	–	High values
Influence of K_m	++	Low K_m reduces time required for assay	++	The higher the K_m the higher S_o that can be assayed
Maximum S_o	++	$S_o < K_m$	++	$S_o < K_m$
Linearity with S_o	++	Linear but may get deviation at low K_{eq}		Nonlinear, therefore calibration curve needed
K_{eq}	+	Problems with low K_{eq} of solubility, inhibition or background absobance	+	With low K_{eq} assay may be too insensitive
Time required	+	Several minutes	+++	Generally <1 minute
Cost of enzyme	–	Relatively high to complete assay in reasonable time	++	Relatively low. Require sufficient to give adequate signal to noise ratio
Cost of cofactors	–	Can be higher than cost of enzyme especially if K_{eq} is low	++	Relatively low. Concentration must not be rate-limiting
Product formation	++	May cause inhibition or reversal of reaction	+++	Too little to cause effect
Unstable product	+	Cannot measure product	++	No problem if assay is rapid
Turbid or coloured sample	+	Blank required	+++	Should not affect reaction rate
Equipment	+++	Sophisticated equipment not required but dedicated equipment being produced	+	Requires sophisticated equipment but this may be used for other analyses

logical material. The value of K_m, relative to the normal concentration if substrate in the assay, may be important in deciding which method of assay is most suitable.

Recently, attempts have been made to extend the useable range of kinetic methods to include substrate concentrations that greatly exceed K_m. One approach is to include competitive inhibitors in the assay since they cause an apparent increase in K_m (Sampson & Baird 1979). An alternative approach has more general applicability. Hamilton & Pardue (1982) have used nonlinear regression analysis to fit data for absorbance (A) and rate (dA/dt) versus time to the Michaelis–Menten equation. This gave linear calibration plots for concentrations from well below K_m to $3.5\,K_m$.

5.2.4 Common indicator species used in routine clinical analysis

Only a limited number of compounds are frequently measured in routine clinical analyses. These indicator species and the properties which make them suitable for sensitive, specific, and yet diverse clinical analyses are also considered below. For a more comprehensive discussion see Carr & Bowers (1980).

5.2.4.1 *Nicotinamide adenine dinucleotides*

Nicotinamide adenine dinucleotide occurs in the oxidized (NAD^+) and reduced ($NADH$) forms, Nicotinamide adenine dinucleotide phosphate also has two forms, $NADP^+$ and $NADPH$. Fig. 5.4 shows their structure and interconversion. NADH absorbs light very strongly at 339 nm ($\epsilon = 6.31 \times 10^3$ litre mol^{-1} cm^{-1}), but the oxidized form does not. $NADP^+$ and $NADPH$ have very similar properties to NAD^+ and $NADH$. Table 5.3 shows the molar absorbance of NADH at particular wavelengths, and indicates how the value changes slightly with alteration in pH, temperature, and ionic strength. The wavelength of maximum absorbance also changes slightly with temperature. The values actually obtained are of course dependent on the type of spectrophotometer used, and especially on the bandwidth of the monochromator.

A simple reaction involving these compounds is the determination of pyruvate using lactate dehydrogenase:

$$\text{Pyruvate} + \text{NADH} + \text{H}^+ \underset{\text{dehydrogenase}}{\overset{\text{Lactate}}{\rightleftharpoons}} \text{L-lactate} + \text{NAD}^+ .$$

The loss of NADH is followed, generally by measuring the absorbance at about 340 nm. Since $K_{eq} = 2 \times 10^4$ at pH 7.0, the reaction will go to completion so long as NADH > pyruvate. The total change in absorbance can then be related to the original amount of pyruvate by use of the Bouguer–Lambert–Beer law, ($A = \epsilon \times c \times l$), where A = absorbance, c = concentration (mol l^{-1}), l = pathlength (cm), and ϵ = molar absorption coefficient (l mol^{-1} cm^{-1}).

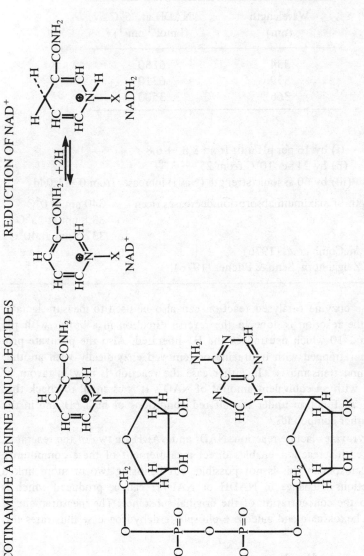

Fig. 5.4 – Structure of NAD(P)$^+$ and the oxidation–reduction reaction.

Table 5.3
Molar extinction of NADH

Wavelength (nm)	NADH at 25°C ($l\,mol^{-1}\,cm^{-1}$)
334	6180
339	6310
366	3500

decreases

    (i) by 16 per pH unit from 8.8. → 6.8
    (ii) by 24 per 10°C from 25° → 35°C
    (iii) by 60 as ionic strength (NaCl) increases from 0 → 1.0 M

The wavelength of maximum absorption decreases from   340 nm at 0°C
                    to   33.8 nm at 25°C
                    to 337.5 nm at 40°C

References: McComb *et al.* (1976).
                Zeigenhorn, Senn & Bücher (1976).

The same enzyme catalysed reaction can also be used to measure lactate. To do this the reaction is done in the reverse direction in a buffer with pH between 9 and 10 which neutralizes the H^+ liberated. Also the pyruvate produced is either trapped with hydrazine or removed enzymically with another enzyme, alanine transaminase. In either case the reaction is now in favour of using lactate with an equivalent amount of NAD^+. It is essential to check that the enzyme can function under the altered conditions of high pH and in the presence of other compounds.

In the pyruvate–lactate reaction, NAD^+ and NADH are two of the reactants, and therefore this reaction enables direct measurement of these compounds. With many substances this is not possible, but by using two or more linked enzyme reactions a change in NADH or NADPH can be produced which is equivalent to the concentration of the original substance. The measurement of glucose with hexokinase and glucose 6-phosphate dehydrogenase illustrates this point.

$$\text{Glucose} + \text{ATP} \xrightleftharpoons{\text{Hexokinase}} \text{Glucose 6-phosphate} + \text{ADP}$$

$$\text{Glucose 6-phosphate} + NADP^+ \xrightleftharpoons[\text{dehydrogenase}]{\text{Glucose 6-phosphate}} \text{6-phosphogluconate}$$
$$+ \; NADPH + H^+$$

Hexokinase has wide specificity for hexoses. Therefore fructose and other naturally occurring carbohydrates are also converted to their phosphates. However, the second enzyme catalysed reaction, which involves $NADP^+$, is very specific for glucose 6-phosphate, thus allowing accurate and specific measurement of glucose. This particular measurement has three disadvantages. Firstly, the coenzyme, $NADP^+$ is more expensive than NAD^+. Secondly, the introduction of two or more enzymes adds to the cost of analysis. Thirdly, there will normally be a lag phase during the initial part of an assay involving two enzymes which requires that there should be a short delay if a kinetic method is used.

An alternative method for measuring NADH or NADPH is to react them with an oxidizing agent, thereby reforming NAD^+ or $NADP^+$. When a tetrazolium salt is used a very intensely absorbing coloured formazan is generated. The chemical reaction only occurs slowly, and therefore an enzyme, for example diaphorase, is normally added. This removal of NAD(P)H gives another way of effectively shifting the equilibrium constant of the reaction (Möllering, Wahlefeld & Michal 1978).

5.2.4.2 *Oxygen and hydrogen peroxide*
Oxygen is used, and hydrogen peroxide formed, in three of the more common enzymic estimations performed in clinical laboratories:

$$Cholesterol + O_2 + H_2O \longrightarrow Cholest\text{-}4\text{-}en\text{-}3\text{-}one + H_2O_2$$
$$Glucose + O_2 + H_2O \longrightarrow Gluconic\ acid + H_2O_2$$
$$Uric\ acid + O_2 + 2H_2O \longrightarrow Allantoin + CO_2 + H_2O_2$$

Hydrogen peroxide is commonly measured with peroxidase by sensitive colorimetric procedures,

$$Chromogen + H_2O_2 \xrightarrow{\ Peroxidase\ } Dye + H_2O$$

Chromogens currently used are o-dianisidine or diammonium 2,2′-azino-bis(3-ethylbenzothiazoline-6-sulphonate). Alternatively, oxidative coupling of two compounds is used. These include phenol with either 4-aminophenazone or 4-amino antipyrine, or the alternative N-ethyl-N-(3-hydroxyethyl)-m-toluidine and 2-hydrazone-2,3-dihydro-3-methyl-6-sulphobenzothiazole (Carr & Bowers 1980). Another system uses the enzyme catalase.

$$CH_3OH + H_2O_2 \xrightarrow{\ Catalase\ } HCHO + 2H_2O$$

$$HCHO + 2\ Acetylacetone + NH_3 \longrightarrow 3,5\text{-diacetyl-}1,4\text{-dihydrolutidine} + 3H_2O$$

Methods which use peroxidase are subject to interference by many compounds including ascorbic acid, uric acid, glutathione, and acetylsalicylic acid, all of which decrease the value obtained. Conversely, the presence of peroxide increases the value. The interference by these compounds is minimized by diluting the plasma samples with a large (> 200-fold) excess of reagents.

The possibility of interference and the cost of second enzyme can both be reduced by measuring one of the other common components in the three reactions. The utilization of oxygen can be measured with an oxygen electrode (Koch & Nipper 1977). This can be done when all the components are in solution. However, the introduction of an immobilized enzyme, glucose oxidase, has produced a more economical, semi-automated process suitable for routine analysis. More recently an H_2O_2-sensitive electrode utilizing immobilized oxidase has come into routine use. These electrodes are discussed in more detail later (see Section 5.4.2).

Those substrates that are measured in clinical laboratories using enzymes as analytical tools are given in Part B, Table 5.1.

5.3 MEASUREMENT OF ENZYMES

Enzymes are measured in body fluids as an aid to diagnosis, and also for monitoring the effect of treatment. The rationale for this is that certain enzymes are located solely or predominantly in a tissue. Also it is known that some enzymes occur in multiple forms, isoenzymes, each of which has a different tissue distribution. When particular tissues are damaged owing to disease, their tissue enzymes and isoenzymes are liberated and their enzyme levels in serum increase.

A more specialized application of enzyme measurements is to determine if a particular enzyme is defective, generally having no or little activity. For these purposes a tissue sample is normally required. These assays are used to establish the cause of an inborn error of metabolism.

Enzymes are measured by kinetic assays, similar to those described for substrates, except that the enzyme, rather than substrate, is rate limiting. There have been attempts to measure the amount of enzyme by immunoassays (Landon et al. 1977), or by titration of the active site (Roth & Selz 1980), but neither of these types of assay is in routine use.

5.3.1 Principles of enzyme assay using coupled enzymes

With only one exception enzyme assays available in kit form, which include other enzymes as analytical tools, all use the $NAD(P)^+/NAD(P)H$ end point. The components and conditions used for this type of assay are summarized in Table 5.4 (Gould 1978).

Serum contains many proteins, some of which will be enzymes, only one of which is to be measured. The activity of this enzyme in the reaction cuvette must be rate limiting. In the case of a serum enzyme this usually involves dilution of the serum by between 1 in 10 and 1 in 30. This dilution also reduces the absorbance at 340 nm due to serum.

Most enzymes have two substrates, and both of these are added at 'optimum concentration'. In practical assays the concentration of substrate may be limited

Table 5.4

Components and conditions for enzyme assays using coupled enzymes

Enzyme sample (e.g. serum)	– rate limiting
	– typical $A_{340} = 0.7–4.6$ (serum)
Coenzyme/co-substrates	– optimal concentration
Coupled enzyme	– > 40-fold excess
NAD(P)$^+$/NADPH	– NAD(P)H > 0.2 mmol/l
Buffer	– pK close to pH optimum of enzyme
Temperature (e.g. 25°C, 30°C, 35°C, 37°C)	– controlled to ±0.1°C
Specific substrate	– start reaction by addition of optimum concentration of specific substrate

Mix, follow conversion of NAD$^+$/NADH at 340 nm

by the solubility of the substrate or by the possibility of inhibition by high S_o, Bergmeyer (1978). If S_o of 19 × K_m can be achieved, then substitution in the Michaelis–Menten equation (5.1) gives

$$v = 0.95\ V$$

that is, the initial velocity, v, is 95% of the maximum velocity. Since equation (5.1) describes a rectangular hypobola (see Fig. 5.2) there is little change of v with small changes in S_o at this level. Therefore this concentration of substrate should give consistent assays so long as the velocity is measured before 10% of the substrate has been converted to product. This should ensure that errors associated with any reverse reaction or inhibition by product are negligible. This concentration of substrate is also sufficient to allow a proportion to be used up in secondary reactions which frequently occur because of the presence of other enzymes in the serum. These reactions should be given time to go to completion, which occurs when the endogenous substrates in serum are used up, before the actual enzyme assay is started by addition of the substrate for which it is more specific.

Aspartate aminotransferase (AST) was the first enzyme to be measured with a coupled enzyme, in this case malate dehydrogenase (MDH). The system used was,

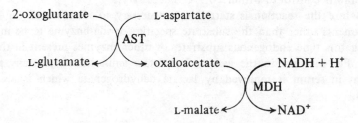

The oxaloacetate formed in the first reaction has to be utilized as soon as it is formed by the reaction catalysed by MDH if measurement of the rate of NADH conversion to NAD^+ is to be an accurate assay for AST. Thus the reaction catalysed by MDH has to operate under the unfavourable condition that the concentration of one of its substrates is virtually zero. To achieve this the total activity of the coupled enzyme must be at least 40-fold greater than the total activity of the enzyme being assayed Bergmeyer (1978). Even then there will be a lag phase as the concentration of oxaloacetate increases to its low steady-state level. This lag phase and the measurements of the velocity should be completed before the reaction deviates significantly from the true rate owing to substrate depletion. If there is more than one coupled enzyme, the lag phase will be increased, and higher total activities of the other enzymes are needed to ensure accurate assays.

The inclusion of large amounts of other enzymes in coupled enzyme systems requires that they are pure, and in particular they must not be contaminated with activity of the enzyme being measured. The use of these purified enzymes is a major contributor to the cost per assay.

The concentration of NAD(P)H present at the start of an assay should not exceed 0.2 mmol/l. At this concentration the absorbance at 340 nm is 1.26 for a 1 cm pathlength. This, added to other absorbances for other light-absorbing species including serum, is close to the upper limit before large errors are likely to be caused by stray light.

Enzyme assays should be carried out at their optimum pH. At this pH small deviations in pH have less effect on the velocity than at other pH's. The optimum pH of an enzyme is dependent on the temperature and concentration of the substrates, and should be determined under the actual conditions to be used in the assay. The components of buffer systems usually inhibit enzyme activity at high concentrations. The concentration of buffer can be minimised if the pK of the buffer is close to the pH used for the assay.

Enzymes do not have an optimum temperature, which accounts for the range of temperatures that have been recommended (see Table 5.4). It is important that all components of the assay mixture should be kept at the temperature selected for the assay, and it is essential that the serum enzyme should not be inactivated at this temperature or during any preincubation step. Changes in activity of enzymes of 5–10% occur per 1°C. Therefore for accurate assays, temperature control to within ±0.1°C is necessary.

Before the reaction is started it is necessary to incubate all the reaction components other than the substrate specific for the enzyme to be measured. During this time endogenous substrates of other enzymes present in the serum react. For instance in the assay of aspartate aminotransferase, any pyruvate present in serum is removed by lactate dehydrogenase which is also in the serum.

$$\text{Pyruvate} + \text{NADH} + \text{H}^+ \xrightarrow{\begin{array}{c} \text{Lactate} \\ \text{dehydrogenase} \end{array}} \text{L-Lactate} + \text{NAD}^+$$

and the assay of aspartate aminotransferase should not start before the above reaction is completed. Some samples of serum contain so much pyruvate that it is now recommended that purified lactate dehydrogenase be routinely included in the assay mixture to hasten its removal (IFCC 1977).

Those enzymes that are measured in clinical laboratories using coupled enzyme systems are given in Part B, Section 5.3 (Table 5.2).

5.4 IMMOBILIZED ENZYMES FOR MEASURING SUBSTRATE CONCENTRATIONS

During the past decade much attention has been focused on the substitution of soluble enzymes by their immobilized counterparts in enzyme-based analyses. This is because of the limited stability of some enzymes under routine operating conditions, and the relatively high cost of many analytically useful enzymes. The exploitation of immobilized enzymes offers the clinical chemist the advantages of re-usability, greater stability, and increased convenience.

The literature contains more than 500 publications describing the use of immobilized enzymes in analytical procedures, and many of these systems have been described and tabulated in several monographs (Carr & Bowers 1980, Bergmeyer 1978, Guilbault 1976) and reviews (Free 1977, Gray, Keyes & Watson 1977, Ngo 1980, Bowers 1982, Mottola 1983). However, relatively few of these analytical techniques have been commercially exploited, and immobilized enzyme preparations have not yet realized their full potential in clinical chemistry.

The successful utilization of immobilized enzymes in clinical chemistry may be divided into three categories: enzyme reactor tubes, bioanalytical probes and dry reagent products. These uses are summarized in Table 5.3 in Part B, Chapter 5, and the principles are discussed more fully in the following paragraphs.

5.4.1 Immobilized enzyme reactor tubes

Early attempts at incorporating immobilized enzymes into automated analytical systems involved the use of packed bed enzyme reactors. These were prepared by packing a column with a gel or finely divided solid which supported the enzyme (Carr & Bowers 1980). These packed bed reactors had the disadvantage of producing a high pressure drop which restricted flow rate and resulted in a slow rate of analysis.

These difficulties have been overcome by immobilizing enzymes on relatively large (20–35 mesh, polyaminostyrene) beads which when packed in Auto-

Analyzer mixing coils could be incorporated into systems based on nonsegmented flow analysis. The size and hydrophobic nature of the beads allowed a fast flow rate with good wash characteristics (Rocks 1973, Miller, Rocks & Thorburn Burns 1976, 1977).

Hornby and co-workers (Sundaram & Hornby 1970, Filippusson, Hornby & McDonald 1972, Hornby, Inman & McDonald 1972) pioneered the first open tubular heterogeneous enzyme reactors (OTHERs) in which the enzyme was covalently bound to the inner surface of a narrow tube. OTHERs have the advantage of being readily incorporated into the widely used air-segmented continuous flow systems without necessitating any major change in existing methodologies. Several commercial manufacturers (see Part B, Table 5.3) now market OTHERs to aid in the measurement of several routinely estimated metabolites. Many of these analytical systems have been evaluated in the literature. These include assays for glucose (Leon *et al.* 1977, Chirillo *et al.* 1979, Werner *et al.* 1979), urea (Collis & Knox 1978, Chirillo *et al.* 1979, Werner *et al.* 1979) and uric acid (Chirillo *et al.* 1979, Werner *et al.* 1979, Leon *et al.* 1982).

Typically a nylon tube of 1 mm bore is used and is internally coated with one or more enzymes using a variety of techniques (Sundaram & Hornby 1970, Horvath & Solomon, 1972, Hornby & Morris 1975, see also Part B, Chapter 5). The operational and storage stability of OTHERs varies with the enzyme involved and also with the method used to coat the inside of the tube. Generally, immobilized glucose oxidase preparations are more stable than immobilized urease and uricase. Shelf life is typically between 1 and 18 months when the tubes are stored filled with buffer at 4°C. When stored dry, loss of enzymic activity is more rapid. In operation between 1000 and 30 000 assays are usually possible before a significant decrease in activity is apparent. Single tubes have been reported to be still analytically useful after a 3-month period of intermittent use (Chirillo *et al.* 1979). The tubes must, however, be protected from extremes of pH, heavy metals, and other enzyme 'poisons' such as cyanide and azide solutions. Longevity of the tubes is increased by the addition of an antimicrobial agent to the buffer solution (Leon *et al.* 1977).

The main advantage in using OTHERs is their re-usability, which minimizes problems arising from the preparation and handling of enzyme solutions. However, contrary to the claims made in many manufacturers' advertising literature, OTHERs do not necessarily result in an economic saving when compared to the use of soluble enzymes. For example, a cost benefit study of replacing the soluble enzyme hexokinase in an AutoAnalyzer II glucose method with its commercially available immobilized counterpart reveals that at least 140 samples per day must be processed before there is an economic advantage in making the change. If only 50 samples per day were analysed the use of the immobilized enzyme tube would be 100% more expensive to use than the free enzyme. On the other hand, if 200 samples per day were to be assayed then the use of the enzyme tube offers a 30% saving compared with the cost of soluble enzyme.

The calculations were based on the assumption that the useful life of the hexokinase tube was 1 month (Technicon publication, July 1981).

Although these enzyme reactors were originally designed for incorporation into systems based on colorimetric determination of the reaction products, they may be interfaced with other appropriate detectors. For example, if the enzymic reaction involves electroactive substrates or products the colorimeter may be replaced by an electrochemical detector, thus rendering the analytical system 'reagentless'. Only a buffer solution is needed (Campbell & Hornby 1977). Another approach is to combine the immobilized enzyme with the transducer in an integral unit to produce a bioanalytical probe — the so called 'enzyme electrode'.

5.4.2 Bioanalytical probes

In principle a bioanalytical probe (enzyme electrode) is a device which consists of a combination of any type of electrochemical sensor in contact with a layer of immobilized enzyme, and which is used to measure the concentration of a substrate. The outer surface of the enzyme layer is exposed to the analyte solution, which is usually stirred or agitated to minimize concentration gradients. The substrate diffuses into the enzyme layer where it is converted to an electroactive compound which is measured by the transducer, either when steady state conditions exist or during the initial stages of the reaction. The latter technique produces a more rapid readout, but is temperature sensitive. The potential or current of the electrode is measured, relative to a reference electrode, and the concentration of the substrate derived. The relationship is logarithmic for a potentiometric electrode and linear for an amperometric electrode.

To increase specificity and to minimize the effect of interfering substances, semipermeable membranes are sometimes placed between the immobilized enzyme and the solution, and also sometimes between the enzyme and the transducer.

The literature contains a multitude of reports on biochemical sensors for the determination of urea, glucose, lactic acid, ethanol, penicillin, cholesterol, amino acids and numerous other substances (see reviews by Moody & Thomas 1975, Carr & Bowers, 1980). Unfortunately the utility of many of these devices is severely limited in practice, particularly in their application to physiological fluids. This is often due to poor selectivity of the transducer. For example, several methods for the analysis of urea use an ammonium ion-sensitive electrode; the electrode is surrounded by immobilized urease which hydrolyses urea molecules as they diffuse into it:

$$(NH_2)_2CO + 2H_2O + H^+ \underset{}{\overset{Urease}{\rightleftharpoons}} HCO_3^- + 2NH_4^+$$

$$\Updownarrow$$

$$NH_3 \text{ (gas)}$$

One of the products of the reaction is NH_4^+ to which the electrode responds. However, the ammonium ion transducer also responds to all other monovalent cations including H^+, and although some selectivity for NH_4^+ over Na^+ and K^+ is displayed, the presence of these ions in blood plasma causes significant interference in the measurement of urea. This particular problem of interfering ions can be eliminated by the use of an ammonia gas-sensitive electrode. In this sensor the enzyme layer is separated from a glass pH electrode by a gas permeable membrane. In alkaline buffer the ammonia gas liberated by the action of urease on urea will diffuse through the membrane and activate the pH sensor. A limitation exists, however, in that the optimum pH for functioning of the NH_3 gas electrode is not the pH of choice for the highest urease activity. Additionally, NH_3 gas electrodes suffer from very slow response and poor washout characteristics.

The glucose analyser, produced by the Yellow Springs Instrument Co., appears to be the only commercial instrument to use an enzyme electrode (Chua & Tan 1978, Spencer, Sylvester & Nelson 1978). The glucose probe (Fig. 5.5) contains a thin layer of glucose oxidase immobilized on glutaraldehyde resin particles and held between two membranes, one of which is placed in contact with a hydrogen peroxide sensitive electrode (that is, a platinum electrode polarised at 0.7 V which acts as the anode of an electrochemical cell). The sample is injected from a special calibrated micro-syringe into a chamber of buffered reagent containing the glucose probe. Glucose which diffuses to the probe from the solution is converted to gluconic acid and hydrogen peroxide by the immobilized glucose oxidase, and the hydrogen peroxide is determined amperometrically at the electrode. The success of the instrument owes much to the nature of the membranes, which not only contain the immobilized enzyme particles but also minimize the amount of interfering substances reaching the electrode. The outer membrane is of a polycarbonate material with a pore size large enough to pass glucose, oxygen, and other small molecules, but too small to allow cells and macromolecules to pass through. The inner membrane is of cellulose acetate with a much smaller pore size (cut off 100 Daltons) which excludes glucose, uric acid, ascorbic acid and most other potentially interfering substances from reaching the electrode, while still allowing some molecules such as hydrogen peroxide to pass through. However, users of this instrument should be aware that some blood preservatives (Hall & Cook 1982, Kay & Taylor 1983) and certain drugs (Lindh *et al.* 1982) do reach the electrode where their oxidation produces erroneous results. The usefulness of enzyme reagents is further exemplified by another feature of the Yellow Springs analyser. The reagent solution contains catalase (soluble form) to scavenge 'stray' hydrogen peroxide which may diffuse back into the probe, causing drifting of the anodic current.

In our laboratory we use a Yellow Springs instrument to analyse urgent blood glucose specimens that arrive at irregular times both day and night; and also as a means of producing rapid results in an out-patient diabetic clinic. The immobilized enzyme—membrane disc is routinely changed at monthly intervals, typically having been used for between 600 and 700 assays. When stored at 4°C the discs have a shelf life of about 1 year.

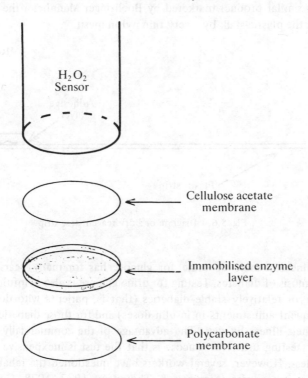

Fig. 5.5 − Exploded view of the Yellow Springs Instrument Co. glucose probe.

Other commerical glucose analysers may also be modified for use with immobilized enzyme preparations. For example, Sokol *et al.* (1980) attached glucose oxidase to a derivatized Teflon membrane which they placed over the tip of the Clark-type oxygen electrode of a Beckman glucose analyser. These authors claim that the immobilized enzyme modification gives similar results to the soluble enzyme normally used in this instrument, and offers the advantage of simplicity, economy of enzyme, and linearity over a greater range of concentration.

5.4.3 Dry reagent chemistry
Free *et al.* (1957) described the first dip-and-read test strip to use enzymes. This was produced by adsorbing the enzyme reagent, glucose oxidase, peroxidase and a chromogen, on to a paper (cellulose) pad. The dried paper responded,

with a colour change when wetted with a solution containing glucose. The concept, under the name   Clinistix, was developed by the Ames Division of Miles Laboratories and is widely used to detect the presence of glucose in urine. The pad containing the dried reagents is cemented to the end of a plastic strip (Fig. 5.6). In a similar product marketed by Boehringer Mannheim the reagent patch is held on the plastic stalk by a very thin nylon mesh.

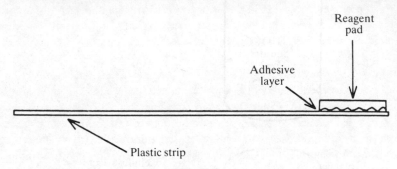

Fig. 5.6 – Diagram of a dry reagent test strip.

Routine screening of urine for glucose has for many years aided in the rapid diagnosis of diabetes. Testing for urine glucose is also helpful in monitoring treatment of relatively stable diabetics (that is, patients who do not need to make frequent adjustments in insulin doses) and for those diabetic patients who control their illness by diet. The advantage of the commercially available dipsticks for testing urine for glucose is that the test is inexpensive, non-invasive, and painless. However, several workers have questioned the reliability of these simple test procedures (Simpson & Thompson 1977, 1978, Gupta, Goyal & Singh 1982, Olesen, Mortensen & Mølsted–Pedersen 1983). Besides differences in user technique, variable concentrations of glucose oxidase inhibitors, normally present in urine, could account for unreliable results.

It is now recognized that maintaining a well-regulated blood glucose concentration can reduce the incidence and severity of the microvascular complications often associated with diabetes. The development of test strips for the determination of glucose in whole blood and serum has made blood assays almost as simple as the detection of urine glucose. In this case the reagent pad is treated with ethyl cellulose which forms a membrane through which small molecules can readily diffuse, but the presence of the coating prevents cells from being adsorbed on the cellulose fibres and allows them to be washed away.

A large drop of blood is applied to the reagent patch and after a given time interval (typically 1 min) excess blood is removed and the colour of the reagent pad compared with a reference colour chart (usually printed on the test strip container wall) and the glucose concentration visually estimated (Kutter 1977,

Sherwood, Warchal & Chen 1983). Reflectance photometers are commercially available to aid with quantifying the colour change; blood glucose measuring kits comprising small portable reflectance meters and reagent stripes are now widely used in hospital side-rooms, diabetic clinics, and by diabetics at home. However, a study in which two of these test systems were compared to a laboratory based AutoAnalyzer method illustrated how inaccurate these simple systems can be (Tomkin & Moore 1977). Although these kits are undoubtedly useful in emergencies, Tomkin & Moore caution medical staff not to rely on results produced by these meters but rather to check with a laboratory estimate as soon as possible.

More recently, Stewart & Kleyle (1983) made a statistical comparison of blood glucose determined by several 'home use' kits and by an established laboratory procedure (hexokinase based). They demonstrated that an experienced operator could obtain reproducible results using the test strip methods, but all the kits had a tendency toward nonlinearity of results at extreme glucose concentrations. All four methods examined gave results that were significantly lower than the laboratory value in the hypoglycemic range. Drucker, Williams & Price (1983) report that results obtained using a portable machine in the laboratory were both accurate and precise, but less satisfactory results were obtained when the tests were performed by doctors and nurses operating outside the hospital laboratory.

The detailed use and limitations of test strips in reaching a rapid clinical diagnosis have been discussed by Kutter (1977). These dry reagent strips must be stored in an atmosphere of very low humidity. This is easily achieved by storing them in a sealed bottle containing a desiccant packet. However, if the container top is not promptly replaced after removing the reagent strip, moisture will enter and may cause rapid deterioration of the product (King, Steggles & Harrop 1982).

Ames have refined this technology, and with the aid of a dedicated microprocessor controlled reflectance photometer — the Seralyzer — offer dry reagent tests for seven different serum constituents. The strips for the analyses of glucose, uric acid, cholesterol and trigylcerides utilize adsorbed enzymes and appropriate chromogens (Zipp 1981). Performing an assay requires the selection and insertion of an appropriate test module; this automatically programmes the instrument for the specified test. After standardization, a test strip is placed on the thermostatted specimen table and a 30 μl volume of diluted serum is manually pipetted into the instrument. When the test has been completed a buzzer sounds and the result is displayed on the instrument panel. In their evaluation of the Seralyzer, Thomas, Plischke & Stortz (1982) reported acceptable precision for glucose and uric acid determinations, but when the results were compared with those obtained by established methods discrepancies were sometimes found. Stevens *et al.* (1983a, b) found the Seralyzer reliable and easy to use and that it met 'state-of-the-art' performance criteria. Clark & Broughton (1983a, b) were less enthusiastic about its performance, and they warn of the pitfalls of dry reagent tests. In particular, they report that sample volume and the time of analysis were critical; poor precision was obtained by an inexperienced operator.

A different approach to the use of dry reagent technology has been pursued by The Eastman Kodak Corporation, based on their vast experience of producing photographic materials. The manufacture of a modern colour film requires that up to 16 thin layers of chemicals are reproducibly coated on a transparent support. Although photographic chemistry is not involved, Kodak have adapted this technology to the production of dry reagent slides. (Curme *et al.* 1978, Shirey 1983). On top of a transparent plastic base are coated one or more reagent layers which contain many or all of the reagents necessary for the analysis. Hydrophillic polymers, such as gelatin or agarose, are used to bind the reagent in these layers. A porous spreading layer is coated on top of the reagent layers. Within this film format, the active area of which is typically 1 cm^2 and 100 μm thick, a variety of physiochemical phenomena can be employed. Fig. 5.7 shows a cross-section of a slide for the analysis of urea. When a drop of serum touches the analysis slide it comes into contact with the porous layer which spreads the sample uniformly and rapidly by capillary action. Below the spreading layer is the reagent layer, which for blood urea determination contains urease and a buffer (pH 8.0) bound in gelatin. Urease catalyses the hydrolysis of urea to ammonia and bicarbonate. Ammonia then diffuses through a semipermeable layer of cellulose acetate butyrate which is interposed between the reagent and indicator layers to minimize diffusion of hydroxyl ions to the dye later. A leuco indicator in a cellulose acetate binder provides the colour forming layer and is coated directly onto a transparent support. After colour development the colour intensity is measured by reflectance through the transparent support against the white spreading layer background (Spayd *et al.* 1978). Similar principles, using other appropriate enzyme systems (see Part B, Table 5.3), have been applied to the assay of glucose (Curme *et al.* 1978), cholesterol (Dappen *et al.* 1982), creatinine (Sundberg *et al.* 1983, Toffaletti *et al.* 1983, Smith *et al.* 1983), and triglycerides (Spayd *et al.* 1978).

Kodak have developed two automatic analysers dedicated to the use of multilayer film technology. The Ektachem Glu/BUN Analyser and the Ektachem 400 Analyser. These instruments contain no pumps or tubing and are designed with the minimum of moving parts. The dry films used for the analyses are stored in the machine as individual slides contained in cartridges. The analysers have built-in microprocessors that control the instrument functions, compute all the data generated during testing, and convert into appropriate units. The Ektachems perform about 120 discrete analyses per hour, and they carry out only those tests that are requested by the operator. Cate *et al.* (1980) reported favourably on their experience with a prototype Ektachem Glu/BUN Analyser; and in an extended clinical trial and evaluation of the method for glucose determination Bandi *et al.* (1981) cite precision, accuracy and reliability as desirable characteristics of the system. Both groups of assessors, however, comment on a pronounced sensitivity to sample viscosity. This results in low values for glucose when the serum contains highly abnormal concentrations of proteins. This effect has been attributed to variation in spot diameters at extreme protein values and to protein interaction with the film matrix (Cate *et al.* 1980). Additionally, aqueous standards cannot be used to calibrate the instrument if serum samples are to be analysed. Lyophilized sera with a special Ektachem value must be used as controls and standards. This characteristic of the Ektachem system is not unique, as other instruments and methods have similar problems.

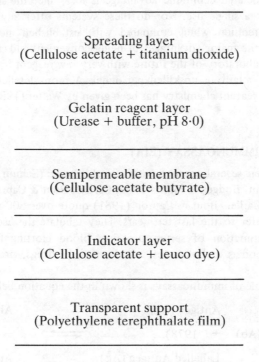

Spreading layer
(Cellulose acetate + titanium dioxide)

Gelatin reagent layer
(Urease + buffer, pH 8·0)

Semipermeable membrane
(Cellulose acetate butyrate)

Indicator layer
(Cellulose acetate + leuco dye)

Transparent support
(Polyethylene terephthalate film)

Fig. 5.7 — Layer representation of Kodak Ektachem dry phase slide for urea assay. The layers are not drawn to scale.

Sundberg *et al.* (1983) have described a novel enzymic method for determining serum creatinine using coated film technology. The method makes use of creatinine iminohydrolase to convert creatinine to *N*-methylhydantoin and ammonia. The ammonia diffuses through a semipermeable layer and is quantitated by reaction with bromophenol blue. The method is claimed to be less affected by interfering substances than the widely used Jaffé procedure (Smith *et al* 1983, Toffaletti *et al.* 1983).

For short term use the reagent slides may be kept at room temperature for up to two weeks (Cate *et al.* 1980). Slides stored for longer periods require to be refrigerated (4°C) in well sealed packages. The limited life of the films probably derives from residual water in the gelatin matrix, and concomitant lability of the trapped enzymes (Greyson 1981).

The quality of results is determined almost entirely by the precision with which the slides are manufactured, and the standard to which this is maintained will largely determine the ultimate success of this approach to clinical analysis.

The use of enzymes in the dry reagent format does not offer re-usability, and so the associated economic advantage is lost, since the slides or strips are discarded after a single use. Nor do these systems offer any improvement in accuracy or precision when compared with established methods. They do, however, have the major advantages of being convenient and simple to use. One can envisage a laboratory of the future with very few reagent bottles or glassware and staffed by relatively unskilled personnel. A more detailed but less critical account of dry reagent chemistry has been given by Walter (1983).

5.5 ENZYME IMMUNOASSAY (EIA)

There have been several recent reviews of this topic (Schuurs & Van Weemen 1977, O'Sullivan, Bridges & Marks 1979, Carlier, Bout & Capron 1981, Blake & Gould 1984). Carlier, Bout & Capron (1981) quote over 600 references, nearly all of which refer to the last ten years. They tabulate the uses of EIA for the quantitative estimation of serum proteins, blood clotting factors, immuno-globulins, antibodies to hormones, a wide range of drugs, viruses, bacteria and parasites.

The principle of immunoassays is shown in the equation below,

$$\begin{matrix} & \text{Antigen (Ag)} & & \text{Ab} - \text{Ag} \\ \text{Antibody (Ab)} & + & \rightleftharpoons & + \\ & \text{Labelled Antigen (Ag}^*) & & \text{Ab} - \text{Ag}^* \end{matrix}$$

Free labelled (Ag*) and unlabelled antigen (Ag) molecules compete for a fixed but limited number of specific binding sites on antibody molecules. After an incubation period the free and bound antigen are normally separated from each other. Then the amount of labelled antigen in either or both fractions is determined. When the quantity of unlabelled antigen is increased, fewer labelled antigen molecules will be bound by antibody. Therefore a standard curve can be produced, and by using this the concentration of antigen in biological samples can be determined. Radioactive labels, hence radioimmunoassay (RIA), have been used since the late 1950s. Enzyme labels became available in the early 1970s.

RIA allows the direct measurement of small quantities of a wide range of compounds present in complex biological samples without prior separation. However, radiolabels have disadvantages. Preparation of labelled antigen may involve a health hazard. The shelf life of the labelled product may be short, for example when γ-emitting isotopes such as iodine-125 are used. Radioactive and organic wastes have to be disposed of safely. The equipment needed to measure radioactive samples is expensive. These disadvantages encouraged research into alternative labels, of these enzymes are the most frequently used. Table 5.5 lists the advantages and disadvantages of enzyme labels.

Table 6.5
Comparison of enzyme − with radio-labels in immunoassays

Advantages of enzyme-labels
    (i) No radiation hazards occur during labelling or disposal of waste
    (ii) Labelled product can have a long shelf life
    (iii) Equipment can be inexpensive and is generally available
    (iv) Homogeneous assays can be completed in minutes

Disadvantages of enzyme-labels
    (i) Plasma constituents may affect enzyme activity
    (ii) Assay of enzyme activity can be more complex than measurement of some types of radioisotopes.

There are two types of quantitative EIA, heterogeneous and homogeneous. In heterogeneous EIA the bound label which is finally measured must be separated from the free label, as is the case with RIA. In homogeneous EIAs the activity of the enzyme label is altered by the antigen and antibody interaction. Therefore separation is no longer necessary since it is the change in enzyme activity that is determined. Homogeneous assays have the advantage, when this is necessary, of rapidity, and they lend themselves to easy automation. But they are also the assays most susceptible to interference by plasma constituents.

5.5.1 Preparation of enzyme labels
Ideally the enzyme labels should be stable, and the chemical reactions should not affect the activity of the enzyme when it is measured. Equally the specific interaction between antibody and antigen or hapten should not be altered by the combination with enzyme. Haptens are small molecules, for example, drugs, steroids which are not usually immunogenic but which become so when coupled to a larger carrier protein.

The reactions used to crosslink enzymes with antibody, antigen, or hapten are similar to those used to form immobilized enzymes. The chemistry of these reactions is discussed in Part A: Chapter 4.

The most common reactions used to form enzyme-labelled antigens or antibodies are:

(i)   glutaraldehyde method (Avrameas 1969) – the dialdehyde links proteins through their amino groups. The main disadvantage is the self-linkage of both types of protein. This is reduced if the reaction is done in two stages instead of one (Avrameas & Ternynck 1971);

(ii) periodate method (Nakane & Kawaoi 1974) – this method is limited to glycoproteins, for example peroxidase. The periodate is used to oxidize the carbohydrate residues of the enzyme; the aldehyde can then react with the amino groups of the antigen protein;

(iii) maleimide method (Kato et al. 1975) – this method joins proteins by combining with sulphydryl groups on both proteins or with an amino group on the antibody (O'Sullivan et al. 1978b). The latter reaction can reduce the amount of self-polymerisation.

Enzyme labelled haptens are formed by a limited number of reactions:

(i)   mixed anhydride method (Erlanger et al. 1959) – a peptide bond is formed between a carboxylic acid residue on the hapten and an amino group on the enzyme;

(ii) carbodiimide method (Tateishi et al. 1977) – this reaction also forms a peptide bond between hapten and enzyme. If necessary, carboxylic acid residues can be introduced into haptens containing hydroxyl groups;

(iii) bifunctional imidate method (O'Sullivan et al. 1978a) – a crosslink is formed between amino groups on the hapten and enzyme.

5.5.2 Heterogeneous EIA

Engvall & Perlmann (1971) pioneered the use of solid-phase supports for heterogeneous assays. These are now the most frequently used method of separation, and are used in all the commercial heterogeneous systems. The solid phases are made from a variety of plastics and are in the form of premoulded multiple sample wells, beads, cuvettes, or tubes. These types of assay are frequently referred to as enzyme linked immunosorbent assays (ELISA). In these assays an antigen or antibody is linked to the surface of the solid phase. It is essential that this linking or coating process is reproducible.

Several types of heterogeneous EIA have been developed. The principles of those systems that are available commercially for routine clinical use are illustrated in Fig. 5.8.

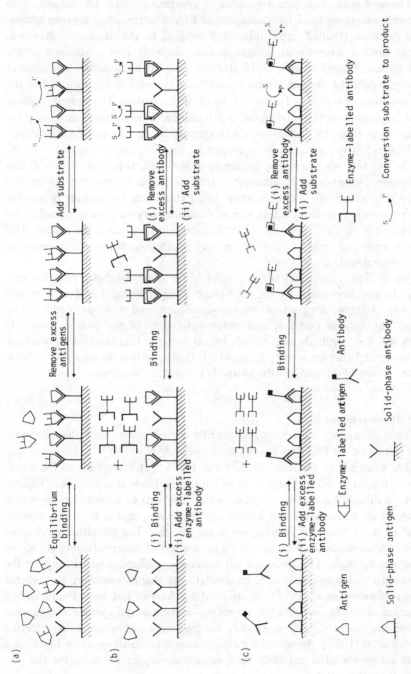

Fig. 5.8 — Principles of commercially available heterogeneous EIA. (a) Competitive EIA for antigen. (b) Direct sandwich EIA for antigen. (c) Direct sandwich EIA for antibody.

Figure 5.8(a) illustrates principles of competitive EIA for antigen. This method is analogous to classical competitive RIA. The first stage involves competition between labelled and unlabelled antigen, in the sample or standard, for the limited quantity of antigen-specific antibody that is attached to the solid surface. Excess antigen molecules are removed by washing. The amount of enzyme-labelled antigen bound by antibody is inversely proportional to the concentration of unlabelled antigen present in the sample. The enzyme activity remaining bound to the solid phase is determined. This method can be used for hapten measurement if hapten molecules are used in place of antigen molecules.

Figure 5.8(b) illustrates the principles of the direct sandwich EIA for antigen. In the first stage antigen, in sample or standard, is incubated with excess solid-phase antigen-specific antibody. After washing, excess enzyme-labelled antibody is added, the excess enzyme being eliminated by washing before the enzyme activity is determined. In this method, the enzyme activity bound to the solid phase is directly proportional to the amount of antigen in the sample. This type of assay can only be used for antigens which possess more than one antigenic determinant.

Figure 5.8(c) illustrates the principles of the direct sandwich assay for antibody. In the first stage excess solid-phase antigen is incubated with sample antibody. After washing, excess enzyme-labelled second antibody is added. This second antibody is raised against immunoglobulins of the animal species in which the first antibody was raised. Excess enzyme is eliminated by washing, and the bound enzyme activity is measured. This activity is directly proportional to the amount of antigen-specific antibody present in the sample.

5.5.3 Homogeneous EIA

The most commonly used homogeneous EIA is the 'enzyme multiplied immunoassay technique' or EMIT ((C), Syva Maidenhead, U.K.).

The principle of this assay is shown in Fig. 5.9(a). It depends on the occurrence of inhibition of the enzyme, to which the hapten is attached, when hapten-specific antibody binds. Free hapten, in the standards or samples, competes for the antibody and relieves this inhibition. Therefore enzyme activity is proportional to the amount of free hapten in the samples. The inhibition of enzyme activity by antibody may be due to (i) prevention of access of substrate to the active site by steric hindrance, or (ii) causing a conformational change in the enzyme, or (iii) prevention of a conformational change necessary for enzyme activity (Rowley et al. 1975). A different mechanism has been found with a thyroxine assay. In this case the thyroxine–malate dehydrogenase conjugate is enzymically inactive, but on binding to thyroxine antibodies it is activated (Ullman et al. 1979). Apparently the thyroxine inhibits the enzyme by binding at the active site. The antibody reactivates the enzyme by removing the thyroxine from the active site.

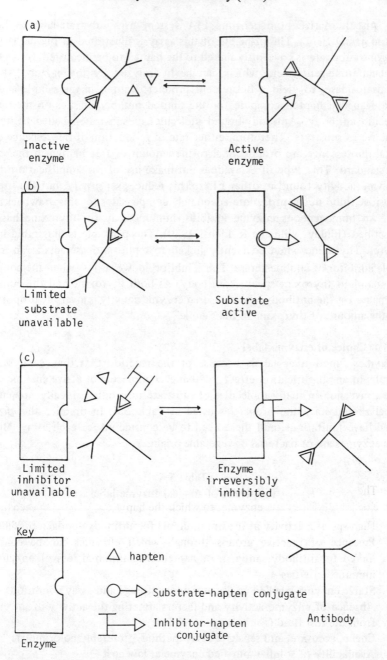

Fig. 5.9 — Principles of commercially available homogeneous EIA. (a) Hapten-labelled EIA. (b) Hapten-labelled substrate EIA. (c) Hapten-labelled with irreversible inhibitor EIA.

An alternative homogeneous EIA system uses substrate-labelled hapten (Burd *et al.* 1977). The principle of this assay is illustrated in Fig. 5.9(b). The enzyme substrate is covalently linked to the hapten to be measured, for example umbelliferone-gentamicin which can be hydrolyzed by the enzyme, that is, β-galactosidase, to yield a fluorescent product. Hapten in the sample and substrate-labelled hapten compete for the limited number of sites on the hapten-specific antibody. Antibody-bound molecules of substrate-labelled hapten are inactive as substrate. Therefore, either rate of production of fluorescence or the total fluorescence are proportional to the amount of free hapten in the sample or standard. This type of assay does not make use of the amplification due to enzyme activity found in other EIAs. This reduces sensitivity, but the use of a sensitive fluorimetric procedure adequately compensates for this drawback.

An homogeneous enzyme inhibitor immunoassay for thyroxine has been described (Finley, Williams & Lichti 1980). The principle is illustrated in Fig. 5.9(c). Thyroxine when covalently linked to a phosphonate acts as an irreversible inhibitor of cholinesterase. This inhibitory power is lost when the conjugate is bound to thyroxine-specific antibody. As both thyroxine and the conjugate compete for the antibody, the measured enzyme activity is inversely proportional to the amount of thyroxine in the sample.

5.5.4 Choice of enzyme label

Enzymes function as labels because of the reaction that they catalyse. This results in an amplification effect, so that each molecule of active enzyme causes the conversion of many molecules of substrate to product. Ideally, the enzyme label should have the properties listed in Table 5.6. In practice the enzymes used have limitations, and these need to be considered for each assay. Most of the enzymes are of bacterial or vegetable origin.

Table 5.6
Properties of an ideal enzyme-label

---

1. High specific activity at the optimum pH for antibody–antigen binding
2. Presence of reactive groups through which enzymes can be covalently linked to antibody, antigen, or hapten with minimal loss of enzyme and immune activities.
3. Stable enzyme-labelled conjugates under storage and assay conditions
4. Absence of enzyme activity and factors affecting the activity of the enzyme from the test fluid (for homogeneous EIA)
5. Cheap, accurate, and sensitive assay method, preferably colorimetric
6. Availability of soluble, purified enzyme at low cost
7. Absence of health hazards attributable to enzyme, substrate(s), or co-factor(s)

The enzymes most commonly used in kit form for heterogeneous assays are listed in Part B, Table 5.4(a). The relevant properties of these enzymes have been thoroughly discussed by Ishikawa (1980) and Yolken (1982). Peroxidase, the cheapest of the enzymes listed, is a glycoprotein which can therefore be conjugated by the periodate method. It is prone to inactivation by preservatives, and by oxidizing and reducing agents. The enzyme is detectable in femtomole amounts by colorimetry. Several chromogens have been used to visualize the enzyme, but they all have disadvantages. The most sensitive, o-phenylenediamine, is photosensitive; 2,2′azino-di-(3-ethylbenzthiazoline-6-sulphonate) is expensive, and the product is unstable; 5-aminosalicyclic acid has low sensitivity; o-toluidine and o-phenylenediamine have been found to be carcinogenic or mutagenic in laboratory animals. Alkaline phosphatase and its conjugates are very stable. The calf intestine enzyme can be detected in femtomole amounts, but the purified enzyme is expensive. Urease, a recent addition to the list, produces a strong purple colour as its end point. It is claimed that this allows for easy visual identification of positive results. Quantitative analyses as sensitive as with other enzyme labels are also claimed. β-Galactosidase can also be detected colorimetrically in femtomole amounts, and it has a potential advantage over peroxidase and alkaline phosphate in that it is absent from animal cells. This accounts for it being the only one of these enzymes that is also used for homogeneous assays. Lysozyme was the first enzyme label, and it is still used for the semi-quantitative assay of a number of drugs pesent in urine in the mg/litre range. More sensitive quantitative assays in serum are possible with glucose-6-phosphate dehydrogenase as the enzyme label. The enzyme is obtained from *Leuconostoc mesenteroides,* as this enzyme uses NAD^+ as one of its substrates. This overcomes the possibility of interference with the equivalent human red cell enzyme which is specific for $NADP^+$. A few urine assays use the more expensive but more sensitive malate dehydrogenase. This enzyme would be susceptible to interference in homogeneous assays of serum constituents.

5.5.5 Assay in EIA
In the assay of enzymes in heterogeneous EIAs the reaction is stopped at a particular time, for example by a pH change, within the linear portion of the progress curve (Fig. 5.1). The whole process of incubation, washing, and enzyme assay is easily automated with the equipment that is now available. Individual assays generally take about 2 hours, but 100 tests can be done at one time.

For homogeneous EIAs a kinetic assay is normally used. An individual assay can usually be done within one to two minutes, although extra time is required to produce a calibration curve. The procedure can be automated, and about 30 assays can be performed at once in about 20 minutes.

The determination of enzyme activity is mainly by photometric methods, often using the same end points as were discussed previously (see Sections

5.2.4.1, 5.2.4.2). β-Galactosidase is measured either colorimetrically or fluorometrically; the latter assay can be about 100 times more sensitive. The assay of lysozyme is more unusual. It is assayed by turbidimetry.

5.5.6 Simultaneous assay of two haptens
A potential advantage of EIA is the relative ease of measuring two haptens simultaneously. This possibility was noted by Wisdom (1976), but the first working system has only recently been published (Blake *et al.* 1982). The advantages of this system compared to separate single-hapten methods, include savings on reagent cost, sample volume, and a reduced overall assay time.

Simultaneous measurement of two haptens using two iodine isotopes I^{125} and I^{132}, was achieved by Mitsuma *et al.* (1972), but the usefulness of this type of assay is restricted by the short half-life of I^{131} and the less than ideal distinction between the γ-emissions of the two isotopes.

The simultaneous EIA of two haptens measured thyroxine and triiodothyronine (Blake *et al.* 1982). Alkaline phosphates, conjugated to thyroxine, was detected at 540 nm by the liberation of phenolphthalein from its monophosphate. The second enzyme was β-galactosidase which was conjugated to triiodothyronine and assayed at 420 nm by the formation of o-nitrophenol from its substrate, o-nitrophenyl-β-galactosidase. The two end points allowed each hapten to be measured in a single tube.

5.6 THE FUTURE
Four aspects of enzymes in clinical analysis have been reviewed. Two of these, the measurement of metabolites and the measurement of enzymes, are well established. The basic methodologies have changed comparitively little in the last ten years, but the availability of increasingly sophisticated machines has considerably increased the number of assays that can be performed both on an individual specimen and in a given time. In the future more sophisticated machines with an ability to do more tests will become available, but they will also be discriminatory, enabling them to do only the tests requested by the physician. The third section of this review dealt with immobilized enzymes. Their main impact is still to come. Dry reagent technology is still in its infancy, but owing to its advantages over conventional wet chemistry, its development should be rapid. Hand in hand with such developments we should also expect to see the introduction of more small portable, simple-to-operate machines for use in the physician's office or by the patients bedside. The final part, on EIA, also described a powerful, versatile technique with considerable potential. The range of available EIAs is steadily increasing, as is the supply of sophisticated dedicated equipment for the automation of both homogeneous and heterogeneous assays. The use of enzymes in clinical analysis will continue, but the range of assays using them will expand because of their convenience, specificity, and reproducibility.

REFERENCES

Avrameas, S. (1969) *Immunochemistry* 6 43–52.
Avrameas, S. & Ternynck, T. (1971) *Immunochemistry* 8 1175–1179.
Bandi, Z. L., Fuller, J. B., Bee, D. E. & James, G. P. (1981) *Clin. Chem.* 27 27–34.
Barker, S. A. & Kay, I. (1975) In: *Handbook of Enzyme Biotechnology,* pp. 89–1.10, (Ed. by Wiseman, A.) Ellis Horwood Ltd., Chichester
Bergmeyer, H. U. (ed.) (1974) *Methods of Enzymatic Analysis,* Vol. 1–4, 2nd ed. Academic Press, New York.
Bergmeyer, H. U. (ed.) (1978) *Principles of Enzymatic Analysis,* Verlag-Chemie, New York.
Bergmeyer, H. U. (ed.) (1983 *et seq.*) *Methods of Enzymatic Analysis,* Vol. 1–10, 3rd ed. Academic Press, New York.
Blake, C., Al-Bassam, M. N., Gould, B. J., Marks, V., Bridges, J. W. & Riley, C. (1982) *Clin. Chem.* 28 1469–1473.
Blake, C. & Gould, B. J. (1984) *The Analyst* 109 533–547.
Bowers L. D. (1982) *Trends Anal. Chem.* 1 191–198.
Burd, J. F., Carrico, R. J., Tetter, M. C., Buckler, R. T., Johnson, R. D., Boguslaski, R. C. & Christner, J. E. (1977) *Anal. Biochem.* 77 56–67.
Campbell, J. & Hornby, W. E. (1977). In: *Biomedical Applications of Immobilised Enzymes and Proteins,* pp. 3–26 (ed. by Chang, T. M. S.). Plenum Press, New York.
Carlier, Y., Bout, D. & Capron, A. (1981) *Bulletin De l'Institut Pasteur* 79 313–382.
Carr, P. W. & Bowers, L. D. (1980) *Immobilized Enzymes in Analytical and Clinical Chemistry,* John Wiley and Sons, New York.
Cate, J. C., Hedrick, R., Taylor, M. & McGlothlin, C. D. (1980) *Clin. Chem.* 26 266–270.
Chirillo, R., Caenaro, G. Paven, B. & Pin, A. (1979) *Clin. Chem.* 25 1744–1748.
Chua, K. S. & Tan, I. K. (1978) *Clin. Chem.* 24 150–152.
Clark, P. M. S. & Broughton, P. M. G. (1983a) *J. Autom. Chem* 5 22–26.
Clark, P. M. S. & Broughton, P. M. G. (1983b) *Ann. Clin. Biochem.* 20 208–212.
Colliss, J. S. & Knox, J. M. (1978) *Med. Lab. Sci* 34 275–285.
Curme, H. G., Columbus, R. L., Dappen, G. M., Eder, T. W., Fellows, W. D., Figueras, J., Glover, C. P., Goffe, C. A., Hill, D. E., Lawton, W. H., Muka, E. J., Pinney, J. E., Rand, R. N., Sanford, K. J. & Wu, T. W. (1978) *Clin. Chem.* 24 1335–1342.
Curtius, H. Ch. & Roth, M. (1974) *Clinical Biochemistry,* Vol. 2, pp. 1186–1304, Walter de Gruyter, Berlin and New York.
Dappen, G. M.. Cumbo, P. E., Goodhue, C. T., Lynn, S. Y., Morganson, C. C., Nellis, B. F., Sablauskas, D. M., Schaeffer, J. R., Schubent, R. M., Snoke, R. E., Underwood, G. M. Warburton, C. D. & Wu, T. W. (1982) *Clin. Chem.* 28 1159–1162.
Drucker, R. F., Williams, D. R. R. & Price, C. P. (1983) *J. Clin. Pathol.* 36 948–953.
Engvall, E. & Perlmann, P. (1971) *Immunochemistry* 8 871–874.
Erlanger, B. F., Borek, G., Beiser, S. M. & Lieberman, S. (1959) *J. Biol. Chem.* 234 1090–1094.
Filippusson, H., Hornby, W. E. & McDonald, A. (1972) *FEBS Lett.* 20 291–293.
Finley, P. R., Williams, R. J. & Lichti, D. A. (1980) *Clin. Chem.* 26 1723–1726.
Free, A. H. (1977) *Ann. Clin. Lab. Sci.* 7 479–485.
Free, A. H., Adams, E. C., Kercher, M. L., Free, H. M. & Cook, M. H. (1957) *Clin. Chem.* 3 163–168.
Gould, B. J. (1975) In: *Handbook of Enzyme Biotechnology,* pp. 5–26 (ed. by Wiseman, A.) Ellis Horwood Ltd, Chichester.
Gould, B. J. (1978) *U.V. Spectrometry Group Bulletin* 6 57–68.
Gray, D. N., Keyes, M. H. & Watson, B. (1977) *Anal. Chem.* 49 1067A–1075A.
Greyson, J. (1981) *J. Autom. Chem.* 3 66–71.
Guilbault, G. G. (1976) *Handbook of Enzymatic Methods of Analysis,* Marcel Dekker, New York.
Gupta, R. C., Goyal, A. & Singh, P. P. (1982) *Clin. Chem.* 28 1724.
Hall, P. M. & Cook, J. G. H. (1982) *Clin. Chem.* 28 387–388.
Hamilton, S. D. & Pardue, H. L. (1982) *Clin. Chem.* 28 2359–2365.
Hornby, W. E., Inman, D. J. & McDonald, A. (1972) *FEBS Lett.* 20 114–116.
Hornby, W. E. & Morris, D. (1975) In: *Immobilized Enzymes, Antigens, Antibodies and Peptides,* pp. 142–169 (ed. by Weetall, H.), Marcel Dekker Inc., New York.
Horvath, C. & Solomon, B. A. (1972) *Biotech. Bioeng.* 14 885–914.
IFCC (International Federation of Clinical Chemistry) (1977) *Clin. Chem. Acta* 80 F21–22.
Ishikawa, E. (1980) *Immunoassay* (Suppl. 1), 1–16.
Kato, K., Hamaguchi, Y., Fukui, H. & Ishikawa, E. (1975) *J. Biochem.* 78 235–237.
Kay, J. D. S. & Taylor, F. (1983) *Clin. Chem.* 29 1558–1559.

King, G., Steggles, D. & Harrop, J. S. (1982) *Br. Med. J.* **285** 1165.
Koch, T. R. & Nipper, H. C. (1977) *Clin. Chim. Acta* **78** 315–322.
Kutter, D. (1977) *Rapid Clinical Diagnostic Tests,* Urban and Schwarzenberg, Baltimore.
Landon, J., Carney, J. & Langley, D. (1977) *Ann. Clin. Biochem.* **14** 90–99.
Leon, P. L., Sansur, M., Snyder, L. R. & Horvath, C. (1977) *Clin. Chem.* **23** 1556–1562.
Leon, P. L., Smith, J. B., Yeung, A. & Yeh, C. K. (1982) *J. Autom. Chem.* **4** 11–16.
Lindh, M., Lindgren, K., Carlström, A. & Masson, P. (1982) *Clin. Chem.* **28** 726.
McComb, R. B., Bond, L. W., Burnett, R. W., Keech, R. C. & Bowers, G. N. (1976) *Clin. Chem.* **22** 141–150.
Miller, J. N., Rocks, B. F. & Thorburn Burns, D. (1976) *Anal. Chim. Acta* **86** 93–101.
Miller, J. N., Rocks, B. F. & Thorburn Burns, D. (1977) *Anal. Chim. Acta* **93** 353–356.
Mitsuma, T., Colucci, J., Shenkman, L. & Hollander, G. S. (1972) *Biochem. Biophys. Res. Comm.* **46** 2107–2113.
Möllering, H., Wahlefeld, A. W. & Michal, G. (1978) In: *Principles of Enzymatic Analysis,* pp. 89–93, (ed. by Bergmeyer, H. U.), Verlag Chemie, Weinheim.
Moody, G. J. & Thomas, J. D. R. (1975) *Analyst* **100** 609–619.
Mottola, H. A. (1983) *Anal. Chim. Acta* **145** 27–39.
Nakane, P. K. & Kawaoi, A. (1974) *J. Histochem. Cytochem.* **22** 1084–1091.
Ngo, T. T. (1980) *Int. J. Biochem.* **11** 459–465.
Olesen, H., Mortensen, H. & Mølsted-Pedersen, L. (1983) *Clin. Chem.* **29** 212.
O'Sullivan, M. J., Bridges, J. W. & Marks, V. (1979) *Ann. Clin. Biochem.* **16** 221–239.
O'Sullivan, M. J., Gremmi, E., Morris, D., Al-Bassam, M. N., Simmons, M., Bridges, J. W. & Marks, V. (1978a) In: *Enzyme-linked Immunoassay of Hormones and Drugs,* pp. 301–310, (ed. by Pal, S. P.), Walter de Gruyter, Berlin and New York.
O'Sullivan, M. J., Gnemmi, E., Morris, D., Chieregatti, G., Simmons, M., Simmonds, A. D., Bridges, J. W. & Marks, V. (1978b) *FEBS Lett* **95** 311–313.
Purdue, H. L. (1977) *Clin. Chem.* **23** 2189–2201.
Rocks, B. F. (1973) *Proc. Soc. Anal. Chem.* **10** 164–165.
Roth, M. & Selz, L. (1980) In: *Trends in Enzymology, FEBS,* Vol. 61, pp. 145–153, (ed. by Vitale, Lj. & Simeon, V.) Pergamon Press, Oxford, New York, Toronto, Sydney, Paris and Frankfurt.
Rowley, G. L., Rubenstein, K. E., Huisjen, J. & Ullman, E. F. (1975) *J. Biol. Chem.* **250** 3759–3766.
Sampson, E. J. & Baird, M. A. (1979) *Clin. Chem.* **28** 1721–1729.
Schuurs, A. H. W. M. & Van Weeman, B. K. (1977) *Clin. Chem. Acta* **81** 1–40.
Sherwood, J. M., Warchal, M. E. & Chen, S. (1983) *Clin. Chem.* **29** 438–446.
Shirey, T. L. (1983) *Clin. Biochem.* **16** 147–155.
Simpson, E. & Thompson, D. (1977) *Lancet* ii 361–362.
Simpson, E. & Thompson, D. (1978) *Clin. Chem.* **24** 389–390.
Smith, C. H., Landt, M., Steelman, M. & Ladenson, J. H. (1983) *Clin. Chem.* **29** 1422–1425.
Sokol, L., Garber, C., Shults, M. & Updike, S. (1980) *Clin. Chem.* **26** 89-92.
Spayd, R. W., Bruschi, B., Burdick, B. A., Dappen, G. M., Eikenberry, J. N., Esders, T. W., Figueras, J., Goodhue, C. T., LaRossa, D. D., Nelson, R. W., Rand, R. N. & Wu, T. W. (1978) *Clin. Chem.* **24** 1343–1350.
Spencer, W. W., Sylvester, D. & Nelson, G. H. (1978) *Clin. Chem.* **24** 386–387.
Stevens, J. F. & Newall, R. G. (1983a) *J. Clin. Pathol.* **36** 9–13.
Stevens, J. F., Tsang, W. & Newall, R. G. (1983b) *J. Clin. Pathol.* **36** 598–601.
Stewart, T. C. & Kleyle, R. M. (1983) *Clin. Chem.* **29** 132–135.
Sundaram, P. V. & Hornby, W. E. (1970) *FEBS Lett.* **10** 325–327.
Sundberg, M. W., Becker, R. W., Esdero, T. W., Figueras, J. & Goodhue, C. T. (1983) *Clin. Chem.* **29** 645–649.
Tateishi, K., Yamamoto, H., Ogihara, T. & Hayashi, C. (1977) *Steroids* **30** 25–32.
Thomas, L., Plishcke, W. & Storz, G. (1982) *Ann. Clin. Biochem.* **19** 214–223.
Toffaletti, J., Blosser, N., Hall, T., Smith, S. & Tompkins, D. (1983) *Clin. Chem.* **29** 684–687.
Tomkin, G. H. & Moore, J. (1977) *J. Irish Med. Ass.* **70** 490–493.
Ullman, E. F., Yoshida, R. A., Blakemore, J. L., Maggio, E. & Leute, R. (1979) *Biochem. Biophys. Acta* **567** 66–74.
Walter, B. (1983) *Anal. Chem.* **55** 498A–514A.
Werner, M., Mohrbacher, R. J., Riendeau, C. J., Murador, E. & Cambiaghi, S. (1979) *Clin. Chem.* **25** 20–23.
Wisdom, G. B. (1976) *Clin. Chem.* **22** 1243–1255.
Yolken, R. H. (1982) *Rev. Infectious Diseases* **4** 35–68.

Zeigenhorn, J., Senn, M. & Bücher, T. (1976) *Clin. Chem.* **22** 151–160.
Zipp, A. (1981) *J. Autom. Chem.* **3** 71–75.

INDUSTRIAL UTILIZATION
OF ENZYMES AND CELLS

CHAPTER 1

Introduction to enzyme utilization

Dr. ALAN WISEMAN, Biochemistry Division, Department of Biochemistry, University of Surrey, Guildford, England

Part B of this book is intended to supply the data (corresponding to Part A chapters) to those interested in the theoretical and practical details of large-scale isolation and purification of enzymes, immobilization of enzymes, and their subsequent use in free and immobilized form in a variety of industrial applications. The scope of these applications has been extended to include the clinical laboratory, where many exciting developments have occured recently.

Very many suggestions for the application of existing commercial enzymes have now been made, and suggestions relating to 'other enzymes' are now too numerous to be completely discussed in a general review of this rapidly expanding field. Major uses of enzymes are nevertheless still relatively few, although useful progress has been clearly achieved in the ten years since the publication of the first edition of this *Handbook*. Much fundamental and applied work remains to be done in manufacturing the required enzyme in appropriate stable physical form. The future lies with 'tailor-made' enzymes prepared through recombinant DNA techniques, and eventually by synthesis of enzymes and enzyme mimics! For review see Wiseman (1984). For important general review see Dunnill, P. (1984) writing on 'Biotechnology and British Industry'. For the use of enzymes in the analysis of foods see Wiseman (1981).

REFERENCES

Dunnill, P. (1984) *Chemistry & Industry,* 2nd July No. 13, 470–475.
Wiseman, A. (1984) In: *Topics in Enzyme and Fermentation Biotechnology* (ed. Wiseman, A.) Volume 9, pp. 202–212. Ellis Horwood Ltd., Chichester.
Wiseman, A. (1981) *Enzymes and Food Processing* (ed. Birch, G. G., Blakebrough, N. & Parker, K. J.), pp. 275–288, Applied Science Publishers, UK.

Practical aspects of large-scale protein purification

Dr. M. D. SCAWEN and Professor J. MELLING, PHLS Centre for Applied Microbiology and Research, Porton Down, Salisbury, England

2.1 INTRODUCTION

The earlier discussion in Part A: Chapter 2 was mainly concerned with the theoretical aspects and principles involved in the various techniques for enzyme purification. This chapter is intended to deal briefly with some of the practical problems involved as well as describing some items of equipment which have been used. This cannot, in the space available, be a comprehensive study, but it is intended as a guide to indicate some of the more important points to consider when undertaking large-scale enzyme purifications. In general, we have only given details of manufacturers and their products for special items of which we have some experience. If any manufacturer or piece of equipment has been omitted this must not be construed as implying any adverse comment, but rather ignorance on the part of the authors.

The details given below refer to enzyme purification in general, but, where the final product has some therapeutic application, special regulations may apply under the Medicines Act 1968. A booklet entitled *Guide to Good Pharmaceutical Manufacturing Practice* (HMSO) is relevant to such therapeutic materials, but many of the points made in it have a general application to the control of any process.

The content of this chapter is intended to complement, so far as possible, the corresponding chapter in Part A, so that information on processes and equipment is more readily accessible.

2.2 ENZYME INACTIVATION

In any large-scale enzyme extraction process, depletion of the product through inactivation rather than physical loss can present an extremely difficult problem. Some aspects of this were considered by Thurston (1972) in relation to the culture situation, but some points are applicable to enzyme purifications.

Losses may be minimized by ensuring that the limits of pH and temperature within which particular enzymes are stable are not exceeded, and hence foaming and high-shear situations are avoided. Even so, losses can occur, and it has become increasingly clear that, particularly in the early stages of a purification process, proteolytic enzymes may frequently be responsible.

The activity of bacterial proteases can be reduced by the selection of suitable conditions of pH and temperature or the addition of inhibitors. The former approach was successful in the purification of a penicillinase from *E. coli* where loss of penicillinase due to alkaline protease activity could be prevented by maintaining the pH at 6.5—7.0 during the first stages of the purification (Melling & Scott 1972). As an alternative, calcium dependent proteases can be inhibited by addition of EDTA, while serine proteases are sensitive to phenyl-methyl sulphonyl fluoride (Gold 1965, Callow *et al.* 1973). However, phenyl-methyl sulphonyl fluoride can react with other hydroxyl groups besides the active serine residue of susceptible proteases. The most important of these is the hydroxyl group of water. It has been shown that in an aqueous environment at pH 8.0 and 25°C the half life of 100 μM phenylmethyl sulphonyl fluoride is about 30 minutes, increasing to some 20 hours at pH 7.6 and 4° (James 1978).

There are a number of metallo-enzymes where the molecule contains one or more metal atoms which are necessary for the maintenance of enzyme activity and also of the protein stability. In such cases the use of chelating agents must be avoided unless they are specific for metals other than the one involved in the protein structure. With this type of enzyme, buffer solutions used during the extraction procedure should usually contain the particular metal ion needed for activity and stability.

2.3 CONTAINERS AND ANCILLARY EQUIPMENT

2.3.1 Glass vessels

Hard glass is an ideal material for enzyme work on the small scale; it is resistant to corrosion, except in strong solutions of sodium hydroxide, and is virtually nontoxic. However, on the large scale, glass vessels have several disadvantages, including lack of structural strength, which may constitute a hazard to workers and a potential embarassment in terms of investement, and a lack of versatility.

2.3.2 Metal vessels

Vessels constructed of mild steel, copper alloys, and aluminium are generally not used, because of corrosion problems. This is common to both fermentation vessels and those used for enzyme isolation. However, there is invariably an exception, and Lingood *et al.* (1955) used a vessel fabricated in aluminium alloy for the production of diptheria toxin because of the need to avoid the presence of iron.

Copper, although very useful in the fabrication of some peripheral items which have no contact with process liquors, should be avoided where contamination of the product could occur. For normal pipework carrying compressed air, water, chilled water, and steam, as well as jacketing of vessels, mild steel is often the material of choice. Used under these conditions the metal may be protected by a coat of zinc, that is, galvanized. These materials are used for the rigid fittings which supply the various services; but to achieve an adaptable multipurpose unit, connections between vessels and other equipment are usually made with flexible tubing such as rubber, polythene, polyvinylchloride, or nylon.

Historically, criteria for determining the materials to be used in vessel fabrication have been drawn from experience in the fermentation industry. Here, the vessels are normally constructed from stainless steel, but, as Solomons (1969) pointed out, there are over 50 different types of stainless steel, not all resistant to the various conditions which may be used. Elsworth *et al.* (1956) listed the stainless steels which are suitable for use in the construction of fermentation vessels. These are the austentic chromium nickel steels BS No. En 58M, En 58B, and En 58A. They are resistant to aqueous solutions of sodium hydroxide, ammonia, phosphoric and nitric acids. However, when working at low pH, that is, pH 1−2, the BS No. En 58J series of steels is found to be more resistant to acids such as citric acid and sulphuric acid. The resistance of stainless steels to corrosion results from the chromium rich oxide layer formed on the surface of the metal.

Stainless steel vessels have been found to be perfectly adequate for use in the purification of enzymes. Possible problems due to leaching of metal ions from the vessels can be avoided by the addition of 2−10 mM EDTA which will effectively remove such ions from solution by chelation. Some advantages of using stainless steel vessels include: ease of cooling and maintaining low temperatures ($+8°C$) by circulation of coolant through the vessel jacket; and robust construction for stirrer attachment and inclusion of pH electrodes. They are also strong, durable and simple to sterilize and clean, capable of functioning under pressure, or maintaining a particular gaseous environment, and easily modified if necessary.

2.3.3 Plastic vessels

Vessels made from polypropylene are of considerable use where corrosive materials, such as ammonium sulphate, are used, or where metal contamination of products is a problem. It is difficult to keep this type of vessel completely metal-free as the stirring mechanism usually has a metal shaft; also the most effective way in which to maintain a low temperature is by inserting metal cooling coils. However, the use of plastic vessels reduces the metal surface area in contact with enzyme solutions and also avoids the possibility of corrosion when handling electrolyte solutions of high concentration, as in ammonium sulphate precipitations.

Although most plastics are inert, nontoxic and can withstand the corrosive effects of many chemicals, they invariably contain plasticizers, extruders and fillers which may be leached out. Therefore in any process utilizing plastic vessels the possibility of enzyme inactivation or contamination due to contact with such materials should be considered before large-scale work is undertaken. Because they are of less robust construction than steel vessels, those made from plastics may need supporting as the capacity increases. The attachment of stirrer equipment also requires some rigid framework. In addition, sterilization is a problem since heating cannot be used and chemical sterilants may produce problems of contamination of the product. Irrespective of the material used in vessel construction, problems of temperature and pH control may be encountered as the vessel size is increased.

Temperature control is often effected by circulation of chilled water or glycol through a vessel jacket so that heat transfer takes place at the vessel surface. However, as the volume of the vessel increases the relative area for heat transfer is decreased ($V \propto r^3$; $A \propto r^2$). This problem can be partly overcome by increasing the flow of coolant and improving the mixing efficiency of the vessel. An alternative method for rapid cooling is to circulate the process liquor through a more efficient heat exchange system such as a plate cooler. Higher temperatures, which may be required for heat precipitation of unwanted proteins, can also be achieved in a similar way.

The efficiency of pH control is determined to a considerable extent by the efficiency of mixing of added acid or alkali within a particular vessel. Mixing efficiency in a stirred vessel is governed by the parameters of vessel size and shape, stirrer speed, impeller design, baffling or other inclusions. Butters (1969) has reviewed the subject of mixing, and this article may be consulted for further details. The viscosity of the product may have an adverse effect on mixing efficiency, and the presence of highly polymerized nucleic acid can severely impair the performance of the stirrer. Poor mixing may result in localized high concentrations of acid or alkali giving rise to enzyme inactivation.

2.4 LIQUID TRANSER

2.4.1 Couplings

On a laboratory scale the transfer of solutions from one container to another is a simple operation. However, when several hundred litres are involved some detailed attention has to be given to the problem. Fig. 2.1 shows a schematic arrangement to allow the maximum flexibility for transfer from vessel to vessel either by gravity feed or pumped flow. Vessels may be connected to each other directly, or via the manifold, such connections being made with flexible tubing. Various ways of coupling the tubing are available including Albany (W. H. Wilcox & Co. Ltd, P.O. Box 23, Southwark Street. London, U.K., SE1 1RX) and Kamlock (Dover Corporation, OPW Division, Cincinnati, USA) fittings which provide quick and simple coupling techniques.

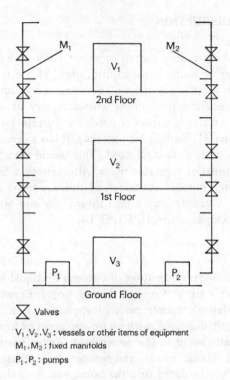

Fig. 2.1 – Schematic arrangement for liquid transfer.

2.4.2 Pumps

A wide variety of pumping equipment is available with flow rates ranging from a few millilitres to hundreds of litres per hour. This area has been reviewed by Solomons (1969), and readers who require details of pump designs and manufacturers are referred to that article.

For rapid transfer of large volumes of process liquors containing enzymes, any pump used should be of a type which produces low shear, and foaming must also be avoided. In addition, the use of a valveless pump prevents problems of seating when particulate suspensions are involved, and in general this type of pump gives rise to few problems. To have a pump which is self-priming is also advantageous.

The addition of specified amounts of reagents can be effected using a metering pump, but it is more common, given the viability of biological materials, to monitor such additions by pH, conductivity, or other relevant parameters.

Peristaltic pumps are widely used in connection with column chromatography, and they provide a simple means of delivering various liquids at low pressure.

2.5 BACTERIAL DISRUPTION

2.5.1 Resuspension

Bacterial pastes, whether deep frozen or direct from a centrifuge, require re-suspending in water or buffer prior to disruption. This may often be achieved by manual or simple mechanical stirring, but occasionally more vigorous methods are needed. A commercial food mixer with a capcity of some 20 litres may suffice, but where greater agitation is needed a Turmix blender may be used (Tech. A.G., Utoquai 31, Zurich, Switzerland). It has a capacity of 30 litres and can be fitted with a stainless steel bowl. This would also be suitable for the comminution of animal or vegetable tissue. Alternatively a Silverson type mixer may be used in any suitable container, as this design of mixer produces no frothing, although it can develop a high shear (Silverson Machines Ltd, Waterside, Chesham, Buckinghamshire, HP5 1PQ, UK).

2.5.2 Liquid shear

For disruption of kilogram quantities of bacteria by liquid shear and Manton–Gualin homogenizer (A.P.V. Co. Ltd, Manor Royal, Crawley, Sussex, UK) is an efficient and relatively simple piece of apparatus. It is important that cell suspensions are evenly dispersed, with no frozen lumps or foreign bodies. Failure can frequently be attributed to the presence of foreign particles and such items as pieces of wood, elastic bands, and washers have caused problems. These materials are usually introduced into the homegenizer via the extract, and cause malfunction of the suction or non-return valve, or blockage at the homogenizing valve and impact ring.

A system of cleaning and maintenance is essential, and after use the machine should be washed through with water and then dismantled. All parts need to be washed individually with hot water and dried. Before reassembling the machine, the piston packing rings should be checked for wear and replaced if necessary, and the cavity above the dampener piston filled with grease. Having reassembled the homogenizer, it should be checked under operative conditions using distilled water.

If the dampener needle value is not screwed in correctly there is a possibility of it being blown our under high pressure. This hazard can be overcome by enclosing the homogenizer in a cabinet, thus protecting the operator and other personnel.

Temperature control can be a problem; we have recorded rises of up to 20°C between inlet and outlet temperatures. Enzyme loss due to this can be reduced by ensuring that the inlet temperature is low, and by connecting the outlet to a plate cooler and then feeding the homogenized suspension into a cooled vessel. Increased viscosity after homegenization is invariably due to DNA, and can be decreased by the addition of a small amount of DNase.

It is of interest that a machine capable of such disruptive power can also be used to remove sex pili from bacterial surfaces without damaging either cells or pili. Operating conditions for cell disruption vary with the type of bacteria, cell concentration, and location of an enzyme within the cell. Unbound intracellular bacteria enzymes can normally be released by a single pass at 55 MPa at bacteria concentrations of 10–20% (w/v). A membrane-bound enzyme, such as cytochrome from *Ps. aeruginosa*, required 3 passes at 55 MPa for release.

Two models of Manton–Gaulin homogenizer are available. The 15M–8BA which can be used either as a single or double stage homogenizer has a throughput of 54 litre/hr and a maximum operating pressure of 55 MPa. A larger version, the K3, is a two-stage homogenizer with a throughput of some 250 litre/hr and a maximum operating pressure of 35 MPa.

Because of their lower operating pressures the larger homogenizers are less efficient at disrupting bacterial cells when used in the single pass mode. The efficiency of operation can be increased by using multiple passes, continuous recycling, or recycling with a continuous bleed (Charm & Mateao 1971). When a homogenizer is operated in this fashion the temperature rise in the homogenate can be very great, so the outlet stream must be fed into a plate heat exchanger or similar high efficiency cooling system.

2.5.3 Grinding

Some organisms, particularly Gram-positive bacteria, have been found difficult to break using liquid shear homogenization, but grinding with glass beads has proved a useful method. The principle of a glass bead mill based on the design of Raheck *et al.* (1969) has been described in Chapter 0.

The smallest Dynomill (Backhoffen, 4000 Basel 5, Schweiz, Switzerland) is the type KDL which may be used either with a batch or a continuous grinding container.

Batch containers with volumes up to 300 ml are available, and hence some 200 ml of bacteria suspension can be treated at a time. It has been found useful to use the batch container to establish conditions of time, bacterial concentration and buffer composition for disruption before using the machine in the continuous mode.

When the continuous grinding container is used the bacterial suspension is pumped in at such a rate that the mean residence time of bacteria in the grinding container is equal to time found to be necessary for disruption when using the batch vessel. Now $D = F/V$, where D = dilution rate, F = flow rate (l/h), V = volume of suspension in grinding container, excluding beads. Then: mean residence time = $1/D$.

It has been found in practice that using 0.1 mm beads the maximum flow rate through the 0.6 litre container is about 12 litre/hr with a 10% (w/v) bacterial suspension. This held for three Gram-positive bacteria and for baker's yeast. It is, however, possible to work with up to 20% (w/v) suspensions.

Larger mills are available. The KD5 (5 litre continuous container) and the

KD15 (15 litre continuous container) have been used for microbial disruption (Dunnill & Lilly 1972). However, larger beads must be used (0.25–0.5 mm) as the separator systems on these larger machines does not retain the smallest beads. Consequently, breaking efficiency for bacteria may be reduced considerably (Hedenskog *et al.* 1969).

Woodrow & Quirk (1982) studied the release of two β-lactamases and a carboxypeptidase G_2 from bacteria in a Dynomill KDL, and found that 0.25 mm beads gave satisfactory release at an agitator speed of 15 m s^{-1} and a 1:22.5 (w/w) cell suspension in all cases. The optimum flow rate for the continuous release of carboxypeptidase was 15 l h^{-1}, giving a residence time of 1.6 min. This result would indicate that this small-scale unit can break up to 4 kg bacteria per hour, and that the larger models could handle up to 100 kg bacteria per hour using a model KD5.

For the small-scale experimental release of enzymes from bacteria, the Bead Beater (Life Science Laboratories, Bedford) proves a useful adjunct to the larger Dynomill type of instrument. Designed for batch operation, it is supplied with a variety of containers of 15–350 ml capacity and can give 95% breakage of most bacteria in 1–3 min.

A Swedish company, Innomed-Konsult A.B., Stockholm, have recently introduced a ball mill for the disruption of bacteria which uses a magnetic coupling between the agitator shaft and drive motor, thereby eliminating the complex separator systems of conventional instruments and permitting the use of 0.1 mm beads to achieve higher efficiencies; this instrument is claimed to be capable of handling up to 50 litre per hour of cell suspension using a 2.2 litre chamber.

The Dynomill KDL has advantages for work with pathogenic bacteria in that its size and design make it possible to enclose it in a safety cabinet (Melling *et al.* 1973).

2.6 CENTRIFUGATION

The principle of centrifugation is well established, and hence it was decided to omit this topic from Part A, Chapter 2, but a few points are worth mentioning:

The rate of settling for spherical particles is given by the equation

$$V = \frac{(P_p - P_L)_g D^2}{18\mu} \times F_s$$

where F_s is a correction factor for particle interaction in hindered settling,

$$F_s = \frac{X_L^2}{10^{1.82}(1 - X_L)} ,$$

D = particle diameter, g = gravitational field, V = terminal settling velocity, X_L = volume fraction occupied by the liquid, μ = viscosity of suspending fluid, P_L = density of liquid, P_p = density of particle.

Now the distance which a particle moves is Vt, where t is the residence time of the particle in the gravitational field. Hence, since particles must impinge on a surface to be removed, only those particles which are within a distance Vt of a surface will be deposited. This has clear implications in centrifuge design. The formation of protein precipitates, and more importantly their removal and recovery from large volumes of process liquor, has been comprehensively reviewed by Bell *et al.* (1983).

2.6.1 Batch centrifuges
A wide variety of batch centrifuges are available with capacities ranging from less than 1 ml up to several litres and capable of applying a centrifugal force up to 100 000 $\times$ g. However, for the deposition of bacterial cells, cell debris, or protein precipitates, forces up to 20 000 $\times$ g are usually sufficient. Machines capable of this and of handling reasonable quantities of material are less numerous.

For instance, the Sorval RC3-B or Beckman J6B centrifuges can hold 6 $\times$ 1 litre containers and give up to 5000 $\times$ g. The Sorval RC5-B or Beckman J2-21M can hold 6 $\times$ 500 ml containers and give up to 13 700 $\times$ g. These and similar machines from other manufacturers are a valuable adjunct to any large-scale enzyme purification plant. Most centrifuges of this type can also use continuous flow rotors, but their capacity is limited to some 300–800 ml of sediment.

2.6.2 Continuous flow centrifuges
In the early stages of any large-scale enzyme purification process it may be necessary to remove solids from several hundred litres of suspension. With this type of centrifuge the deposited solids are retained in the centrifuge bowl and the clarified supernatant is continuously discharged.

There are three basic designs of centrifuge which can retain several kilograms of solid and operate at sufficiently high flow rates; these are the disc type centrifuge, the hollow bowl centrifuge, and the basket centrifuge.

Disc type centrifuge
A disc type centrifuge is shown in Fig. 2.2, and in this particular case the deposited solid can be discharged intermittently without stopping the machine. However, limited experience with a centrifuge of this type has indicated some losses of enzyme activity when the solid is discharged, possibly as a result of high shear forces set up during this process. Nevertheless, disc type centrifuges without solid discharge have proved most efficient for clarification.

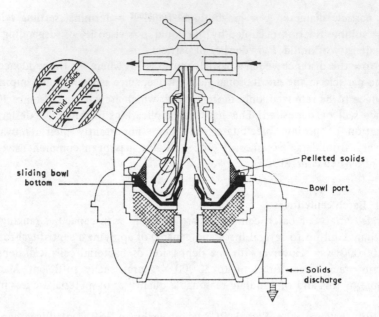

Fig. 2.2 – Continuous disc type centrifuge (Charm & Matteo 1971; by courtesy of Academic Press Inc.

The bowl used with these machines contains a central stack of coned discs which ensures an almost constant length of flowpath, and the deposition of solids at the bowl wall produces only a minor reduction in the flowpath. Thus, there is little loss of efficiency as the process proceeds. However, cleaning is laborious, and some loss of deposit is difficult to avoid. Similarly reassembly is time consuming.

Disc type centrifuges are available from De Laval Separator Co., New York, USA; Westfalia Separator Ltd., Wolverton, Bucks, UK; Bird Machine Co., South Walpole Massachusetts, USA. These have solid capacities of up to 20 kilograms and operate at about 8000 × g. The selection of a suitable flow rate is essentially empirical and will vary greatly with the nature of the feed material.

Hollow bowl centrifuges
These have a tubular section bowl in order to obtain a sufficient length of flow path, and solids are deposited onto the bowl (Fig. B2.3); thus as centrifugation continues the effective bowl diameter, and hence the centrifugal force, are reduced. However, such a centrifuge is easily cleaned, and by using a bowl liner which can be withdrawn, the recovery of deposited material is virtually 100%. Likewise reassembly is a quick and simple procedure. The Pennwalt centrifuges are perhaps the best known of this type, and details are given in Table 2.1, although they are also available from Carl Padberg, Lahr, West Germany.

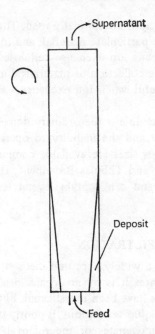

Supernatant

Deposit

Feed

Fig. 2.3 – Hollow bowl centrifuge.

Table 2.1

Characteristics of Pennwalt tubular bowl clarifiers

Model	Bowl capacity (g wet deposit)	Maximum RCF	Bowl weight
T-41-24Y	200	50 000 or 13 000	1.4 kg
A.S.16	3500	13 000	27 kg
A.S. 26	5200	16 000	63 kg

Flow rates must again be determined empirically, but for deposition of cell debris after bacterial disruption flow rates of about 50 litre/hr have been used with the two larger machines.

Basket centrifuges

These are designed to operate at much lower g forces than the centrifuges described above (1000 rpm compared with 10 000 rpm), and they are in essence centrifugal filters, the bowl being perforated to allow egress of filtrate. Porous bowl liners of some filter cloth are normally used. The main purpose of these machines is to collect large particulate material, and in the context of enzyme purification this usually means ion exchange cellulose or resins. Basket centrifuges may be useful for the collection of materials during batch adsorption of enzymes; they are also useful when ion exchangers are being regenerated or equilibrated.

Perhaps the simplest example is the ordinary domestic spin dryer. Its main drawbacks are low capacity and the inability to operate on continuous flow. Basket centrifuges of various sizes are available commercially (from Carl Padberg, Lahr, West Germany and Thomas Broadbent, Huddersfield, UK) which will operate continuously and can contain several tens of kilograms of ion exchangers.

2.7 TANGENTIAL FLOW FILTRATION

Although centrifugation is a widely accepted method for the removal of cell debris and protein precipitates it is by no means ideal for large-scale use. For this reason other techniques have been investigated. Filtration, although used in many industries for solid—liquid separations, is poorly suited to the clarification of bacterial or animal homogenates or the removal of protein precipitates. These are often gelatinous and slimy in nature and filter very poorly unless large quantities of filter-aid are used, or the precipitation conditions can be adjusted to give a flocculant precipitate.

An alternative method of preventing the blocking of the filtration membrane is to use tangential flow, in which the liquid flow is at right angles to the direction of filtration. By using a sufficiently high flow rate, blocking of the filter can be minimized. The use of this technique for the removal of bacterial cells from a culture has been demonstrated on a small scale (Valeri *et al.* 1979; Tanny *et al.* 1980). Quirk & Woodrow (1983) using a Millipore Pellicon cassette system demonstrated the feasibility of tangential flow filtration in the clarification of two bacterial extracts: a *Pseudomonas fluorescens* containing aryl acyl amidase (Hammond *et al.* 1983) and a *Pseudomonas* sp. containing carboxypeptidase G_2 (Baird *et al.* 1976). These results showed that both the organism and the method of cell breakage had a marked effect on the filtration rate. In addition the isotropic membranes used were very prone to blocking by proteins and debris and by the build-up of a secondary membrane.

The recovery of enzyme could be improved by diluting the extract and by washing the membranes. However, these steps lead to a considerable increase in volume.

A new type of membrane, which has an asymmetric structure (Asypore™), has recently been made available by Domnick Hunter Filters Ltd (Birtlry, Co. Durham. UK). These membranes are much less prone to fouling and hence can handle much higher concentrations of solids. Using the aryl acyl amidase from

Pseudomonas fluorescens the effects of pressure, feed velocity, and membrane configuration were investigated using a unit containing a total of 128 cm² of membrane (Le *et al.* 1984a). These same membranes were also tested on a larger scale employing *Erwinia chrysanthemumi*.

A 1 m² membrane assembly was used to harvest the cells from 100 litres of culture fluid in 2.5 h. The solids concentration in the retentate increased from 0.55% dry weight to 22% dry weight. This same membrane assembly was then used to clarify the extract produced by tha alkali lysis of these bacteria (Le *et al.* 1984b).

The authors estimate that to harvest the bacteria from a 500 litre culture in 2.5 h would require a total of 7.5 m² membrane. The economics of the membrane filtration process compare very favourably with centrifugation, with the membrane process costing an estimated 25% less than centrifigation. In addition the problems of aerosol generation are effectively eliminated, which may be an important consideration when pathogenic or genetically engineered bacteria are employed.

2.8 CONCENTRATION

2.8.1 Ultrafiltration

There are some four basic designs of ultrafiltration units available. These are: stirred cells, thin channel systems, cartridge membranes, and hollow fibres.

Stirred cells

A stirred cell consists of a cylindrical vessel, the base of which is formed by the ultrafiltration membrane resting upon a rigid support. The unit is capable of pressurization, and the contents are stirred with a magnetic stirrer to prevent concentration polarization at the membrane. These units are available with capacities ranging from a few millilitres up to some 2 l (Amicon Ltd, Stonehouse, Gloucestershire, UK; Chem. Lab. Ltd, Ilford, Essex, UK; and Millipore, Harrow, Middlesex, UK).

Although not suitable for large-scale work the units are useful for preliminary trials as well as concentration of column eluates.

Thin Channel systems

These also use flat membranes which are sandwiched in a filter press arrangement. The material which is to be concentrated flows across the membrane through narrow channels designed to produce a laminar flow of liquid and thus cut down concentration polarization. The concentrate is usually recirculated from a reservoir, and the whole system can be pressurized to increase the ultrafiltration rate. Ultrafiltrate is collected separately as it emerges from the 'filter press'.

Units of this type are available from Amicon; Sartorius, Gottingen, West Germany, and Millipore. Linko *et al.* (1973) have compared the performance in thin channel systems of Amicon, Millipore, and Sartorius membranes with cellulose acetate membranes of their own manufacture.

The capacity of these units is about 20 litres, which can be increased if additional reservoirs are included in series. Additional reservoirs may also be used to dialyse material, by replacing the ultrafiltrate with fresh buffer solution.

Thin channel ultrafiltration has been criticized on the grounds that high liquid shear generated by movement of solution through the channels may damage some enzymes (Charm & Matteo 1971). A number of bacterial enzymes including L-asparaginase, several penicillinases, and cytochrome oxidase were not adversely affected, but a DNA-dependent RNA polymerase showed considerable loss of activity. It is clear, therefore, that care is needed in deciding whether or not an enzyme can be subjected to this process.

The concentration rates are highly variable, depending upon such factors as protein concentration, salt concentration, applied pressure, nature of the material, and number of membranes used. It is therefore extremely difficult to make valid comparisons, and potential users are most strongly recommended to make their own evaluations for their particular system.

The main disadvantage of these systems is that assembly and subsequent cleaning are very time consuming, but such apparatus has been found most useful at the final stages of concentration to reduce volumes from 10 to 20 litres down to some 1 to 2 litres.

Cartridge membranes

To obtain a larger surface area without making large and unwieldly stacks of flat membrane, one method has been to mould the membrane into a tube and then incorporate a number of such membrane tubes into a cartridge; liquid can then be recirculated through the cartridge and sufficient pressure for ultrafiltration obtained by restricting the flow downstream of the cartridge. In some systems the membrane may be moulded round a plastic rod which is grooved to produce a thin channel effect; it has been suggested that this is more efficient in reducing concentration polarization than a hollow membrane tube.

The cartridge system has been used in various large-scale units with membrane areas of up to several square feet. The manufacturers of such systems, include Amicon, Babcock & Wilcox, P.C.I. Ltd, Whitchurch, Hampshire, UK; Romicon Inc., Rotterdam, The Netherlands.

Molecular weight cut-off values range from 500 to 300 000.

Hollow fibres

These are microtubular membranes about 0.2–1.1 mm diameter and of variable length. They can be mounted into cartridges for easy manipulation, and provide large membrane areas within a small unit volume.

Amicon produce a laboratory model CH4 concentrator having a membrane area of 830 cm². Hollow fibre cartridges are available in a range of molecular weight cut-offs from 1000–100 000 and with fibre diameters from 0.2–1.1 mm. The larger fibre diameters 0.5 or 1.1 mm are most suitable for general work with protein solutions. With this model, typical ultrafiltration rates of 1–2 litre/h are obtained.

Larger versions of these concentrators are also available. The Amicon DC 30EM takes cartridges of 0.45 m² surface area and can be adapted to take from 3–7 such cartridges. The nominal molecular weight cut-offs range from 1000–100 000. This type of concentrator is ideally suited to pilot plant scale operation, and typical ultrafiltration rates of 50–150 litre/hr are obtainable. For larger scale operations Amicon produce a model DC 120EM which uses cartridges of 27 m² area and can give ultrafiltration rates up to 600 litres/hr.

Besides concentration, ultrafiltration can also be used for the rapid removal of salts from a protein solution, either by concentration, dilution, and concentration steps, or by the continuous replacement of the ultrafiltrate volume with fresh buffer in diafiltration.

2.8.2 Dialysis

Dialysis of protein solutions, using dialysis tubing, which retains molecules having molecular weights greater than 5000, has been used as a concentration step for many years on the laboratory scale: the surrounding liquid consisting of an aqueous solution of some inert high molecular weight solute such as polyethylene glycol. Although this approach is possible on the large scale it is unpractical and expensive. Even for simple desalting, without concentration, such a process is probably only suitable for volumes of a few litres. For dialysis of even this volume, and certainly for greater volumes, 10–50 litres, this is totally unpractical.

For dialysis (desalting or buffer exchange) on this scale, gel filtration or ultrafiltration can be employed. However, an alternative and perhaps cheaper, albeit slower, approach is to use an artificial kidney dialysis cartridge. These are readily available in a range of sizes from 0.6–2.5 m² surface area (for example, C-D Medical Systems Ltd, Slough, Berkshire, UK) and require no more than two pumps: one for the protein solution which is pumped through the fibres, and one for the buffer which is circulated around the fibres. This arrangement is simple to operate, and reliable, and by using a suitable size of reservoir it is possible to desalt 20 litres of protein solution overnight.

2.9 CHROMATOGRAPHY

2.9.1 Columns

The principles of small-scale column chromatographic techniques have been described in detail elsewhere (for example, Fischer 1969, Petersen 1971). These principles are equally important in large-scale chromatography (Janson & Hedman 1982). Thus the columns should be constructed so as to have the minimum dead volume above and below the packing, and the end pieces should be designed so as to ensure an even distribution of material over the entire surface area of the column, which may be 10 000 cm², or more, in cross-sectional area.

Large-scale columns of conventional glass or plastic construction are available from Amicon–Wright Ltd (Stonehouse, Gloucestershire, UK), and from Pharmacia Fine Chemicals Ltd. Both of these manufacturers also supply short, sectional 'stack' columns for applications requiring the use of soft gels. The Pharmacia stack column is 37 cm diameter and has a volume of 16 litres. Up to ten of these columns can be connected in series, giving a column of 160 litre capacity but which has most of the flow characteristics of a column only one-tenth that capacity.

Amicon–Wright manufacture a range of sectional 'stack' columns from 18 to 44 cm diameter and capacity from 4.6 to 30 litres. These columns feature an adjustable end piece to compensate for changes in bed height. As with the Pharmacia, system up to ten columns can be connected in series. The use of 'stack' columns can be advantageous in a manufacturing environment, because if one section becomes contaminated or fouled it can rapidly be removed and replaced without seriously affecting the running of the column.

Both of these manufacturers also supply stainless steel columns, Pharmacia covering the range 75–500 litres and Amicon–Wright 39–1100 litres. These columns have the disadvantage of being opaque, but offer many advantages, in the industrial situation, of strength and ease of cleaning and sterilization. Amicon–Wright also offer a range of high-performance stainless steel columns, up to 11 litres capacity that can withstand pressures up to 200 kg/cm², although of course the larger the column the lower is the maximum operating pressure.

The packing of such large columns needs to be carried out with care so as to avoid stratifying particles of varying sizes or the inclusion of cavities in the gel bed. Pharmacia give detailed instructions for packing their large columns, which usually involves swelling the dry gel *in situ*. However, the newer, more rigid materials are supplied in a wet state and are much less demanding of correct packing conditions. Indeed they can in many cases be packed very rapidly by the application of a slight positive pressure to the top of the column.

Large-scale columns can be run under gravity flow, but it is often easier and more convenient to use pumped flow, particularly if salt gradient elution is used. In our laboratories we routinely use roller type peristaltic pumps of the type manufactured by Watson Marlow Ltd (Falmouth, UK); these pumps offer a range of flow rates, up to 800 litre/hr, which are adequate for most types of chromatographic column. Alternatively, lobe rotor or centrifugal pumps can be used. The choice of pumps for large-scale chromatography has been discussed by Janson & Hedman (1982).

Large-scale gradient makers can be readily constructed from pairs of identical plastic or stainless steel tanks of the desired capacity, connected either by tubing at the base or by a siphon. The initial buffer tank is stirred with an overhead stirrer. With this simple equipment, linear concentration gradients up to several hundred litres can be prepared.

2.9.2 Gel chromatography

There are three main suppliers of materials for gel chromatography: Pharmacia Fine Chemicals Ltd, LKB Instruments Ltd, and Bio-Rad Laboratories Ltd. They all produce ample literature in various aspects of gel chromatography. The Sephadex (dextran) and Bio-Gel P (polyacrylamide) series of gels are well known and have been used for the large-scale purification of numerous enzymes. They are available in a wide range of pore and particle size (Tables 2.2, 2.3). However, these types of gel become increasingly less rigid as the pore size increases, such that the gels capable of fractionating proteins of molecular weight greater than 70 000 are too soft for routine use in large-scale chromatography. They are, however, excellent for desalting operations. Agarose gels are traditional gels which offer a greater rigidity, although they also have a larger pore size than the dextran or polyacrylamide gels (Tables 2.4, 2.5), making them suitable for the fractionation of large proteins, DNA, RNA, and viruses.

Table 2.2

Characteristics of various Sephadexes (by courtesy of Pharmacia (GB) Ltd)

Sephadex type		Dry bead diameter (μm)	Fractionation range (Daltons)		Bed vol. ml/g dry Sephadex
			Peptides and globular proteins	Dextrans	
Sephadex G-10		40–120	−700	−700	2–3
Sephadex G-15		40–120	−1500	−1500	2.5–3.5
Sephadex G-25	Coarse	100–300	1000–5000	100–5000	4–6
	Medium	50–150			
	Fine	20–80			
	Superfine	10–40			
Sephadex G-50	Coarse	100–300	1500–30 000	500–10 000	9–11
	Medium	50–150			
	Superfine	10–40			
Sephadex G-75		40–120	3000–80 000	1000–50 000	12–15
	Superfine	10–40	3000–70 000		
Sephadex G-100		40–120	4000–150 000	1000–100 000	15–20
	Superfine	10–40	4000–100 000		
Sephadex G-150		40–120	5000–300 000	1000–150,000	20–30
	Superfine	10–40	5000–150 000		18–22
Sephadex		40–120	5000–600 000	1000–200 000	30–40
	Superfine	10–40	5000–250 000		20–25

Table 2.3

Characteristics of the Bio-Gel P series of beaded polyacrylamide gel chromatography media (by courtesy of Bio-Rad Ltd)

Product		Diameter of hydrated beads (μm)	Fractionation range (Daltons)	Bed volume ml/g dry gel
Bio-Gel P-6DG	Desalting gel	90–180	1000–6000	7
Bio-Gel P-2	Fine	40–80	100–1800	3.5
	Extra fine	<40		
Bio-Gel P-4	Coarse	150–300	800–4000	5
	Medium	80–150		
	Fine	40–80		
	Extra fine	<40		
Bio-Gel P-6	Coarse	150–300	1000–6000	7
	Medium	80–150		
	Fine	40–80		
	Extra fine	<40		
Bio Gel P-10	Coarse	150–300	1500–20 000	9
	Medium	80–150		
	Fine	40–80		
	Extra fine	<40		
Bio-Gel P-30	Coarse	150–300	2500–40 000	11
	Fine	80–150		
	Extra fine	<80		
Bio-Gel P-60	Coarse	150–300	3000–60 000	14
	Fine	80–150		
	Extra fine	<80		
Bio-Gel P-100	Coarse	150–300	5000–100 000	15
	Fine	80–150		
	Extra fine	<80		
Bio-Gel P-150	Coarse	150–300	15 000–150 000	18
	Fine	80–150		
	Extra fine	<80		
Bio-Gel P-200	Coarse	150–300	10 000–200 000	25
	Extra fine	<80		
Bio-Gel P-300	Coarse	150–300	60 000–400 000	30
	Fine	80–150		
	Extra fine	<80		

Unfortunately, agarose gels are chemically and physically unstable, so the gels cannot be readily sterilized or used with strongly denaturing solvents, at pH values less than 4 or greater than 9 or above 40°C. By crosslinking agarose gels with dibromopropanol, beaded crosslinked agarose can be prepared. This retains the original macroporous structure of the parent agarose coupled with a much greater chemical and thermal stability (Table 2.5). Such gels are stable in all but the harshest of solvents, and can be repeatedly sterilized by autoclaving.

Table 2.4

Characteristics of the Bio-Gel A series of beaded agarose chromatography media
(by courtesy of Bio-Rad Ltd)

Product		Diameter of hydrated beads (μm)	Fractionation range (Daltons)	% agarose in gel
Bio-Gel A-0.5 m	Coarse	150–300	<10 000	10
	Medium	80–150	to	
	Fine	40–80	500 000	
Bio-Gel A-1.5 m	Coarse	150–300	<10 000	8
	Medium	80–150	to	
	Fine	40–80	1 500 000	
Bio-Gel A-5 m	Coarse	150–300	10 000	6
	Medium	80–150	to	
	Fine	40–80	5 000 000	
Bio-Gel A-15 m	Coarse	150–300	40 000	4
	Medium	80–150	to	
	Fine	40–80	15 000 000	
Bio-Gel A-50 m	Coarse	150–300	100 000	2
	Fine	80–150	to	
			50 000 000	
Bio-Gel A-150 m	Coarse	150–300	1 000 000	1
	Fine	80–150	to	
			150 000 000	

Table 2.5

Characteriatics of the Sepharose gels (by courtesy of Pharmacia (GB) Ltd)

Sepharose type	Agarose conc. %	Wet bead diameter (μm)	Fractionation range	
			Proteins	Polysaccharides
Sepharose 2B	2	60–20	$7\times10^4 - 40\times10^6$	$20^5 - 20\times10^6$
Sepharose CL-2B	2	60–200	$7\times10^4 - 40\times10^6$	$10^5 - 20\times10^6$
Sepharose 4B	4	60 140	$6\times10^4 - 20\times10^6$	$3\times10^4 - 5\times10^6$
Sepharose CL-4B	4	60–140	$6\times10^4 - 20\times10^6$	$3\times10^4 - 5\times10^6$
Sepharose 6B	6	45–165	$10^4 - 4\times10^6$	$10^4 - 10^6$
Sepharose CL-6B	6	45–165	$10^4 - 4\times10^6$	$10^4 - 10^6$

For the chromatography of proteins, two new materials have been introduced in recent years, both of which are composite gel materials and are considerably more rigid than conventional media. Sephacryl is prepared by crosslinking dextran with N,N-methylene bis acrylamide, and it gives a range of gels
as shown in Table 2.6. The largest pore size of Sephacryl is suitable for the
chromatographic separation of cells. Ultrogel is a polyacrylamide–agarose
mixture, and is available in a wide range of porosities as shown in Table 2.7. It

offers useful chromatographic properties and intermediate stability. Trisacryl GF 05 is a totally synthetic gel designed for the large-scale desalting operations. As such it offers extreme chemical and physical stability.

Table 2.6

Properties of the Sephacryl range of gel chromatography media
(by courtesy of Pharmacia (GB) Ltd)

Sephacryl type	Wet bead diameter (μm)	Fractionation range (Daltons) Proteins	Polysaccharides
Sephacryl S-200	40–105	$5\times10^3 - 2.5\times10^5$	$1\times10^3 - 8\times10^4$
Sephacryl S-300	40–105	$1\times10^4 - 1.5\times10^6$	$2\times10^3 - 4\times10^5$
Sephacryl S-400	40–105	$3\times10^4 - \ 8\times10^6$	$1\times10^4 - 2\times10^6$
Sephacryl S-500	40–105	–	$4\times10^4 - 20\times10^6$
Sephacryl S-1000	40–105	–	$5\times10^5 - <10^8$

Table 2.7

Characteristics of the Ultrogel and Trisacryl series of gel chromatography media
(by courtesy of LKB Instruments Ltd)

Gel type	Wet bead diameter (μm)	Fractionation (proteins) (Daltons)	Composition
Ultrogel AcA22	60–140	$10^4 - 1.2\times10^6$	2% acrylamide 2% agarose
Ultrogel AcA34	60–140	$20\times10^3 - 350\times10^3$	3% acrylamide 2% agarose
Ultrogel AcA44	60–140	$10\times10^3 - 130\times10^3$	4% acrylamide 4% agarose
Ultrogel AcA54	60–140	$5\times10^3 - 70\times10^3$	5% acrylamide 4% agarose
Ultrogel Aca202	60–140	$1\times10^3 - 20\times10^3$	20% acrylamide 2% agarose
Trisacryl GF05	40–80	$300 - 2000$	Synthetic polymer
Ultrogel A2	60–140	$12\times10^4 - 25\times10^6$	2% agarose
Ultrogel A4	60–140	$5.5\times10^4 - 9\times10^6$	4% agarose
Ultrogel A6	60–140	$2.5\times10^4 - 2.4\times10^6$	6% agarose

All of these newer materials are available in a particle size comparable to that of the finest grades of the conventional gels. As a result they offer potentially higher resolution which can in many cases be traded off for higher flow rates and hence greater productivity.

The newer chromatographic materials are all supplied as ready swollen suspensions in water and therefore need no pretreatment other than equilibration with the desired buffer. The older materials, Sephadex and Bio-Gel P gels, are supplied as dry powders, and to ensure good flow rates with these materials the correct pretreatment is essential.

These gels can be swollen in the required solvent at room temperature or by boiling. A table of pretreatment times has been published by Pharmacia for Sephadex G (Table 2.8). The advantage of pretreatment by boiling is minimum time loss with simultaneous gel degassing. We have found similar conditions to be suitable for pretreatment of the Bio-Gel P series.

Table 2.8

Pretreatment requirements of Sephadex (by courtesy of Pharmacia (GB) Ltd)

	Minimum swelling time	
Type of Sephadex	At room temperature	On boiling water bath
G-10, G-15, G-25, G-50	3 hours	1 hour
G-75	24 hours	3 hours
G-100, G-150, G-200	3 days	5 hours
LH-20	3 hours	—

Having pretreated the gel and allowed it to equilibriate at the chosen working temperature, the column can be packed. The gel is poured into a column as a fairly thick slurry (50–75% by volume) and allowed to settle for a few minutes. Any air bubbles are removed by stirring with a glass or plastic rod, and the bed is then allowed to pack with a flow of buffer. To stabilize the column, about three bed volumes of equilibriation buffer should be passed through. However, in practice one volume is often sufficient when running preparative columns. The homogeneity of the bed should always be checked by running a coloured substance such as blue dextran through the column, as there are few things more frustrating, or expensive, than trying to rescue several grams of enzyme from a poorly packed column.

It is particularly important with the soft, large-pore gels that the column is not too long, or the hydrostatic pressure too high, otherwise a severe reduction in flow rate can occur owing to compression of the beads. The use of several columns connected in series may therefore be a solution in some cases (see above).

The newer, more rigid chromatographic materials, for example Sephacryl, Sepharose CL, and Ultrogel, because they can withstand a greater operating pressure, can with advantage be packed under constant flow conditions which are slightly greater than those to be finally employed. This gives a more evenly packed bed than is obtainable if gravity flow alone is used.

The eluant flow is best maintained by the use of a pump, and a variety of pump types are available. In laboratory columns a peristaltic pump is commonly used, and these are available in a wide range of sizes able to cope with very large columns. Tubing wear can be a critical factor on a large peristaltic pump, and the tubing should be closely watched. Alternatively, a regular programme of replacement can be implemented.

A great advantage of peristaltic pumps is that the liquid only contacts the tubing, so cleanliness and sterility are easily maintained.

Stainless steel lobe and gear pumps or centrifugal pumps can also be used, but these should be carefully chosen to ensure there is no possibility of lubricant or metal particles reaching the column, and that they can easily be cleaned.

The sample is best applied via the pump or via a second pump dedicated to sample loading. Care must be taken to prevent air bubbles reaching the column. This can best be performed by using a three-way valve on the column inlet coupled with a bubble-trap on the buffer pump side of the valve.

The fraction from a large chromatography column may be several litres in volume, and, until fairly recently , the collection of such fractions presented a problem. With columns running at low flow rates, manual changing of bottles is feasible, particularly so if a 24 h shift system is in operation. However, in recent years a number of fraction collectors have been introduced which are capable of collecting fractions of any size in almost any container. These are manufactured by Pharmacia as the PF 30 or FRAC 300, and by LKB as the Multi-Rac or SuperRac fraction collectors. These offer varying degrees of sophistication and control and are an invaluable aid to the running of large chromatography columns.

Monitoring the eluant flow from a column is equally important. UV monitors for protein are available which can handle high flow rates, as are monitors for pH, conductivity, flow rate, and pressure. The sophisticated monitoring and automation of chromatography systems is something which is not normally done in small-scale operation; but in large-scale, particularly industrial, operations it is essential because of the value of the product. Some control and monitoring equipment is available from Pharmacia, although in many cases it may be desirable to construct suitable equipment in-house. Thus it is possible to have a fully automated, continuously operating chromatography system which loads the sample, collects fractions, and regenerates the column, and has various fail-safe devices to protect the column and product should a leak develop, for instance.

2.9.3 Ion exchange chromatography
There are three types of ion exchanger available for large-scale purification of enzymes. These are the ion exchange resins, ion exchange celluloses, and ion exchange large-pore gels.

Because of their stability, ease of sedimentation, and high porosity when packed into columns, ion exchange resins exhibit useful properties particularly for large-scale purification of enzymes. However, the structural integrity of enzymes can be affected adversely when in proximity to a matrix having a high charge density, such as is the case with ion exchange resins, where a high degree of protein binding occurs. In addition, protein binding is restricted to the surface of resin particles because of the high crosslinked hydrophobic matrix. However, as indicated in Chapter 0, resins do have some uses in enzyme purification where the stability of the enzyme permits.

Ion exchange celluloses can be used in both column and batch processes, although for large-scale work difficulty has been encountered in using large columns of these materials. The cellulose is easily compressed, and this leads rapidly to a reduction in flow rate. Thus the main application of ion exchange cellulose in large-scale enzyme purification has been for batch adsorption and batch elution.

Details of commonly used ion exchange celluloses are given in Table 2.9.

Table 2.9

Properties of some commonly used commercially available
cellulose ion exchangers

Ion exchanger	Total capacity meq.g^{-1}	Available capacity (mg/g)	Physical form
Cellex D[a]	0.4	–	Fibrous
	0.7	–	Fibrous
	0.9	–	Fibrous
Whatman DE23[b]	1.0	450 (albumin)[d]	Fibrous
Whatman DE51[b]	0.22	150 (albumin)[d]	Microgranular
Whatman DE52/(32)[b]	1.0	660 (albumin)[d]	Microgranular
Whatman DE53[b]	2.0	800 (albumin)[d]	Microgranular
DEAE Sephacel[c]	1.4	160 mg/ml (albumin)[e]	Bead
Cellex CM[a]	0.7	–	Fibrous
Whatman CM23[b]	0.6	150 (γ-globulin)[f]	Fibrous
Whatman CM52/(32)	1.0	400 (γ-globulin)[f]	Microgranular

Produced by [a]BioRad Laboratories Ltd; bWhatman Chemical Separation Ltd; [c] Pharmacia Fine Chemicals Ltd.
Measured in [d]0.005 M phosphate buffer, pH 8.5; [e]0.01 M Tris-HCl buffer, pH 8.0; [f]0.08 M phosphate buffer, pH 3.5.

There are several factors which influence the ability to carry out satisfactory preparative batch adsorption of enzymes. However, assuming the correct conditions of ionic strength, pH, and preparation of the ion exchange cellulose, the next parameter of importance is the vessel in which the process is carried out. The ability to maintain a correct temperature is essential, therefore a jacketed steel vessel connected to chilled water or a plastic tank with cooling coils is required. The methods for maintaining the cellulose in a suspended state are restricted to mechanical stirring or sparging with air. A large number of enzymes will adsorb to the exchanger within 10–20 minutes, but in our experience others require up to 24 hours for adsorption to take place. It is with the latter that agitation is of importance.

A mechanical stirrer is the most convenient method, as it enables work to be carried out under both aerobic, and even anaerobic conditions if a closed vessel is used. Stirring can create the serious problem of fines production if carried out too vigorously, but if a variable speed motor is installed the problem can be overcome by reducing the stirrer speed to the point at which the cellulose may be kept in suspension without excessive fines production.

Recovery of ion exchange celluloses can be effected after a suitable settling time by removal of the supernatants by syphoning or pumping, the settled slurry being finally collected via a valve at the base of the vessel. Removal of the remaining liquor from the slurry by using the conventional Buchner funnel/filter paper technique can be tedious. However, this problem may best be overcome by using a basket centrifuge (see above) fitted with a close weave cotton bag. Subsequently the basket centrifuge can be used in the elution process.

Ion exchange celulloses are simple to regenrate. When using a Buchner funnel and filter, 2 kg of cellulose can take two days to regenerate, but with the introduction of a basket centrifuge 20 times this amount can be treated. Anionic exchangers for regeneration are suspended in 0.5 N HCl, and when the pH drops below 2 the cellulose is removed and suspended in 0.5 N NaOH, checking that the pH is above 12. For cationic exchangers this process is reversed. The cellulose is washed several times in water to remove salt formed during the regeneration process. Equilibration of the cellulose is best carried out by a single suspension in a concentrated buffer, followed by several washings in the final equilibration buffer and checking that the correct pH and conductivity have been reached. Reproducible results are frequently only observed after an exchanger has undergone several adsorption and regeneration processes.

Relatively small quantities of cellulose may be used, for example when adsorbing extracellular enzymes from culture supernatants. In such cases, although the cellulose may be collected as described above, it may be reasonable to elute the enzyme after packing the collected cellulose into a column (see above for column details). This allows a much closer control of the elution process than is achieved using batch elution.

The Sepahdex ion exchangers (Table 2.10), particularly those based on Sephadex G50, are not readily suited to large-scale chromatography, because of their low rigidity and tendency to shrink and swell with changes in the ionic strength or pH of the buffer. The Sephadex G25 based exchangers are considerably more rigid, but do not offer a very great capacity for most proteins because binding is restricted to the surface of the beads.

For large-scale ion exchange chromatography the ideal materials are the newer exchangers based on crosslinked agarose or Trisacryl (Table 2.11). These offer extremely good flow rates and capacities and show no tendency to change in volume with variations in ionic strength or pH. They can also be regenerated *in situ,* a considerable labour-saving advantage, which can lead to a semi-continuous automated process. Pharmacia have recently introduced a variant of the ion exchange Sepharoses, called Fast Flow, which was specially developed for large-scale ion exchange chromatography.

Table 2.10
Properties of Sephadex ion exchangers (by courtesy of Pharmacia (GB) Ltd)

Ion exchanger		Total capacity[a] meq per 100 ml	Available capacity[b] for haemoglobin (g per 100 ml)	Molecular weight limit
DEAE-	A-25	50	7	3.5×10^4
Sephadex	A-50	17.5	25	2.5×10^5
QAE-	A-25	50	5	3.5×10^4
Sephadex	A-50	10	20	2.5×10^5
CM-	C-25	56	5	3.5×10^4
Sephadex	C-50	17	35	2.5×10^5
SP-	C-25	30	3	3.5×10^4
Sephadex	C-50	9	27	2.5×10^5

Table 2.11
Properties of ion exchangers based on Sepharose CL-6B and Trisacryl M
(by courtesy of Pharmacia (GB) Ltd. and LKB Instruments Ltd)

Ion exchanger	Bead size (μm)	Exclusion limit for proteins molecular weight	Total capacity (meq per 100 ml)	Available capacity for haemoglobin (g per 100 ml)
DEAE-Sepharose CL-6B	45−165	10^6	15 ± 2	10
DEAE-Trisacryl M	40−80	10^7	30	8−9
CM-Sepharose CL-6B	45−165	10^6	12 ± 2	10
CM-Trisacryl M	40−80	10^7	20	9−10

2.10 AFFINITY CHROMATOGRAPHY

As discussed in Part A: Chapter 2, affinity chromatography is not used extensively for large-scale enzyme purification. When affinity chromatography is employed the user has the choice of preparing his own affinity matrix by activating a support, and then reacting it with the ligand, using a ready-activated matrix or using a ready-prepared affinity matrix.

The major suppliers of chromatography media prepare a range of activated supports suitable for the coupling of affinity ligands, as shown in Table 2.12. These same manufacturers also prepare a wide range of immobilized ligands including dyes, lectins, polynucleotides, nucleotide cofactors, protein A, phenyl boronic acid, heparin, gelatin, avidin, calmodulin, and phenothiazine.

Table 2.12

Properties of some commercially available activated affinity gels

(by courtesy of Pharmacia Fine Chemicals Ltd, LKB Instruments Ltd, and BioRad Ltd)

Affinity matrix	Activator, spacer arm and functional group		Ligand specificity
CNBr Sepharose 4B	CNBr	—	$-NH_2$
AH Sepharose 4B	CNBr	Hexamethylene diamine	$-COOH$
CH Sepharose 4B	CNBr	6-amino caproic acid	$-NH_2$
Activated CH Sepharose 4B	CNBr	N-hydroxysuccinimide ester of 6-amino caproic acid	$-NH_2$
Epoxy activated Sepharose 6B	1,4-bis-(2,3-epoxypropoxy-)butane		$-NH_2$, $-OH$, $-SH$
Activated thiol Sepharose 4B	Glutathione-2-pyridyl disulphide		$-SH$, $-C=O$, $-N=N$ alkyl or aryl halides
Act.-Ultrogel AcA 22	Glutaraldehyde		$-NH_2$
AC-Ultrogel AcA 34	H_2N-NH_2	6-amino caproic acid	$-NH_2$
Affi-Gel 10	N-hydroxysuccinimide ester on 10-atom hydrophilic spacer		$-NH_2$
Affi-Gel 15	N-hydroxysuccinimide ester on 15-atom hydrophilic spacer		$-NH_2$
Affi-Gel 102	NH_2 on 6 atom hydrophilic spacer		$-COOH$
Affi-Gel 202	COOH on 10 atom hydrophilic spacer		$-NH_2$
Amino ethyl Bio-Gel P2 or P150	Diaminoethane		$-COOH$
CM Bio-Gel A	Carboxymethyl		$-NH_2$

In general these ready prepared matrices and coupled ligands are available only in small quantities; in many cases the user requiring large amounts often has no choice but to synthesize his own. However, recently the need for large amounts of some affinity matrices has been recognized, and it is now possible to obtain large quantities of certain matrices from the manufacturers.

Examples of the large-scale application of affinity chromatography are given in Part A, Chapter 2, and in two recent reviews: Hill & Hirtenstein (1983) and Janson (1984).

REFERENCES

Baird, J. K., Sherwood, R. F., Carr, R. J. G. & Atkinson, A. (1976) *FEBS Lett* **70** 61–66.
Bell, D. J., Hoare, M. & Dunnill, P. (1983) *Advances Biochem. Eng.* **26** 1–72.
Butters, J. R. (1969) *Process Biochem.* **5** 25–27.
Callow, D. S., Atkinson, A. & Melling, J. (1973) *FEBS Abs.* No. 54, Dublin.
Charm, S. E. & Matteo, C. C. (1971) *Methods Enzymol.* **22** 476–556.
Dunnill, P. & Lilly, M. D. (1972) In *Enzyme Engineering*, pp. 101–113 ed. by L. B. Wingard. Wiley, New York.
Elsworth, R., Meakin, L. R. P., Pirt, S. J. & Capel, G. H. (1956) *J. Appl. Bact.* **19** 264–278.
Fischer, L. (1969) In *Laboratory Techniques in Biochemistry and Molecular Biology*, vol. 1, pp. 151–396, ed. by T. S. Work & E. Work. North Holland Publishing Co., Amsterdam and London.
Gold, A. M. (1965) *Biochemistry* (Washington) **4** 897–901.
Hammond, P. M., Price, C. P. & Scawen, M. D. (1983) *Eur. J. Biochem.* **132** 651–655.
Hedenskog, G., Enebo, L., Vendlova, J. & Prokes, B. (1969) *Biotech. Bioeng.* **XI** 37–51.
Hill, E. A. & Hirtenstein, M. D. (1983) In: *Advances in Biotechnological Processes* pp. 31–66. A. Mirzrahi & A. VanWeszel, eds. Alan R. Liss, Inc., New York
James, G. T. (1978) *Anal. Biochem.* **86** 574–579.
Janson, J. C. (1984) *Trends Biotechnol.* **2** 1–8.
Janson, J. C. & Hedman, P. (1982) *Adv. Biochem. Eng.* **25** 43–99.
Le, M. S., Spark, L. B. & Ward, P. S. (1984a) *J. Membrane Sci.* (in press).
Le, M. S., Spark, L. B., Ward, P. S. & Ladwa, N. (1984b) *J. Membrane Sci,* (in press).
Lingood, F. V., Matthews, A. C., Pinfield, S., Pope, C. G. & Sharland, T. R. (1955) *Nature* **176** 1128.
Linko, M., Wallinder, P. & Linko, Yu-Yen (1973) *FEBS Abs.* No. 59, Dublin.
Melling, J., Evans, C. G. T., Harris-Smith, R. & Stratton, J. E. D. (1973) *J. Gen. Microbiol.* **77** xviii.
Melling, J. & Scott, G. K. (1972) *Biochem. J.* **130** 55–62.
Petersen, E. A. (1970) In *Laboratory Techniques in Biochemistry and Molecular Biology,* vol. 2, pp. 223–396, ed. by T. S. Work & E. Work. North Holland Publishing Co., Amsterdam and London.
Quirk, A. V. & Woodrow, J. R. (1983). *Biotechnol. Letts.* **5** 277–282.
Rahacek, J., Beran, K. & Bicik, V. (1969) *Appl. Microbiol.* **17** 462–466.
Solomons, G. L. (1969) *Materials and Methods in Fermentation,* Academic Press, London.
Tanny, G. B., Mirelman, D. & Pistole, T. (1980) *Appl. Environmental Microbiol.* **40** 269–273.
Thurston, C. F. (1972) *Process Biochem.* **8** 18–20.
Valeri, A., Gazzei, G. & Genna, G. (1979) *Experientia* **35** 1535–1536.
Woodrow, J. R. & Quirk, A. V. (1982) *Enzyme Microb. Technol.* **4** 385–389.

CHAPTER 3

The applications of enzymes in industry

PETER S. J. CHEETHAM, Tate & Lyle plc, Group Research and Development, Philip Lyle Memorial Research Laboratory, PO Box 68, Reading, England

GLOSSARY OF TERMS

7ACA	7-aminocephalosporonic acid
7ADCA	7-aminodesaceloxycephalosporonic acid
6APA	6-amino penicillanic acid
BOD	biological oxygen demand
DE	dextrose equivalent
DEAE cellulose	diethylaminoethyl cellulose
DNA	deoxyribonucleic acid
DNP	2,4-dinitrophenol
HFCS	high fructose corn syrup (isoglucose syrup)
NAD	nicotinamide adenine dicucleotide
NADP	nicotinamide adenine dinucleotide phosphate
RNA	ribonucleic acid.

3.1 INTRODUCTION

The industrial use of enzymes is by no means new. Many relatively poorly understood traditional processes have been evolved, sometimes over thousands of years of continuous usage by a trial and error procedure. Such processes are usually characterized by a very slow rate of development such that they have not always adopted recent scientific and technological advances to the same extent as other more dynamic and innovative industries, such as the chemical industry. They also overcome the economic problems inherent in biotechnology through unique advantages peculiar to each particular process. These problems arise from the low aqueous concentrations of the products formed, their instability, and the complex reaction mixtures produced including unused substrates and side-products, from which the desired product must be recovered and concentrated. The peculiar advantages of existing biotechnologies include the exceptional organoleptic and nutritional properties of beer, cheese, and bread, the exceptional activities of interacting communities of microorganisms in sewage treatment, and the high value and low bulk of antibiotics. However, in the last few decades our recently acquired knowledge concerning the biocatalytic capabilities of enzymes and microorganisms has made possible the creation of a new generation of rationally researched and developed biologically based processes and products.

Currently, the majority of the enzymes used in industry are used in food processing, for instance three of the four enzymes most commonly used in industry, α-amylase, glucoamylase, and glucose isomerase, are employed for the production of glucose and fructose syrups. This emphasis is hardly surprising, since most foods are derived from natural sources, which are synthesized enzymatically in the plant, animal, or microorganism of origin. Furthermore, enzymes are present and active in many foods, for instance stored meat becomes more tender owing to the action of enzymes, particularly lysosomal proteases (cathepsins) and collagenases; and we use enzymes to break down our food in our digestive tracts. However, there is every likelihood that the industrial uses of enzymes will increase in the future, considering that it has been estimated that about 40% of the manufacturing output in developed industrialized countries is biological in nature and origin (Dunnill 1981).

In this chapter, I have tried to present the general principles and concepts concerning the applied use of enzymes in industrial and other applications, and to give representative examples of their use. In many cases this includes the use of enzymes still associated with their parent cells, as opposed to soluble enzymes. Comprehensive coverage is not possible in a work of this size, and full technical details are often not available, particularly when the process has genuine commercial importance or promise. Many laboratory and to a lesser extent pilot-scale processes are described in the academic and patent literature, but it is often very difficult to discern the extent to which these processes have acquired commercial importance.

Biotechnology has been defined as the application of scientific and engineering principles, especially those from microbiology, biochemistry, genetics and biochemical and chemical engineering, to the processing of materials by biological agents to provide goods and services (Bull *et al.* 1982). Enzymes are frequently used for process improvement, for instance to enable the utilization of new types of raw material or for improving the physical properties of a material so that it can be more easily processed, which may take the form of increasing the solubility and decreasing the viscosity of a material so as to facilitate filtration or pumping during processing. Secondly, enzymes are used for product improvements, such as changes in the colour, aroma, texture, taste, or shelf life of a foodstuff so as to make it more acceptable to the retailer or customer.

To be useful and thus commercially usable, enzymes must enable a product to be produced that has some of the following advantages: (i) is of better quality than the traditional product, (ii) is of improved utility in the applications intended for the product, (iii) is cheaper, which of course may be achieved indirectly — for instance by a decrease in the cost of the labour and/or machinery required in the manufacturing process, (iv) lastly and most importantly, enzymes may enable products to be produced which were not previously available or only available in limited quantities owing to a restricted supply from the natural source.

The usefulness and therefore profitability of an enzyme mediated process or product may, at least initially, be only marginally advantageous when compared with the existing competing processes or products. For instance, the production of high fructose syrups using immobilized glucose isomerase only really took off during a period of exceptionally high sucrose prices during the mid-1970s, caused by shortages in supply due to hurricanes, pest damage, and an attempt to establish a sugar producers' cartel; but it has subsequently displaced sucrose from a number of large volume uses such as soft drinks, such that several million tonnes of high fructose syrups are now sold per annum.

Enzymically produced products can be divided into three main categories: firstly, those that exactly simulate the products traditionally formed by other means. Secondly, there are those products that simulate the traditional products but with some differences and/or improvements to the process and/or the product. Thirdly, there are novel products which were not previously available until enzymic production was possible.

To be commercially useful, enzymes do not have to occupy the central role in a process, actually producing the compound(s) of interest. A range of important applications exist in which enzyme treatments have been found to be very useful in improving the quality or the ease of production of a product, or in producing an intermediate in a process that is difficult to synthesize chemically, often in processes which have been traditionally carried out successfully without the use of any such enzyme treatments. For instance, addition of lipases has been found to accelerate the ripening of cheeses, and pectinases and amylases are employed to reduce the viscosity of fruit juices so as to facilitate the subsequent clarification of the juice by filtration or flocculation.

3.2 PRODUCTION OF ENZYMES

Whenever possible it is advantageous to use thermostable enzymes, as by carrying out the reaction at higher temperatures faster reaction rates, decreased substrate viscosities, increased reactant solubilities, and, especially, decreased microbial contamination, can be obtained. If used at ambient temperatures these enzymes are often more stable than the corresponding enzymes derived from mesophilic strains of cells.

Organisms suitable for use in the production of enzymes should possess a number of useful attributes. These include easy and rapid growth in large fermenters on comparatively cheap and simple nutrients without the need for inducers. A high yield of enzyme should be obtained in a form that is easy to isolate, purify, and concentrate without the formation of toxic or immunogenic metabolites. The organism should also have stable physiological characteristics and be readily acceptable to the food and drug authorities (for a review see Barfoed (1981)).

Selection for strains which hyperproduce enzymes by forming mutants which lack the requirement for inducers due to inactive repressor sites, which exhibit low catabolite repression, which do not carry out feedback repression, or for enzymes which are relatively resistant to end-product inhibition, can also be a profitable strategy. For instance, the glucosidase (naringinase) selected for commercial use in the debittering of grapefruit and oranges was specially selected to be resistant to inhibition by the glucose formed from the bitter flavinoid glycoside by the action of the enzyme (Sakaguichi *et al.* 1971). In fact, despite the dramatic advances made recently in the field of genetic engineering, classical enzyme selection techniques are still very productive, particularly when novel and/or exotic environments are examined, for microbial cells progressing enzymes with exceptional properties. Examples are the range of chlorinating enzymes obtained from marine microorganisms (Neildlelman & Geigert 1983),and an enzyme which produces a novel sugar, isomaltulose, very productively from a bacterial pathogen of rhubarb (Cheetham *et al.* 1982). Even more exceptional discoveries have been made very recently. Microorganisms that grow at about 250°C in deep water volcanic vents must possess extremely heat stable enzymes, and a microorganism from the intestine of a wood boring worm that both degrades cellulose and fixes nitrogen has been found (Waterburg *et al.* 1983).

In short, the majority of enzymes in current industrial use are of microbial origin and are produced in conventional aerobic submerged fermentations, which allows greater control of the conditions of growth than solid state fermentations. Much is often known about the selection of strains of organisms and culture conditions, but little about the regulation of the synthesis, degradation, and excreation of the enzyme by the producer organism. These enzymes are often extracellular, facilitating isolation and purification. Otherwise the cells are disrupted and the enzyme extract purified and supplied in a concentrated liquid form. Alternatively the enzyme can be precipitated and supplied as a dry powder mixed with an inert diluent such as lactose. Note that *B. coagulans* glucose isomerase is one of the few commercial enzymes to be manufactured using continuous fermentation.

In the future the dramatic advances currently being made in genetic engineering techniques offer the prospect that it may be possible to easily produce an enzyme irrespective of its catalytic function or its origin in the same quantities and at the same prices that a few enzymes, such as proteases and the carbohydrases used in sugar and starch processing, command today. This may be achieved by increasing the number of gene copies in an organism and thus the concentrations and amounts of enzyme produced or by interstrain genetic transfer techniques. These techniques are protoplast fusion and what is loosely referred to as 'genetic engineering' whereby restriction endonucleases are used to selectively dissect the DNA, which is then joined together in different combinations by DNA ligase. These techniques have the advantage that they work well with cells of very different characteristics and phylogenetic origins, and that the progeny cells are usually not severely disabled by the accumulation of deleterious mutations as is the case with the more conventional mutagenic techniques. However, expression of the introduced genetic material is uncertain and is probably the major problem in this field. The expression of enzymes in host microorganisms is affected by many variables, several of which are poorly understood. These factors include the choice of host strain, ribosome binding site, plasmid copy number, and amino acid sequence at the N-terminal end of the enzyme. For instance, in one of the best understood hosts, *E. coli,* foreign genes often give rise to large deposits of non-native insoluble protein products. Since genetically engineered cells can often be metabolically disabled such that their rate of growth is lower than that of the native cells, it would appear that their use in the immobilized cells may be advantageous since they do not need to grow when used in this form.

An exciting application of genetic manipulation techniques has been described by Winter *et al.* (1982). They converted the cysteine 35 at the active site of the tyrosyl tRNA synthetase of *Bacillus stearothermophilus* into serine, by cloning the gene for the enzyme into a vector so as to facilitate mutagenesis with mismatched synthetic olionucleotide primers. The recombinant clone expressed high levels of the enzyme in the *E. coli* host, the altered enzyme being less active owing to its lower K_m for ATP. This method promises to be a general means of engineering novel enzyme activities by redesigning existing enzymes, for instance so that they have different substrate ranges, pH and temperature optima and also so that they are secreted from, rather than retained in, their parent cells. Site-specific modifications will to some extent replace traditional mutagenesis and screening procedures, particularly when there is not an obvious characteristic that makes screening easy. However, such precise modification techniques presume a knowledge about the relationship between the primary structure of the enzyme under investigation and its activity, stability and other characteristic properties that is not established for any industrially important enzymes; or any enzymes at all for that matter. Genetic cloning techniques involving manipulation of the promotor regions can be used to amplify the expression of certain genes so that their protein products may represent an appreciably larger proportion of the cells protein. Genetically recombined cells can also be used in an immobilized form. For instance immobilized *B. subtilis* carrying a plasmid for rat proinsulin produced proinsulin continuously when cell growth, but not protein synthesis or excreation, was inhibited by the addition of an antibiotic such as novobiocin which inhibits DNA replication (Mosbach *et al.* 1983). In this study the cells were immobilized in 100–300 μm diam. agarose beads formed using a soya oil as the suspending medium (Wikström *et al.* 1982).

3.3 USE OF ENZYMES – GENERAL COMMENTS

The historical developments in the availabilities of commercial enzymes are depicted in Table 3.1. The most used enzymes are α-amylase, glucoamylase, glucose isomerase and various proteases. Only about 20 enzymes are used in appreciable quantities, so that the entire enzyme industry has only a relatively small total annual turnover. However, several enzymes are of sufficient industrial interest to be traded on commodity markets. Note also that the very high tonnages of malt amylases used are deceptive, because this is mostly malted barley. As in any commercial field, some products are increasing appreciably in usage, for instance rennet; whereas others have a more static market such as the amylases and proteases used in textile desizing and leather baiting.

Table 3.1
The most important type of industrial enzymes

Source	Name	Commercially available before			Current production (tons of enzyme protein per year)
		1900	1950	1976	
Animal	Rennet	X			2
	Trypsin		X		15
	Pepsin		X		5
Plant	Malt amylase	X			10 000
	Papain		X		100
Microbial	Koji	X			?
	Bacillus protease		X		500
	Amylogucosidase			X	300
	Bacillus α-amylase		X		300
	Glucose isomerase			X	100
	Microbial rennet			X	10
	Fungal α-amylase	X			10
	Pectinase		X		10
	Fungal protease	X			10

(Aunstrup 1977).

There is a large variety of commercially available enzymes differing in biological source, activity, purity, physical form, and characteristics such as pH and temperature optima. In particular, a considerable number of amylases and proteases of consistent quality and activity are offered by enzyme suppliers (Table 3.2). This wide choice is often useful when the requirements of particular applications are considered: for instance the presence of contaminating enzymes may have a beneficial effect. Thus certain commercial preparations have become associated with particular industrial applications and in due course have been 'tailored' even more closely to these requirements; for example, products containing α-amylase designed for different applications. These include a protease-free α-amylase from *A. oryzae* for regulating diastatic levels in flour, a blend of bacterial proteases and α-amylase for use in the production of crackers, and α-amylase preparation designed to regulate diastatic levels in bread and rolls and for improving the elasticity of gluten in flour, another α-amylase containing product designed to delay the onset of stailing and moisture loss in cakes, cake-mixes, brownies and bread plus others with application in the brewing, detergent, pharmaceutical, starch processing, paper, textile, leather, wine and fruit juice industries. Note that immobilized enzymes can also be bought, the best known being glucose isomerase manufactured by Novo. Others include trypsin, urease, ribonuclease, peroxidase, papain and other proteases, amylase, asparaginase, alcohol dehydrogenase, alkaline phosphatase, and glucose oxidase.

Table 3.2
Some industrial applications of enzymes

Enzyme (trivial name)	Main source	Main applications
Alanine and formate dehydrogenases	*Bacillus subtilis* and *Candida boidinii*	Alanine formation from pyrivic acid
Aminoacylase	*Escherishia coli*	Resolution of racaemic amino acid mixtures
Aspartase	*Escherischia coli*	Production of aspartic acid
Aspartate decarboxylase	*Pseudomonas dacunhae*	Alanine production
Bacterial α-Amylase (α-Glucanase)	*Bacillus subtilis licheniformis* and *amyloliquefaciens*	Pre-thinning of starch during glucose syrup formation, starch hydrolysis during brewing and before distillation; paper, textiles, cleaning, pharmaceutical and animal feed operations
Fungal α-Amylase	*Aspergillus oryzae* and *niger*	Starch liquification, fruit, vegetable, brewing, baking, confectionery and paper manufacture
Bacterial α-Amylase	*Bacillus cereus, circulans, megateriam* and *polymyxa* and *Streptomyces* sps	Starch degradation into maltose and glucose syrups for use in food and beverages products and in alcohol production
Bacterial isoamylase	*Bacillus cereus*	Ditto
Plant α and β amylases	Barley and soya beans	Production of a range of malt extracts containing glucose and maltose for brewing and baking applications
β-Amylase	*A. niger* and *oryzae* and *Bacillus* sps	Ditto
Amyloglucosidase	*Aspergillus niger, awamori* and *oryzae; Rhizopus niveus, oryzae* and *delemar* and *Trichderma viridae*	Glucose syrup production from pre-thinned starch, for use in brewing, alcohol, baking, textiles, paper, fermentation, pharmaceuticals, soft drinks and confectionery
Anthrocyanase	*Aspergillus niger*	Decolorization of red grapes
Catalase	*Aspergillus niger, Penicillium* sps, *Micrococcus lysodeikticus* and bovine liver	Removal of hydrogen peroxide in the milk, cheese and egg processing industries. It is also used in sterilization, oxidation and foam, plastic and rubber production

Table 3.2 – *continued*

Enzyme (trivial name)	Main source	Main applications
Cellulase	*Trichoderma reesei, Aspergillus niger, oryzae, phoenicis* and *wentii* and *Mucor miehei*	Fruit and vegetable processing
Cyclodextrin glucosyl transferase	*Bacillus macerans* and *megaterium*	Formation of cyclic dextrins
Dextranase	*Klebsiella aerogenes, Penicillium funicolosum* and *liacnium* and *Fusarium* and *Flavobacterium* sps	Polysaccharide hydrolysis during sugar production and in food processing
Diacetylreductase	*Aerobacter aerogenes*	Removal of diacetyl from beer, which gives the beer an off-flavour
Epoxysuccinate hydrolase	*Nocardia tartaricus*	Formation of tartaric acid
Eumarase	*Brevibacterium ammoniagenes*	Malic acid production
α-galactosidase	*Aspergillus niger, Mortierella vinacea Saccharomyces cerevisiae*	Oligosaccharide hydrolysis during sugar refining and soya bean milk production
β-galactosidase (lactase)	*Escherichia coli, Bacillus* sps *Aspergillus niger* and *oryzae, Klebsiella fragilis* and *Saccharomyces fragilis* and *lactis,* also *Kluyveromyces lactis* and *fragilis*	Hydrolysis of lactose in milk and other dairy processing applications, especially whey hydrolysis, also pharmaceuticals
β-glucanase	*Bacillus subtilis* and *circulans, Aspergillus niger* and *oryzae, Penicillium emersonni* and *Saccharomyces cerevisiae*	Polysaccharide hydrolysis during brewing and the extraction of fruit juice and other products from plant materials, e.g. flavours
Glucose isomerase	*Actinoplanes missouriensis Bacillus coagulans, Streptomyces albus, olivaceus, olivochromogenes* and *phaeachromogenes* and *Arthobacter* sps	Isomerisation of glucose into high fructose syrups
Glucose oxidase	*Aspergillus niger, Penicillium glaucum, notatum* and *chrysogenium*	Antioxidant, e.g. in fruit or albumin preservation and in the control of colour in wine, and formation of gluconic acid
Histidine ammonia lyase	*Achromobacter liquidium*	Urocanic acid production
11 α-hydroxylase	*Rhizopus arrhizus* and other microbial sources	11 α-hydroxylation of progesterone leading to the synthesis of cortisone hydrocortisone, prednisolone, and prednisone
Inulinase	*Kluyveromyces fragilis* and *Aspergillus* and *Candida* sps	Sweetener formation

Table 3.2 – *continued*

Enzyme (trivial name)	Main source	Main applications
Invertase	*Kluyveromyces fragilis, Saccharomyces carlsbergensis* and *cerevisiae* and *Candida* sps	Production of invert sugar for confectionery and humectant uses, and in brewing and artificial honey manufacture
Isomaltulose synthase	*Erwinia rhaponticii* and *Protaminobacter rubrum*	Formation of isomaltulose
Lipase/esterase	*Aspergillus niger* and *oryzae, Mucor javanicus, pusillus* and *miehei, Rhisopus niveus* and *lipolytica* and *cylindracea* and calf, kid, lamb and pig pancreas	Leather and wool processing, cheese and butterfat flavour modification, fat and oil modification, waste treatment
Lactoperoxidase		Cold sterilization of milk
Lipoxygenase	Soya flour	Whitens bread and oxidises oils
Malic acid decarboxylase	*Leuconostoc oenos*	Beverage production
Naraginase	*Aspergillus niger*	Debittering of citrus fruit juice
Nisinase	*Bacillus cercus*	Removal of nisin (an antibiotic) from milk
Pectinase/pectin esterase	*Aspregillus niger, ochraceus* and *oryzae, Rhizopus oryzae Trichoderma reesei* and *Penicillium simplicissium*	Extraction and clarification of fruit juices for use in soft drinks, beer, and wine. Also extraction of spices and coffee
Penicillinase	*Bacillus lichenformis* and *cereus*	Removal of antibiotic from milk
Penicillin amidase (acylase)	*Bacillus megaterium, Escherichia coli* and *Basidiomycetes* and *Achromobacter* sps	Formation of 6-aminopenicllinic acid for the synthesis of semi synthetic antibiotics
Phenylalanine ammonia lyase	Yeast	Formation of phenylalnine
Phytase	*Aspergillus ficcium*	Removal of phytic acid from cereals
Plant proleases (papain, ficin, bromelain etc.)	*Papayi latex, Ficus carica* and *Bromus* sps	Yeast extract production, chill proofing of beer, baking, leather, textiles, pharmaceuticals and human and animal food processing including meat tenderization
Animal proteases including trypsin chymotrypsin etc.	Cattle, sheep and pigs	Leather and pharmaceutical industries food processing especially protein hydrolysis such as the hydrolysis of cheese whey protein, peptide synthesis
Microbial proteases	*Aspergillus niger*	Cheese, meat, fish, cereal, fruit, beverages and baking industries
Microbial proteases	*Aspergillus flavus*	Food processing
Microbial proteases	*Aspergillus oryzae* (acid protease) (neutral protease)	Protein hydrolysis, especially meat and fish processing, brewing, baking
Microbial proteases	*Aspergillus melleus, Endothia parasitica, Mucor miehei* and *pusillus*	Cheese manufacture (milk coagulation)

Table 3.2 – *continued*

Enzyme (trivial name)	Main source	Main applications
Microbial proteases	*Bacillus licheniformis* and *subtilis alkaline proteases*	Detergent and leather industries, meat, fish and dairy produce processing
Microbial proteases	*Bacillus subtilis* neutral protease	Beverage production and baking
Microbial proteases	*Bacillus cereus*	Beverage and baking
Pullulanase	*Klebsiella aerogenes* and *Bacillus* sps	Debranching of starch during glucose syrup formation and brewing
Rennin (Chymosin)	Fourth stomach of un-weaned calves or lambs	Coagulation of milk during chees-making
Ribonucleases	*Penicillium citrinum,* *Aspergillus oryzae* and *Streptomyces griseus*	Formation of nucleotides for use as flavouring agents
Sulphydryl oxidase	Milk whey	Reduction of the cooked flavour of milk
Tannase	*Aspergillus oryzae* and *niger*	Beverage processing, e.g. tea and beer
Thermolysin	*Bacillus proteolyticus*	Production of the high intensity sweetener aspartame (aspartic acid-phenylalanine methyl ester)
Tryptophase		Tryptophane formation
Xylanase	*Streptomyces* sps *Aspergillus niger* and *oryzae* and *Sporotin-chium dimorphosporium*	Processing of cereals, tea, coffee, cocoa and chocolate
Enzymes used as complex mixtures		
Pectinase, hemicellulases and proteases	*Aspergillus, Rhizopus* and *Trichoderma* sps	Fruit processing (extraction and clarification)
Glucose oxidase and catalase	*Aspergillus niger*	Antioxidants uses, e.g. in soft drinks and butter and gluconic acid production
Amylase, β-glucanase protease and cellulase	*Bacillus subtilis*	Brewing
Mixtures of enzymes from Actinomyces, Achromo-bacter and Pseudomonas sps		Cell lysis preparations
Proteases, amylases and also lipases	*Bacillus subtilis*	Detergents
Mixed enzymes and bacteria		Waste treatment (cleaning agents)

About 60% of the enzymes used are proteolytic and 30% are carbohydrases, the total amounts used remain low as enzymes are usually used at concentrations well below 1% by weight of the substrate being processed (Tables 3.1 and 3.2 and Fig. 3.1).

Commercially available enzymes can be divided into three classes in terms of availability, price and purity. Enzymes used on a large scale such as glucose isomerase are relatively cheap and are available in bulk quantities, but are often used in a form that is relatively impure by biochemical standards. Many enzymes that are widely used in smaller quantities, particularly in clinical analysis, such as glucose and cholesterol oxidases, are reasonably pure but are available in relatively small quantities by comparison with industrial enzymes. They are also fairly expensive because of the high capital and labour costs associated with the need to use a series of purification techniques of comparatively low capacity and efficiency. Thirdly, speciality enzymes which are usually produced for research purposes, often on an *ad hoc* basis, are of variable purity, of very limited availability, and are usually extremely expensive so that large-scale trials are prohibitively expensive. This is a factor which often militates against the development of novel enzyme based products and processes, since enzyme suppliers are usually reluctant to make large quantities of enzyme available at a reasonable cost until a sizeable commercial demand exists, but of course the commercial demand usually cannot exist until the process can be scaled-up, tested in a commercial environment, and reasonable quantities of product produced for testing in applications and for toxicology.

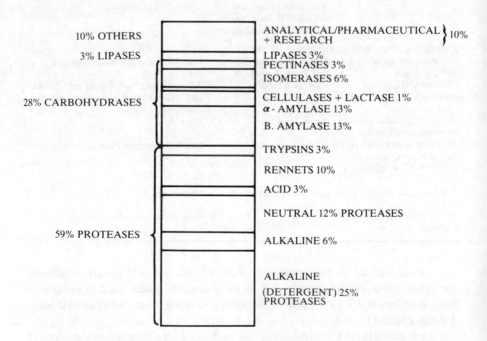

Fig. 3.1 – The relative usage of the various industrially important enzymes (Godfrey & Reichelt, 1983).

3.3.1 The characteristics of industrial enzymes

Enzymes, unlike many other substances, are recognized and sold by their activity rather then their weight, so that the stability of enzyme preparations during storage is of prime importance. The enzymes used in industry are very rarely crystalline, chemically pure, or even single protein preparations. These impurities need not interfere with the activity of the enzyme. However, enzyme impurities can catalyse the formation of side-products, or present a toxicological hazard. For instance the potent carcinogen aflatoxin can contaminate extracts from *Aspergillus flavus*. In such cases their presence does prevent the enzyme from being used. As regards toxicological hazards, the potential allergenic activity of enzymes should be recognized, and the presence of toxic metabolites, often derived from the parent cell, should be guarded against. The most common contaminating material in enzyme preparations is often inactivated molecules of the enzyme itself.

An enzyme that is commercially useful in food processing must be cheap *vis a vis* the overall processing costs, and must be active under the physical conditions prevalent during traditional food processing steps. That is, it is preferable to screen different enzymes for activity under these conditions rather than to manipulate an established process or product so as to accommodate a potentially useful new enzyme. The enzyme must also be stable, many industrially used enzymes operate at temperatures in excess of $50°C$; it must be available in sufficient quantities, and it must be safe. Since the costs of petitioning regulatory authorities are so great, whenever possible it is much easier to make use of an enzyme already approved for food use rather than to obtain legal status for a new enzyme. It is also noticeable that the same enzyme may be useful in very different applications, for instance α-amylases are used in both brewing and baking instructions, and proteases in brewing, baking, cheese-making, and in the tenderization of meat.

The majority of enzyme preparations perform their technological function during the preparation and processing of food rather than in the final product itself. Thus the use of enzyme are advantageous because they operate under conditions of pH, temperature, etc. that are consistent with the retention of the desired structure and other properties of the food, and minimize the energy requirements during processing, whereas the high temperatures and pressures associated with the use of chemical catalysts would often be detrimental to the product.

Frequently the residues of enzyme remaining in the final product stream are very small indeed, such that if the product is intended for food or pharmaceutical use the enzyme can be regarded as a processing aid, although in some cases they do come within the definition of food additives, especially as they can become active during subsequent processing. In this respect the use of immobilized enzymes is additionally advantageous because the immobilized enzyme or cell is usually completely recoverable from the reaction mixture. Thus it can be re-used repeatedly without any contamination of the final product and without the need to heat the product so as to denature the enzyme, a procedure which cannot always be used because of the temperature sensitivity of the product. Also, larger concentrations of immobilized rather than free enzyme can be utilized because the immobilized enzyme can be recovered and re-used, resulting in a shortening of reaction times and/or the size of vessel needed to carry out the reaction, and a virtual absence of enzyme from the final product such that it only has to be approved as a food processing aid and not as a food additive.

3.4 SOURCES OF ENZYMES (see Table 3.2)

Microbial cells are the usual sources of enzymes for industrial use except for traditionally used plant and animal enzymes such as the plant proteases papain, ficin, and bromelain which are used for meat tenderization, and rennin which is used in cheese manufacture. The vast majority of microbial enzymes are produced from only about 25 organisms, including a dozen fungi; indeed it has been estimated that only about 2% of the world's microorganisms have been tested as enzyme sources.

Microbial enzymes are more useful than enzymes derived from plants or animals because of the great variety of catalytic activities available, because cheap, regular, abundant supplies of even quality can usually be developed, occasionally by surface culture but usually by the deep culture aerobic fermentation techniques widely used in the production of antibiotics; also because microbial enzymes are usually more stable than the corresponding plant and animal enzymes, and because microbial enzyme production is usually a more convenient and safer procedure than production from plant and animal sources. Genetic and environmental manipulation to increase the yield of cells (Demain 1971), to increase the enzyme activity of the cells by making the enzyme of interest constitutive; or by inducing it; or to produce altered enzymes (Betz *et al.* 1974) may be employed easily using microbial cells because of their short generation times, their relatively simple nutritional requirements, and since screening procedures for the desired characteristic(s) are easier. These techniques are especially important in the case of activity or stability limiting enzymes in multi-enzyme conversions and when the normal regulatory controls on the synthesis of a desired enzyme needs to be removed, when it is desirable to lessen or remove substrate or product inhibition, the latter being a natural mechanism which exists in order to prevent the overproduction of the product of a metabolic pathway, or to acquire some other favourable characteristic. An interesting example of the latter is the improvement in amyloglucosidase production following the genetic recombination of high and low yielding *Aspergillus niger* strains to give a high yielding strain with the same good broth filtration characteristics of the low yielding strain (Ball *et al.* 1978). Generally the objective is to maximise the rate of enzyme formation and the concentration of enzyme in the cells or in the fermentation broth so as to minimize the production costs of the enzyme. This second term is especially important since it also determines the amounts of material which must be processed in order to obtain the required amount of enzyme. This yield equals the mass of cells obtained multiplied by their enzyme activity, or in the case of an extracellular enzyme the concentration of enzyme obtained multiplied by its volumetric activity. It is therefore desirable to use the optimum combination of a selected strain of microorganism, optimal fermentation and recovery conditions and the most appropriate equipment in good working order. Usually the cells are grown in submerged, well agitated and aerated fermenters; however a significant number of industrially important enzymes are grown in solid or semi-solid state fermentations. These include lactase, α-amylase and a protease from *A. oryzae,* pectinase and a protease from *A. niger,* α-amylase from *Rhizopus,* and rennet from *Mucor pusillus.*

Wherever possible it is preferable to use non-spore forming strains which do not form toxins. It is also advantageous to use constitutive mutants in the case of cells which normally require an inducer to produce a particular enzyme, and catabolite resistant mutants so that glucose and other sugars can be used in the growth medium without causing repression of the desired enzyme(s).

Higher plants are not generallly satisfactory sources of enzymes for industrial and clinical uses because they accumulate waste products in vacuoles. These compounds are often enzyme inhibitors and/or toxins, particularly when oxidized by contact with air. These wastes, many of which are phenols, are then released when the cell is disrupted, contaminating and often inactivating any enzymes that are extracted.

Despite the phylogenetic and biochemical diversity of microorganisms only a comparatively restricted number are used as sources of enzymes, often because of legislative restrictions, but this tendency does result in some economies as the purification of several enzymes from a single organism is more likely to be carried out; the isolation and purification of the enzyme, especially on a large scale, usually being the major element in its final cost to the producer. *Aspergillus niger, A. oryzae* and *Bacillus subtilis* are the most useful and well known enzyme producer organisms (Table 3.1). In general the fungal enzymes have neutral or acidic pH optima and are not thermostable, whereas the bacterial enzymes tend to have neutral or alkaline pH optima and are often thermostable.

Up to the present day, extracellular enzymes have been most commonly used because there is no need to disrupt the cells and separate the cells from the cell debris, because yields of up to 1% (v/v) can be obtained from fermenters, which are higher on a volumetric basis than can be achieved for intracellular enzymes, because extracellular enzymes are usually available in a purer form than extracellular enzymes, and because extracellular enzymes are usually more stable than enzymes with an intracellular location, possibly because they usually contain disulphide bonds. However, some intracellular enzymes, such as glucose isomerase, glucose oxidase, penicillin acylase, and asparaginase have found important commercial applications, although extracellular versions of such enzymes would almost certainly be very useful commercial products.

Synthesis of intra- or extracellular enzymes is an aspect of cellular metabolism that is of great interest since it influences the type and yield of enzymes obtained in fermentations. Partly because of the complexity of the processes involved, optimization has tended to be empirical. However, Terui *et al.* (1967) have produced a model which assumes that the rate of enzyme synthesis is directly proportional to the amount of mRNA present, and that it is the quantity of mRNA present which is the rate of limiting step. This model was found to be consistent with the experimental results obtained for several fermentation processes for the production of enzymes.

Current ideas suggest that enzymes destined for secretion are synthesized on membrane-bound polysomes, the precursor proteins containing hydrophobic amino acid sequences (signal peptides) which are cleaved off by specific proteases during the secretion of the enzyme (Ramalay 1979, Priest 1983). Yields of secreted enzymes can be increased by incorporating surfactants into the culture medium, presumably by causing faster release of enzyme. Similarly, yields can be increased by the use of morphological mutants which are hyperenzyme producers, for instance α-amylase, amyloglucosidase, invertase and trehalase can be obtained in greater amounts from morphological variants of *Neurospora* (Gratzner 1972).

One of the very few membrane-bound enzymes to have been commercially exploited so far as cholesterol oxidase which is used in a clinical assay for blood cholesterol, it is unusual in being located on the outside of the parent cell, such that it can be liberated by detergents without causing disruption of the cell, and the consequent release of intracellular proteins to contaminate the enzyme operation.

For detailed information on the sources, production and uses of microbial enzymes see Halpem (1981), Gutcho (1974) and Duffy (1980).

3.5 THE ISOLATION, PURIFICATION AND FORMULATION OF ENZYMES
(see Part A, Chapter 2 and Part B, Chapter 2)

Most of the enzymes in current use in bulk quantities have extracellular locations, whereas the majority of enzymes are intracellular and/or membrane-bound. Therefore, if the full potential of enzymes is to be exploited cheap efficient methods of extracting and purifying enzymes will have to be discovered which take into account the low concentrations of enzymes that are typically found in the cells, which are usually less than 1% (w/w). For comprehensive reviews, see Darbyshire (1981), Bruton (1983) and Lambert & Meers (1983).

Intracellular enzymes are usually released from inside the cell by mechanical or chemical disruption, or by permeablization of the cell membrane so as to allow leakage. Common methods of cell disruption include high pressure homo-genizers, ball mills, freeze-thawing, pH, temperature, or osmotic shocking, or enzyme digestion such as with lysozyme or the mixture of enzymes derived from *Micrococcus lysodeikticus*. The resulting crude enzymically active prepara-tions are then clarified free of remaining cells and cell debris by filtration or centrifugation. Centrifugation is the most common method, one or more of the large-scale designs of centrifuge being used. These include tubular bowl, disc bowl, basket and scroll centrifuges which besides their large capacities may have the additional advantage of being usable continuously or semicontinuously. (Fig. 3.2, Higgins *et al.* (1978)).

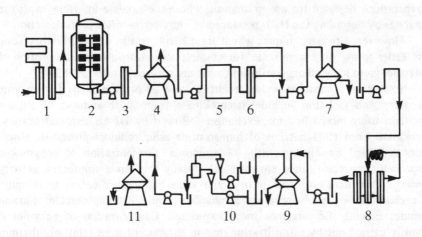

Fig. 3.2 – Flow diagram for the continuous isolation of β-galactosidase from *E. coli* (Higgins *et al.* 1978) 1, continuous-flow plate sterilizer, 2, 1000 (fermenter, 3.6, heat exchangers, 4, 7, 9, 11, disc-bowl centrifuge, 5, homogenizer, 8, heat exchanger and holding coil, 10, continuous-flow mixer.

The resulting cell-free extracts can then be purified by a consecutive series of fractionation and concentration steps. These include ion-exchange chromato-graphy and membrane techniques, such as ultrafiltration and reverse osmosis, although concentration polarization is a serious problem in the latter techniques. Precipitation of the enzyme is also a commonly used method, since many treatments are reversible, easy, and cheap to carry out on a large scale, and are relatively non-denaturing (Wang *et al.* (1979)). Precipitants include organic solvents, polyethylene glycol, dextran, polyacrylic acid, or ammonium sulphate which acts by salting the proteins out of solution. Coagulation or flocculation are also used on occasions. The ease of extraction and yield of an enzyme can be influenced by the fermentation conditions used, the phase of growth the cells are harvested in, and by any delays in processing.

Comparatively little work has been done on the development of large-scale methods for the purification and isolation of enzymes, such that often workers resort to scaled-up laboratory methods despite their diffficulties, inefficiencies, and costs. Thus if techniques such as gel filtration or affinity chromatography need to be used to produce very pure enzyme preparations, then such enzymes can only be produced in small quantities and will be expensive. For larger scale applications these techniques require the development of new support materials with greater mechanical strength. The need for the development of new commercially important techniques are also apparent, for instance for the removal of catalase by an affinity chromatography method. This technology could then be used for a number of applications where catalase must be removed from enzyme preparations destined for use in analysis, where the catalase interferes with the analyses by degrading the H_2O_2 produced by the enzyme prior to its detection.

However, a few techniques which are difficult to use on a laboratory scale are easier to use on a large scale, for example the adsorption and desorption of enzymes from ion-exchange resins in basket centrifuges (Bruton, 1983).

Examples of batchwise ion exchange and adsorption steps for enzyme recovery and isolation include streptokinase from 2501 aliquots of culture medium using mixed bed ion exchanger followed by DEAE chromatography, urokinase from 1001 batches of human urine using cellulose phosphate absorbent followed by elution with 3% ammonia. Concentration of enzymes is especially important since enzymes are usually sold on a volumetric activity basis, rather than on their purity, and since many purification techniques, especially those involving chromatography, result in considerable dilution besides making the enzymes more expensive. Concentration of enzymes is usually carried out by ultrafiltration, but in the case of some relatively thermostable enzymes, such as amyloglucosidase, concentration can be by evaporation.

The enzyme is usually marketed as a concentrated liquid often containing preservatives, or as a solid powder diluted with bulking agents such as lactose; although solid preparations can present a health hazard due to the dust generated. Good storage stability is of great importance, owing to the obvious difficulties in marketing a product of continuously decreasing activity.

3.6 LEGISLATION ON THE USE OF ENZYMES

Traditionally, enzymes have been regarded as being safe, owing to their description as natural extracts; but this status is being eroded worldwide, although actual toxic enzymes are very rare. In the UK at present there are no specific controls on the use of enzymes in food production, in which they are classed as food processing aids.

The use of enzymes in foods and in their manufacture is regulated in most countries, with local regulations varying somewhat. For a review of enzyme regulations see Roland (1981) and Denner (1983). Introduction of a new enzyme therefore requires a demonstration that it is not toxic, by means of animal feeding and other trials. Note also that the legislative status of immobilized cells is not clear, especially when the viability of the cells is retained during operational use.

The following list includes microorganisms that have been traditionally used in foods or food processing, and so are generally thought to be harmless by enzyme producing companies, having stood the test of time over many generations:

Bacillus subtilis (including strains known under the names *B. mesentericus, B. natto* and *B. amyloliquefaciens*).

Aspergillus niger (including strains known under the names *A. awamori, A. foetidus, A. phoenicis, A. saitoi* and *A. usumii*).

Aspergillus oryzae (including strains known under the names *A. sojae* and *A. effusus*).
Mucor javanicus
Rhizopus arrhizus
Rhizopus oligosporus
Rhizopus oryzae
Saccharomyces cerevisiae
Kluyveromyces fragilis
Kluveromyces lactis
Leuconostoc oenos.

The food Additives and Contaminants Committee (FACC) of MAFF (The Ministry of Agriculture, Fisheries and Food) has classified enzymes into five classes, A–E, on the basis of their safety for use in foods and in the manufacture of foods. This classification takes into account the nature of the enzyme itself, including its catalytic activities, any immunogenic, allergenic or toxic effects elicited by the enzyme or by contaminants contained in the enzyme, for instance derived from the fermentation broth (HMSO, 1982). The definition of each class is as follows:

Group A: Substances that the available evidence suggests are acceptable for use in food.

Group B: Substances that on the available evidence may be regarded as provisionally acceptable for use in food, but about which further information must be made available within a specified time for review.

Group C: Substances for which the available evidence suggests possible toxicity and which ought not to be permitted for use in food, until adequate evidence of safety has been provided to establish their acceptability.

Table 3.3

Classification of enzyme preparations by Ministry of Agriculture Fisheries and Food (1982)

Source	Enzyme preparation
(i) Group A	
Ananas comosus	– bromelain
Ananas bracteatus	– bromelain
Carica papaya	– papain
	– chymopapain
Edible oral or forestomach tissues of the calf, kid or lamb	– triacylglycerol lipase
Porcine or bovine pancreatic tissues	– triacylglycerol lipase
	– α-amylase
	– trypsin
Porcine gastric mucosa	– pepsin A
	– pepsin B
	– pepsin C
Abomasum of calf, kid or lamb	– chymosin (rennet)
Adult bovine abomasum	– bovine pepsin A
	– bovine pepsin B
Bovine liver	– catalase
(ii) Group B	
Aspergillus niger	– α-amylase
	– immobilized and non-immobilized *exo*-1, 4-α-D-glucosidase (glucoamylase)
	– cellulase
	– β-D-galactosidase (lactase)
	– *endo*-1, 3(4)-β-D-glucanase
	– glucose oxidase
	– catalase
	– pectinesterase
	– pectin lyase
	– polygalacturonase
Aspergillus oryzae	– α-amylase
	– neutral proteinase
Bacillus coagulans	– immobilized and non-immobilized glucose isomerase
Bacillus licheniformis	– α-amylase
	– serine proteinase
Bacillus subtilis	– α-amylase
	– *endo*-1,3(4)-β-D-glucanase (laminarinase)
	– neutral proteinase
Endothia parasitica	– endothia carboxyl proteinase
Klebsiella aerogenes	– pullulanase
Mucor miehei	– acid proteinase
Mucor pusillus	– acid proteinase
Penicillium emersonii	– *endo*-1,3(4)-β-D-glucanse
Penicillium funiculosium	– dextranase
Penicillium lilacinum	– dextranase
Saccharomyces cerevisiae	– β-D-fructofuranosidase (invertase)
Streptomyces fradiae	– serine proteinase
Streptomyces olivaceous	– immobilized glucose isomerase
Trichoderma viride	– cellulase

Reproduced with the permission of the Controller of Her Majesty's Stationery Office.

Group D: Substances for which the available information indicates definite or probable toxicity and which ought not to be permitted for use in food.

Group E: Substances for which inadequate or no toxicological data are available and on which it is not possible to express an opinion as to their acceptability for use in food.

In this review of enzymes, MAFF recommended certain enzymes (Table 3.3) for inclusion in a list of enzymes acceptable or provisionally acceptable for use in foods; with the data on the enzymes in Group B to be reviewed after 2 years. It is to be expected that following comment and representations, regulations will follow along the lines suggested by the FACC.

The dextranases derived from *Penicillium funiculosom* and *Penicillium lilacinum* were also permitted, but were restricted to use during the early stages of sugar refining.

Permitted enzyme preservatives included sorbic acid, SO_2, and the methyl, ethyl and propyl esters of 4-hydroxybenzoate. Note that MAFF classify the food products of enzyme action into three classes, additives, flavours modifiers and bulking agents.

3.7 ENZYME MANUFACTURERS

Table 3.4 gives a comprehensive list of companies that sell enzymes, of which about half-a-dozen are obviously dominant in terms of the variety, quality and quantity of enzymes retailed. A trade association exists called 'The Association of Microbial Food Enzyme Producers'. Table 3.2 provides a summary of the present-day uses of enzymes in industry. Frozen cell pastes are sold as sources of enzymes by the PHLS, Porton Down.

Table 3.4

Some industrial enzyme suppliers (with addresses)

A.B.M. Chemicals Ltd,
Poleacre Lane, Woodley, Stockport, Cheshire, SK6 1PQ, UK
Advance Biofactures Corporation,
35 Wilbur Street, Lyncbrook, NY 11563, USA
Aktieboluget Montoil,
Fack S 100 55, Stockholm, Sweden
Akzo Chemie,
Statwinstraat 48, P.O. Box 247, The Netherlands
Alltech Inc,
271 Gold Rush Road, Lexington, Kentucky 40503, USA
Alpha Color S.P.A.,
Via Valtellina 48, 20159 Milan, Italy
Amano Pharmaceutical Co. Ltd,
1-21-chome, Nishiki, Naka-Ku, Nagoya, Japan
A.P.I. Laboratory Products Ltd,
Unit D2, Grafton Way, Basingstoke, Hants, UK

Table 3.4 — *continued*

Armour Pharmaceutical Co. Ltd,
Jampden Park, Eastbourne, Sussex, BN22 9AG, UK
B. D. H. Chemicals Ltd,
Broome Road, Poole, Dorset, BH12 4NN, UK
Beckman Instruments, Inc. (Microbics),
6200 El Camino Real, Carlsbad, CA92008, USA
Biddle Sawyer and Co. Ltd.
P.O. Box 170, 3 Lovat Lane, London EC3P 3EX, UK
Biochemi G.m.b.H,
Bulkmarketing,
Kundle, Tyrol A-6250, Austria
Biocon Ltd,
Kilnagleary, Carrigaline, Co. Cork, Eire
Biotec, Inc,
2800 Fish Hatchery Road, Madison, WI 53711, USA
Laboratories Biotrol,
Div: Clinical Diagnostic Div, 1, rue de Foin, 75140 Paris Cedex O3, France
Biozyme Laboratories Ltd,
Unit 6, Gilchrist-Thomas Estate, Blaenavon, Gwent, NP4 9RL, UK
Böhringer Ingelheim,
Bdrreich Chemikalien, Ingleheim am Rhein 6507, W. Germany
Boehringer Co. (London) Ltd
Bilton House, 54/58 Uxbridge Road, Ealing, London W5 2TZ, UK
Boehringer Mannheim G.m.b.H,
P.O. Box 51, D-6800 Mannheim 31, W. Germany
Calbiochem,
P.O. Box 12087, San Diego, Ca92112, 10933 N. Torrey Pines Road, La Jolla CA 92037, USA
Cambrian Chemicals Ltd,
Suffolk House, George Street, Croydon, CR9 3QL, UK
Cambridge Medical Diagnostics, Inc,
575 Middlesex Turnpike, Billerica, MA 01865, USA
Chr. Hansen Laboratories A/S
3 Sankt Annae Plads, DK-1250 Copenhagen, Denmark
CIBA-Geigy Ltd
Basle, Switzerland
Collaborative Research Inc,
Research Products Div, 1365 Main Street, Waltham, MA 02154, USA
Corning Biosystems,
Corning Glass Works, Croning, New York 14830, USA
Dairyland Food Laboratories Inc,
Progress Avenue, Waukesha, Wisconsin 53187, USA
Daiwa Kasei KK,
3—11 Vehonmachi-5-chome, Tennoji-ku, Osaka, Japan
Diamalt,
Friedrechstasse 18, D-800 Munchen 40, Postfach 400 469, W. Germany.
Diamon Shamrock,
620 Progress Avenue, Waukesha, Wisconsin 53186, USA
Diosynth BV,
P.O. Box 20, Oss, The Netherlands
Enzyme Center, Inc,
33 Harrison Ave, Boston, MA02111, USA
Enzyme Development Corporation,
2 Penn Plaza, New York, New York 10121, USA
Fermco Biochemics Inc,

Table 3.4 – *continued*

2638 Delta Lane, Elk Grove Village, Illinois 60007, USA
Genzyme Biochemicals Ltd,
Springfield Mill, Maidstone, Kent, ME14 2LE, UK
Genzyme Corporation,
1 Bishop St., Norwalk, CT 06851, USA
Gist Brocades NV,
P.O. Box 1, Wateringseweg 1, Delft 2600MA, The Netherlands
Glaxo Operations UK Ltd,
Ulverton, Cumbria, LA12 9DR, UK
Godo Shusei Co. Ltd,
6-2-10, Ginza, Chuo-ku, Tokyo, Japan
Grindestedvaeket A/S,
38 Edwin Rahrs Vej, 8220 Braband, Denmark
Grindstedvaerket, G.m.b.H,
D-2 Hamburg 54, Kellerbleek 3, Postfach 54097, W. Germany
Grinsted Products Ltd,
Northern Way, Bury St. Edmunds, Suffolk, 1P32 6NP, UK
Hankyu Kyoei Bussan Co. Ltd,
5/6 chome, Tenjin Bashi-suji, Oyodo-Ki, Osaka, Japan
Hayashibara Shoji Co. Ltd,
198 Shimoishii, Okayama City, Japan
Henkel KG a.A,
d-4000 Dusseldorf 1, Postfach 1100, W. Germany
Hoechst,
D-6000 Frankfurt/M80, Postfach 800320, W. Germany
Hopkins and Williams Ltd,
P.O. Box 1, Romford, Essex, RM1 1HA, UK
Hughes and Hughes (Enz.) Ltd,
Elms Industrial Estate, Church Road, Harold Wood, Romford, Essex, RM3 0HR, UK
Kingsbridge Industrial Inc,
P.O. Box 24, 200 Tapei, Taiwan
Kyowa Hakko Kogyo Co. Ltd,
Ohtemachi Building, 6–1, Ohtemachi, 1-chome, Chiyoda-ku, Tokyo, Japan
Laboratories Sanders SA
47–51 rue Henri Wafelaerts, Brussels 6, Belgium
Dr. Madis Laboratories Inc,
375 Huyler Street, S. Jackensack, New Jersey 07606, USA
Meiji Seika Kaisha, Ltd,
Kyobashi, Chuo-Ku, Tokyo, Japan
E. Merck,
Frankfurther Str. 250, Postfach 4119, D-6100 Darmstadt 1, W. Germany
Miles Kali-Chemi G.m.b.H,
3, Hannover-Kleefeld, Hans-Buckler Allee 20, Postfach 690307, W. Germany
Miles Laboratories, Inc,
Research Products Div, P.O. Box 2000, 1127 Myrtle Street, Elkhart, IN 46515, USA
Miles–Seravac (Pty) Ltd,
Moneyrow Green, Holyport, Maidenhead, Berkshire, UK
Millipore U.K. Ltd,
Millipore House, 11–15 Peterborough Road, Harrow, Middlesex, HA1 2YU, UK
Mitsui and Co. Ltd,
P.O. Box 822, Tokyo Central, Japan
Munton and Fison Ltd,
Cedars Factory, Stowmarket, Suffolk, IP14 2AG, UK
Murphy and Son Ltd,
Wheathampstead, St Albans, Hertfordshire, UK

Table 3.4 − *continued*

Naarden Internation NV,
Postbus 2, Naarden-Bussum, The Netherlands
Nagase and Co. Ltd,
Konishi Building, 2,2-chome Honcho, Nihonbashi Chuo-ku, Tokyo, Japan
New England Biolabs, Inc,
32 Tozer Road, Beverley, Ma 01915, USA
New Englans Enzyme Centre (Tufts),
Tufts Medical School, 136 Harrison Avenue, Boston, MA02111, USA
Otto Norwald KG,
2 Hamburg 50, Heinrichstrasse 5, W. Germany
Novadel Ltd,
12/14 St Ann's Crescent, Wandsworth, London, SW18 2LS, UK
Novo Industri A/S,
Novo Alle, DK-2880 Bagsvaerd, Denmark
Ikasa Industriae Comercia de Diastase Ltda,
Rue Araraquara, 41-Diadema-SP, Brazil CEP 09900, Caixa Postal 352, Brazil
Organnon Laboratories Ltd,
Newhouse, Lanarkshire, ML1 5SH, UK
Oriental Yeast Co.
Enzyme Development Centre, 4−1 Minamisuta 4-chome, Suita, Osaka 564, Japan
Pfizer Inc,
World Headquarters, 235 East 42nd Street, New York, New York 10017, USA
Pharmacin State Economic Trust,
15 Iliensko Chaussee, Sofia, Bulgaria
PHLS Centre for Applied Microbiology and Research,
Porton Down, Salisbury, Wiltshire, SP4 0JG, UK
P.L. Biochemicals Inc,
1037 West Mckinley Avenue, Milwaukee, Wisconsin 53205, USA
Powell and Scholefield Ltd,
38 Queensland Street, Liverpool, L7 3JG, UK
Premier Malt Products Inc.,
1137 North 8th Street, Milwaukee, Wisconsin 53210, USA
Rohm G.m.b.H,
Kirshenallee, Postfach 4−42, D-6100 Darmstadt 1, W. Germany.
Rohm and Haas Co,
Independence Mall West, Philadelphia, Pennsylvania 19105, USA
Royal Netherlands Fermentation, Ind. Ltd,
P.O. Box 1. Delft, The Netherlands
Schering AG,
1-Berling 65, Mullerstr 170−172, W. Germany
Schmitt-Jourdan,
22 Rue de la Tourellw, 92100 Boulogne-Billancourt, Paris 8, France
G. D. Searle and Co,
2634 South Clearbrook Drove, Arlington Heights, Illinois 60005, USA
Sigma Chemical Company INc,
St. Louis, Morrisvilles, Missouri, USA
Societe Rapidase,
15 rue des Comtesses, 59113 Seclin, France
Societa Prodotti Antibiotici,
20143 Milano, Via Biella, 8, Italy
Sturge Chemicals, Denison Road, Selby, North Yorkshire, YO8 8EF, UK
Sumitomo Shoji Kaisha Ltd,
2 Nishikicho Building 24-1 Kandanishikicho 3-chome, Chiyoda-ku, Tokyo, Japan
Swiss Ferment Co,
Vogesenstrasse 132, 4056 Basel 13, Switzerland

Table 3.4 – *continued*

Tanabe Seiyaku Co. Ltd,
via Siber Hegner Benelux BV, Postbus 414 Rotterdam, Westersingel 107, Japan and The
    Netherlands
Ubichem Ltd,
281 Hithermoor Road, Stanwell Moor, Staines, TW19 6AZ, UK
United States Biochemical Corporation,
P.O. Box 22400, Cleveland, OH44122, USA
Vifor SA,
48 route de Drize, 1227 Carouge, Geneva, Switzerland
W.B.E. Ltd,
Sandyford Industrial Estate, Foxrock, Dublin 18, Eire
Windsor Laboratories Ltd,
Bedford Avenue, Slough, Berkshire, UK
Worthington Diagnostics,
P.O. Box 650, Hall Mill Road, Freehold, NJ 07728, USA
Yakult Biochemical Co. Ltd,
8–21 Jingikan-Machi, Nishinomiya-shi, Hyogo, Japan

The most rapid growth in terms of the formation of new companies would appear to be in the use of enzymes as analytical devices. This is probably because extensive testing prior to regulatory approval is not needed, and new ideas can be commercialized comparatively rapidly without the need to scale-up via pilot size equipment, as in most other areas of industrial enzymology.

3.8 BIOCHEMICAL APPLICATIONS

Enzymes are one of the central areas of study in biochemistry. Free (soluble) enzymes are usually used, although the immobilization of enzymes is a useful methodology for solving problems in academic biochemistry. These include the following-

1.   By being able to remove an immobilized enzyme from its reaction mixture very rapidly, by filtration, centrifugation, or other means, for instance using a magnet in the case of enzymes immobilized to magnetically susceptible supports, the reaction can be easily stopped. When using soluble enzymes this is not a procedure that is easy to perform without recourse to methods that denature the enzyme. Furthermore, following its recovery the immobilized, but not the free enzyme can be easily and rapidly transferred to a different vessel, and thus different reaction conditions can be employed. Thus Zingard & Uziel (1970) used immobilized alkaline phosphatase for the sequential analysis of t-RNA by endonucleases, and Knorre *et al.* (1973) used phosphomonoesterase and snake venom immobilized into DEAE-Sephadex with a triazine dye for determining the nucleotide composition of oligonucleotides.

2. By co-immobilizing enzymes to form novel metabolic pathways, faster rates of reaction can be obtained than when the corresponding enzymes are used together in free solution. This is because the diffusion layers of the enzymes can overlap when co-immobilized, or be in close proximity to each other. Thus metabolic intermediates produced by one enzyme have a shorter distance to diffuse before becoming the substrate for a second enzyme than if the enzymes were used in free solution.

3. The subunit structure of enzymes can be investigated by immobilization; for instance an enzyme could be immobilized by one subunit only, the other subunits dissociated, and then reassociation of hybridization experiments using the immobilized subunit and free subunits studied. (Swaisgood et al. 1976, Horton & Swaisgood, 1976).

4. Immobilized enzymes have been used widely in the field of affinity chromotography to isolate enzyme inhibitors and labelled peptides. Trypsin, trypsin—plasmin, and trypsin—kallikrein inhibitors have been isolated by Fritz et al. (1969) by passing crude extracts of the inhibitor down a column containing the appropriate immobilized enzyme, and eluting the inhibitor by rupturing the immobilized enzyme/inhibitor complex with acidic buffer. The proteolytic inhibitor in wheat germ has been isolated by a similar method (Fritz et al. 1970). Givol et al. (1970) prepared a column of ribonuclease convalently bound to Sepharose with cyanogen bromide. They then labelled soluble ribonuclease uniquely at the tyrosyl residue by reaction with a substrate analogue, the diazonium derivative of 5′-aminophenylphosphonyl)-uridine 2′-(3′)-phosphate, and digested the labelled enzyme by carboxymethylation and trypsin hydrolysis. The digest was passed down the column of immobilized ribonuclease, which had an affinity for the substrate analogue, and the peptide possessing the substrate analogue was thus separated from the rest of the digest. Similarly, DNP-labelled peptide was isolated by employing an anti-DNP Sepharose column. Isoleucine-t-RNA synthetase immobilized on the same support has been used for the isolation of t-RNA specific for isoleucine (Denburg & Deluca (1970)).

5. In some respects immobilized enzymes are useful models of membrane-bound enzymes as both are active while attached to a solid substratum. This is especially true of membrane-bound enzymes which have been extracted and then immobilized to a more defined support, liposomes being a favourite support (Gregoriadis 1976).

A good example is the work of Storelli et al. (1972) who used the sucrase—isomaltase complex as a model active transport system. The enzyme was incorporated into a mixture of purified phospholipids which spontaneously form a single lipid bilayer called a black lipid membrane, when spread over a small hole in a Teflon plate separating the two compartments of the apparatus. Sucrose added to one compartment disappeared rapidly, and fructose appeared in the second compartment at rates several orders of magnitude faster than could occur by diffusion of the sucrose or fructose alone, either when no enzyme was incorporated in the membrane or when the enzyme was inhibited. No transport of sucrose took place when the enzyme was added to preferred membranes.

6. Immobilization is a method of maintaining the reaction conditions of an enzyme constant. Thus *in vivo* the concentrations of reactants are constantly changing, and often the enzyme(s) is being actively synthesized and/or degraded during the course of the reaction. Also in batch laboratory experiments substrate concentrations fall, product(s) accumulates, and the enzyme cannot be exposed to steady state conditions. However, immobilized enzymes can be maintained in essentially constant reaction conditions by using immobilized enzyme in a packed bed or continuous stirred tank reactor in which fresh substrate is continuously supplied to the enzyme at a constant rate and products continuously removed.

7. Enzyme activities are frequently stabilized by immobilization, different methods of immobilization frequently having very differing stabilizing effects (Table 3.5).

Table 3.5

The effect of the immobilization technique used on the activity and stability of *Erwinia rhapontici* cells producing isomaltulose.

Immobilization technique	Activity [g product/ g immobilized cell/h]	Half life (hours)
Calcium alginate	0.325	8500
DEAE cellulose	0.583	400
Polyacrylamide	0.13	570
Glutaraldehyde aggregated cells	0.153	40
K-carrageenan-locust bean gum	0.263	37.5
Bone char	0.01	25
Agar	0.34	27
Xanthan—locust bean gum	0.10	8

(from Cheetham & Bucke 1982).

For details of the immobilization methods see *Methods in Enzymology* **44** (ed. Mosbach, K.) Academic Press (1976).

3.9 USE OF ENZYMES IN ANALYSIS (see also Part B, Chapter 5)

3.9.1 General
The best known uses of enzymes in analysis are as soluble enzyme 'kits' especi-

ally useful in clinical biochemistry, such as for measuring blood glucose, using glucose oxidase and catalase; and cholesterol using cholesterol oxidase and cholesterol esterase. More recently the use of enzyme electrodes in which enzymes are bound to transducers making use of potentiometric or amperometric detection have become widespread. The first to be described used glucose oxidase to measure glucose amperometrically (Clark & Lyons 1962). Thus the advantageous properties of enzymes such as their substrate stereospecificity and high affinity for their substrate can be made use of in 'reagentless' analytical devices which can, for example, provide proof of the optical purity of a chemical as well as its chemical identity and concentration. Such analytical applications of immobilized enzymes are especially useful when automatic, closed system continuous flow analysers are to be used.

Indirect enzyme electrodes involve the detection of an enzyme reactant by a specific, usually ion-selective electrode. Direct enzyme electrodes incorporate redox enzymes coupled to an electrode with a flow of electrons between the cofactor involved in the enzyme reaction and the electrode. Transducer bound enzymes consist of a thin layer of enzyme in close contact with a transducer, such as oxygen electrode, together with a thin semipermeable membrane to secure the enzyme in place. Substrate then diffuses from the bulk solution that the electrode is placed in, undergoes reaction, and then the transducer measures the appearance of the product. pH electrodes and ammonium ion selective electrodes are also commonly-used potentiometric electrodes. Amperometric transducers in contrast usually measure the flux of electroactive species.

The first successful automated analytical application using immobilized enzymes was reported over a decade ago by Hicks & Updike (1969). Using polyacrylamide entrapped lactate dehydrogenase and glucose oxidase, lactate and glucose were quantitated by spectrophotometric detection of a coupled dye. Subsequently a very wide range of practical devices have been developed including sensors for BOD, biomass, carcinogens, creatinine, α-fetoprotein, gastrin, glucose, glutamate, human chorionic gonadotropin, human serum albumin, NADH, sodium ions, sucrose, sulphate, syphilis bacteria, and urea (Table 3.6).

Table 3.6
Enzyme electrode probes

Type	Enzyme	Sensor
Acetic acid, formic acid	Alcohol oxidase	Pt
Acetylcholine	Acetylcholinesterase	Choline
Acetyl-β-methylcholine	Acetylcholinesterase	Acetylcholine
Adenosine monophosphate (AMP)	5-Adenylate deaminase	NH_4^+

Table 3.6 — *continued*

Type	Enzyme	Sensor
Alcohols[a]	Alcohol dehydrogenase	Pt
	Alcohol dehydrogenase/diaphorase	Pt
	Alcohol oxidase	Pt
D-Amino acids[b]	D-Amino acid oxidase	NH_4^+
L-Amino acids[c] (general)	L-Amino acid oxidase	Gas (NH_2)
		NH_4
		Pt
	Decarboxylases	CO_2
L-Arginine	Arginase	NH_4
L-Asparagine	Asparaginase	NH_4^+
L-Cysteine	*Proteus morganii*	H_2S
L-Glutamine	Glutaminase	NH_4^+
L--Glutamic acid	Glutamate dehydrogenase	NH_4^+
	Glutamate decarboxylase	CO_2
L-Histidine	Histidinase	NH_4^+
L-Lysine	Lysine decarboxylase	CO_2
L-Methionine	Methionine ammonia lyase	NH_3
L-Phenylalanine	Phenylalanine ammonia lyase	NH_3
L-Tyrosine	Tyrosine decarboxylase	CO_2
	Tyrosinase	Gas (O_2)
Amygdalin	β-Glucosidase	CN^-
Butyryl thiocholine	Cholinesterase	Pt(SCh)
Cholesterol	Cholesterol esterase/ cholesterol oxidase	$Pt(H_2O_2)$
		$Pt(O_2)$
	Cholesterol oxidase	$Pt(O_2)$
Creatinine	Creatininase	NH_4^+
	Creatininase (purified)	NH_3
Glucose	Glucose oxidase	pH
		$Pt(H_2O_2)$
		Pt(quinone)
		Pt(DCIP)
		$Pt(O_2)$
		1^-
		$Gas(O_2)$
	Glucose oxidase/peroxidase	Pt
Lactic acid	Lactate dehydrogenase	Pt $[Fe(CN)_6]$
		C(NADH)
	Cytochrome-b_2	Pt
Lactose	β-Galactosidase/glucose oxidase	Gas (O_2)
Maltose	Maltase/glucose oxidase	Gas (O_2)
NADH	Alcohol dehydrogenase	Pt
	Mitochondria	Gas (O_2)
Nitrate	Nitrate reductase (nitrite reductase)	NH_4^+
Oxalic acid	Oxalate decarboxylase	Gas (CO_2)
Penicillin	Penicillinase	pH
Peroxide	Catalase	$Pt(O_2)$
Phosphate	Phosphatase/glucose oxidase	$Pt(O_2)$
Sucrose	Invertase/mutarotase/glucose oxidase	$Pt(O_2)$
Succinic acid	Succinate dehydrogenase	$Pt(O_2)$
Sucrose	Invertase/glucose oxidase	$Pt(H_2O_2)$
Sulphate	Aryl sulphatase	Pt

Table 3.6 — *continued*

Type	Enzyme	Sensor
Thiosulphate	Rhodanease	CN^-
Urea	Urease	NH_4^+
		pH
		Gas (NH_3)
		Gas (CO_2)
Uric acid	Uricase	Pt (O_2)

[a]Responds to methanol, ethanol, allyl alcohol.
[b]Responds to D-phenylalanine, D-alanine, D-valine, D-methionine, D-leucine, D-norleucine, D-isoleucine.
[c]Responds to L-leucine, L-tyrosine, L-phenylanine, L-tryptophan, L-methionine.
(Guilbault 1980). Reprinted by permission of the publishers, Butterworth & Co. (Publishers) Ltd. ©

An application that illustrates the great potential of immobilized enzymes in trace analyses involves the determination of parts per thousand-million levels of nitrate in environmental water samples. The nitrate was reduced to nitrite in the presence of methyl viologen, that is,

$$NO_3^- + 2H^+ + 2MV^+ \xrightarrow[\text{reductase}]{\text{nitrate}} 2MV^{2+} + NO_2^- + H_2O \ ,$$

where MV^+ and MV^{2+} represent the oxidized form of methyl viologen and its one electron reduction product respectively. The nitrite produced in the reaction was determined by the classical azo dye reaction monitored at 543 nm. This sytem is rapid and simple and showed improved specificity over other more commonly used nitrate methods. The results were significant enough for the US Environmental Protection Agency to choose it as a standard technique (Senn *et al.* 1976).

Several enzyme electrodes based on potentiometric cation and ammonium ion specific electrodes have been developed. Guilbault *et al.* (1970) used the production of ammonium ions by immobilized urease to determine urea, but Na^+ and K^+ were found to interfere with the assay. L-amino acid oxidase has been immobilized by several methods and used with cation electrodes sometimes in conjunction with immobilized catalase (Guilbault *et al.* 1970). Guilbault & Nagy (1973) also described two enzyme electrodes for the determination of L-phenylalanine. The first consists of L-amino acid oxidase and horseradish peroxidase immobilized in a polyacrylamide gel and used with an ion selective iodide electrode, and the second consists of the same enzyme system in conjunction with a silicone rubber based support. Similar systems have been described using D-amino acid oxidase (Guilbault 1971) and asparaginase (Guilbault 1971). Enzyme electrodes can, therefore, be used for a wide range of assays, and their accuracy and practicality make them highly suitable for the routine clinical monitoring of physiological fluids.

An extremely sensitive method for determining copper in solution was reported by Stone & Townshend (1973). Copper was removed from polyphenol oxidase which had previously been bound to a polyacrylamide gel by the method of Barker *et al.* (1970), and the resulting apoenzyme was reproducibly and selectively reactivated by incubation with the copper solution to be determined. Immobilized enzyme was used because it made the removal of copper and handling of the apoenzyme easier, showed increased stability on storage, and could be reused. Another interesting example is the use of an immobilized enzyme to measure the freshness of fish. In this case the accumulation of inosine monophosphate, inosine, and hypoxanthine is the indicator of freshness, and these substances are measured by 5'-nucleotidase, nucleoside phosphorylase, and xanthine oxidase co-immobilized to cellulose triacetate membrane and connected to an oxygen electrode which detects the consumption of oxygen as reaction takes place.

Enzyme based analytical devices are characterized by their sensitivity, response times and limits of detection, but are limited by the stabilities of the enzymes involved. Other problems are the high cost of these devices, their slow response times usually due to internal diffusional restrictions on reactants moving in and out of the immobilized enzyme/cell preparations, and the need to match the effective range of concentrations to be measured with the linear portion of the Michaelis—Menten curve of the enzyme in use. Thus a range of enzymes differing in K_m values are needed in order to cover the different concentrations of reactants that must be measured.

Owing to the slow rates of electron transfer between reactants and electrode even when mediators such as 4 4' bipyridyl are used, much attention has been devoted to enzyme electrodes which measure the heat of the enzyme reaction. Heat can be measured, either directly by thermistors, or indirectly by measuring the potential generated by the heat of reaction via thermopiles. For instance, Mattiasson *et al.* (1977) were able to continuously monitor the concentration of cyanide present in blast furnace waste waters down to 20 μM by means of the heat generated during reaction with the enzymes rhodanese and injectase; with a response time of 2—3 min.

Although enzyme electrodes have received most of the attention in immobilized enzyme based analysis, several other types of applications have been presented; for example, solid surface fluorescence has been developed. In this technique, enzymes and cofactors are lyophilized onto a silicone rubber pad. When an analysis is required, the pad is reconstituted and the analyte solution added to the 'immobilized enzyme' spot. Quantitation is accomplished by measuring the rate of production of fluorescence in a fluorometer. The only advantage that this technique has in common with other immobilized enzyme techniques is the increased reagent stability (Reitz & Guilbault, 1975).

A second extremely versatile application using an immobilized enzyme is the immobilized enzyme stirrer. The prototype stirrer was a Teflon stir bar with immobilized urease held in place by a nylon net. Used in conjunction with an ammonia ion selective electrode, initial response times and sensitivity were reported as 2 sec and equivalent to 70% conversion (Guilbault & Stokbro, 1975).

A currently promising development involves field effect transistors. These devices are based on conventional silicon gate technology, and rely on the electrode potential created by ions or by reactions involving whole molecules. The devices usually comprise a silicon chip embedded in an unreactive material such as epoxy resin. Surrounding the chip is a pocket of sensing fluid that carries current to the chip's surface. A membrane made of a polymer or a gel surrounds the fluid. These materials have specially selected properties; some are punctured with holes designed to admit molecules up to a certain size. Others are impregnated with, or have covalently attached to them, organic substances, such as enzymes or cells. These are called BioF.E.T.s, and they react with the substance(s) under analysis setting up electrical potentials which the microchip detects. For instance, the concentration of small amounts of penicillin in fermentors can be achieved using β-lactamase; alternatively, whole bacteria can be incorporated into the gel overlaying the chip.

Another promising development is chemiluminometric assays which utilize luciferinase enzymes to carry out very sensitive analysis of enzymes or substrates which are involved in reactions utilizing NADH, NADPH, or ATP. The NAD(P)H oxidoreductase from *Beneckea harveyi* catalyses the oxidation of the former coenzymes with the emission of light, while the firefly luciferinase converts ATP to ADP + light.

3.9.2 In clinical assays (see also Parts A and B, Chapter 5)

Enzymes are now accepted as standard laboratory reagents in clinical biochemistry laboratories for the diagnosis of diseases, for following the course of diseases and the response of patients to therapy, and for identifying and monitoring the concentration of drugs or their metabolites in blood or other body fluids (Table 3.6). For reviews see Price (1983) and Atkinson (1983). An important new development are enzyme linked immunoassay (ELISA) techniques in which specific antibodies are linked to enzymes such as peroxidase or β-galactosidase whose reaction generates the chromophore by which the extent of antibody binding is measured. Thus excellent sensitivities and low thresholds are obtained without recourse to the use of radioactivity as in radioimmunoassay. Enzymes are routinely used to diagnose hepatic, myocardial, pancreatic, and prostatic diseases, anaemias, leukaemias, muscular dystrophy, tumours and toxaemias of pregnancy. These applications are based on the general phenomenon that diseased cells become leaky and lose a proportion of their enzymes which eventually find their way into the blood. The extent to which the serum enzyme activity is elevated above the normal level, depends primarily on the extent and severity of the damage to the cells. Thus, in liver diseases glutamate— pyruvate transaminase and glutamate—oxaloacetate transaminase are measured, as they give very high peaks of activity of $\cong$ 600 and 400 mU/ml during the first week of jaundice. In myocardial infarction the above two enzymes plus lactate dehydrogenase and creatine phosphokinase are measured.

Enzymes are also used in drug detection, antibiotic assay, antibody or antigen detection, or in the detection of enzymes and metabolites in extracellular fluids, and are usually quicker to use than immunoassays, more specific than photometric assays, and cheaper than chromatographic or radio-labelled immunoassays. At the present time, enzymes obtained from animal sources predominate in use, although the trend is towards the use of microbial enzymes and particularly the use of thermostable enzymes; for instance, thermostable glycerokinase from *B stearothermophilus,* because many of the enzymes used at present are rather unstable, even under normal ambient conditions. Immobilization of enzymes is another approach to the stabilization of enzymes for clinical use. Particularly suitable and convenient methods of immobilization include enzymes permanently attached to the inside of coiled tubes which can be fitted directly into autoanalysers, for instance L-asparaginase has been immobilized to the inside of nylon tubing, and enzymes have been impregnated into paper strips which are simple and robust enough to be used even in the surgery and the home. Advances are likely to result from the application of thin-film technology to this field.

Two recent developments in the use of microbial enzymes are, firstly, the estimation of highly toxic chemotherapeutic drugs in serum, such as the antifolate drug methotrexate which is used in cancer chemotherapy, since it inhibits dehydrofolate reductase and so reduces the supply of reduced folates to the cancer cells. Methotrexate is assayed spectrophotometrically by its inhibition of *Lactobacillus casei* dehydrofolate reductase (Falk *et al.* 1976). Secondly, there is the development of a rapid and sensitive enzyme mediated colourimeter assay for the toxic analgesic paracetamol which depends on the liberation of aminophenol from the paracetamol by a purified enzyme derived from *Pseudomonas fluorecens* (Atkinson *et al.* 1980). Microbial proteins other than enzymes have also proved useful; for instance protein A from *S. aureus* binds to the Fc region of IgG molecules such that fluorescent, radioactive and enzyme conjugates of protein A have been used to replace the second antibody in radioimmunoassay.

An assay to determine the concentration of a particular substrate molecule should fulfil the following requirements: For simplicity, the total change in reaction rather than the initial velocity should be measured. The enzyme should be present in excess so that the reaction reaches completion, preferably equilibrium, rapidly. The presence of high concentrations of enzyme also has the effect of minimizing the effect of any inhibitors in the assay mixture. The enzyme should also be as pure as possible and be specific for the compound under test so as to avoid false positive reactions. The substrate concentration should be low, to give quicker conversion, but high enough to give reliable results. The range will depend upon the sensitivity of the monitoring equipment.

Other factors such as pH and temperature should be carefully controlled. The total change in some parameter, for example, light extinction, fluorescence, or gas uptake or output, is measured from the start to the completion of the reaction and compared with the change obtained under similar conditions with a range of known concentrations of the pure substrate of the enzyme by means of a standard curve.

If the enzyme reaction does not produce an easily measurable change, then a second (indicator) enzyme is frequently used. This reacts with the product of the original enzyme to give an easily measured product; for example,

$$\text{glucose} + H_2O + O_2 \xrightarrow{\text{glucose oxidase}} \text{gluconic acid} + H_2O_2$$

$$H_2O_2 + \begin{array}{c}\text{dye}\\ \text{(reduced)}\end{array} \xrightarrow{\text{peroxidase}} H_2O + \begin{array}{c}\text{dye}\\ \text{(oxidized)}\end{array}$$

Several different dyes can be used; in each case the amount of oxidized dye is measured. Examples of metabolites assayed enzymatically include diphosphoglycerate, glucose, cholesterol and trigylcerides.

If a large number of analyses need to be carried out for the same compound, automated techniques for doing the reaction and following its progress are worth considering. These include the Technicon autoanalyser systems widely used in clinical biochemistry for routine analysis of samples. An objection to such techniques is that they waste expensive pure enzyme. The development of simple techniques for immobilized enzymes should avoid some of this waste. The immobilized enzyme may be attached to a piece of tubing and so form part of the analytical system. The possibilities of using enzyme electrodes are also very numerous.

As mentioned earlier, enzymes are commonly used in clinical and research laboratories in coupled enzyme systems to determine the activity of another enzyme. A commonly used coupled enzyme is:

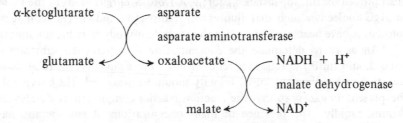

The enzyme to be assayed, asparate aminotransferase, produces oxaloacetate which is a substrate for the second enzyme, malate dehydrogenase. This oxidizes its other substrate, reduced NAD, so that the reaction is readily followed by the decrease in extinction at 340 nm. If the assay is to be accurate the oxaloacetate must not accumulate; only if it is used as soon as it is formed will the rate of formation of NAD be a true measure of the activity of asparate aminotransferase. Since the second enzyme has to function with a minute substrate concentration, a great excess has to be present in the assay mixture. Bergmeyer (1953) calculated that even with a 100-fold excess of the second enzyme, compared to the first, there was likely to be a 4% error in the initial rate measurements. Thus these coupled enzyme systems are likely to be very expensive, and it is essential that the second enzyme be free from any contaminating activities that might interfere.

Initial rate measurements can also be used to determine the concentration of compounds other than substrates, for example, inhibitors, activators and prosthetic groups (Townsend 1973). Thus the concentrations of various pesticides, drugs, and metal ions can now be determined with enzymes. Perhaps the most surprising application of enzymes is in the estimation of compounds for which the enzyme has no special affinity. Thus Rubenstein *et al.* (1972) have been able to detect levels of morphine as low as 3×10^{-8} M using lysozyme. They covalently attached morphine to lysozyme, the compound remained active, and they also prepared anti-morphine γ-globulin. Addition of this antibody to the morphine–lysozyme caused 98% inhibition of enzyme activity. When the antibody was added to a mixture of morphine and morphine–lysozyme there was competition for the binding sites on the antibody. Thus the amount of enzyme activity found was related to the free morphine concentration.

Even whole cells can prove useful; for instance the hydrogen producing bacterium *Clostridium butyricum* has been immobilized and used as a BOD sensor (Matsunaga *et al.* 1980). Immobilized *Pseudomonas fluorescens* cells have been used as a glucose sensor (Karube *et al.* 1979); and co-immobilized invertase, mutarotase, and glucose oxidase used as a sucrose electrode (Sato *et al.* 1976).

A good example of an enzyme used on a large scale in clinical biochemistry is cholesterol oxidase, used for the estimation of serum cholesterol levels, it is derived from a number of organisms such as *Mycobacterium, Nocardia, Proactinomycetes, Streptomyces* and *Corynebacterium,* a 3β specific, extracellular form of the enzyme being obtained from *Schizophylleum* (Sugiura *et al.* 1977). Incorporation of a non-ionic detergent in the culture medium makes the enzyme effectively extracellular; the detergent must be nontoxic and must not produce toxic degradation products such that the Triton X-100 often used to activate the isolated enzyme in the clinical assay for serum cholesterol, and used to solubilize the enzyme from harvested cells after growth has ceased, cannot be used. The presence of yeast extract, inducer, and detergent in the growth medium provided to the cells has a synergistic effect on the cholesterol oxidase activity of the cells. After extraction of the enzyme, detergent can be removed by treatment with a water-miscible solvent or with a macroreticular resin (Cheetham 1979) and the enzyme purified free of interfering catalase by DEAE chromatography before concentration by ultrafiltration (Buckland *et al.* 1974).

The other enzyme used in the serum cholesterol assay, cholesterol esterase, is produced by *Pseudomonas fluorescens* when supplied with an inducer. The enzyme accumulates both intra- and extracellularly and is then purified (Terada & Uwajima 1977). Cholesterol esterase can also be obtained from *Candida rugosa*, being purified free of lipase and protein by hydrophobic chromatography.

Uricase is important in the biochemical diagnosis of gout and some forms of rheumatism by the detection of uric acid in serum urine. Uricase is obtained from *Micrococcus biteus* which secretes the enzyme into the culture medium and is then purified by ammonium sulphate precipitation and column chromatography (Snoke *et al* 1977). In particular, chromatography on Sepharose derivatized with a hydrophobic ligand such as 1,6 hexanediamine to remove contaminating catalase is used, the catalase being retained more strongly than the uricase. Uricase can also be obtained from *Aspergillus flavus* and *oryzae*, the enzyme being extracted by pulverizing the frozen mycelia followed by purification by column chromatography, and from *Streptomyces gannmycicus* (Nakanishi & Shigemsa, 1981).

Creatinine amidohydrolase and creatine amidohydrolase are used in sequence to convert creatinine to creatine, and thence to urea and sarcosine. Cultivation of the producer organisms (*Flavobacterium, Micrococcus,* or *Corynebacterium*), production of the enzyme, and the accumulation of the enzymes in the medium, can be obtained in one step, and the enzymes can then be separated by DEAE chromatography.

NADH-peroxidase is used in the enzymic assay of many substances where H_2O_2 is generated, for instance by coupling the oxidase reaction leading to the formation of H_2O_2 with the NADH reaction. NADH peroxidase is obtained by digesting *Streptococcus faeralis,* or by treating it with a surface-active agent, and then purifying the extracted enzyme (Röder *et al.* 1980)

Dehydrogenases such as glucose-6-phosphate dehydrogenase are important in the analysis of many compounds using enzymes, such as glucose by glucose-6-phosphate dehydrogenase. The enzyme is then purified by affinity chromatography (Röder *et al.* 1976).

Phosphodiesterase is useful for experimental diagnoses and has possible pharmaceutical uses. This enzyme cleaves the cyclic ring of cyclic AMP which acts as a 'second messenger' in biochemical reactions. A suitable enzyme is obtained from *Dictyostelium discoidium* which is useful because of its low K_m; a low K_m being required because of the low concentrations of a c AMP present in biological materials (Gerisch 1974).

Lactate oxidase is used for the clinical assay of serum lactate levels. A highly specific and stable lactate oxidase from *Streptococcus facaelis* is used, the intracellular enzyme being extracted and purified prior to use (Esders *et al.* 1979).

Glucose oxidase is used to assay glucose and to inhibit the growth of tumours. Before use it must be separated from contaminating catalase which would otherwise catalyse the degradation of H_2O_2. This can be achieved by exploiting the fact that glucose oxidase contains carbohydrate residues which can be degraded by dextranase or sodium peroxide, so reducing the extent to which the catalase interacts with the glucose oxidase, and enabling separation of the two enzymes by column chromatography. Alternatively the catalase can be precipitated by treatment with acridine base or removed by ion-exchange chromatography (Keyes 1980). Before use the enzyme can then be stabilized by freeze drying in the presence of polymers such as polyethylene oxide polymer.

Other enzymes useful in clinical analysis include glycerphosphate oxidase (Misaki *et al.* 1981), glycerol kinase (Imamura *et al.* 1982), L-amino acid oxidase (Yoshino *et al.* 1982), galactose oxidase (Terado & Aisaka 1982), xanthine oxidase & Machida 1982), acyl-CoA synthetase (thiokinase) (Yamada *et al.* 1981), alcohol oxidoreductase (Eggeling *et al.* 1981), choline oxidase (Nakanishi & Machida, 1981) and glycerol dehydrogenase (Atkinson *et al.* 1982).

3.10 MEDICAL USES OF ENZYMES

L-Asparaginase is used for treating leukemias and disseminating cancers which require asparagine for growth (Mauer & Simone 1976). It is produced from *E. coli, Serratia marcesens,* and various *Erwinia* species, although *B. coagulans* produces an asparaginase which has no anti-tumour activity (Christie *et al.* 1974). *Pseudomonas fluorescens* produces an interesting protein which has identical asparaginase and glutaminase activities. Incorporation of an amino acid in the culture medium, such as L-glutamic acid, stimulated enzyme production which is released by lysing the cells with alkalis and removing contaminating endotoxins and pyrogens with basic ion exchange resins (Grabner *et al.* 1975), or purified by precipitating with glycerol and ethyl alcohol.

Trypsin or collagenase is used to remove dead tissue from wounds, burns, ulcers, etc. so as to speed the growth of new tissue and skin grafts, as well as to inhibit the growth of some contaminant organisms (Sizer 1972). Collagenase is obtained from a non-flagellated form of *Clostridium histolyticium,* and contaminating proteases may be removed by ion exchange chromatography (Chiulli & Wegman 1974). It prefers to act at —gly in helical regions in the native collagen —Z—Pro—X—gly—Pro—X—. (Note that there is a great need for a collagenase derived from a safer producer organism than Clostridia).

Keratinases may be used similarly to remove skin callouses or the excess keratin formed in conditions such as psoriasis as well as in the dehairing of hides during the preparation of leather and wool. The enzyme is obtained from *Trichophyton mentagrophytes.* Bromelain can also be used as a debriding agent (Galbraith, 1981).

Pterin deamidase deaminates pterin, pteroic acid and folic acid and has anti-tumour activity. The enzyme is widely distributed among moulds and has been purified from an *Aspergillus* source (Kusakabe *et al.* 1976). Superoxide dismutase is used as an anti-inflammatory agent in veterinary applications and streptokinase for the dissolution of blood clots. Urokinase is also used for its fibrinolytic/thrombolytic activity. Urininase is currently extracted from human urine, but more efficient sources of supply from cultured cells and engineered organisms are under development. Streptokinase is secreted from haemolytic streptococci and so is readily and cheaply available (Huna *et al.* 1983). Immobilized froms of these enzymes have been used (Everse *et al.* 1981).

Lastly, an enzyme which prevents bloat in animals can be obtained from *Streptomyces griseus.* It appears to act by reducing the viscosity of the gastric mucous (Hahn *et al.* 1975).

3.11 THE USE OF ENZYMES AS CATALYSTS IN ORGANIC CHEMISTRY

3.11.1 Introduction

Enzymes have great potential for use as reagents in chemical syntheses since they catalyse rapid, stereospecific reactions under mild conditions and because a wide range of substrates are transformed with great fidelity of reaction such that little or no side-product(s) are formed. However, enzyme mediated syntheses have not yet been widely accepted as routine procedures in organic chemistry, perhaps because of the unfamiliarity of chemists with the use of water as a solvent, especially as larger quantities of water need to be removed once the reaction is completed. Nevertheless, many examples exist where novel synthetic procedures can be adopted by the use of enzymes or existing procedures can be greatly simplified, for instance by the elimination of protective groups and through the formation of chemically and especially optically, pure products. Such reactions include reductions, since dehydrogenases such as alcohol dehydrogenase have a wide substate specificity; condensations, for instance in the synthesis of amino acids using pyridoxyl phosphate dependant enzymes; and functionalisations of nonactivated carbon atoms such as in the hydroxylation of steroids and the use of monooxygenases (Jones, 1980).

In general an enzyme is much more specific than the corresponding chemical reagents, being confined to structurally closely related molecules, but are very much more reaction specific irrespective of whether an isomerization, oxidation, transfer, or condensation reaction is taking place. Chemical reagents sometimes exhibit substrate specificity, but this is usually due to steric exclusion of molecules larger than a certain size, whereas enzymes discriminate more precisely, selectively reacting with a particular molecule even when smaller chemically similar substrates are present in excess. Substrate specificities can on some occasions be so wide that synthetic compounds which do not occur in nature are often successfully transformed.

Some asymmetric chemical synthesis are possible, for instance the L-dopa used for the treatment of Parkinsons disease is made commercially with a chiral rhodium catatlyst. However such non-enzymic reactions are usually rather difficult and give low yields of product and lower optical purities than when the same compound is synthesized enzymically. Since enzymes are usually strict chiral reagents, as befitting asymmetrical catalysts with such a pronounced asymmetrical structure, the optical purity of the products of enzyme reactions is likely to prove increasingly attractive. especially if regulatory authorities begin to require the production of optically pure materials, and as the demand for new, more sophisticated, drugs, hormones, and antibiotics increases.

The structural and stereochemical specificities of enzymes need to be well documented before they are used in chemical syntheses, as such information enables the structures and configurations of the products to be predicted with confidence even for previously untried non-physiological substrates. Chemists also prefer the enzyme to be functionally pure, that is, not containing impurities which might interfere with the desired reaction, and also to be stable, cheap, easy to use, and commercially available, since they are rarely willing to produce and purify their own enzymes (Jones, 1980).

The structural specificity of enzymes can be exploited to effect selective or regiospecific reactions on molecules possessing two or more chemically similar functions. The ease with which this can be done can often be extremely difficult to duplicate chemically. Some enzymes, particularly dehydrogenases, kinases and synthetases, are very specific as to the substrate acted on, whereas others will carry out a defined reaction, such as oxidases and isomerases which act on a range of usually structurally related substrates. In other cases, enzymes are highly specific for one substrate, often a coenzyme, but are nonspecific for the cosubstrate.

As with all reactions, it is the thermodynamically preferred products which normally accumulate in enzyme-catalysed processes. However, since such reactions are microscopically reversible, catalysis in thermodynamically less preferred directions can be induced by appropriate selection of the reaction conditions. This is illustrated by the synthesis of cephalothin by acylation of 7-aminocephalosporonic acid. In this process exploitation of the plug flow kinetics of a continuous flow immobilized enzyme column resulted in increased yields of product, since the cephalothin tended to be removed before the newly synthesized amide bond could be hydrolysed (Belg. Patent No. 803,832; 1973).

Polynucleotide synthesis is another example where the use of enzymes has overcome serious deficiencies in the chemical procedure. For instance, see the synthesis of polyriboinosinic (poly I) and polyribocytidylic (poly C) acids (Hoffman *et al.* 1970), and the synthesis of the trinucleotides U–A–A, U–A–G and U–G–A Gassen & Nolte 1971). The polymer Poly IC is a potent inducer of interferon, the antiviral compound, while the trinucleotides act as protein synthesis termination codons. In the latter synthesis the use of an immobilized enzyme in plug flow columns prevented the normal phosphate hydrolysis reaction from becoming appreciable such that the thermodynamically less preferred products accumulated.

Ⓟ ⓅOCH₂ ... Inosine or Cytosine $\xrightarrow[\text{from } M. \text{ luteus}]{\text{Polynucleotide phosphorylase}}$ Poly-I or Poly-C

HOCH₂ ... + A - A or A - C or G - A $\xrightarrow{\text{Ribonuclease}}$ U - A - A or U - A - C or U - C - A

3.11.2 Stereospecificity of enzymes

The absolute stereospecificity of enzymes is their most remarkable feature, and their ability to discriminate between enantiomers and between stereo-heterotropic groups and faces is probably of greatest utility and value to synthetic organic chemists. The opposite isomer is not acted on at all when it is the asymmetric group itself that is undergoing reaction; however, when the asymmetric group is at a site in the substrate molecule remote from where the enzyme acts, both isomeric forms of the substrate can be attacked although there is usually a marked preference for one isomeric form. In fact catalysis yielding only one enantiomer can be achieved using any chiral reagent, for instance L-proline has been used considerably (Eder *et al.* 1971, Hajos & Parrish 1974). Note that the ability of an enzyme to catalyse the transformation of only one of two enantiomeric forms is due to the difference in energies between the two possible enantiomeric transition states, and that the unattached isomer often bonds to the active site, thus behaving as a competitive inhibitor.

The first example of the use of an enzyme to carry out an industrially useful chemical synthesis is an acyloin condensation involved in the asymmetric synthesis of D(−) ephedrine. The reaction is carried out in high yield by a number of *Saccharomyces* sps. (Rose 1961). Resolution of racaemic mixtures of amino acids using pig kidney acylase were another important early application of enantiomer separation using enzymes (Greenstein 1954, Jones & Beck 1976). L-amino acids find important applications in the food, pharmaceutical and cosmetic industries and large-scale acylase catalysed resolution of L-alanine, L-valine, L-tryptophan and L-phenylalanine using microbially derived enzymes have been in full commercial operation in Japan for a decade. L-lysine can be obtained from racemic α-amino-ε-caprolactam in a similar way by using yeast mediated hydrolysis (Fukumura 1976) − see p. 351−352.

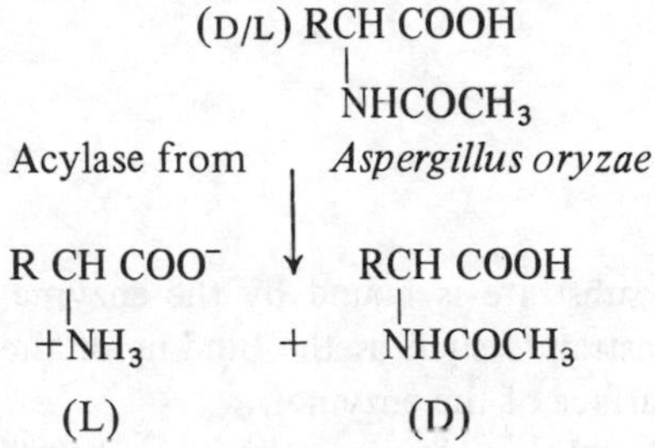

$$(\text{D/L}) \; \underset{\overset{|}{\text{NHCOCH}_3}}{\text{RCH COOH}}$$

Acylase from   *Aspergillus oryzae*

$$\underset{\overset{|}{+\text{NH}_3}}{\text{R CH COO}^-} \quad \downarrow \quad \underset{\overset{|}{\text{NHCOCH}_3}}{\text{RCH COOH}}$$

$$\text{(L)} \qquad\qquad + \qquad \text{(D)}$$

Proteolytic enzyme mediated resolutions are usually affected via hydrolysis of peptide, or equivalent, bonds. Enantiometric discriminations in the opposite, thermodynamically unfavourable, peptide bond formation direction can also be exploited as illustrated in the following reaction where the equilibrium is displaced in the desired direction by the insolubility of the p-tolumide product (L-form) in the aqueous reaction medium (Huang & Niemann 1951).

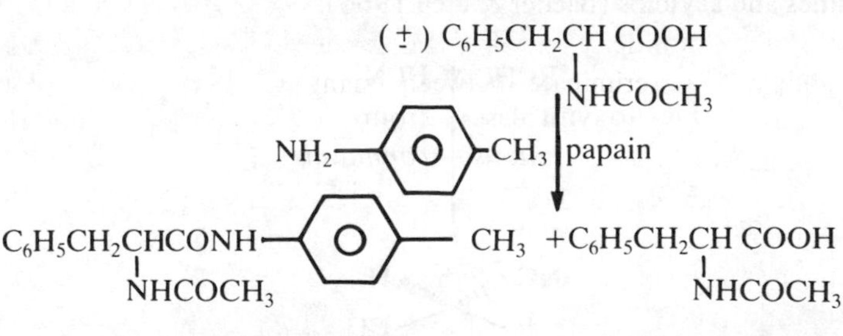

$$(\pm) \; \underset{\overset{|}{\text{NHCOCH}_3}}{\text{C}_6\text{H}_5\text{CH}_2\text{CH COOH}}$$

NH$_2$—⟨O⟩—CH$_3$ | papain

$$\underset{\overset{|}{\text{NHCOCH}_3}}{\text{C}_6\text{H}_5\text{CH}_2\text{CHCONH}}—⟨O⟩—\text{CH}_3 \; + \underset{\overset{|}{\text{NHCOCH}_3}}{\text{C}_6\text{H}_5\text{CH}_2\text{CH COOH}}$$

L -enantiomer   D - enantiomer
(insoluble)   (soluble)

3.11.3 Prochiral stereospecificity
Commercially one of the most useful enzymic reactions is 11 α-hydroxylation of steroids by *R. arrhizus* enzyme. This is currently used in the manufacture of

very large amounts of cortisone, prednisone, and related steroids. This route involves only 11 steps as compared with the 37 steps required by the previous chemical process (Ahrnowitz & Cohen 1981).

While the abilities of enzymes to discriminate between enantiomers is very important in resolutions and asymmetric synthesis, the use of racemic mixtures has the disadvantage that only half the substrate has the desired configuration, and the residual undesirable enantiomers must be recycled or even discarded. Exploitation of the prochiral stereospecificities of enzymes overcomes this problem because conversion of all the substrate to product becomes possible when an enzyme operates stereospecifically on a symmetrical molecule. That is, the enzyme is able to distinguish between chemically identical groups such as X and X′ in symetrical molecules such as

$$\begin{array}{c} X \\ | \\ C \\ X^{\diagup} {\overset{|}{\underset{Y}{}}} {\diagdown} Z \end{array}$$

presumably because the substrate is bound by the enzyme in a manner which imposes severe steric constraints such as the binding of the groups in the substrate to a plane on the surface of the enzyme.

For instance a C=O addition reaction which can be effected stereospecifically on a practical scale is the oxynitrilase catalysed addition of HCN to aldehydes. An inexpensive and readily available enzyme extracted from almonds reacts exclusively with the Si-face of the carbonyl group of a broad range of different aldehyde substrates. It is best used in an immobilized form in a continuous flow system. The R-cyanohydrine thus obtained can then be converted into other useful asymmetric compounds such as α-hydroxy acids, substituted ethanolamines and acyloins (Becker & Pfeil 1966).

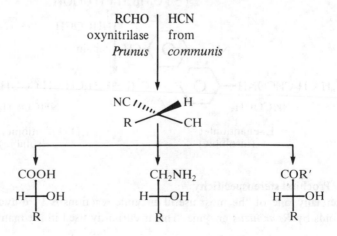

3.11.4 Combinations of stereospecificity

A significant example of the industrial use of an enzyme which can make pro-chiral distinctions is the stereospecific reduction of just one of the enantiotopic ketone groups of a pentane dione steroid to give the required enantiomer of the monohydroxy product in a 70% yield. The reaction is catalysed by an enzyme associated with *Rhizopus arrhizus* which is doubly enantiotopically specific, the hydrogen being presented to the Si-face of the *pro*-R carbonyl group (Bellet & van Thuong 1969) and the acyclic keto group being unreactive.

One of the most striking illustrations of the degree to which the various specificaties of an enzyme can be exploited is provided by the abilities of the pyridoxal phosphate-dependent enzymes β-tyrosinase and tryptophanase to produce L-tyrosine, L-DOPA and L-5-hydroxytryptophan (a serotonin precursor) in high yields from their cheap achiral precursors, ammonia and pyruvic acid, together with the appropriate side chain source (Fukui *et al.* 1975, Abbott 1976). The process operates using columns of enzymes from *E. coli* immobilized on Sepharose or in polyacrylamide:

$$RH + CH_3COCOOH + NH_3 \xrightarrow[\text{Tryptophanase}]{\beta\text{-Tyrosinase or}} RCH_2\underset{\underset{NH_2}{|}}{C}HCOOH$$

3.11.5 Multiple-step reactions

Although the use of enzymes can be expensive, these costs can be more than offset by the ease of operation and high yields characteristic of enzyme reactions, particularly if a number of sequential reactions can be performed in one container simply by adding the appropriate enzymes. This feature relies on the substrate specificity of the enzymes concerned, and can only be exceptionally performed with chemical reagents because of the very different conditions required and their non-specificity which results in the generation of numerous side-products and low yields of the desired product. Thus product formation can often be maximized by using it as the substrate for a second enzyme without the need to isolate the intermediate, such that the equilibrium of the first reaction is effectively displaced in favour of product formation. Examples include the formation of prednisolone from Reichstein compound S via hydrocortisone, secondly formation of L-citrulline from ATP, ammonium carbonate and ornithine and thirdly coenzyme A from cysteine and pantothenic acid. All three examples are best carried out in the form of immobilized enzymes in column reactors so as to maximize yields of product.

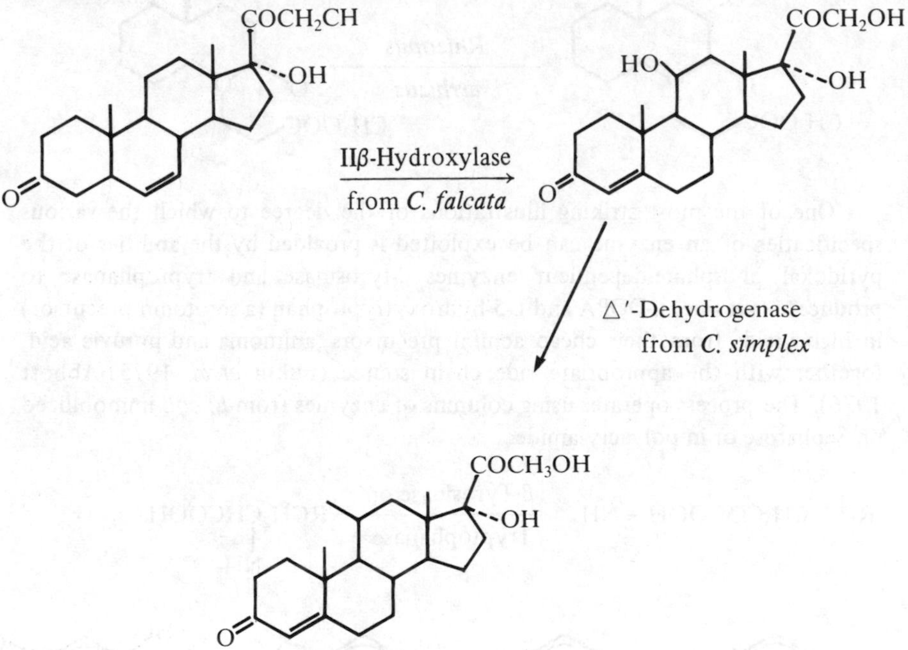

$$\xrightarrow[\text{from } C.\ falcata]{\text{II}\beta\text{-Hydroxylase}}$$

$$\xrightarrow[\text{from } C.\ simplex]{\triangle'\text{-Dehydrogenase}}$$

Conversion of Reichstein's compound S to prednisolone is by sequential operation of two enzymes in a flow system.

ATP + (NH$_4$)$_2$CO$_3$

+

NH$_2$(CH$_2$)$_3$CHCOOH
|
NH$_2$
L-ornithine

$$\xrightarrow[\text{and}]{\substack{\text{Immobilized carbamyl} \\ \text{phosphate synthetase} \\ \\ \text{ornithine transcarbamylase}}}$$

NH$_2$CONH(CH$_2$)$_3$CHCOOH
|
NH$_2$

L-citrulline

In a two-step reaction with both enzymes in the same reaction medium, carbamyl phosphate is first formed from ATP and ammonium carbonate. It then acts as a co-substrate with L-ornithine in the carbamylase-catalysed formation of L-citrulline (Miura *et al.* 1975).

Similarly synthesis of coenzyme A involves five sequential enzyme-catalysed steps. These are pantothenic acid → phosphopantothenic acid → phosphopathenyl cysteine → phosphopantotheine → dephospho coenzyme A → coenzyme A (Shimuzu *et al.* 1975).

$$\underset{\underset{\text{L-cysteine}}{\overset{|}{NH_2}}}{HSCH_2CHCOOH} \quad + \quad \underset{\underset{\text{Pantothenic acid}}{\overset{|}{HO}\quad\overset{|}{CH_3}}}{HOOC(CH_2)_2NHCOCH\ \overset{\overset{\displaystyle CH_3}{|}}{C}\ CH_2OH} + ATP$$

Multiple enzymes
from *B. ammoniagenes*

Coenzyme A

Other examples of the use of enzymes in chemical synthesis include the regeneration of NAD(P)H cofactors by coupling with alcohol dehydrogenase and aldehyde dehydrogenase, which converts ethanol to acetate; or by using these enzymes together with formate dehydrogenase which converts methanol to CO_2. The most important feature of these enzymes is the use of ratios of enzymes such that the first step, the formation of the aldehyde, is the slow step such that the concentration of aldehyde in the reactor is as low as possible in order that the enzymes are protected from inactivation (Wong & Whitesides 1982).

More complex examples of the use of enzymes in chemical synthesis include the regeneration of NAD(P)H cofactors, using glucose-6-sulphate and glucose-

6-phosphate dehydrogenase. Although the enzyme was less active when glucose-6-sulphate rather than glucose-6-phosphate was used this cofactor was considerably more stable, and also the glucose-6-sulphate was more easily prepared (Wong *et al.* 1981). The same approach can also be extended to other systems such as the enzyme catalysed synthesis of N-acetylglucosamine with the *in situ* regeneration of uridine-diphosphate glucose and uridine – diphosphate galactose (Wong *et al.* 1982).

    Archer *et al.* (1975) have carried out the complete enzymatic synthesis of a cyclic decapeptide antibiotic Gramicidin S which is used as a growth factor in animal feed. Enzymes from three microorganisms were used; two enzymes from *Bacillus brevis* catalyse the synthesis of Gramicidin S from its constituent amino acids. Adenylate kinase from *Saccharomyces cerevisiae* and acetate kinase from *Escherichia coli* were used to regenerate *in situ* the ATP needed to synthesize the peptide bonds.

3.11.6 Synthesis of radioactive compounds

Enzymic syntheses are often faster and give higher yields of non-racemic products than most chemically catalysed reactions. These advantages can be of particular importance when using expensive and/or unstable isotopes. Use of immobilized enzymes enables rapid recovery of the product free of antigenic or pyrogenic substances. For instance pharmaceutical grade ^{13}N-L-alanine has been synthesized in 4 min, using immobilized enzymes, whereas the same synthesis using free enzymes took much longer, the shorter reaction period being especially important since ^{13}N has a half-life of only 10 min. Higher yields were also obtained using the immobilized system, which involved the formation of ^{13}N-L-glutamic acid from ^{13}N-ammonia and α-ketoglutarate using glutamate dehydrogenase followed by addition of pyruvate and treatment with immobilized glutamate transaminase (Cohen *et al.* 1974).;

    Another important application is the use of co-immobilized glucose oxidase and lactoperoxidase to iodinate proteins with ^{125}Iodine. In this method exposure of the protein to extreme oxidizing conditions is prevented by the addition of glucose such that the immobilized glucose oxidase generates H_2O_2, which is then used by the lactoperoxidase in the labelling reaction. This method is much more gentle than chemical iodination, so that denaturation of the protein and oxidation of amino acids is prevented. Also the reaction can be easily controlled since the immobilized enzymes can be rapidly removed and the reaction terminated by centrifugation or application to a gel filtration column.

3.12 RESTRICTION ENDONUCLEASES

Restriction endonucleases are sequence-specific RNAses which prevent the expression of foreign DNA, for instance derived from bacteriophages, by cleaving the unprotected invasive DNA into fragments. Almost without exception,

Table 3.7

Sequence specificities of some restriction endonucleases

Designation	Occurrence	Sequence
Producing cohesive ends		
EcoR1	*Escherichia coli* RY13	$5' .. $ G AATTC $.. 3'$
EcoR11	*Escherichia coli* R245	C$\overset{\circ}{C}$ (^A_T) GG
Bam H1	*Bacillus amyloliquifaciens* H	G↓GATCC
Bgl 11	*Bacillus globiggi*	A↓GATCT
Hind 111	*Haemophilus influenzae* R$_d$	A↓AGCTT
Pst 1	*Providencia stuartii* 164	GTGCA↓G
Sac 1	*Streptomyces achromogenus*	CAGCT↓C
Zma 1	*Xanthomonas valvacearum*	C↓CCGGG
Sal 1	*Streptomyces albus*	G↓TCGAG
Xho 1	*Xanthomonas holcicola*	C↓TCGAG
Ava 1	*Anabaena variabilis*	C↓PyCGPuG
Hae 11	*Haemophilus aegyptius*	PuGCGC↓Py
Kpn 1	*Klebsiella pneumoniae*	GGTAC↓C
Sst 11	*Streptomyces stanford*	CCGC↓GG
Hinf 1	*Haemophilus influenzae* R$_T$	G↓ANTC
Ava 11	*Anabaena variabilis*	G↓G(^A_T)CC
Asu 1	*Anabaena subcylindrica*	G↓GCC
Taq 1	*Thermus aquaticus* YT1	T↓CGA
Hpa 11	*Haemophilus parainfluenzae*	C↓CGG
Hap 11	*Haemophilus aphrophilus*	C↓CGG
Mno 1	*Moraxella nonliquifaciens*	C↓CGG
Hha 1	*Haemophilus haemolyticus*	$5' .. $ GCG↓C $.. 3'$
Mbo 1	*Moraxella bovis*	↓GATC
Producing blunt ends		
Pvu 1	*Proteus vulgaris*	$5' .. $ CGA↓TCG $.. 3'$
Sma 1	*Serratia marcescens* S$_b$	CCC↓GGG
Pvu 11	*Proteus vulgaris*	CAG↓CTG
Hpe 1	*Heamophilus parainfluenzae*	GTT↓AAC
Hind 11	*Haemophilus influenzae* R$_d$	GTPy PUAC
Bal 1	*Brevibacterium albidum*	TGC↓CCA
Hae 1	*Haemophilus aegyptius*	(^A_T)GG↓CC(^T_A)
Hae 111	*Haemophilus aegyptius*	GG↓$\overset{\circ}{C}$C
Bsu R1	*Bacillus subtilis* X5	GG↓CC
Alu 1	*Arthrobacter luteus*	AG↓CT

The abbreviations for restriction endonucleases were devised by using the first letter of the generic name and the first two letters of the species name of the organism from which the enzyme was isolated. The position of the cleavage in the nucleotide chain is symoblized by the arrows; °; indicates the methylated nucleotides in the modification reaction. Nucleotide bases: C, cytosine; G, guanine; A, adenine; T, thymine; Pu, G or A; Py, C or T.

(Szalay *et al.* 1979). Reprinted by permission of the publishers, Butterworth & Co. (Publishers) Ltd. ©

all restriction endonucleases recognize short, non-methylated DNA sequences (Table 3.7). They can be divided into two groups based on the position of the cleavage site relative to the recognition sequence. Class 1 restriction endonucleases cleave double stranded DNA at positions outside the recognition sequence and generate randomly sized fragments of DNA. The second type is of greater use to molecular geneticists because it is extremely specific and breaks the DNA precisely within the sequence that it recognizes. The cleavage sites of Class II restriction endonucleases are located, in most cases, within the recognition sequence. Most of the Class II restriction endonucleases recognize 4, 5 or 6 base pair palindromes and generate fragments with either flush ends or staggered ends. The DNA fragments with staggered ends containing 3, 4, or 5 nucleotide singlestranded tails are said to have 'sticky ends'. DNA fragments produced by Class II restriction endonuclease cleavage can be separated on gels according to their molecular weight, then used for sequence analysis to elucidate genetic information stored in DNA. Furthermore, isolated DNA fragments can be inserted into a small extrachromosomal DNA, for example, plasmid, phage, or viral DNA, and its replication and expression can be studied in clones of prokaryotic or eukaryotic cells. Thus restriction endonucleases and cloning technology are very powerful modern tools for attacking genetic problems in medicine, agriculture and industrial microbiology.

The number of restriction endonucleases that have been described now approaches 400, although only about 70 are available commercially. The best known as EcoRI. It is of great importance in DNA analysis because of its high substrate specificity, recognizing a particular sequence of double stranded bases, and produces overlapping single stranded ends which are capable of reuniting easily.

Most restriction endonucleases are specific for guanosine—cytosine sites, and only one endonuclease, Aha 111, is reliably known to attack adenosine—thymine areas. this enzyme being obtained from a saline and high temperature tolerant algae by affinity chromatography (for a review see Szalay *et al.* 1979). Note that in order to conserve these expensive enzymes, immobilization is an obvious opportunity, and Mosbach and co-workers have recently successfully immobilized EcoRI, see also Chrikjian (1982).

3.13 BIOCHEMICAL PROCESSING

In the majority of cases the use of biological catalysts in the form of enzymes or cells is greatly facilitated by their immobilization and subsequent use in enzyme reactors. It is impossible to separate the design of the reactor from that of the catalyst since the properties of the immobilized enzyme or cell will greatly affect the reactor design. Immobilized biocatalysts are at present utilized in various industrial processes. Most of these are on a relatively small scale by the standards of the chemical industry. There are two obvious potential candidates for a second large-scale process in succession to glucose isomerase, immobilized lactase, and amyloglucosidase (glucoamylase).

Before launching into examples or enzymes used in biochemical processing and details of processes employing enzymes, immobilized enzymes and cells, and enzyme reactors, the following general points which help the successful application of biological catalysts in industry should be emphasized, as they arise from a consideration of the useful features of enzyme-based processes which have achieved, or are likely to achieve, commercial success.

1.   The process should be simple to use and robust, such that the conditions of use do not have to be finely tuned, and that the enzyme(s) involved are very stable. In this connection the use of thermostable enzymes is advantageous as they allow processes to be operated at elevated temperatures such that microbial contamination is discouraged and high solute concentrations can be used without having to process unduly viscous substrate and product streams. Furthermore, thermostable enzymes are often found to be more resistant to other denaturants than the corresponding mesophilic enzymes.

2.   It is advantageous to use enzymes approved for use by the government regulatory authorities and which are readily available in large quantities and which can be produced cheaply and abundantly.

3.   The process should be cheap in relation to the enhancement in value affected in the substrate; that is, more expensive enzyme preparations can be tolerated when high value low bulk products such as pharmaceuticals are being formed than when comparatively low value high bulk products such as sugars or solvents are being produced.

4.   The use of high substrate concentrations are preferred so as to decrease the volumes of fluid which must be processed and the volumes of solvent which must be removed prior to the recovery of the product.

5.   It is advantageous if the maximum possible extent of conversion of substrate into product, that is, equilibrium conversion, can be maintained throughout the life-time of the reactor, so as to minimize wastage of substrate and to facilitate isolation of the products. Thus use of enzymes in which the point of equilibrium greatly favours the formation of products is thus very desirable.

6.   Highly active enzyme preparations are desirable so as to decrease the sizes of reactors needed to achieve a required productivity and to decrease the reaction times required.

7.   The production, purification, and isolation of the product is simplified if products or side-products are formed which are in a different phase from the other reactants, or which are formed under conditions of high or low pH or high temperatures, or which themselves help to maintain the sterility of hygienic quality of the reactor during operation, such as antibiotics or organic acids.

8. Process parameters such as residence time, pH, substrate concentration, etc. are frequently interdependent. These interactions can profoundly affect the performance of the process and the quality of the product, such as for example the activity and stability of glucose isomerase (Fig. 3.3), and so should be carefully assessed. It is important to note that the conditions that are optimal at the beginning of the use of a biocatalyst do not necessarily prove optimal once the biocatalyst has been in use for a long period of time.

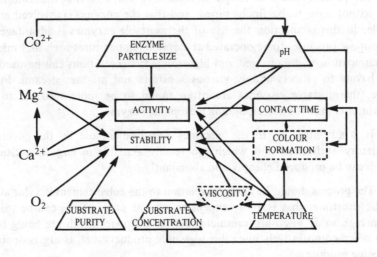

Fig. 3.3 – Interrelationships between various process variables in the case of immobilized glucose isomerase (Poulson & Zittan 1976).

Points 4–7 can be described as the intensification of the process, that is the facilitation of the easy formation of a cheap, defined product using the smallest reactor volume and energy input possible. As with many other topics in bio-technology the industrial use of enzymes requires a multidisciplinary eclectic approach if success is to be achieved. The interdependence of all the unit opera-tions which go to make up a process should be recognized. That is, a process engineering approach is very valuable, and the importance of the entire process, rather than any individual step or operation, however, scientifically novel it may be, should be given priority. Process engineering aspects should be taken into account at an early stage of research, and the project optimized and assessed on an economic basis. For instance, if an expensive reagent is used to immobilize an enzyme, it may be desirable to replace it with a less satisfactory but cheaper reagent.

3.14 APPLICATIONS OF ENZYMES IN THE FOOD INDUSTRY

The enzyme industry has now recovered from the discontinuation of protease-based detergents, caused by the fear of side-effects, and now some detergent enzymes are being reintroduced after favourable investigations. The starch processing industry is the largest user of enzymes, mainly owing to the success of the α-amylase/amyloglucosidase based process for forming glucose syrups. Success has continued with the high fructose corn syrups formed by immobilized glucose isomerase.

3.14.1 Polysaccharide processing

Starch consists of a mixture of amylose (15–30%) and amylopectin (70–85%). The latter is a branched polymer containing α1–6 branch points every 24–30 glucose residues. However in 'waxy' cereal starches such as from certain species of maize, rice and sorgham, amylopectin is the only component of the starch.

Glucose is mainly produced by the enzymatic hydrolysis of starch according to the reactions:

$$\text{starch} \xrightarrow[\alpha\text{-amylase}]{\text{bacterial}} \text{dextrins} \xrightarrow{\text{amyloglucosidase}} \text{glucose}$$

Traditionally, glucose syrups were produced by acidic hydrolysis, but it was not possible to achieve dextrose equivalents greater than about 55 without generating off-tastes. α-amylase and glucoamylase (amyloglucosidase) are probably the most used enzymes in the starch industry, and are relatively inexpensive. Immobilized forms of these enzymes must thus provide very clear economic or technical advantages before commercialization will be considered, especially in the case of bacterial α-amylase which must act on macromolecular starch.

3.14.2 Bacterial α-amylase (EC 3.2.1.1)

Prior to extensive hydrolysis the starch must be solubilised. This is achieved by heating the starch slurry such that the starch granules burst and the starch becomes dispersed and geletinized, and associated protein coagulates. This operation is carried out in a jet-cooker where mechanical shear is applied, steam is injected and pressure applied so that a temperature of 100–105°C is experienced for 5–10 min. Usually the starch suspensions contain 30 or more % dry solids starch, and so form very viscous suspensions. This process, and more recently α-amylase, is used as a thinning agent. Enzyme thinning has become very widespread since 1973 when *Bacillus lichiniformis* α-amylase has been available; because this enzyme is partially resistant to such extreme conditions. Following geletinization and thinning by the limited action of the α-amylase possible in such short periods, the starch is cooled and the α-amylase activity allowed to continue at about 95°C hours, yielding a product with a DE of around 12.

α-Amylase hydrolyzes the internal α-1-4 glycosidic links in amylose and amylopectin at random to produce liquid, less viscous, lower molecular weight products limited by the α-1.6-glycosidic bonds which forms the branch points in the native starch (amylopectin) molecule. The products of hydrolysis have an α configuration at the reducing glucose end. It is believed that the enzyme acts by the combined effect of carboxyl and histidine groups in the active site. α-Amylases derived from *Bacillus subtilis* var amyloliquifaciens contain tightly bound Ca^{2+} ions (one per molecule) which are required for activity and which greatly stabilize the activity to heat. It is conveniently employed as soluble enzymes at about 85°C and pH 5.5−7.0 for 2 h, as an alternative to the use of hydrochloric acid for the solubilization of starch which leads to excessive colour and reversion product formation. Owing to the intractability of potato and waxy starch to gelatinization, a two-stage procedure for these raw materials is used, in which a second dose of enzyme is added and incubation continued until the desired glucose content is reached.

Large quantities of α-amylase are produced from *B. amyloliquefaciens*. Improvements of yield have been obtained by mutation, by limiting aeration, and by increasing the CO_2 concentration. The enzyme has a requirement for Ca^{2+} ions, a pH optimum of 6, a molecular weight of 10−20 000, and consists of four subunits bound together by one zinc atom, but polymers of six to sixteen subunits have been found in crystalline α-amylase. *B. licheniformis* α-amylase has a greater thermostability than the *B. amyloquefaciens* enzyme, and so can be used at temperatures of up to 110°C in continuous starch liquefication processes. It is also less dependent on Ca^{2+} ions for stabilization requiring only 5 ppm Ca^{2+} rather than the 150 ppm required by *B. amyloliquifaciens* α-amylase. The *B. lichenformis* enzyme produces mainly maltose, maltotriose and maltopentaose with the maltohexaose formed initially being completely hydrolysed. By contrast the *B. amyloliquefaciens* amylase produces malto-hexaose as the major product (Norman 1981). Following thinning of the starch, continued action of the α-amylase forms maltodextrins which then pass on to the subsequent saccharification stage, although very minor quantities can be purified and used in coffee whiteners and high fat products such as mayonnaise salad dressings and ice creams where they help to reduce the calorific content. α-Amylase also finds application in sugar cane processing, since small quantities of starch occur in sugar cane, and thus in sugar juice. This starch can be hydrolyzed prior to evaporation by *Bacillus licheniformis* α-amylase at 85−95°C, such that invertase is destroyed, and no inversion of the sucrose can take place.

Some small-scale immobilization studies have been conducted; for instance, α-amylase from *Bacillus subtilis* has been immobilized on cyanogen bromide activated carboxymethyl cellulose and used in a stirred reactor for the hydrolysis of wheat starch (Linko *et al.* 1975). The initial activity of the immobilized enzyme was lower than that of the soluble enzyme, but because of thermal inactivation of the soluble enzyme, the productivity of the immobilized enzyme was greater. The immobilized α-amylase produced relatively more glucose and maltose than the soluble enzyme, owing to the limited diffusion of larger substrate molecules such that small fragments of the polysaccharide can more easily diffuse into the interior of the polymer where they are completely hydrolyzed. No external diffusion restrictions existed, and the immobilized α-amylase was used for several consecutive batch reactions.

3.14.3 Amyglucosidase (EC 3.2.1.3)

Amyloglucosidases (glucoamylase, α-amylase or α-1-4 glucan glucohydrolase) catalyze the stepwise hydrolysis of α-1-4 links in starch and oligosaccharides releasing β-glucose molecules from the non-reducing end of the molecule. The α-1-6 branch links are also hydrolyzed but very much less rapidly, a glucose syrup with a dextrose equivalent of 97–98 with a glucose content of 95–97% w/w and 3–5% higher saccharides being formed. (Note that the enzyme is also capable of slowly hydrolyzing α-1-3 bonds). The glucose is either used as a syrup or crystallized to give pure solid glucose. Amyloglucosidase is used batchwise on a large scale in the starch processing industry in stirred tanks at 55–60°C and pH 4.5 for 48–92 h incubation periods to hydrolyze the dextrins formed by α-amylase to glucose. Calcium ions stabilize it against heat or alkali denaturation. Hydrolysis can be carried out cheaply using the soluble enzyme which is added to the starch when 15–30% of the possible hydrolysis has been achieved by the action of α-amylase. Removal of the enzyme activity once the reaction is completed is achieved by heat denaturation or ion exchange chromatography. In industrial practice the maximum possible hydrolysis is required, using substrates as concentrated as possible in order to minimize the subsequent evaporation costs. In practice, 30–40% w/w solutions are employed, and the equilibrium reaction mixture contains 94–97% glucose. A particular advantage of using enzymes rather than HCl for the hydrolysis of starch is that less pure grades of starch can be used as the protein contaminants are not hydrolyzed to amino acids, which then cause browning reactions. Following the saccharification by amyloglucosidase, the syrup is filtered to remove protein and fat, and purified with carbon columns followed by ion exchange resins.

Glucoamylase is an extracellular enzyme produced by *Aspergillus* or *Rhizopus* species. Amyglucosidase from Rhizopus can be separated into isoenzymes with properties similar to those of the isoenzymes from *A. niger*. The enzyme is a glycoprotein containing mannose, glucose, galactose and uronic acid, and has a moleculer weight of 60–100,000 and a pH optimum of 4.3–4.5. One of its disadvantageous features is its relative inactivity towards α-1-6 bonds resulting in the use of high enzyme doses and/or long incubation times in order to achieve the desired degree of hydrolysis. Therefore the combined use of amyloglucosidase and the 'debranching' enzyme pullulanase has been proposed, for instance by Hurst (1975) who achieved a 1–2% increase in dextrose content despite having to carry out the reaction at pH 5.5–6.0 where the amyloglucosidase is less active. Although isoamylases have a more suitable pH optimum for this task, they cannot be used because of their thermolability.

Immobilization of extracellular enzymes such as amyloglucosidase by entrapment within the parent cells is obviously not possible, but practically every other method of immobilization has been applied. Amyloglucosidase is a relatively cheap enzyme, so an immobilized enzyme would need very superior properties as compared to the soluble enzyme in order to find wide industrial uses. Nevertheless, many attempts have been made to use immobilized amyloglucosidase so as to benefit from the advantages given by a continuous process. These include logistic convenience, and lower capital and energy costs. Other advantages include a reduction in the reaction time from about 75 h in the conventional batch process to a residence time of less than 1 h using the immobilized enzyme. Because of the shorter reaction times fewer side-reactions occur, so that refining is cheaper and the volume of plant required is much smaller than when using the soluble enzyme (Table 3.8).

Table 3.8

A comparison of soluble and immobilized amyloglucosidase for saccharification of starch

		Soluble	Immobilized
Feed stock	– concentration	35% (w/w)	30–35% (w/w)
	– pH	4.5	4.5
	– DE	10–15	30–40
Reactor size (for 25 000 ty^{-1})		60 000 l	5000 l
Residence time		48–72 h	0.5 h
Colour formation		considerable	nil
Approximate maximum final DE		95–97	94

Immobilized amyloglucosidase preparations have half-lives measured in months, while operating at temperatures high enough to eliminate microbial contamination. However, it has usually proved impossible to achieve the same degrees of hydrolysis using immobilized glucoamylase as with the enzyme in free solution. This is because the enzyme is reversible and is able to catalyze the formation of α-1-4 and α-1-6 linked oligomers from glucose. van Beynum et al. (1980) studied the equilibrium reaction mixture formed by glucoamylase, concluding that the reversal of the α-1,6 glucosidase causes a build-up of isomaltose. In immobilized preparations the inevitable internal diffusion limitations increase the contact time of the reactants with the enzyme so that the formation of the side products isomaltose, maltose and other saccharides is favoured such that glucose concentrations as high as those achieved using the soluble enzyme cannot be attained (Table 3.8).

Fig. 3.4 — An industrial scale immobilized amyloglucosidase system produced by Tate & Lyle, Plc.

Immobilized glucoamylase has been studied on a pilot plant scale by Lee *et al.* (1976). The immobilization of the enzyme was actually carried out in the packed column by recycling a solution of glucoamylase through a bed of glutaraldehyde treated alkylamine silica beads. The resulting immobilized enzyme column was then operated continuously for 70 days at 38°C with 30 wt% dextrin feed at flow rates 250—500 kg per day, producing a syrup with a dextrose equivalent DE of 92—93. The reactor system consisted of a number of parallel columns maintained at constant conversion by decreasing the flow rates through the column as the activity of the immobilized enzyme decreased. The useful life of a column was estimated as two half lives, and it was concluded that at least seven columns would be needed to keep variations in the production rate to within 10%.

Recently Tate & Lyle have succeeded in developing a commercially feasible immobilized amyloglucosidase, its chief advantage being its high volumetric activity, and thus the high throughputs that can be used, such that the size of plant required is very considerably less than when the soluble enzyme is used (Fig. 3.4). The amyloglucosidase is immobilized as a gel of enzymically active protein by simultaneous precipitation with acetone and crosslinking with glutaraldehyde in the presence of bone-char which acts as an inert mechanical support for the enzyme gel (Daniels and Farmer 1981). This method has been successfully scaled-up to a large scale, and is beginning to be used commercially in US corn processing factories.

Immobilized amyloglucosidase has also found applications in the saccharification of the low DP polysaccharide stream resulting from the chromatographic separation of fructose syrups in order to enhance their fructose content (Poulsen *et al.* 1980), and to hydrolyze residual maltose molecules following the alkaline isomerization of maltose (Walon 1980).

3.14.4 Maltose syrups

Two principal types of maltose syrup are produced (Table 3.9), one contains 30—50% maltose and 6—10% glucose, and has a dextrose equivalent of 42—49, which is used in jams and confectionery because it is resistant to colour formation, is nonhygroscopic, and does not crystallize as readily as glucose syrups. The second contains 30—40% maltose and 30—50% glucose, and has a dextrose equivalent of 63—70. It has a high content of fermentable sugars, but is stable on storage, and so is used in the brewing of beer and in bread making (Maeda & Tsao 1979). It is sometimes referred to as 'brewers adjunct'.

Fungal α-amylases derived from *A. niger* and *oryzae* are much less heat stable than bacterial α-amylases, but are widely used because they produce large amounts of maltose and maltotriose from liquefied starch. The enzyme acts by hydrolyzing the penultimate α-1-4 glycosidic link at the non-reducing end of the amylose or amylopectin chains, liberating maltose units in a sequential fashion until the action of the enzyme is limited by an α-1-6 bond. A 10—20 dextrose equivalent (DE) syrup containing about 50% maltose is formed in stirred tanks, there being the usual relationship between enzyme concentration, substrate concentration, and the incubation time. The process is terminated by heat denaturation of the enzyme at 80–85°C, but for complete hydrolysis of starch to maltose an additional enzyme such as pullulanase is required.

Table 3.9

The composition of glucose syrups

Sample	Dextrose equivalent	% Saccharides (carbohydrate basis)						
		DP[a] 1	DP 2	DP 3	DP 4	DP 5	DP 6	DP 7
Corn syrup AC[b]	27	9	9	8	7	7	6	54
Corn syrup AC	36	14	12	10	9	8	7	40
Corn syrup AC	42	20	14	12	9	8	7	30
Corn syrup AC	55	31	18	12	10	7	5	17
Corn syrup HM, DC	43	8	40	15	7	2	2	26
Corn syrup HM, DC	49	9	52	15	1	2	2	19
Corn syrup DC	65	39	31	7	5	4	3	11
Corn syrup DC	70	47	27	5	5	4	3	9
Corn syrup DC, E	95	92	4	1	1	sum of DP 5, 6, 7		2

[a] DP = degree of polymerization.
[b] AC = acid conversion, DC = dual conversion (acid–enzyme), HM = high maltose, E= enzyme conversion.

From: Maeda & Tsao (1979).

β-Amylase (EC 3.2.1.2) is widely distributed in higher plants, especially in barley and soya beans. It is an exoenzyme, producing β-maltose, is inhibited by sulphydryl reagents, and has a pH optimum of 6.5–7.0. β-Amylase has no activity towards α-1-6 bonds, and so complete hydrolysis of the amylopectin molecules in starch is not achieved. Thus it can produce a 80% maltose, 20% dextrin syrup when incubated with starch liquefied either with acid at elevated temperatures or bacterial α-amylase.

Maltose syrups can be manufactured using fungal α-amylase, or plant or microbial α-amylase, to a low dextrose equivalent only, so as to minimize subsequent maltotriose formation. Maltose is then formed by the maltogenic α-amylase or β-amylase, and the syrup then heated to deactivate the enzyme and clarified, decolourized and concentrated by evaporation. Barley and soya β-amylases have been used extensively, but because they are relatively expensive, microbial enzymes are increasingly used. Unfortunately, β-amylases caanot act on the α1−6 bonds in starch so that β-limit dextrins are formed and the final maltose content is limited to 60%. Therefore the combined use of β-amylase and debranching enzymes such as pullulanase have been proposed.

Amyloglucosidase and fungal α-amylase have also been used in combination to give highly converted syrups containing typically 30−35% glucose and 40−45% maltose. This is because both enzymes have similar pH optima. The syrups are widely used and are resistant to crystallization even at low temperatures and high solids concentrations.

Microbial β-amylases are also found for instance in *B. megaterium, circulans cereus* and *polymyxa,* and *Pseudomonas* and *Streptomyces* spp. These microbial enzymes have the advantage of being potentially cheaper and more abundant than the plant enzymes.

A potentially important observation is that β-amylase and α-1-6 glucosidases can be formed simultaneously by certain *Bacillus* spp. This is an advantage because previously two organisms had to be cultured separately, and because of the differing pH and temperature requirements of the two enzymes it has been difficult to react both enzymes on starch at the same time, resulting in a two-step procedure for the manufacture of maltose. α-1-6 Glucosidase formation was markedly stimulated by the presence of Mn^{2+} in the media, which did however tend to suppress formation of the β-amylase, so that the time and amount of Mn^{2+} added had to be carefully chosen. The enzymes are extracellular, are precipitated by salts or organic solvents, and are isolated by adsorption by starch, carbon, or diatomaceous earth. A β-amylase active under alkaline conditions and stable in the presence if chelating agents has been isolated from *B. subtilis* (Boyer & Ingle 1977), and an acid pH optimum β-amylase has been obtained from *Streptomyces* (Koaze *et al.* 1975).

Production of maltose syrups by an immobilized two-enzyme system has been proposed, (Mårtensson 1974) for instance, by co-immobilizing β-amylase and pullulanase, a debranching enzyme, in a crosslinked copolymer of acrylamide and acrylic acid. As these two enzymes act sequentially on the same substrate, a pronounced advantageous microenvironmental effect can be expected for the co-immobilized system. Increased operational stability was also observed when compared to the soluble enzymes.

Similarly, β-amylase has been immobilized together with an α-1, 6-glucosidase by adsorption to activated coal. Both enzymes were obtained from *Bacillus cereus.* This immobilized two-enzyme system hydrolyzed starch at 50°C to a product composition of 90.5% maltose, 7.5% maltotriose, and 2% other oligosaccharides (Takasaki & Takahara 1976).

3.14.5 Glucose isomerase (EC 5.3.1.5)
Glucose has almost 65% of the sweetness of sucrose on a weight basis. Fructose has 120−180%, depending on the conditions of testing. Thus a means of con-

verting glucose, which is available readily and very cheaply when corn starch is abundant, to fructose, increases its sweetness, and thus its value. Alkaline isomerization is possible, but produces excessive colour and side products. D-Xylose keto isomerases have been found which will also isomerize glucose when supplied in high concentrations, and these now produce all the high fructose corn syrup (HFCS) now produced, being the largest volume use of an immobilized enzyme in the world, with several million tonnes of product produced per annum. The availability of HFCSs has radically altered the carbohydrate sweetener market in the USA, where they have acquired many of the markets previously held by sucrose syrups. However, operations located within the EEC have experienced considerable difficulties since 1977, because a high import tariff has been introduced to safeguard the production of domestic sucrose from sugar beet.

The first glucose isomerase to be described was by Marshall & Kooi in (1957). Glucose isomerase is well suited to use in an immobilized form since it is intracellular, stable at elevated temperatures sufficient to discourage microbial contamination, and because all the reactants are small molecules so that there are few diffusional problems. Traces of magnesium, supplied as magnesium sulphate, are required to activate the enzyme. Magnesium ions are competed with by calcium ions such that the magnesium ions must be present in approximately a 20-fold excess. Mn^{2+} ions are required for the isomerization of xylose, but Co^{2+} ions for the isomerization of both xylose and glucose. However, the Co^{2+} ions are tightly bound to the enzyme and so do not have to be added to the substrate.

Side product formation during isomerization is a function of pH, temperature, and the time of reaction. Use of immobilized enzyme columns enables much shorter contact times to be used, compared with batch reactions such that colour formation is greatly reduced, particularly in comparison with early products formed by soluble enzyme in batch processes where slightly alkaline conditions had to be maintained for considerable periods if complete isomerization was to be achieved.

The approaches used to immobilize glucose isomerase were many and varied, both whole cells and cell-free preparations being used (Table 3.10). The earliest attempts consisted of the enzyme from *Streptomyces* species, partly purified and adsorbed onto DEAE cellulose. The first commercial process, which was introduced in 1974, was operated batchwise. Since then, the mechanical properties of the immobilized glucose isomerase preparation have been improved such that generally packed bed reactors are used (Table 3.11). Generally a minimum of 8 packed-bed reactors are used in parallel so as to maintain a constant overall productivity. In the batch process the reaction conditions were not optimal because excessive colour and by-products were formed during the long reaction periods. In the packed bed process, the reaction time is relatively short, and optimal isomerization conditions can be employed and a purer product formed. Note that the exhausted immobilized cells can be used as cattle feed. Later plants used preparations consisting of whole cells of *Bacillus, Streptomyces, Actinoplanes,* or *Arthrobacter* combined into stable particles with good hydraulic properties by a variety of methods including flocculation, heat treatment and crosslinking with glutaraldehyde or similar agents (Bucke & Wiseman 1981). The great bulk of the high fructose glucose syrup manufactured at present is made using Novo's whole cell preparation of *Bacillus coagulans* employed in several plants in the USA and also some in Europe, Japan and South Korea, with design capacities of up to 1200 tons dry weight (as dry substance). The total capacity of these plants can be estimated to be several million tons per year.

The glucose isomerase associated with *Bacillus coagulans* cells are centrifuged out of the culture medium, crosslinked with glutaraldehyde, broken into small particles and dried to improve the mechanical strangth of the pellets. Alternative procedures include freezing the crosslinked precipitate prior to drying, allowing the centrifuged cells to autolyse before crosslinking and then flocculating and drying, or disrupting the cells and then flocculating and crosslinking the homogenate. The immobilized enzyme is then used in column reactors at about 60°C to produce syrups containing approximately 42–43% (w/w) fructose and with a low colour and psicose content (Amotz *et al.* 1976) (Fig. 3.2).

A glucose isomerase derived from a strain of *Actinoplanes missouriensis* is used by Gist Brocades. The cells are cultivated on a corn steep liquor medium and salts with the yield of enzyme being improved by the use of molasses as a carbon source. The mycelium-fixed enzyme is entrapped in gelatin and crosslinked with glutaraldehyde. This system of immobilization was chosen because the carrier material and coupling reagents are permitted in foodstuffs or as food processing aids. The entire procedure of immobilization is continuous and

Table 3.10

Immobilization methods that have been used for the commercial exploitation of glucose isomerase

Company	Immobilization method†
Novo Industry	*Bacillus coagulans* cells lysed, crosslinked with glutaraldehyde and granulated
ICI Americas Inc.	*Arthrobacter* cells flocculated, heated and then granulated
Gist–Brocades	*Actinoplanes missouriensis* cells entrapped in gelatine, crosslinked with glutaraldehyde and granulated
Clinton Corn Processing Co.	Enzyme extract adsorbed to ion exchange resins
Miles Labs. Inc.	Cells crosslinked with glutaraldehyde and granulated
Miles Kali-Chemie	Cells crosslinked and then granulated
CPC Int. Inc.	Enzyme extract adsorbed to granular ceramic supports
Sanmatsu	Enzyme adsorbed to ion exchange resins
Snam Progetti	Cells entrapped in cellulose acetate fibres

†Unless otherwise stated, cells of *Streptomyces* species are used.

applicable to other enzymes applied in food processing. The enzyme preparation consists of uniform spherical particles of 1 mm diameter. The halflife of such immobilized enzyme preparations is reported to be 430 h in batch processes in the presence of cobalt, and 500 h for column operation in the absence of cobalt.

R. J. Reynolds Tobacco Co. utilizes a strain of *Arthrobacter* for production of glucose isomerase. The cells are aggregated and further processed before use in a fixed bed reactor (US Patent 3645848, 1972). Baxter Laboratories has a process based on enzyme from *Streptomyces phaeochromogenes,* which is immobilized on an anion-exchange cellulose support (Brit. Patent 1274158, 1972). Clinton Corn Products, one of the pioneers in the field, introduced an industrial

Table 3.11

The production of glucose and high fructose syrups from starch

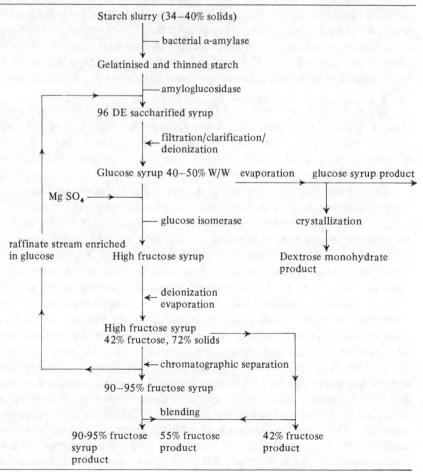

process based on immobilized glucose isomerase (isolated from a *Streptomyces*) in 1972 (Thompson *et al.* 1974). The enzyme was adsorbed on DEAE-cellulose and used in a continuous computer controlled process. The reactor consisted of several shallow beds placed in series with each bed containing material of a different age. The isomerization was carried out at 60°C and pH 7.0–7.5, using as substrate a 73% glucose solution of 30–35% solids. The product contained 42% fructose.

Glucose isomerases can also be obtained from *Corynebacterium*, *Curtobacterium*, *Bacillus stearothermophilus* and *Flavobacterium arborescens*, which is exceptional in that maximal enzyme activity was obtained when the cells were grown on lactose.

The list of immobilized glucose isomerase preparations can be extended considerably. For instance, purified glucose isomerase has been immobilized on DEAE–cellulose porous glass, phenylformaldehyde resins, and also entrapped in cellulose fibres and hollow fibres. Whole cells with isomerase activity have been immobilized by entrapment in cellulose acetate fibres as well as polyacrylamide and collagen membranes. The potential use as industrial catalysts of most of these preparations appears, however, to be rather limited, especially since there are a number of processes as described above already on the market (Brodelius 1978).

There are signs, however, of a tendency to return to immobilized preparations in which partly purified, soluble enzymes are immobilized onto stable inorganic supports with excellent flow characteristics. The greater cost of these is balanced by up to 10-fold increases in volumetric activities which allow more compact plant to be used, thus lowering the initial capital costs of the process.

Despite the dramatic commercial success of glucose isomerase there is room for improvement in glucose isomerases. An enzyme which is not inhibited by calcium would find a ready market, as all the refining could be left until after the enzyme steps, thus eliminating at least one refining stage. The glucose isomerases currently used operate satisfactorily at 55–60 °C, but an even more temperature stable enzyme would be welcomed, as would an enzyme capable of operating at a pH below 7, where colour formation resulting from fructose destruction is low. It is unlikely that anything can be done to improve the fructose content of the high fructose syrup. At equilibrium, the product contains about 55% fructose, but in practice it is uneconomical to take the reaction beyond 42% fructose at which level the syrup is 'isosweet' on a solids basis with sucrose, except when it is used in acid beverages such as cola. In such beverages fructose appears less sweet and so 55% fructose syrups are needed to replace sucrose in cola-type beverages. Such syrups are now made employing chromatographic separators to produce a fructose-rich stream ($\cong$ 95% fructose) and a glucose-rich 'raffinate'. The former is blended back with the feedstock to enrich its fructose content, the latter may be recycled . Similarly, 95% fructose syrups can also be manufactured (Bucke & Wiseman 1981). Note that a glucose and oligosaccharide rich 'raffinate' stream is produced by the separator, and this can be further saccharified with amyloglucosidase prior to further isomerisation (Table 3.11).

Many new improved forms of glucose isomerase are being developed. By way of example, Miles Labs. Inc. (Elkhart, In., USA) produce a new form of immobilized GI by heat treating *Flavobacterium arborescens* cells, crosslinking with polyethylenamine, chitosan and glutaraldehyde, shaping, and then drying. The resulting immobilized enzyme has a broad pH range of 6.0–8.5, is relatively insensitive to heavy metal inhibition, and has a half-life of 58 days while producing 42% fructose at 60°C from 40% glucose syrup. The initial activity was 2.0–2.5 empty column volumes h^{-1} for 42% fructose, about double the activity of the existing commercially available immobilized glucose isomerases.

Two interesting attempts have been made to produce a high fructose syrup by means other than glucose isomerase. Heady (1980) has invented a method involving the polymerization of the fructose derived from sucrose into a levan polymer using the fructosytransferase levan-sucrase. The polymer is then degraded into a fructose syrup containing 66–75% fructose by treatment with dilute acid or acidic ion-exchange resins. The residual glucose is then isomerized to fructose by glucose isomerase so as to increase the yield of fructose, or fermented to ethanol.

The second process uses a glucose-2-oxidase (EC 1.1.3.10) derived from *Basidiomycetes* which oxidises glucose to glucosone, which on catalytic hydrogenation produces fructose:

$$\text{Glucose} + O_2 \; - - - \rightarrow \; H_2O_2 + \text{glucosone}$$

$$\text{Glucosone} + H_2 \; - - - \rightarrow \; \text{fructose}$$

An elegant feature of this method is to employ the hydrogen peroxide produced as a side-product to oxidize olefins to epoxides and glycols. That is:

$$\text{Ethylene} + \text{HCl} + H_2O_2 \xrightarrow{\text{Chloroperoxidase}} \text{2-chloroethanol}$$

$$\text{2-chloroethanol} \xrightarrow{\text{NaOH}} \text{ethylene epoxide}$$

$$\text{ethylene epoxide} \xrightarrow{\text{Epoxidase}} \text{ethylene glycol}$$

The thermal stability of the 2-oxidase was substantially increased by the modification of lysine residues with ethyl acetimidate.

3.14.6 Inversion of sucrose
Sucrose can be hydrolyzed to invert sugar by acids or enzyme (Invertase, EC 3.2.1.26). The product consists of a 1:1 mixture of glucose and fructose and is somewhat sweeter than sucrose. It is mainly used in food and confectionery products as a humectant to hold moisture and prevent drying. Inulinases and

invertases can be used, the enzymes usually being derived from *Aspergillus* or *Saccharomyces* spp in which they are chiefly associated with the cell wall. Applications of invert sugar include artificial honey, the brewing industry, in jams since it crystallizes less readily than sucrose, in the recovery of scrap confectionery, and in the formation of liquid centres in confectionery where a hard sucrose centre containing invertase is given a surface coating of chocolait.

Invertase (β-fructofuranosidase) yields a syrup of higher purity than acid hydrolysis of sucrose. Although the production of invert sugar with soluble enzyme is relatively inexpensive, attempts have been made to immobilize the enzyme for continuous production of invert syrups. Invertase is a relatively stable enzyme especially when used in the presence of high concentration of sucrose and immobilization results in preparations with very long lifetimes. For instance, invertase from yeast was entrapped in cellulose triacetate fibres. In this immobilized form the enzyme was found to have satisfactory activity and stability under operational conditions, especially in high sucrose concentrations where the free enzyme showed severe substrate inhibition which can be correlated with the reduced water activity of the substrate. It was concluded that the observed differences between free and entrapped invertase were due to diffusional effects (Marconi *et al.* 1974).

3.14.7 Sugar refining
3.14.7.1 *Raffinase*
Raffinose, an α-galactoside, accumulates in sugar beet, especially in cold conditions. It interferes with the crystallization of sucrose from sugar beet molasses. The yield of sucrose can be increased by the hydrolysis of raffinose by α-galactosidase (EC 3.2.1.22). This increased yield is partly due to the actual formation of sucrose from the hydrolysis of raffinose, and partly to the elimination of a molassogenic agent. The mould *Mortierella vinacea raffinosutilizer* proved to be the best source of this enzyme, a major factor being that it produces no invertase. When grown under controlled conditions, mycelial pellets of uniform size (20–30 mesh) containing the α-galactosidase are produced. These pellets are conveniently used as a preparation of immobilized α-galactosidase in a continuous process.

An industrial process involving pellet-bound α-galactosidase was introduced in 1968 by the Hokkaido Sugar Co. Ltd, Japan (Obara *et al.* 1976/7). A similar process was designed for the Great Western Sugar Co., USA, which began operation in 1974. The enzyme reactor consists of a number of U-shaped open vessels each equipped with agitators and a replacement screen at the top for removal of catalyst from the product stream. The molasses to be treated in the reactor are continuously adjusted to $\cong 30\%$ (w/w), 48–52 °C and pH 5.0–5.2 with sulfuric acid. The optimum pH for the α-galactosidase reaction is 4.0, but the pH is maintained at 5.2 so as to avoid acid catalysed inversion of the sucrose. The reaction time is 1.5–5 h, but longer residence times can be used without trouble if molasses of higher raffinose content needs to be treated. The enzyme pellets are used for about 25 days before they are discharged. Since the introduction of the enzyme process the capacity for beet processing has been increased by about 10% and the amount of sucrose obtained by 3%, with a consequent reduction in the amounts of molasses formed.

The enzyme has also been insolubilized within the cells of *M. vinacea* var *raffinosutilizer* with glutaraldehyde. In a batch reactor, beet sugar solutions containing raffinose were treated with the cells at 50 °C. After more than 250 cycles over a 3-week period, the treated cells retained 72% of original activity whereas the untreated cells retained only approximately 5% (Saimura *et al.* 1975).

α-Galactosidases from *Circinella muscair*, *Absidia griseola*, *Bacillus stearothermophilus*, and *Aspergillus niger* have also been used, the latter for the hydrolysis of the α-galactosides such as melibiose and raffinose present in soybean milk and oil seeds.

3.14.7.2 *α-amylase*
α-Amylase finds occasional use in sugar refining since small quantities of starch can occur in sugar cane. This starch can be hydrolysed prior to evaporation by *B. licheniformis* α-amylase. It is essential that no inversion of the sucrose takes place.

3.14.8 Dextranase (EC 3.2.1.11)
Dextran is an α-1-6 linked glucose polymer formed by *Leuconostoc* spp. that grow in mechanically damaged sugar cane. Dextranases (1,6 α-glucan-6-glucanohydrolases) are used to degrade this polymer into isomaltose and isomaltotriose at $\cong$ 50–60 °C and pH 4.5–6.0, so preventing the undesirable increases in viscosity in the cane juice that makes the clarification and crystallization of sucrose by evaporation difficult. Dextranases have also been suggested as a component of toothpastes in order to help prevent dental caries by dissolving the dextrans produced by oral bacteria, and to produce dextran used as a blood substitute. In the former application the dextranase was formed by *Fusarium monififorme* which has a high affinity for hydroxyapetite. Endo-dextranases are formed by *Penicillium lilacinum*, *lutem* and *funiculosam*, *Klebsiella aerogenes*, *Fusarium monoliformi* and *Flavobacterium* sps. Exo dextranases are formed by *B. coagulans* and *Arthobacter globiformis*.

3.14.9 Debranching enzymes
Some enzymes act specifically on the α-1-6 bonds of the amylopectin of liquefied starch and produce maltose and maltotriose from starch. These enzymes are pullulanase and isoamylase. Pullulanase (EC 3.2.1.41) is more correctly named pullulan-6-glucanohydrolase, so called because it hydrolyzes the linear polysaccharide pullulan. *Klebsiella pneumonia* is the commercial source of pullula-

nase from which it is secreted during growth, although a portion of the enzyme appears to be cell-bound. Starch with a high amylopectin content is used as the substrate to induce the enzyme. Other sources include *Bacillus* spp and *Aerobacter aerogenes*. The *Klebsiella* enzyme consists of at least three isoenzymes of varying molecular weight that can be interconverted proteolytically. The best known commercial application is in brewing.

Recently Novo have developed a pullulanase-like debranching enzyme from *Bacillus acidopullulyticus* (Nielson *et al.* 1982). Thus it should be possible to use the new pullulanase and glucoamylase together under conditions that are optimal for both enzymes, so as to produce glucose syrups of > 96 DE without having to excessively dilute the substrate or use large quantities of the glucoamylase. An additional potential application is use in combination with β-amylase to produce high maltose groups.

Other debranching enzymes which do not hydrolyse pullulan and have only a limited action on α-limit dextrin, are classified as isoamylases. This enzyme can be obtained from *Flavobacterium* and *Pseudomonas deramosa* and acts on α-1-6 links only if they are situated at a branch point and not if they are found in a linear chain as in pullulan.

3.14.10 Cyclodextrin glucosyltransferase and other amylases

Cyclodextrin glucosyltransferase forms circular cyclodextrins from starch. They consist of 6.7 or 8 glucose units (termed α, β and γ cyclodextrins respectively). It can also transfer the glucose from starch to the glucose residue of sucrose. The enzyme is obtained from the alkophile *Bacillus macerans, B. stearothermophilus* or *B. magaterium* which tends to produce β-cyclodextrin only (Shiosaka 1976). The enzyme is extracellular and can be purified by adsorption to starch. The enzyme from *B. macerans* has been used to produce dextrins containing fructose by transglucosylating starch hydrolysates of DE over 10 with sucrose or invert sugar. β-Cyclodextrin is produced most cheaply and in by far the largest quantities because it is very much less soluble than the α and γ forms, a property which facilitates its manufacture by causing it to precipitate from solution so enhancing yields. Cyclodextrin glucosyltransferase has been immobilized recently by Kato & Horikoshi (1984). *Pseudomonas stutzeri* produces an exoamylase, which hydrolyzes starch to maltohexaose. Another exoamylase hydrolyzes starch to maltohexaose (Robyt & Ackerman (1971). Similar exo-amylases are produced by *Klebsiella pneumoniae* (Kainama *et al.* 1972) and another enzyme obtained from *B. lichiniformis* produces predominantly maltopentaose (Saito 1973). Lastly, isopullulanase derived from some spp. of *A. niger* degrade pullulan to isopanose (Sakano *et al.* 1972).

The bitter taste of the high intensity sweetener stevioside has been successfully removed by using a glucosyltransferase enzyme such as dextran-sucrase, α-amylase, or α-glucosidase to transfer glucose from starch or sucrose, forming α-glucosyl stevioside (Kaisha *et al.* 1980).

3.14.11 Cellulase (EC 3.2.1.4)
The enzymatic hydrolysis of cellulose to glucose is one of the most exciting new developments in enzyme engineering. Cellulose is the most abundant organic material, and unlike some other potential energy sources, it is renewable. Very large quantities of cellulose waste are produced per annum in the form of straw, saw-dust etc. and the development of simple, efficient and economical processes for the conversion of such waste cellulose to glucose is of global interest. Glucose produced in such a way could be used as a sweetener, may serve as an energy source for single cell protein production, or can be converted to valuable chemicals such as ethanol. The hydrolysis of cellulose is catalyzed by cellulase (from, for example, *Trichoderma viride*), which is a complex mixture of several enzymes with various hydrolytic activities, prepared by semi-solid or submerged fermentation in which Sophorose acts as an inducer.

The mode of action on cellulose is still not clearly understood. The current status is that the cellulose complex contains several *endo-β-1,4-glucanases* (Fig. 3.5), one of which may be the enzyme that acts first on cellulose. The enzyme acts synergistically with the *exo-β-1,4-glucancellobiohydrolase*. The enzyme system furthermore comprises an *exo-β-1,4-glucanglucohydrolase* and cellulase. Cooperation of several of the enzymes is necessary for optimal hydrolysis. Most of the enzymes are glycoproteins of MW 40–75 000. One endoglucanase has a MW of 12 000.

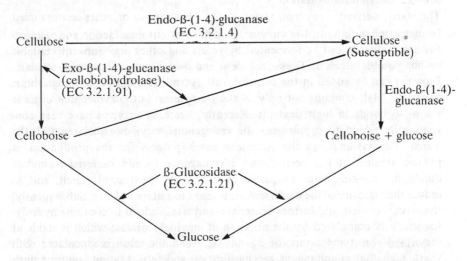

Fig. 3.5 – The mode of action of cellulase on cellulose-cellulase enzyme systems (Ryu & Mandels 1980). Reprinted by permission of the publishers – Butterworth & Co. Ltd.

The cellulase system of *Trichoderma viride* is induced by sophorose and repressed by glucose; the cellobiase activity is unstable below pH 4. This enzyme is important for glucose production during cellulose hydrolysis so as to relieve end-product inhibition by cellobiose.

Other cellulases that have been investigated include those from the wood rotting fungus *Sporotrichum pulverulentum*. Thermophilic species are especially interesting, for instance those from *Sporotrichum thermophilic, Chaetomium thermophile, Thermomospora curvata, Clostridia* spp, and *Aerobacter* spp. which have temperature optima of 45, 77, 65, 67 and 50 °C respectively. A very interesting example is *Streptomyces flavogriseus* which produces cellulase and a glucose isomerase simultaneously (Huepfel *et al.* 1980). For general reviews, see Ryu & Mandels (1980) and Enari (1983).

Present commercial applications are on a small scale, and include use as digestive aids, to degrade cellulose in foodstuffs, for instance in garlic and mushrooms, and in domestic wastes. For instance the hemicellulase impurities in some cellulases are used in the degradation of wheat flour pentosans, in order to improve baking quality. Similarly, the hemicellulases present in impurities in commercial pectinases are used to liquefy mucilage in the preparation of liquid coffee. Hemicellulase preparations available commercially come from *Aspergillus* spp and also contain xylanases, mannanases and arabinofuranases. Finally, Gist-Brocardes sell a multi-enzyme mixture derived from *B. subtilis* or *A. oryzae* for the treatment of silage or cattle fodder.

3.14.12 Ethanol fermentation

The starch derived from maize, potatoes, barley cassava or other sources must be pretreated with hydrolase enzymes, which carry out liquefaction and saccharification, before it can be fermented by yeasts and other organisms into potable or non-potable alcohol. These include α and β-amylases and amyloglucosidase. Enzymes can be added in the form of malt, (germinated barley) although this is expensive. Malt contains not only α- and β-amylases, but enzymes that degrade the α-1-6 bonds in limit dextrins. Recently, bacterial enzymes have been conventionally added to supplement the endogenous enzymes associated with the starch. For instance, in the American batch process for the production of potable alcohol, it has been found advantageous to add bacterial α-amylase during the cooking stage to partly hydrolyse the gelatinized starch, and to reduce the viscosity of the mash, making it easier to agitate and mix. Subsequently, the mash is cooled, and further bacterial α-amylase is added to continue hydrolysis, which is completed by the addition of amyloglucosidase which is stable at the new lower temperature of 55–60 °C. Then the mesh is inoculated with yeast, such that simultaneous saccharification and fermentation continue until the dextrose is exhausted, and the alcohol distilled off.

By contrast, in the German batch process α-amylase is not added until the mashing stage, which is carried out in two stages: a high temperature liquefaction using bacterial α-amylase (80 °C) followed by a lower temperature liquefication at 55–60 °C, using fungal α-amylase. Amyloglucosidase and then yeast are added to complete the conversion of the starch into ethanol (see Godfrey & Reichelt (eds.), 1983). The commercial production of non-potable alcohol for use as a fuel for internal combustion engines has increased rapidly in the last few years, particularly in Brazil which relies on the fermentation of sugar cane or cassava. Use of immobilized cells has also been proposed; the most advanced effort appears to be that of the Kyowa Hakko Kogyo Co. who use columns of *Saccharomyces cerevisiae* entrapped in calcium alginate pellets. Ethanol has been produced from nonsterilized diluted cane molasses for over half a year. The productivity was 0.6 kl ethanol/kl of reactor volume/day, 8.5% (v/v) ethanol being produced in 95% of the theoretical yield, and a semi-commercial plant consisting of two 10 kl reactors has been operated since May 1983. Another new technical development is the use of aqueous two-phase systems for the direct conversion of starch to glucose and ethanol (Mattiasson & Larsson 1984).

3.14.13 Brewing

In the brewing of beer, enzymes are used at two stages, mashing and conditioning (Fig. 3.6). In mashing, proteases, pullulanase, α-amylase, and β-glucanase are added to the barley malt, which has been finely ground to increase hydration and enzyme penetration together with malt extract, cereal adjuncts, and hot water to form the mash. During mashing, starches in the barley are degraded by the enzymes to form the sweet wort. Fermentable sugars are produced, plus limit dextrins, which are unfermentable, and persist in the final beer, adding to the calorific value. The β-glucanase (EC 3.2.1.6) derived from *B. subtilis, B. amyloliquifaciens,* or *A. niger,* degrades β-glucan polymers (barley gums) that increase the viscosity of the wort, making it difficult to process and especially to filter. A considerable degree of degradation of the β-glucan is achieved because the microbial enzyme is thermostable such that some activity may survive the mashing process and so exert an influence later in the process. (Note that this action of β-glucanase is also employed in the preparation of chicken feed, since β-glucan cannot be digested by chickens, and it coats the metabolizable materials reducing their nutritional value).

In infusion mashing the temperature is kept constant at about 65°C, the enzymes being stabilized by using as concentrated a mash as possible, whereas in decoction mashing the temperature is gradually raised, so favoring the activity of the more stable enzymes such as proteases, β-glucanases and β-amylases. The temperature of mashing influences the amounts of residual dextrins, a higher percentage of fermentable sugars being present at the lower temperatures. Any small starch granules that survive malting are resistant to enzymic saccharification, probably because of the presence of a coating of polymeric materials.

Hops are next added to the wort, which is boiled so as to inactivate the enzymes. Papain is the protease most commonly used to degrade the protein component of haze, a complex of polyphenols, carbohydrate and protein, thus chill-proofing the beer and extending its shelf life. Treatment is at 4 °C for 3–4 days, and must be controlled, since excessive proteolysis can affect the foaming and other qualities of the beer. Other enzymes that have been used for chill-proofing include tannase from *A. niger* which degrades the polyphenol component of haze (Beck & Scott 1974); this enzyme also having been proposed as a means of solubilizing solids in instant tea (Scott 1975). One advance that may occur in the future is continuous conditioning of beers, as this step appears to be amenable to the use of continuous immobilized enzyme reactors. Amyloglucosidase has on occasion been added actually during fermentation. The spectrum of carbohydrates produced is altered, and more glucose is probably made available (Pfisterer & Wagner 1975). However, conventionally amyloglucosidase is used later in the process after fermentation. In this case the reason for adding amyloglucosidase is to degrade residual dextrins, so as to ensure a maximum alcohol content in the beer, to eliminate the need to add priming sugar to sweeten the beer or to produce a beer that is low in assimilatable sugars, and that is possibly better tolerated by diabetics (Woodward & Bennett 1973). In the brewing of 'live' beers, amyloglucosidase is not added, as its continued presence would be deleterious, and pasteurization is not carried out. In the case of 'dead' beers, the proteases and amyloglucosidase added are partly inactivated by the pasteurization that this product undergoes.

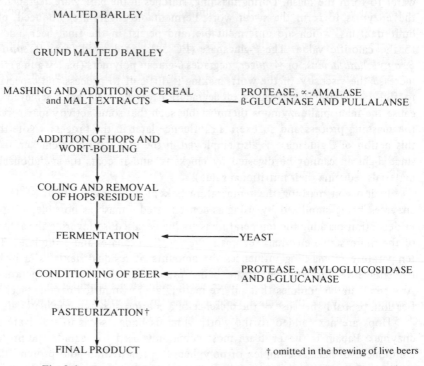

MALTED BARLEY

GRUND MALTED BARLEY

MASHING AND ADDITION OF CEREAL   PROTEASE, α-AMALASE
and MALT EXTRACTS ◄———————— β-GLUCANASE AND PULLALANSE

ADDITION OF HOPS AND
WORT-BOILING

COLING AND REMOVAL
OF HOPS RESIDUE

FERMENTATION ◄——————— YEAST

CONDITIONING OF BEER ◄——————— PROTEASE, AMYLOGLUCOSIDASE
                                      AND β-GLUCANASE

PASTEURIZATION†

FINAL PRODUCT   † omitted in the brewing of live beers

Fig. 3.6 – Beer manufacture – a flow diagram of the main operations.

There are a number of other processes in the brewing industry where immobilized enzymes have a potential use; immobilized glucoamylase has been used for the hydrolysis of dextrins in beer and immobilized β-glucanases has been proposed to facilitate filtration of the yeast (Linko *et al.* 1981). If the treated beer contained residual yeast cells the glucose produced could be utilized in the second process. Fully fermented yeast-free beers have been treated with the insoluble glucoamylase at 5 °C for 3 days to produce glucose at 0.6%. These beers remained clear, and were of satisfactory flavour. Diacetyl (2,3-butanedione) which imparts a butter-like flavour, has been removed from beer, using diacetyl reductase, especially that associated with yeast cells (Thompson *et al.* 1970) or *Aerobacter aerogenes* (Scott, 1975). The beer that was formed is better quality than that treated with absorbents rather than enzyme.

By adapting and perhaps modifying processes developed for the sugar industry, many glycolytic enzymes might be useful in beer production, provided that the organoleptic properties of the product are acceptable to the customer. Immobilized microbial α-amylase could be added to assist or even replace the natural amylases of malt. Production of a highly fermentable sugar wort could be achieved by the action of immobilized α- and β-amylases and glucoamylase. In addition to this, immobilized β-glucanase might be useful in adjustment of beer viscosity, since β-glucan contributes to the viscosity of beer. All these potential applications are only possibilities of future developments. Non-alcoholic beer-type beverages are produced, using mash recipes similar to those used for conventional alcoholic beers but with the addition of extra maltogenic amylases and proteases so as to extract more flavour and 'body'-giving components. Often the mash is then fermented as usual, and the ethanol then removed under vacuum.

3.14.14 Baking

Traditional methods of baking are based on the presence of endogenous enzymes which catalyze natural changes during growth, ripening and storage. They also carry out the saccharification of starch prior to fermentation and the degradation of the gluten which is a very important determinant of the rheological properties of the dough. Fungal proteases derived from *A. oryzae,* including both *exo-* and *endo*-peptidases are used because they have a low thermal stability and are inactivated during baking. The texture of the dough can thus be improved, so facilitating processing and decreasing energy inputs. Loaf volumes are substantially increased, and many other improvements are claimed (see below).

Saccharification can take place using endogenous α- and β-amylases acting synergistically on the starch made available by the rupture of starch granules during milling, the β-amylases forming maltose and limit dextrins and the α-amylases glucose, maltose and low molecular weight oligosaccharides containing α-1-6 bonds. Their combined effect is to make maltose available to the yeast maltase which forms glucose, which can then be used for fermentation, and thus gas production; and to make the dough easier to work by degrading the starch and altering the rheological properties of the dough. The ideal situation is to have sufficient enzyme activity such that fermentable sugars are produced at the same rate as their utilization by the yeast cells.

The activity of β-amylases is nearly the same in ungerminated and germinated flour; but α-amylase activities can be very low in ungerminated cereals, and develop more slowly even in germinated cereals. Thus its level is very dependent on the conditions of growth and harvesting, as in dry conditions little germination will occur. Wheat α-amylase also has the disadvantage of being relatively heat stable so that it continues to act during the baking process, causing overproduction of dextrins with the consequent formation of a sticky, moist, inelastic crumb structure. Endogenous α-amylases can be destroyed by treating with superheated steam or it can be diluted out with flour low in α-amylase. A low α-amylase content results in a low concentration of dextrins and poor gas formation, and thus a poor quality bread of low volume is formed. Supplementation of flour with exogenous fungal α-amylases is common, particularly as higher activities are required for the modern, rapid, and continuous baking processes that are used for the production of most of present day bread (Fig. 3.7). α-Amylase supplementation of flour is reported to not only enhance the rate of fermentation and reduce the viscosity of the dough, resulting in improvements in the volume and texture of the product, but also to generate additional sugar in the dough which improves the taste, crust colour and toasting qualities of the bread.

α-Amylase was originally added in the form of malt. Now α-amylase derived from *A. oryzae* predominates since malt α-amylase alters the colour of the bread, and has a higher protease content. Also malt α-amylase levels are variable and are more thermostable, sometimes leading to the formation of sticky bread. The use of the microbial   enzyme is especially useful, since although its activity increases as the temperature of the dough rises as baking commences, it is inactivated at about 60 °C before appreciable amounts of starch are gelatinized, so making control of the reaction easier. Also it has a negligible protease content and appears to aid the even distribution of the gas in the dough, thereby imparting favourable textural qualities and an increased loaf volume. Advantages in colour and shelf life have also been claimed for amylase supplemented bread. α-Amylases are not added when an unleavened product is to be manufactured, the dough in this case being composed of gluten, low molecular weight dextrins, and a much higher sugar content than is customary in fermented products. These proteins are essential to give baked materials a good structure by retaining gas. As gluten is found only in wheat, only this type of cereal is used to make baked products.

In some applications such as biscuit manufacture a flour containing high molecular weight dextrins but with extensively degraded gluten is required in order for it to be sufficiently plastic to enable easy processing. Many flours do not naturally have this property, and possess only very low protease activities, so proteases are added to degrade the gluten. Specific thermostable endo-peptidases obtained from *B. subtilis* are used, and they give a product with an enhanced elasticity which is important so that the dough can be spread thinly during manufacture. Use of enzyme for gluten degradation is superior to chemical methods such as bisulphite treatment.

There are several other minor applications of enzymes in the baking process. Lipoxygenase added in the form of soya flour has some use as a whitener. It acts by generating lipohydroperoxides that in the presence of oxygen oxidise gluten and bleach cereal pigments such as carotene. Bread with a fine crumb and expanded volume is formed. This process is still carried out as first described by Haas & Bohn (1934), (Emken 1977). Some work has also been carried out on the enzymic hydrolysis of pentosans in wheat and rye flours, and fungal phytase from *A. ficcum* has been used to treat wheat, removing phytic acid, and anti-nutritional factor (Liener 1977).

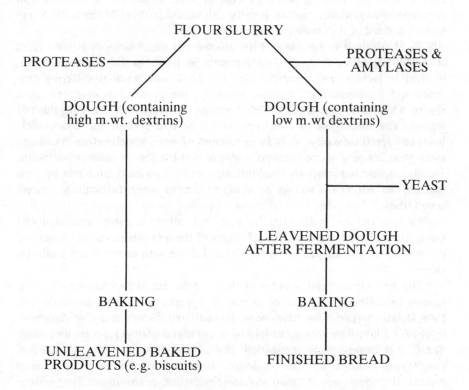

Fig. 3.7 – Manufacture of baked priducts – a flow diagram of the principal steps involved: to show the use of enzymes.

3.14.15 The dairy industry

Milk and its derivative products are an important nutritional source of carbohydrate, protein, fat, minerals, and vitamins. The dairy industry utilizes enzymes in many processes, and new applications are apparently under development.

3.14.15.1 Lactose hydrolysis

One of the most important new developments is the use of immobilized β-galactosidase (lactase EC 3.2.1.23) for the hydrolysis of lactose in milk products. Lactose or milk sugar is a disaccharide found in the milk of mammals (40% solids in cows). Lactose itself is less sweet and soluble than sucrose, and it cannot be absorbed directly from the intestine. Lactase hydrolyzes lactose into glucose and galactose, which have a combined sweetening power of about 0.8 relative to sucrose; the hydrolyzed product is also 3–4 times more soluble than lactose, and the monosaccharides are easily and directly absorbed from the intestine.

Lactase is found in the intestine of young mammals, and normally disappears after the weaning period, except in some humans. It is also found in certain microorganisms. Lactase activity also occurs in most of the lactic *Streptococci* utilized in the cheesemaking process.

The hydrolysis of lactose to form glucose and galactose is of interest from several points of view. A significant percentage of the world population is intolerant to lactose, and therefore may not drink milk without suffering diarrhoea and gastrointestinal distress. Secondly, the production of cheese gives rise to whey as a byproduct, which contains relatively high concentrations of lactose. The utilization of this byproduct is limited because the disaccharide tends to crystallize easily. Very large volumes of whey are therefore discharged every year, causing waste disposal problems and the loss of valuable nutrients. Thirdly, lactose is a relatively insoluble sugar and causes many problems by crystallizing out during the storage or transportation of whey, particularly concentrated whey.

Lactose can be hydrolyzed by acid, but colour is easily generated and acidic ion-exchange resins are fouled. Many of these problems can be eliminated or reduced by the enzymatic hydrolysis of lactose with immobilized β-galactosidase.

The best commercial sources of this enzyme are as *Saccharomyces lactis,* moulds like *Aspergillus niger* or *oryzae,* or bacteria such as *Escherischia coli.* Pure lactase preparations have been derived from *Escherichia coli, Saccharomyces* (or *Kluyveromyces* according to recent nomenclature) species, and fungi. The *E. coli* lactose finds application only in analytical chemistry. Of the *Saccharomyces* lactases, the lactase derived from the dairy yeast *S. lactis* is at present the most widely used commercial lactase preparation. The optimal process conditions for this lactase (35–40 °C, pH 6.6–6.8) are close to the natural temperature and pH of milk, and so it is most useful in the treatment of milk and sweet cheese whey. *S. lactis* lactase, however, also shows an appreciable activity at lower temperatures, down to about 4°C, where the growth of spoilage bacteria is very slow. Thus milk (or whey) can be treated during the overnight storage period that is customary in dairies.

The second most well-known commercial lactase preparation is a fungal lactase derived from *Aspergillus niger*. Its optimal process conditions are around 50 °C and pH 3.5–4.5. Application of this lactase is therefore limited to acid whey. Enzyme from both sources are inhibited by a product, galactose, such that complete hydrolysis is difficult to achieve.

Lactases find their application in several areas:

Milk for people with milk intolerance problems, especially infants.

Milk destined for cheese and yogurt making.

Whey or lactose for the production of sweeteners and soluble hydrolyzed whey syrups.

Concentrated milk products such as condensed milk, and for preventing the gritty texture of ice-cream made with lactose.

Batch hydrolysis is not a commercially viable approach except in the case of milk where lactases can be added to containers of milk during packaging, the lactose being hydrolyzed during the transportation and storage of the milk. Therefore considerable effort has been devoted to the immobilization of lactase. β-Galactosidase from the above and other sources have been immobilized on a variety of supports on laboratory-scale inorganic carriers, collagen, and agarose, as well as polyacrylamide gels and cellulose triacetate fibres. Whole microbial cells exhibiting β-galactosidase activity have been immobilized in polyacrylamide gels. The choice of enzyme source is influenced by the application of the preparation. Hydrolysis of lactose in acid whey, for instance, is conveniently achieved with the enzyme from *A. niger* which has the pH optimum of about 4.0, while for other substrates with a more neutral pH, such as milk, the enzyme from *S. lactis* or *E. coli* is more suitable.

Some of the preparations of immobilized β-galactosidase listed above have been evaluated for large-scale operation, or investigated in a pilot plant. Probably the first pilot-plant sized lactose hydrolyzing process was by Snam Progetti in a Dairy plant in Milan, Italy. Yeast β-galactosidase was entrapped in cellulose triacetate fibres and used to hydrolyze lactose in sterilized skim milk. After 50 cycles. 10 000 l of milk had been processed. No problem with bacterial contamination was experienced if appropriate disinfection was carried out, and the loss of enzymatic activity was very low such that after 50 cycles only 9% of the initial activity had been lost. Furthermore, the use of fibres of higher activity led to a marked improvement in hydrolysis, thereby reducing the fibre-to-milk ratio. In fact, the results of the pilot plant experiments were so encouraging that an industrial plant with the capacity of 10 t of treated milk per day was built in Milan and was ready for use in 1977 (Pastore *et al.* 1974).

A second process for the production of lactose hydrolyzed whey has been developed by Corning Glass Works, and tested on a pilot-plant scale. A column with the capacity of 6800 l per day of liquid whey was utilized. Before the whey was passed through the enzyme column dissolved protein was removed by ultra-filtration. The immobilized enzyme is estimated to have a life of at least 4000 h. The hydrolysis of lactose in whey has also been carried out with β-galactosidase immobilized on alumina particles on a pilot plant scale. The process was oper-ated as a fluidized bed in a column of 3 inches diameter. During use a layer of protein was formed on the enzyme particles, which could be removed by soni-cation. The tracer experiments revealed that plug flow conditions were not approached.

In general, a major problem in the treatment of whey is the fouling of the immobilized enzymes by undissolved solids which necessitates expensive filtra-tion or centrifugation prior to treatment. Continuous fluidised beds have been used to try to avoid this problem (Cheryan et al. 1975).

3.14.15.2 *Cheese manufacture*

a) *Coagulation.* The first step in cheese manufacture is the coagulation of milk (Fig. 3.2). Coagulation can be divided into two distinct phases, enzymic and then non-enzymic. In the primary enzymic phase a proteolytic enzyme such as chymosin (rennet), or less effectively, pepsin, carries out an extremely specific and limited proteolysis, cleaving a phenylalanine–methionine bond of K-casein, making the casein micelle metastable. In the second non-enzymic phase, the milk gels owing to the influence of calcium ions. The temperature coefficients for these reactions are 2 and 10–12 respectively, so that the coagulation of milk can be selectively retarded in immobilized enzyme reactors by lowering the temperature to inhibit the non-enzymic phase, but allowing completion of the enzymic phase; with coagulation taking place when the eluate from the reactor is warmed. It is very important to carry out the coagulation at the correct pH and temperature, and with the optimum dose of rennet, in order to form curd with the correct firmness, so that whey separation is made very easy.

Rennet (EC 3.4.23.4) is made by a saline extraction of the abomasum (fourth stomach from unweaned calves) and contains several gastric enzymes, the principal milk-clotting enzyme being an acid protease, chymosin (EC 3.4.4.3). Because of the relative scarcity of rennet, because veal consumption is decreasing at the same time that cheese consumption is decreasing, replacements have been developed from more convenient sources, the first to be discovered with a sufficiently high milk coagulating to general protease ratio being from *Mucor pusillus* (Arima et al. 1968). Subsequently *Mucor miehei* and *Endothia parasitica* were also found to produce a milk coagulating enzyme (EC 3.4.23.6) (Aunstrup 1976), which can replace rennet in the production of cheese, or more specifically in the coagulation of milk proteins, so as to achieve separation of the casein from the whey. These enzymes, unlike other proteases, have similar specificities but not identical properties to that from *Mucor pusillus*. The quality of this enzyme can be improved by inactivating the cellulase impurity by heat treatment of the enzyme preparation in the prescence of sodium chloride. A similar enzyme can also be obtained from the yeast *Cryptococcus albidus* and other similar organisms. It is produced extracellularily and precipitated with ammonium sulphate and organic solvents. Microbial coagulants are now so useful that they are responsible for about one third of all the cheese produced worldwide, but suffer from the disadvantage of being too stable and so are threatened commercially by improved methods of producing chymosin by recombinant DNA technology. The use of thermally destabilized microbial rennets results in residual enzyme levels in the milk product similar to or below those encountered when calf rennet is used (Branner-Jorgensen 1983). An unexpected benefit has been an increase on some occasions of the specificity of the microbial enzyme, making it virtually indistinguishable from the action of calf rennet. Also some microbial rennets help impart a flavour that is popular with consumers.

b) *Flavour development.* Lipolytic enzymes (EC 3.1.1.3) are produced from a variety of organisms, including *Mucor, Miehei* and *Myriococcum, Aspergillus, Rhizopus* and *Bacillus* spp. These preparations have both lipase and esterase activities, and are used in the food industry to produce desirable flavours in cheese. The enzyme hydrolyzes triglycerides to form free fatty acids which can then be converted to ketones. The enzyme can be prepared by absorbing onto diatomaceous earths at low pH, followed by desorption at higher pH, and imparts a flavour characteristic of the animal from which it is derived. Neutral proteases have also been found to accelerate cheese ripening by increasing the development of flavour, whereas acid and alkaline proteases caused bitter tastes (Law & Wigmore 1982). The acceleration of cheese ripening provides a major opportunity for enzymes in the dairy industry in the future.

In the cheesemaking process, rennet, pepsin, and/or microbial milk coagulation enzymes can all be added to the milk, together with the starter culture of *Lactobacillus*. A curd is then formed by coagulation of the casein, which is separated from the whey and pressed. Then proteases and lipases are added together with the moulds appropriate to the type of cheese being formed, in order to develop the texture and flavour of the cheese. An alternative approach is to encapsulate the cheese ripening enzymes and the milk components they act on in milk fat microcapsules which are then added to the pasteurized milk before it is made into cheese.

The use of immobilized enzymes in cheesemaking is attractive, as a continuous process would then be possible, and because milk-clotting enzymes that are in short supply can be used most economically by re-using them. Thirdly, proteases which are not suitable for use in a soluble form because of self-proteolysis, can be employed in an immobilized form in which this effect is prevented. Finally, use of an immobilized milk-clotting enzyme would allow the use of a second protease selected solely for its beneficial effects on cheese ripening, thereby avoiding a compromise between milk ripening and clotting activities.

c) *Other applications*

In the dairy industry up to 0.05% H_2O_2 is conventionally used to 'cold pastuerize' milk destined for cheese manufacture, and in the preservation of milk in some countries lacking refrigeration. The H_2O_2 is destroyed by catalase after the 'pasteurization' (Ohlson & Richardson 1974). This method is, however, limited by the high cost of catalase, which could be used more economically in an immobilized form, and by the tendency of the hydrogen peroxide to reduce the food value of the milk or cheese by oxidizing some amino acids, particularly methionine. Lysozyme has been reported to be added to cows' milk in order to help 'humanize' it, one of the major differences between human and bovine milk being their lysozyme contents.

Hydrogen peroxide and thiocyanate are both found naturally in milk, and they result in the formation of an antibacterial agent by the lactoperoxide catalyzed oxidation of the thiocyanate. This addition of hydrogen peroxide to milk activates the lactoperoxide system, resulting in enhanced 'self-pasteurization'. Note that lactoperoxidase activity is regularly assayed in milk as it is a good indicator of the efficiency of pasteurization.

Milk is also treated with penicillinase prior to further use in order to remove the residues of penicillin left in the milk of cows treated with penicillin to prevent mastitis. The penicillin is removed, as it could be harmful to people allergic to penicillin and could affect microorganisms involved in cheesemaking.

Trypsin has been successfully used to solubilize water-soluble heat denatured cheese whey protein, the enzyme being recovered after the reaction is completed by a cellulose based affinity absorbent (Monti & Jost 1978). The treatment of milk with trypsin also inhibits the development of an oxidized flavour on storage. Trypsin immobilized on porous glass has been used to remove such flavours, and a packed bed reactor was found to be more efficient than a batch reactor or fluidized bed.

A similar serious problem in the form of a cooked flavour arises in the production of ultra high temperature sterilized milk. This off-flavour is connected with the exposure of sulphydryl groups due to the thermal denaturation of proteins in the milk. A sulfhydryl oxidase with broad substrate specificity has been isolated from milk immobilized on porous glass with a view to reoxidizing these sulphydryl groups. (For reviews see Swaisgood 1977, 1978, 1980).

Lastly, note that lipases are also used in the generation of flavours in butterfat products in a similar manner to their use in cheese flavour development, and the enzyme nisinase derived from *B. cereus* is used to remove the polypeptide antibiotic nisin which is produced or added in cheese products.

3.14.16 Organic acids

α-Keto acids can be used in the treatment of chronic uremia. Chemical synthesis is often difficult and oxidase enzymes are expensive, so that use of immobilized enzymes for their synthesis is preferred.

The sun-screening agent urocanic acid is produced from L-histidine using L-histidine ammonia lyase from, for instance, *Achromobacter liquidium* entrapped in polyacrylamide gel. The preparation was heat treated at 70 °C to denature urocanase, which degrades the urocanic acid. This procedure and/or the immobilization appeared to lyase the cells, making the enzyme more available to the substrate. The immobilized cells are used in columns and supplemented with Mg^{2+} ions to maintain operational stability (Yamamoto *et al.* 1974).

A similar procedure for the production of malic acid has been developed using *Brevibacterium ammoniagenes* immobilized in polyacrylamide. The process has been successfully operated by the Tanabe Seiyaku Co. since 1974, the product being used in the treatment of hepatic malfunctioning. The immobilized cells have an operational halflife of 525 days at 37°C, while achieving an equilibrium (80%) conversion of 1.0 M fumaric acid into malic acid with a 70% yield of product. Fortuitously it was discovered that addition of bile salts suppressed the side reaction of fumaric into succinic acid besides breaking down the membrane of the cells which acts as a barrier to the passage of reactants (Yamamoto *et al.* 1976, 1977).

3.14.17 Amino acids
3.14.17.1 *Introduction*
Amino acids are used as food supplements, medicinal agents, and for the synthesis of other substances. Synthesis by enzymes is most appropriate as it is the natural L-amino acids that are biologically most active. Chemical synthesis yields a racemic mixture of L and D amino acids which can then be resolved using aminoacid-acylase.

3.14.17.2 *Enzymic resolution*
Resolution of racemic mixtures of L and D amino acids is achieved by the use of acetylated amino acid DL mixtures of which only the L-form is hydrolyzed to the free L-amino acid by aminoacylase. Industrial scale production has been carried out for about 10 years by the Tanabe Seiyaku Co. of Japan, using the aminoacylase from *Aspergillus oryzae* immobilized to DEAE-Sephadex and used in column reactors (Chibata *et al.* 1976). This process is the first example of the use of an immobilized enzyme on a commercially viable scale; this resolution system is especially used for methionine, but resolution of phenylalanine, tryptophan and valine mixtures have also been described (see p. 313).

Considerable effort has been expended on optimizing the process, taking into account factors such as the ease and cost of immobilization, the activity and stability of the immobilized enzyme, the practicability of regenerating the activity of the immobilized enzyme, and the effect of the flow-rate of substrate on the rate and extent of reaction. Using a 1000 l column reactor and obtaining 100% deacylation of the L-isomer, a yield of 6.4–21.5 tons of product was predicted, depending on the amino acids being formed, the actual yield being some 70–90% of this value. During continuous operation at 50°C a half life of 65 days was estimated, although after 30 days use the column was regenerated *in situ* by the adsorption of a quantity of fresh enzyme corresponding to that lost over the 30-day period of operation. Changing the dimensions of the column or the direction of flow of substrate had no effect on the performance of the column, and the pressure drop developed in the column was directly proportional to the flow rate of substrate and the length of the column. The DEAE-Sephadex support proved very stable, and no deterioration in absorptive capacity or physical properties was apparent over 5 years of use.

The operating costs of the above process are approximately 60% of those of the corresponding process using the soluble enzyme, because of reduced labour costs, improved yields, and easier isolation of the product. A variety of other immobilization techniques have been attempted, and a number of other enzymes tested in an attempt to achieve these ends. For instance, racemic amino acid mixtures can be resolved, as several proteases and peptidases specifically hydrolyse the L-isomers of amino acid esters; however, the D esters are usually strong inhibitors of such enzymes. Also, D-amino acid oxidase could be used to specifically transform the D-amino acid to a α-keto acid, and α-amino acyl amidase obtained from *Pseudomonas* spp which produces L-α-amino acids and D-α-amino amides from DL amino acid mixtures could be used.

3.14.17.3 *Enzymic production of amino acids*

Aspartic acid is used in medicines and as a food acidulant. It has been produced on a commercial scale since 1973 by the Tanube Seiyaku Co. in Japan, producing about 1700 kg of product/m^3 column/day. The aspartase activity of *E. coli* cells immobilized initially in polyacrylamide and more recently in carageenan is used, the preparations are heat shocked to denature the contamination fumarase activity that catalyzes unwanted side reactions. Comparatively little enzyme activity is lost upon immobilization, the pH optimum was decreased to 8.5, the enzyme lost its sensitivity to Mn^{2+} ions, and the immobilized cells had a much greater operational stability than either the immobilized enzyme or free cells, a half life of about 120 days being obtained at 37°C. A 1 M solution of ammonium fumarate supplemented with 1 mM Mg Cl_2 is used as substrate, and a 95% degree of conversion is conventionally obtained. This process replaced earlier ones using soluble aspartase and fermentation, being some 60% cheaper because of 10-fold reduction in the cost of cell production, even though recovery of the product by crystallization is very similar, irrespective of the type of catalyst used (Sato *et al.* 1975, Chibata *et al.* 1976).

D-Phenylglycine used in the synthesis of the semisynthetic penicillin ampicillin, by coupling to 6-aminopenicillinic acid, is also produced by enzyme mediated resolution of D, L-phenylglycine amide. An aminopeptidase enyme derived from *Pseudomonas putida* specifically hydrolyzes the L-phenylglycine amine to L-phenylglycine and ammonia, leaving the D form unchanged. As the free amino acid has a very low solubility it precipitates, leaving D-phenylglycine amide which is 99.99% optically pure. The L-phenylglycine is then chemically precipitated, hydrolyzed, racemized and then recycled (Nielsen 1980). D-phenylalanine has non-addictive analgesis properties probably by inhibiting enzymes responsible for degrading encephalins. Optically pure N-acyl L-D-phenylalanine esters can be formed from racemic mixtures using serine proteases that are not inhibited by the L-isomer (Bauer 1981).

Recently the Mitsui Toatsu Chemical Co. have announced the completion of a 200 tonne/yr plant for the production of the essential amino acid L-tryptophan from indole, and DL-cysteine and DL-serine are produced as side products. The method of production has not been disclosed, but may be based on the US Patent by Asai *et al.* (1982) which describes the reaction of L-serine with indole by tryphophan synthetase or tryptophanase, the L-serine being generated from DL or D serine by serine racemase.

Also, it appears that the Tanabe Seiyaku Co. have been operating a process since 1982 for the production of L-alanine from fumaric acid using the aspartase activity of immobilized *Escherichia coli* cells to form L-aspartic acid which is then converted to the L-alanine by the L-aspartate β-decarboxylase activity of immoblized *Pseudomonas dacunhae* cells. Use of mixtures of separately immobilized cells in the same reactor proved disadvantageous, and the sequential use of the two immobilized enzymes was found to be most efficient, with the pH of the reactants having to be adjusted from pH 8.5 to pH 6.0 between the columns. The formation of the side products L-malic acid and D-alanine by fumarase and alanine racemase respectively were eliminated by incubating the cells at an acid pH before immobilization (Takamatsu *et al.* 1982) (see Table 3.12). Also, because of the poor flow pattern of substrate in the column and changes in pH caused by the generation of CO_2 by the immobilized aspartame decarboxylase, this column is operated at high pressures to keep the CO_2 in solution. The same company also have a process for the production of L-citrulline from L-alanine using immobilized *Pseudomonas putida* with L-arginine deimidase activity (Chibata *et al.* 1975).

The third and probably the most interesting new commercial process for producing amino acids originates from Degussa AG (F.G.R.). It involves the continuous conversion of α-ketoacids into the corresponding amino acids by a substrate specific dehydrogenase, and NADH and alanine dehydrogenase in an ultrafiltration reactor. The coenzyme is retained in the reactor by linking it to a water soluble polyoxyethylene polymer which is continuously regenerated by formate dehydrogenase when supplied with formate ions. L-Alanine would appear to be the preferred product (Wandrey *et al.* 1981). This process is notable as the first commercial scale process to use an enzyme ultrafiltration reactor, which has often been proposed, but has never progressed further than laboratory studies before. Similarly cofactor regeneration as used in this process has never been utilized commercially before.

Table 3.12

The enzyme activities of immobilized preparations for alanine production with or without pH treatment of microorganisms

		Enzyme activities (μ mole/h $\times$ g of gel used)			
Microorganisms	pH treatment	Aspartase	L-Aspartate β-decarboxylase	Alanine racemase	Fumarase
E. coli	–	7510	–	89.4	1310
	+	7470	–	1.0	1.4
P. dacunhae	–	–	1140	28.0	745
	+	–	1030	0.3	0.5

Before immobilization with carrageenan, *E. coli* and *P. dacunhae* cells were treated at low pH, pH 5.0 at 45°C for 1 h, + pH 4.75 at 30°C for 1 h respectively (Takamatsu *et al.* 1982).

3.14.18 Antioxidants

3.14.18.1 *Introduction*

The spontaneous oxidation of foodstuffs by atmospheric or dissolved oxygen is often a problem, causing losses in nutritional value and organoleptic quality, and the formation of toxins; for instance the oxidation of unsaturated lipids is a common problem. Enzymes can be used as antioxidants; for instance glucose oxidase, superoxide dismutase, catalase, glutathione peroxidase and cholesterol oxidase have been proposed as antioxidants, or for removing oxygen or other reactive species from foods.

3.14.18.2 *Glucose oxidase* (EC 1.1.3.4)

Glucose oxidase is the commonest enzyme used as an antioxidant to prevent changes in the colour and flavour of foods during processing, transportation, and storage; for instance in citrus fruit based drinks, beers and wines, sauces and dressings, and a vast variety of dried foods such as cake mixture and instant soups. *A. niger* and *Penicillium glaucum* and *notatum* are good sources of glucose oxidase, and also the catalase which improves its stability by consuming the H_2O_2 produced during the oxidation reaction, which would otherwise denature the enzyme.

　　Glucose oxidase is also used to remove residual glucose from eggs and milled egg whites, so preventing browning (Maillard) reactions and the development of off-flavours. The enzyme is generally combined with catalase and FAD and so has to be extracted from cells before being used. However, this enzyme is relatively expensive, and therefore an immobilized enzyme system would be beneficial. In a series of experiments using immobilized glucose oxidase and catalase in a continuous packed bed reactor, plugging and gel formation occurred. Precipitation and removal of mucin protein eliminated these complications, but this solution was considered unsatisfactory by the egg industry.

Glucose oxidase can also be used in enzyme-based assays for glucose, and in the production of gluconic acid from glucose. If a high fructose syrup or invert sugar is treated with glucose oxidase, a mixture of fructose and gluconic acid is obtained, which can then be separated by ion exchange chromatography. Gluconic acid is useful as a complexing agent for metal ions in foods and pharmaceuticals, and also as a sweetener for diabetes. The shelf life of sea foods can be prolonged by treating with a glucose oxidase–catalase mixture. Gluconic acid is formed which lowers the pH, so inhibiting microbial spoilage without affecting the organoleptic properties of the food.

3.14.19 Protein processing
3.14.19.1 Introduction
Enzymes are widely used to increase the value and availability of proteins, for instance by enabling the recovery of protein hydrolysate from otherwise scrap bones or fish. Proteases differ widely in their substrate specificity, so that combinations of different enzymes can be used to increase the degree of hydrolysis of a protein, and also differ in factors such as their pH optima, so that they can be used over a wide range of operating conditions. Although a host of microbial proteases are known, proteases derived from plant and animal sources are most widely used in industrial applications. These include papain, ficin and bromelain from plants and trypsin and pepsin and rennin from animal sources. Problems associated with protease action are bitterness due to the release of hydrophobic amino acids and difficulties in controlling the extent of reaction. Note that although a very wide range of proteases from various sources with varying specificities and properties are available, only a comparatively small range of peptidases can be obtained. Applications of proteases include hydrolysis of gelatine in photographic films to enable recovery of silver, Korosi (1976) leather treatments Monsheimer & Pffeiderer (1981) , and use as cleaning agents. Applications in the food industry enable the intrinsic functional properties of proteins such as viscosity, whipping ability and emulsifying power to be optimized by controlled proteolysis. Examples include the enzymic modification of soy and whey proteins, so as to make them functionally more suitable for food applications, brewing, hydrolysis of wheat gluten, the production of yeast extracts, the production of gelatine from collagen, and in the preparation of peptones, which are hydrolyzed proteins used in microbiological growth media. Proteases also prove useful in the recovery of protein from the blood, offal and

bones from slaughter-houses, the decolorization of haemoglobin so as to make this abundant source of protein more acceptable, and from fish processing wastes, the enzyme apparently becoming adsorbed onto the surface of the substrate (for review see Ward 1983). Several reports of the use of proteases in the extraction of gelatine from bones have been made. Since fresh bones contain a significant lipid content, some workers have found that its removal using *Rhizopus arrhizus* lipase prior to gelatine extraction was advantageous (Laboureur & Villabon 1972).

Many of the characteristic properties of proteins such as solubility, emulsifying activity, water and fat binding capability and elasticity can be lost or greatly reduced during the processing of foods. These losses are currently especially important, since many modern foodstuffs require proteins with a high degree of functionality. For instance, proteins could be enzymically phosphorylated, glycosylated or hydrolyzed to improve their solubility without the losses in the other useful properties caused by the restoration of solubility by enzymatic hydrolysis of the proteins; which requires careful control of the degree of hydrolysis. Similarly, thio-disulphide isomerase, sulphydryl oxidase, or disulphide reductase could possibly yield useful results in meat and bread, particularly as they may eliminate the need to add chemical flour oxidizing agents to the latter (Whitaker 1977).

One of the most promising applications is the use of iso-electric soluble soy protein hydrolysate for fortification of soft drinks in tropical countries; its good solubility also facilitates incorporation into meat products. Several reports of the use of proteases in the extraction of gelatine from bones have been made. Semifresh bones contain a significant lipid content and some workers have found that its removal using *Rhizopus arrhizus* lipase prior to geletine extraction was advantageous (Laboureur & Villalon 1972).

Two other very interesting applications of proteases are firstly the use of a protease to convert porcine insulin into human insulin (Obermeier & Seipke 1984), and secondly the use of aminopeptidase to produce the pharmaceutically useful substance D-phenylglycine by selectively hydrolyzing the L-phenyl glycinamide in a racemic mixture.

A process for decolorization of a haemoglobin has been developed. After enzyme hydrolysis, the unwanted haem colour is adsorbed onto active carbon, and removed by centrifugation and filtration. Uses of the new almost colourless globin hydrolysate are at present under investigation.

Proteases such as papain are reputed to be widely used to tenderize meat, papain being especially useful as it is relatively heat stable, so that its action continues during the early stages of cooking. The best method of administration appears to be to inject the enzyme into the jugular vein of the cow prior to slaughtering. This method appears to overcome one of the chief problems associated with the enzyme tenderization of meat, that is achieving a uniform distribution of the enzyme such that even hydrolysis occurs. Pesin, trypsin and chymotrypsin have been used clinically as a digestive aid, especially when enterically coated, and also in the preparation of precooked cereals and baby foods. Proteases have also been used medically for fibrinolysis, as anti-inflammatory agents in the potentiation of drug activity, and in the treatment of cystic fibrosis, burns, ulcers and acne. (For a review, see Christie 1980). Lastly, papain has been used as a shrink-proofing agent for wool, where some degree of surface attack can confer a permanent anti-shrink finish.

3.14.19.2 *The plastein reaction*
This reaction involves the hydrolysis of a protein by a protease followed by a change in the reaction conditions such that the protease works 'in reverse', resynthesizing pepticle bonds, but with the amino acids and peptides in a new primary structure (sequence). This reaction can be used to combine amino acid esters into proteins, to fortify soy bean protein with methionine, to increase their solubility by the incorporation of glutamic acid, and also to remove undesirable residues. For instance, phenylalanine can be removed from protein destined for consumption by phenylketonuric patients (Yamashita *et al.* 1975, 1976).

3.14.19.3 *Aspartame*
Aspartame is a synthetic dipeptide ester, L-asp-L-phe-OMe which is about 200 times as sweet as sucrose. It has recently been released for sale in N. America and Europe by G. D. Searle. It was originally synthesized chemically and reported by Mazur *et al.* (1969). Subsequent improved methods of synthesis have been developed which involve the use of metaloproteases such as thermolysin 'in reverse'. Metaloproteases are used because, unlike the more common proteases, they have no esterase activity. High yields of product, and thus the commercial practicability of the process, are dependent on the tendency of the aspartame molecule to form a 1:1 addition compound with excess phenylalanine methyl ester substrate molecules in which it is in effect simultaneously purified, isolated and concentrated in one step (Isowa *et al.* 1979). One enzymic route to aspartame synthesis is the condensation of N-benzyloxycarbonyl-L-aspartic acid with L-phenylalnine methyl ester to form N-benzyloxy carbonyl-L-aspartyl-L-phenylalanine methyl ester using thermolysin. It has been described in detail by the Toyo Soda Co. (Oyama *et al.* 1981). Unlike chemical methods, such enzymic routes produce no detectable β-peptide linked aspartame, which has a bitter taste and so has to be separated from the α-isomer.

Aspartame can also be produced by alkylation of the phenylalanine by the esterase activity of subtilisin in the presence of an alcohol (Davino 1981).

3.14.19.4 *Others*
A number of other examples of the use of proteases to produce medically or industrially important substances have been reported recently (for a review see

Chaiken *et al.* 1982), for instance the synthesis of an angiotensin-11 derivative
(Isowa *et al.* 1977), and the synthesis of Leu and Met-enkephalins (Kullman
1980). Angiotensin is produced as the protected valine-5-amide derivative
using papain, subtilisin, or microbial metalloproteases. The reactants used were
t-butyloxycarbonyl peptides and peptide ethyl esters for the carboxyl and amine
components respectively. Papain catalyzed the reactions without exhibiting any
esterase activity, provided that the reaction mixture contained a high methanol
concentration.

The formation of each bond of the pentapeptide enkephalins were catalyzed
by either papain or α-chymotrypsin using the t-Boc amino acids and peptides
or their esters as the carboxyl component and amino acid and peptide hydra-
zides as the amine component.

Peptide semi-synthesis, where large peptides are joined together using
proteases to carry out the condensation reactions, have been reported for
several natural compounds such as insulin (Morihara *et al.* 1979), staphylococcal
nuclease-T (Komoriya *et al.* 1980), proteinase inhibitor (Sealock & Loskowski
1969), and ribonuclease-S (Homandberg *et al.* 1982).

Morihara *et al.* (1982) took porcine insulin and replaced the β-chain terminal
alanine with threonine to form human insulin. The alanine was removed by
carboxypeptidase A and the threonine attachment catalyzed by trypsin. A large
excess of the amine component was required and the use of organic solvents also
improved the yields. Trypsin was also used to form the staphylococcus nuclease.
Synthesis of the ribonuclease and protease inhibitor were promoted by using a
molecular trap, that is in the former case the product of the reaction is a small
fragment of ribonuclease non-covalently bound to the completed ribonuclease
molecule, which constitutes a new molecular species and thus promotes the
formation of the enzyme by mass action effects.

3.14.20 Flavouring agents
In the formation of soya sauce from soya beans, proteases are added to the
fermentation so as to reduce its duration. Cultures of *A. oryzae* and *A. sojae* can
be immobilized as complex mixtures of *endo* and *exo*-proteases with both acidic
neutral and alkaline pH optima. In addition, treatment with carbohydrases is
carried out to degrade undesirable sugars, including stachyose and raffinose.
Also, neutral proteases have been used to remove the characteristic 'bean flavour'
in the manufacture of soya milk.

5′-Phosphodieesterases from *Aspergillus* and *Penicillium* species are used to
degrade yeast RNA into its constituent ribonucleotides. In a second step, adenosine monophosphate is converted into inosine monophosphate by a 5-AMP
deaminase obtained from *A. oryzae*. These guanosine, adenosine and inosine
mononucleotides are useful flavouring agents which enhance the flavour of
monosodium glutamate when added in equimolar amounts.

A sour dough flavour suitable for baking and that can be spray dried can be

prepared from rennin treated curd and lactic ac
lated with organisms that produce extracellular e
and incubated. A continuous hydrolysis process usin
also been developed.

The use of cell-wall degrading enzymes in the extrac
resin is widespread, because of the resulting reduced extra
creased yields of product. It is interesting that one of the roles o
to increase the rate of rehydration of dried plant tissues. Cellulas
lases and proteases are especially common active components in such p
Pseudomonas putida immobilized in agar gel has been used to degrade   ine
on a laboratory scale (Middlehoven & Bakker 1982). However, production of
decaffeinated coffee by treatment with supercritical carbon dioxide appears to
be preferred in practice.

3.14.21 Fruit processing

A very wide range of temperate and tropical fruits, both with and without
stones, and including apples, pears, berries, citrus fruits and vegetables are processed by present-day industry. Mixed preparations of pectinases, hemicellulases
and cellulases derived from *Aspergillus, Rhizopus* or *Trichoderma* spp. are added
to fruit and grapes during maceration, so as to enhance the extraction of juice,
colour and flavours by supplementing the action of the endogenous enzymes
present in the fruit, and also to reduce the times required for fermentation and
clarification (Fig. 3.8). These enzymes act by reducing the integrity of the
structural elements of the fruit, and by reducing their water-retaining capacity.
In unripe fruit, pectin is present in an insoluble form and so helps to maintain
the hardness of the fruit, but when the fruit ripens the pectin becomes partially
degraded by endogenous emzymes making the fruit softer. Pectic enzymes
include pectin lyases which are found exclusively in fungi and which act on high
methoxypectins; pectate lyases which occur chiefly in bacteria and which act
on low methoxy or de esterified pectins; pectin esterases which demcthylate
the esterified polygalacturonic acids; and endo and exo poly galacturonidase
(pectinases) which are possessed by fungi, plants and some bacteria and
which hydrolyse polygalacturonic acid residues (Pilnik & Rombouts 1981).
The most commonly used enzyme for fruit treatment is pectinase (polygalactu-
ronidase EC 3.2.1.15), which together with hemicellulases, cellulases and/or
amylolytic enzymes are added to the juice during mashing. The pectinase degrades $1,4$-α-D-galacturonic bonds in pectin, a structural component of fruit.
Pectin is a heteropolysaccharide found as a structural component in various
fruits and vegetables, can become at least partly solubilized during processing, so spoiling the appearance and impairing the flocculation and/or the
filterability of the juice and reducing the yield of juice. Thus, pectinase treatment reduces the viscosity of the juices and enables the production of a more
stable and concentrated product. Pectinase preparations are usually a mixture

and *exo*-polymethylgalacturonidases and are obtained from *Aspergillus* pp (for review see Fogarty 1983).

Pectin forms a gel when heated and then cooled and so is useful in the manufacture of jellies and jams. The carboxyl groups of the polymer can be naturally esterified to form methoxy groups. When a pectin with a high methoxy-pectin content is used, as is usually the case, the gelation process requires the use of high concentrations of sugar and acids and precise control of conditions; whereas when low methoxy pectin is used, sugar and acid are not always required, and gelation proceeds easily in the presence of ions such as calcium. This form of pectin is also useful as a film-forming agent, for instance in yogurt and mayonnaise. Pectin esterase (or more properly pectin methyl esterase) is used to de-esterify natural high methoxy content pectin. Such enzymes have been obtained from *Aspergillus, Penicillium* and *Fusarium* spp (Ishii *et al.* 1980), and are usually used as contaminants of pectinase preparations.

The amylases and cellulases that are also added degrade any residual starch that is often present in immature fruit, as this can impart a haze to the product (see below) and also facilitate masceration and colour extraction respectively.

A specific $\beta 1,3/\beta 1,6$ glucanase has been recently specifically developed so as to facilitate the clarification and filtration of wines made from grapes infected with *Botrytis cinerea* or 'noble mould'. *B. cinerea* produces a $\beta 1,3$ linked glucan which is not susceptible to hydrolysis by dextranases, and the β-glucanases useful for the hydrolysis of the $\beta 1,4$ linked glucans from barley that are encountered in beers. Destruction of the $\beta 1,3$ glucan also has the advantage of preventing the suspension of contaminant microorganisms in the wine, resulting in considerable savings in the SO_2 needed to prevent undesirable infections (Dubourdieu *et al.* 1981). Similarly amylogucosidase is used to remove starch haze from the juice produced from immature apples.

The use of immobilized proteases to prevent hazes in wine caused by proteins coming out of solution has been proposed, but the stabilities of such preparations were comparatively low, owing to inhibition by phenolics, iron ions, and of course ethanol. Immobilized cells have also been used to deacidify wines after fermentation. This is important because the malolactic fermentation, a conversion of malic to lactic acid with an accompanying decrease in acidity, is undesirable if it takes place in the bottle but this secondary fermentation is very desirable if it occurs prior to bottling as it increases the quality of the wines flavour. For instance *Leuconostoc oenos* immobilized in alginate have been used by Gestrekius & Kjaer (1983) to deacidify wines. Viable cells immobilized in alginate and stored in a resting medium of sterile grape juice containing 5–12% alcohol were preferred.

Tartaric acid is an important byproduct of wine manufacture. Traditionally, dextrorotatory tartaric acid has been obtained from the tartar in wine lees. However, this is an unstable and restricted source, so that the preferred method is the chemical conversion of the readily available substrates maleic anhydride or maleic acid to *cis*-epoxysuccinic acid followed by their bioconversion to tartaric acid by the *cis*-epoxysuccinate hydrolase of *Nocardia tartaricans*. Growing cells, dried or pulverized cells, or even a purified cell extract can be used (Miura *et al.* 1977).

Another important use of enzymes in fruit processing, is the use of naringinase and limonase to hydrolyze the bitter component of some citrus fruits, especially grapefruit. Rhamnose and prunin are formed without the loss of the natural yellow colour of the fruit. Naringinase is often used in an impure form, and it is a common contaminant of pectinases (Chandler & Nichol 1975). Debittering of citrus juices has been successfully carried out on a laboratory scale using *Arthobacter globiformis* cells immobilized in acrylamide gel. The juice remained otherwise unaltered, and the activity of the cells was retained after many uses. 17-dehydrolimonoate A-ring lactone was identified as the major metabolite and the reaction was mediated by limonin D-ring hydrolase and limonoate dehydrogenase (Hasegawa *et al.* 1982). A related triterpene, nomilin, that has been involved in the delayed bitter taste of certain citrus fruits is also debittered by the *A. globiformis* cells (Hasegawa & Pelton 1983).

3.15 USE OF ENZYMES IN THE EXTRACTION OF NATURAL PRODUCTS

There have been a number of reports of the use of enzymes in the extraction of oils from sources such as fish, rapeseed, yeast, palms, and soya beans. Cellulases and pectinases are used in palm oil extraction. In soya bean and fish, much oil has been found to be associated with protein, so that addition of proteases increases the yield of oil and protein. Use of thermostable proteases is preferred, but in general the use of enzymes is limited by the minimal water contents of the various process streams. *Trichoderma uride* and *A. niger* cellulases, hemicellulases and proteases have been used to extract hydrocarbons from *Euphorbia* plants (Weil *et al.* 1982) and similar enzymes used to extract sapogenins from *Helleborus* (Isacc 1977).

A process has been developed for loosening the shells of shrimps and for extracting the viscerol mass of clams, using a mixture of carbohydrases and cellulases derived from *Aspergillus niger* (Fehmerling 1970). Glucomananase is used to reduce the molecular weight of soluble gums present in coffee extracts prior to their concentration and peptido glutanase is used to increase the amounts of L-glutamic acid or glutamylpeptides in foods and beverages which are prepared by the hydrolysis of peptide containing raw materials. The enzyme is obtained from *B. circulans* which releases the enzyme during autolysis (Kikuchi *et al.* 1974).

Endogenous microbial enzymes are sometimes utilized to break down their parent cells, and thus extract valuable intracellular materials. For instance, in the production of yeast extract cells are allowed to autolyse at about pH 5 and 550°C. Proteases are probably the most important class of enzymes involved in autolysis, although others such as glucanases, lipases and nucleases also have some activity. Attempts have been made to increase the rate of extraction and the final yield by the addition of exogenous proteases such as papain.

Mixtures of lytic enzymes are obtained from *Actinomyces, Achromobacter* and *Pseudomonas* spp. Cells of species liable to lysis are used as inducers for the lytic enzymes during growth of the parent cells (Shinkarenke *et al.* 1976). Especially interesting is a *Streptomyces* preparation intended to lyse dental caries producing organisms such as *Streptococcus mutans* (Yoshimura *et al.* 1975). These enzymes can after be stabilized by incorporating carboxymethyl cellulose, glycerol or monoethanolanine.

3.16 DETOXIFYING ENZYMES

Toxins or anti-nutritional compounds can often be formed by endogenous enzymes acting on foodstuffs, particularly during the processing of the foods when their structural integrity is diminished, for instance following the rupturing of cellular materials. Alternatively, toxic endogenous substances can be degraded enzymically to nontoxic materials. Therefore the control of such processes can be of great benefit, for instance by heat denaturation of undesirable enzymes. Three of the most important substances are, firstly, glucosides of toxicants such as the goitogenic thiocyanates, isothiocyanates and oxazolidine-thiones. Secondly, oligosaccharides such as stachyose, raffinose and lactose which can result in abnormal carbohydrate metabolism. Thirdly phosphate esters such as nucleic acids and phytic acid which cause the deposition of uric acid with gout-like symptoms and complex with essential dietary metals and proteins respectively. Nucleic acids can be hydrolyzed by endogenous yeast ribonuclease which is activated by heat shocking the cells (Maul *et al.* 1970) and phytic acid is partly hydrolyzed by endogenous phytase but also requires the addition of extra phytase in the form of wheat germ (Kon & Ohlson 1973).

Pollutants can be natural compounds such as lignified woods, or artificial, such as organophosphorous insecticides. Microbial enzymes are a potentially effective means of degrading pollutants because many have actually evolved over the course of millions of generations to carry out this role in their natural environment, either to protect the parent cells from toxins or to actually use the pollutant as a carbon or energy source. Detoxification may result in complete mineralization of the pollutant or in partial decomposition associated with the loss of undesirable biological activities. Thus despite the variety of potential toxins produced, comparatively few are recalcitrant, such that they are resistant to biological action and will persist for long periods in the environment.

The use of enzymes still associated with whole cells rather than isolated enzymes for the degradation of wastes would appear to be most advantageous because the enzymes are at least partly protected from denaturants by being retained inside a substantially intact cell wall and membrane. Furthermore, the enzymes concerned are often difficult to extract in a fully active form.

A considerable amount of research has been carried out recently into the enzymic and microbiological degradation of pesticides. For instance, *Pseudomonas putida* has been found to possess dehalogenases which will degrade mono- and dichloroacetates and propionates derived from pesticides and herbicides. Especially impressive is the observation that Dalapon degrading activity was stable over 20 000 h in a chemostat (Slater *et al.* 1979, Berry *et al.* 1979). An enzyme extract obtained from a mixed microbial population adapted to grow on Parathion was found to hydrolyze a number of organophosphate insecticides. This enzyme could also be successfully immobilized to facilitate its used, having an operational halflife of 280 days (Mannecke 1976, 1977).

α-Amylase has been used to clarify colloidal starch—clay suspensions, commonly referred to as 'white waters', that are produced in large volumes by paper mills. After enzyme treatment the solids can then be flocculated easily by the addition of alum. It was advantageous to immobilize the enzyme so that it could be re-used and so that the water did not have to be treated to destroy the enzyme prior to recycling of the clarified water and its use in paper manufacture (Smiley 1974). Other applications include the removal of carcinogenic aromatic amines including benzidine, napthylamines and aminophenyl from waste waters derived from the coal, resin, plastics and textile industries, using hydrogen peroxide and peroxidase to enzymically crosslink the pollutants (Klibanov & Morris 1981). Also cyanide can be detoxified using immobilized (*Stemphylium loti* (Nazaly & Knowles 1981).

The brown, partly chlorinated lignin derivatives commonly known as 'Kraft lignins' present in pulp mill effluents can be decolorized by using the white rot fungus *Coriolus versicolour* entrapped in calcium alginate beads and supplemented with sucrose as a carbon and energy source (Livenocke *et al.* 1981) or the lignolylic fungus *Phanerochaete chrysosporium* (Kirk & Yang 1979). This biological method is an alternative to the chemical bleaching of the lignin, and is especially important as the chlorinated lignins produced during bleaching have been shown to be mutagenic.

Pseudomonas denitrificans, again immobilized in alginate, has been used to denitrify drinking water. The cells are provided with an exogenous carbon source and they oxidize nitrates and nitrites completely to gaseous products. When operated in continuous reactors the activity of the cells decayed with a halflife of 30 days, and after incubation with full nutrients at intervals, fresh cell growth occurred and increases in the rate of reaction were observed (Mattiasson *et al.* 1981, Nilsson & Ohlson 1982).

The degradation of phenols remaining in the waste waters from hospitals, laboratories and after coal processing to coke and removal of H_2S and NH_3, can be achieved using the fungus *Aureobasidum pullatans* adsorbed to fibrous asbestos (Takahashi *et al.* 1981) by using cells of a *Pseudomonas* sp adsorbed to anthracite coal, or by cells entrapped in alginate gel (Hackel *et al.* 1975). This treatment is advantageous in that it avoids damage to conventional biological waste water treatment systems. Finally, enzymes can even be used to detoxify nerve gases. The US Army have used an enzyme derived from the nerve fibres of squid and immobilized on agarose beads to render harmless obsolete stocks of the nerve gas soman (1,2,2 trimethyl propylmethyl phosphorofluoridate) and its derivatives.

3.17 ENZYME-BASED DETERGENTS

Massive quantities of enzymes are used in detergents. Detergent proteases are typically alkalis tolerant (pH 9–13), thermostable, detergent resistant serine proteases secreted by *Bacillus subtilis* and *Bacillus lichenformis*. Because detergents contain sequestering agents to overcome hardness in water supplies, metallo proteases cannot be used. Enzymes derived from alkalophilic *Bacillus* spp. are especially tolerant of alkaline conditions and can grow in the presence of surface active substances. Often, neutral proteases are produced as well which can cause substantial loss of alkaline protease activities. Usually the enzyme is salted out of solution and recovered by filtration. Such alkaline proteases work well even at the lower temperatures experienced during presoaking and at mildly alkaline pH also. The use of successful enzyme based detergents dates from the mid 1960s when the first alkalis tolerant protease, Alcalase, was introduced by Novo.

There is also a potential for the use of proteases and α-amylases in detergents in order to hydrolyze food residues containing protein and starch. However, practical application of α-amylases is limited by their instability in the presence of chlorine which is present in most machine dishwashing formulations to act as a bleach.

Trends are to use enzyme based detergents in pre-wash soaking or 'spotting' treatments, and to wash at lower temperatures so as to conserve energy. Use of liquid detergent formulations is also likely to be increasingly prevalent. The market for enzyme containing detergents collapsed during the 1970s owing to problems with the allergic response of some workers and consumers. These problems have now been largely overcome by the formation of dust-free preparations in which the enzyme is granulated and often covered by an inert layer of wax, cellulose. polyalkylene glycol or dextran. In its most sophisticated form, the so-called 'prilling' process, enzyme is mixed with a filler, a binder and water; extended as spheres and coated in wax and spray-cooled to form 0.5–1.0mm diameter granules.

3.18 USE OF ENZYMES AS CLEANSING AGENTS

Enzymes are included in cleaning preparations for the dispersion of solids and films in and on pipework, heat exchangers, tanks, etc. The preparations compromise mixtures of *Bacillus subtilis* spores and amylases, cellulases, lipases and proteases and appear to act even in systems containing high contents of undissolved solids such as in sewerage systems.

Fouling of industrial filters is an especially costly and inconvenient phenomenon even when back flushing and/or the circulation of cleaning solutions is possible. Examples are numerous and include the filtration of cheese whey to recover protein and the filtration of fruit juices and wines. Cleaning solutions generally comprise an enzyme and detergent. Precise formulation depends on the application in mind; for instance, trypsin and papain are used for fouled dairy filters, α-amylases and β-glucanoses in yeast and cereal uses, and cellulases and pectinases for wines and fruit juices. The Genex Corporation in particular are concentrating on household and janitorial applications, for instance for the removal of bathroom mould, because existing products are hazardous or ineffective and because such products do not require the extensive testing prior to sale that pharmaceutical and food products are subject to.

3.19 THE LEATHER INDUSTRY

Skins are soaked initially to clean them and to allow rehydration. This latter process is aided by the addition of very low concentrations of proteases. Use of pancreatic trypsin is especially favoured since contamination lipases tend to solubilize fats and gums, further improving water uptake.

Dehairing is carried out using alkaline proteases such as subtilisin in a very alkaline bath. Alkaline conditions tends to swell the hair roots, so easing the removal of the hair by allowing the proteases to selectively attack protein in the hair follicle. Other specific enzymes are used for skins from particular species.

Dewooling of skins is conventionally carried out by applying a 'paint' containing typically lime, sodium chlorite and a protease to the 'flesh-side' of the skin followed by incubating. For finer wooled skins, less active enzyme is required and can be applied as a powder. Papain has also been reported to be used to cause shrinkproofing of wool by carrying out limited surface proteolysis.

Baking of hides involves degradation of some elastin and keratin, removal of hair residues and deswelling of collagen. A variety of proteases are used by the industry, depending upon the type and quality of hide. The process remains a highly skilled craft requiring intuitive manipulations of conditions such as pH, temperature, enzyme concentration and incubation time if the best quality leathers are to be produced.

3.20 TEXTILES

Fabrics are normally 'sized' by the application of adhesive starch to increase strength during weaving. Subsequent desizing by the liquefaction of the starch by bacterial α-amylases is carried out because the sized fabric is less absorbent to liquids, so that operations such as dyeing and bleaching are more difficult (Underkofler 1976). Bacterial α-amylase is most useful because desizing can be accelerated by operating at elevated temperatures and because of the resistance of the bacterial α-amylase to inhibitors and temperature. The starch and dextrins appear to exert a stabilizing effect. Reaction can be carried out at 50–75°C and at pH 5.5–7.0 in the presence of surfactants so as to ensure an even action of the bacterial and fungal enzymes, but temperatures as high as 105–110°C and reaction times as low as 1–2 minutes can be achieved using *B. licheniformis* α-amylase.

3.21 PAPER MANUFACTURE

Starch is used to size the surface of paper, and as a coating binder, but must first be reduced in viscosity before the benefits in strength and stiffness can be obtained. Hydrolysis can be carried out by acidic hydrolysis or oxidation with sodium hypochlorite. Another method uses α-amylases and involves heating the starch with the enzyme, such that the starch granules swell and then gel (Kalp 1975). During the subsequent period the enzymes thin the starch, and then finally rapid heating with steam denatures the enzyme and completely solubilizes the starch which is then diluted prior to use. When potato starch is used, fungal α-amylase is used at 50°C; but when maize starch is used, bacterial α-amylase must be used, as the fungal α-amylase is active only at relatively low temperatures at which the starch is below its gel point. For a review of new developments see Clayton *et al.* (1984).

3.22 ANTIBIOTICS

About 150 antibiotics are produced on a commercial scale and find a variety of applications in medicine and agriculture. Virtually all are produced in conventional fermentations, and some are subsequently enzymically modified (Brodelius 1978). Immobilized whole cells have a potential for achieving a more efficient production of existing antibiotics and the formation of novel antibiotics especially, when the antibiotic is naturally secreted into the culture medium. For instance, Morikawa *et al.* (1979) reported the production of penicillin G from glucose using *Penicillin chrysogenum* immobilized in polyacrylamide. The same group also used Bacillus cells in a similar manner to form bacitracin (Morikawa *et al.* 1980). The activity of these cells had a halflife of a week when used in a reactor, but productivity was limited by cell growth at the surface of the cells.

3.23 PENICILLIN ACYLASE (amidase or deacylase, EC 3.5.1.11)

The antibiotic activity or toxicity of the penicillin molecule is markedly influenced by the chemical nature of the acyl side chain substituent at the 6-amino group. In practice a variety of semisynthetic penicillin antibiotics are synthesized from 6-amino penicillanic acid (6-APA) by chemical substitution. The parent benzyl or phenoxy methyl penicillin molecules (penicillins G and V respectively) are produced by fermentation and then treated with immobilized penicillin amidase to remove the side chain without degrading the sensitive β-lactam part of the molecule. Thus benzyl penicillin forms 6-APA and phenylacetic acid.

This process is the best known use of an immobilized enzyme in the pharmaceutical industry. Several companies, including Beechams, Bayer, Snam Progetti, Tanabe Seiyaki, Squibb, and Astra, have developed their own industrial scale processes (Lagerlof *et al.* 1976).

The enzyme is usually obtained from *E. coli* and used at pH 4.5–5.5. This enzyme has a molecular weight of 70,000 and is inhibited by phenylmethansulphonylfluoride, but not by di-isopropyl-fluorophosphate, indicating that serine is not in the active centre. Sodium dodecyl-sulphate dissociates the enyme reversibly to units with a molecular weights of 20,500. Usually, isolated enzyme is used, although the Tanabe Co. process uses entrapped whole cells. Other organisms have also been used as the source of penicillin acylase; for instance, *Bacillus megaterium* by Squibb, *Bovista plumbea* by Biochemia, and an *Achromobacter* sp. by the Toyo Jozo Co.

A shift of pH optimum of the enzyme can occur when the enzyme is immobilized onto a charged support. For example, enzymes immobilized to positively charged supports are likely to have their apparent pH optimum shifted to lower pH values owing to repulsion of hydrogen ions from the inside of the support — the local pH is therefore higher than outside the support. This is an advantage in the synthesis of penicillin where the pH optimum for synthesis is 3.6 pH units above the optimum pH for catalytic activity of the enzyme in solution. Immobilized enzymes are highly desirable in the penicillin industry and in other pharmaceutical applications because the enzyme does not contaminate the product if it has been properly attached to the support by covalent bonding. Also, use of a column packed bed reactor of immobilized enzymes helps to take the reaction to completion by removal of product.

In the Astra process, penicillin acylase purified from *Escherichia coli* was immobilized on cyanogen bromide-activated Sephadex G200 Logerlof *et al.* 1976. Supports used by other companies include cellulose triacetate fibres, bentonite plus filter aid and DEAE cellulose. After coupling, 48% of the enzyme activity was recovered on the polymer, which had an activity of about 225 units per g of wet polymer. The enzyme preparation has been used in a batch process since 1973 for the industrial production of 6-APA. In a batch 100 kg of penicillin G was hydrolysed by 16.5 kg of wet Sephadex–enzyme complex (corresponding to 3.7×10^6 units) at 35°C and pH 7.8. The immobilized enzyme could be used for more than one hundred batches without addition of fresh enzyme, provided that the operations were carefully performed. Special care must be taken in the recovery and recharging of the enzyme to avoid losses of activity. This reactor design had some additional advantages such as a higher production capacity and lower costs. Compared to the old method using whole microbial cells in suspension, this new process has resulted in higher yields of a purer product, easier handling of the catalyst, and better economy. In the recirculation process, operated at 37°C and pH 7.8, 6-APA of 98% purity was obtained in 90% yield.

It was concluded that no leakage of protein from the support occurred, and therefore 6-APA and semisynthetic penicillins of hypoallergenic quality could be produced. Normally allergenic proteins have to be removed from the 6-APA in a special purification step.

New side chains are placed on the 6-aminopenicillanic acid nucleus using acid chlorides of suitable grouping or by the use of the amidase in reverse (at pH values just above 7). The latter is not easy in some cases (for example in ampicillin production) owing to difficulty in taking the enzymic conversion towards completion. Nevertheless, the Toyo Jozo Co. Ltd (Japan) has developed a process involving cells of *Bacillus megaterium* or *Achromobacter* sp. adsorbed to DEAE-cellulose for the production of ampicillin (Fuji *et al.* 1973). The penicillin was, produced in 54% yield by passing a solution of 6-APA (0.3%) and D-phenyl-glycine methyl ester (0.09%) through a column containing the immobilized cells of *Achromobacter* sp. at a flow rate of 0.5 column volumes/h. The Otsuka Pharmaceutical Co. Ltd has used succinoylated penicillin acylase adsorbed on DEAE-Sephadex to synthesize ampicillin. To minimize enzyme leakage, the enzyme was succinoylated to introduce a number of carboxylic groups, thereby increasing the binding force between the enzyme and solid support. The enzyme-carrier complex was used in a column of 37°C, and the yield of ampicillin was 67%, which after ten runs had decreased to 58% (Kamogashira *et al.* 1972).

Other potential uses of biocatalysts include the use of cell-free extracts of *Cephalosporium acremoniam* to racemise isopenicillin N to penicillin N (Demain 1981).

3.24 CEPHALOSPORINS

Cephalosporin antibiotics can be formed directly by fermentation or by enzymic deacylation to form the 7-aminocephalosporanic acid nucleus (7-ACA), followed by acylation to introduce the desired side chain. Alternatively, the five membered ring of penicillins can be chemically enlarged to the six membered dihydro-thiazine ring of cephalosporins and the side chain removed enzymically without the need for protective agents for reactive groups.

Typically, the Toyo Jozo Co. have used a deacylating enzyme from *Bacillus megaterium* to produce 7-aminodesaceloxycephalosporonic acid (7-ADCA) from phenylacetyl-7-ADCA (Fuji *et al.* 1976). The enzyme was used adsorbed to celite in a packed column reactor and a 85% yield obtained. A similar enzyme from *Proteus rattgeri* was used in the same manner, and a 90% yield was obtained. Abbot & Berry (1982) have used an enzyme from *Streptomyces capillispira* to de-esterify cephalosporin methyl esters.

In all reactions it is the thermodynamically preferred product(s) which normally accumulates; however, since all reaction are microscopically reversible, the reverse reaction can be carried out by selecting the appropriate conditions. Thus the immobilized *B. megaterium* enzyme could also be used to carry out the reverse reaction, the acylation of 7-ADCA or 7-ACA. Essentially similar procedures have also been developed by other companies, often using enzymes from different microbiological sources and other immobilization supports.

Enzyme action is also fundamentally involved in the mode of action of many drugs, the drugs acting for instance as an enzyme inhibitor. Thus clavulinic acid derived from *Streptomyces clavuligerus* inhibits the β-lactamase enzymes that inactivate penicillins. Thus co-administration of the penicillin and β-lactamase inhibitor makes a more efficient combination; for example, Beechams sell a mixture of potassium clavulanate and amoxycillin under the trade name 'Augmentin'. A similar principle has been followed in the case of another antibiotic potentiator, 7-hydroxytropolone, which inhibits the bacterial enzymes that inactivate aminoglycosides antibiotics such as gentamycin by adding adenyl residues onto the antibiotic molecule. Furthermore, Bayer sell an inhibitor of small-intestinal α-glucosidase called 'Acarbose' which can be used to treat diabetics by delaying carbohydrate reabsorption.

Immobilized enzyme has been used to phosphorylate aminoglycoside antibiotics such as kanamycins (UK Patent 1 485 797, 1977).

Lastly, a β-transcarbamoylase enzyme has been used to transfer a carbamoyl group to a 3-hydroxymethyl cephalosporin to produce the semisynthetic cephalosporin antibiotic 3 carbamoyloxy methyl cephalosporin. The enzyme is derived from *Streptomyces* spp. which are disrupted in a Dynomill and purified before use. Deo & Gaucher (1983) managed to stabilize the formation of the antibiotic patulin by *Penicillium urticae* immobilized in K-carageenan by allowing the immobilized cells to grow and then using them batchwise with nitrogen-free media. The half life was three times greater than that obtained for free cells or cells grown in a free form and then immobilized. For further information on the preparation of pharmaceutical compounds by immobilized enzymes, see Abbot (1976).

3.25 MISCELLANEOUS USES OF BIOCATALYSTS

A variety of applications of biocatalysts exists that are not easily classified under the above headings. These include the enzymic synthesis of esters and lactones from carboxylic acids for use in flavours (Gatfield & Sand 1984), the use of pancreatic trypsin in the extraction of papyrus manuscripts from Egyptian mummy's, the use of di-isopropylphosphofluoridase from squids to detoxify obsolete stocks of nerve gases and the use of haemoglobin entrapped in polyurethane as an 'artificial gill' to extract oxygen directly from sea water. Heparin immobilized onto Sepharose has been used in extracorporeal shunt to prevent blood clots formed during treatment in heart-lung machines and methanol dehydrogenase from *Methyl lobacterium organophilum* has been used to oxidize glycerol to glutaraldehyde for use in the manufacture of cosmetics (Wolf 1982). Laccase, a polyphenol oxidase from *Allemaria tenius* or *Coriolus versicolor* is used to accelerate the drying of traditional Japanese paints (Urushi laquers) and a mixed preparation of β-glucanase, cellulose and pentosanase is included in pelletized barley feed for poultry so as to improve its digestibility.

Glucose oxidase and amyloglucosidase are included in at least one commercially available toothpaste where the hydrogen peroxide produced by their combined activity on starch exerts a bacteriocidal action. Immobilized tannase has been used to solubilize 'tea cream' (Weetall & Detar, 1974; Coggan *et al.* 1975). This is a complex of phenols and caffeine that precipitates when tea is allowed to cool and stand and so is a problem in the manufacture of instant tea products.

Lastly, an interesting application of enzyme membranes in photography is based on the finding that the stable chymotrypsin derivative *cis*-4-nitroconnamoyl-α-chymotrypsin, which is enzymically inactive, undergoes on UV-irradiation a *cis-trans* isomerization with the subsequent formation of a labile *trans*-4-nitroconnamoyl-α-chymotrypsin derivative. The latter is rapidly hydrolyzed to yield an active enzyme which may, in the presence of a substrate, bring about substantial chemical change in the system and thereby intensify the effect of light.

3.26 CONCLUSION

To date enzymes have made a respectable if modest impact in industry. The 1981 estimate for the world market for industrial enzymes was about 65 000 tonnes valued at about £400 million. This is estimated to represent some 10% of the world market for catalysts. Predictions for future markets vary greatly, but are generally optimistic, one reason being that the costs of enzymes are increasing at less than the rate of inflation. In the short term, the use of immobilized lactase to hydrolyse whey to a monosaccharide syrup which is more nutritionally acceptable both to lactose intolerant humans and animals and which is easier to transport and store than the original whey, is a distinct commercial possibility. In the medium term, conversion of the galactose thus produced to the more useful monosaccharide glucose would be advantageous as would the discovery of forms of glucose isomerase which have a lower pH optimum so as to reduce colour formation, is resistant to inhibition by Ca^{2+} ions, and/or which produces a syrup containing a much greater fructose content than is achieved at present. In the long term, a commercially viable method of achieving the enzymic hydrolysis of cellulose to glucose syrups would be an enormously significant breakthrough not only to the food industry but also for the production of fuel ethanol by fermentation. Major advances are most likely to arise from the discovery of new enzymes rather than through modifications of reactor design, although subtle effects may be important. An ability to specify the properties of enzymes through a combination of protein and genetic engineering could have long-term future applications, not least in microelectronics by making possible the development of newer and smaller-scale integrated circuit devices.

Such potential process illustrate the multidisciplinary skills required to bring them to fruition and the need for the correct economic and political climate to enable commercial as well as scientific and technological success to be achieved. The future for the applied use of enzymes appears bright and must remain so if many of our economic and environmental problems are to be solved. The only certainty is that we will continue to be astonished by the imagination, skill, and ingenuity displayed by new advances in the field of enzyme technology.

NOTE IN PROOF

A good indication of the speed with which biocatalysts are being applied in industry and elsewhere is the number and diversity of new emerging applications of biocatalysts that have become apparent during the preparation of this chapter. This is especially true of applications in the chemical industry. Use of biocatalysts in this sector of industry has been slow. This is probably because of the comparatively small value of many products, as compared with pharmaceuticals for instance, and because of the biological nature of most of the reactants and the need to work in non-aqueous solvents. Nevertheless, progress is taking place as is illustrated by the following examples.

Acrylamide
The Nitto Chem. Ind. Co. have introduced an enzyme-based process for the production of acrylamide from acrylonitrile. A 4000 t.p.y. plant is scheduled for completion in 1985. Hitherto acrylamide production has required the use of expensive inorganic catalysts. The enzymic reaction employs an nitrilase immobilized in acrylamide. It is advantageous because 100% yields are obtained when the reaction is carried out at less than 10°C, rather than the 80–140°C used in the chemical process. Thus, energy savings are made. Also, because the product is so pure, it can be polymerized without further purification. The enzyme is derived from *Corynebacterium* or *Nocardia*. The addition of polymerization inhibitors to prevent premature polymerization of the product adversely affected the enzyme. However, treatment with glyoxal or glutaraldehyde reversed this effect (Watanabe *et al.* 1981).

Propylene oxide

Propylene oxide formation from propylene is an important conversion in the chemical industry; De Bont *et al.* (1981) have shown that this reaction can be carried out by the ethylene mon-oxygenase of *Mycobacterium*. Also, Hou *et al.* of Exxon have used the methane mon-oxygenase of methanotrophic bacteria, such as *Methylosinus* sp, immobilized on glass beads for the epoxidation of gaseous propylene. The reaction required the regeneration of $NADH_2$ and the humidification of the propylene. The cells could be successfully regenerated by supplying methanol. This is a rare example of a gas-phase reaction in biotechnology.

In addition, Exxon have used the propane mon-oxygenase of resting cells of newly isolated bacteria grown on propane. Both reactions with propane and propylene appear to be mediated by a single enzyme that can be obtained as a cell-free soluble fraction (Hou *et al.* 1983a, b).

Vinyl chloride

Biocatalysts also have considerable potential for the detoxification of wastes from the chemical industry. For instance a *Mycobacterium* that degrades the carcinogen vinyl chloride has been isolated from an environment close to a PVC producing factory. It has been used in an immobilized form and degraded 99% of the vinyl chloride supplied. This method has considerable potential as existing methods of removing vinyl chloride by combustion are expensive (De Bont *et al.* 1981).

Other notable recent advances have occurred in the following areas:

Biosensors

Recent advances in this technology have involved an extension of the applications of such devices and especially their miniaturization. For instance a semiconductor consisting of a Si_3N_4-gated pH-sensitive field-effect transistor (FET) has been made sensitive to penicillin by depositing a penicillinase-albumin membrane onto it (Matsuoka *et al.* 1984). Another novel development involves the use of *Salmonella typhimurium* revertants, immobilized to an oxygen electrode to measure the potential carcinogenicity of test samples and so carry out a rapid automated AMES mutagenicity test (Karube *et al.* 1979). For a review of this topic see Gronow (1984). The ultimate potential of the synergistic combination of biocatalysts and microelectronics is indicated by explanatory research taking place into the development of computers with biological components.

Amino acids

Another area where rapid advances in biocatalysis have taken place is in amino acid production, stimulated in part by the need for large quantities of aspartic acid and phenylalanine for Aspartame manufacture. Phenylalanine production has been especially stimulated because it has until recently only been available in small quantities and high prices. The Tanabe Seiyaku Co have developed a

method of converting acetamidocinnamic acid into L-phenylalanine in 94%
yield using whole cells of *Alcaligenes faecalis* and *B. sphaericcus* strains selected
from soil sources and sewerage sources (Nakanischi *et al.* 1981 & 1982). The Genex
Corporation use phenylalanine-ammonia lyase to produce L-phenylalanine
from trans-cinnamate and ammonium ions (Swann, 1984). The same company
also have a process for producing L-serine from glycine and formaldehyde using
serine hydroxymethyltransferase which has been cloned into *E. coli* and other
host cells (Anderson *et al.* 1984).

P. S. J. Cheetham
5.12.84

REFERENCES

Abbott, B. J. (1976) *Adv. Appl. Microbiol.* **20** (ed. Perlman, D.) Academic Press, 203–257.
Abbott, B. J. & Berry, D. R. (1982) US Patent 4,316,955 assigned to Eli Lilly & Co.
Ahronowitz, Y. & Cohen, G. (1981) *Sci. Amer.* **245** 141–152.
Allen, D. H. (1980) *Economic Evaluation of Projects,* Inst. Chem. Engs.
Amotz, S., Nielsen, T. K. & Thiesen, N. O. (1976), U.S. Patent 3,980,521 assigned to Novo
　　Ind., A/S, Denmark.
Anderson, D. M., Hsiao, H.-Y., Somerville, R. L. & Herrmann, K. M. (1984) UK Pat. Applic.
　　GB 2 130 216 A.
Archer, M. C., Colton, C. K., Cooney, C. L., Demain, A. L., Wang, D. I. C. & Whitesides,
　　G. M. (1975). In: *Enzyme Technol. Grantees – Users Conf.* (Ed. Pye, E. K.), pp.
　　1–8. Natl. Sci. Found. (USA).
Arima, K., Yu, J., Iwasaka, S. & Tamara, G. (1966) *Appl. Microbiol.* **16** 1727.
Asai, Y., Shimada, M. & Soda, K. (1982), U.S. Patent 4,335,209, assigned to the Mitsui
　　Toatsu Chemicals Inc.
Atkinson, A., Bruton, C. J., Comer, M. J. & Sharp, R. J. (1982) US Patent 4, 342, 827.
Atkinson, T. (1983) *Phil. Trans. Royal Soc. B* **300** 399–410.
Atkinson, A., Hammond, P. M., Price, C. P. & Scawen, M. D. (1980), U.K. Patent Applic.
　　No. 8038634.
Aunstrup, K. (1976), U.S. Patent 3,988,207 assigned to Novo Terapeutisk Lab. A/S, Den-
　　mark.
Aunstrup, K. (1977) In: *Biotechnological applications of proteins and enzymes* (Eds.
　　Bohak, Z. & Sharon, N.) Academic Press, pp. 39–49.
Ball, C., Lawrence, A. J., Butler, J. M. & Morrison, K. B. (1978) *Eur. J. Appl. Micro. &
　　Biotechnol.* **5** 95–102.
Barfoed, H. L. (1981) in *Essays in Appl. Microb.* (ed. Norris, J. R. & Richmond, M. N.)
　　J. Wiley Ltd.
Barker, S. A., Somers, P. J., Epton, R. & McLaren, J. V. (1970) *Carbohyd. Res.* **14** 287–
　　296.
Bauer, D. P. (1981) US Patent 4,262,092.
Beck, C. I. & Scott, D. (1974) In: *Food related enzymes.* (Ed. J. R. Whitaker). Advs. in
　　Chem. Series No. 136, pp. 1–30, Am. Chem. Soc.
Becker, W. & Pfeil, E. (1966) *J. Amer. Chem. Soc.* **88** 4299–4300.
Bellet, P. & van Thuong, T. (1969), U.S. Patent 3, 432, 393.
Bergmeyer, H.-U. (1953) *Biochem. Z.* **324** 408–432.
Berry, E. K. M., Allison, N. & Skinner, A. J. (1979) *J. Gen. Microbiol.* **110** 39–45.
Betz, J. C., Brown, P. R., Smyth, M. J. & Clark, P. H. (1974) *Nature* **247** 261–264.
Boyer, E. W. & Ingle, M. B. (1977), U.S. Patent 4,061,541, assigned to Miles Labs. Inc.
　　USA.
Branner-Jorgensen, S. US Patent 4,386,160.
de Bont, J. A. M., Tramper, J. & Luyten, K. Ch. A. M. (1981) *Abs. Commun.* 2nd Eur.
　　Cong. Biotechnol. S.C.I. pp. 158.
Brodelius, P. (1978) *Advs. Biochem. Eng.* **10** 75–130.
Bruton, C. J. (1983) *Phil. Trans. Royal Soc. B* **300** 249–260.
Bucke, C. (1983) In: *Microb. Enz. in Biotechnol* (Ed. Fogarty, W. M.) Appl. Sci., pp. 93–
　　129.
Bucke, C. (1977) In: *Topics in Enzyme and Fermentation Biotechnology* (Ed. Wiseman,
　　A.), **1** Ellis Horwood, pp. 147–168.

Bucke, C. & Wiseman, A. (1981) *Chem. and Ind.* 7 234–240.
Buckland, B. C., Richmond, W., Dunnill, P. & Lilly, M. D. (1974) In: *Industrial aspects of Biochemistry* (Ed. Spencer, B.) FEBS pp. 65–79.
Bull, A. T., Holt, G. & Lilly, M. D. (1982) *Biotechnol. Int. Trends and Perspectives,* OECD.
Chaiken, I. M., Komoriya, A., Ohno, M. & Widmer, F. (1982) *Appli. Biochem. & Biotechnol.* 7 385–399.
Chandler, V. V. & Nicol, K. J. (1975) *CSIRO Food Res.* Q. 35 79–88.
Charney, W. & Herzog, H. L. (1967) *Microbial Trans. of Steroids,* Academic Press, 728 pp.
Cheetham, P. S. J. (1979) *Anal. Biochem.* 92 447–452.
Cheetham, P. S. J., Imber, C. E. & Isherwood, J. (1982) *Nature* 299, 628–631.
Cheryan, M., van Wyk, P. J., Olson, N. F. & Richardson (1975) *Biotechnol. Bioeng.* 17 585–598.
Chibata, I., Tosa, T., Sato, T. & Yamamoto, K. (1975) US Patent 3,886,040.
Chibata, I., Tosa, T., Sato, T. & Yamamoto, K. (1975). Japanese Patent Kokai 100, 289/75.
Chibata, I., Tosa, T. & Sato, T. (1976). In: *Methods in Enzymology* (Ed. Mosbach, K.) Academic Press 44, 739–746.
Chibata, I., Tosa, T., Sato, T. & Mori, T. (1976). (n: *Methods in Enzymology.* (Ed. Mosbach, K.) 44 Academic Press, pp. 746–759.
Chiulli, A. J. & Wegman, E. H. (1974). U.S. Patent 3,821, 364, assigned to Adv. Biofactures Corp, USA.
Chirikjian, J. G. (1982) US Patent 4,342,833 assigned to Betlesda Res. Labs.
Christie, A. A., Wood, C., Elsworth, R., Herbert, D. & Sargeant, K. (1974), U.S. Patent 3,843,445, assigned to Sec. State Soc. Services (GB).
Christie, R. B. (1980) *Topics in Enzyme and Fermentation Biotechnology* 4 25–83.
Clarke, L. C. Jr. & Lyons, C. (1962) *Ann. N.Y. Acad. Sci.* 102 29–45.
Coggan, P. & Sanderson, G. W. (1975), U.K. Patent 1 413 351.
Clayton, D. W. Jurasek, L., Paice, M. G. & O'Leary, S. (1984) *Proc. Biotech '84,* pp. 623–636.
Cohen, M. B., Spolter, L., Chang, C. C., MacDonald, N. S., Takahashi, J. & Bobinet, O. D. (1974) *J. Nucl. Med.* 15 1192–1195.
Daniels, M. J. & Farmer, D. M. (1981). U.K. Patent Applic. GB. 2 070 022 B.
Darbyshire, J. (1981). In: *Topics in Enzyme and Fermentation Biotechnology* 5 (Ed. Wiseman, A.), pp. 147–183.
Davino, A. A. (1981) US Patent 4,293,648.
De Fines, J. (1969). In: *Fermentation Advances* (Ed. Perlman, D.), Academic Press, p. 385.
Demain, A. L. (1971) *Adv. Biochem. Eng.* 1 113–118.
Demain, A. L., Kornomi, T., Baldwin, J. E. (1981) US Patent 4,307,192.
Denburg, J. & Deluca, M. (1970) *Proc. Natl. Acad. Sci* 67 1057–1062.
Denner, W. H. B. (1983) in *Ind. Enzymol.* (Ed. Godfrey, A. & Reichelt, J.) Macmillan Publishers Ltd. pp. 111–169.
Deo, Y. M. & Gaucher, G. M. (1983) *Biotechnol. Lett.* 5 125–130.
Deus, B. & Zenk, M. H. (1982) *Biotechnol. Bioeng.* 24 1965–1974.
Dubourdieu, D., Villetta, J. C., Desplanques, C. & Ribereau-Gayon, P. (1981) *Connaissance de la Vigne et du vin.*
Duffy, J. I. (1980) *Chemicals by enzymatic and microbial processes.* Noyes Data Corp., New Jersey, USA.
Dunnill, P. (1981) *Chem and Ind.* April 204–247.
Eder, U., Sauer, G. & Wiechert, R. (1971) *Angew Chem. Int. Edn.* 10 496.
Eggeling, L., Paschke, M. & Sohm, H. (1981) US Patent 4,250,261.
Emken, E. A. (1978) *J. Amer. Oil Chem. Soc.* 55 416–421.
Enari, T-M. (1983). In: *Microb. Enzs. and Biotechnol.* Appl. Sci. publishers, pp. 183–223.
Esders, T. W., Goodhue, C. T. & Schubert, R. M. (1979). U.S. Patent 4,166,763, assigned to Eastman Kodak Co. USA.
Everse, J., Everse, K. E. & Merrer, L. C. (1981) US Patent 4,305,926.
Falk, L. C., Clark, D. R., Kalwan, S. M. & Long, T. F. (1976) *Clin. Chem.* 22 785.
Fehmerling, G. B. (1970). U.S. Patent 3513071.
Fogarty, W. M. (1983) in *Microb. Enzs. & Biotechnol.* Appl. Sci. Pub. pp. 1–92.
Fogarty, W. M. & Kelly, C. T. (1983) *ibid,* pp. 131–182.
Fritz, H., Gebhardt, M., Fink, E., Schramm, W. & Werle, E. (1969) *H.S.Z. Physiol. Chem.* 350 129–138.
Fritz, H., Gebhardt, M. & Meister, R. (1970) *H.S.Z. Physiol. Chem.* 351 571–574.
Fukui, S., Ikeda, S., Fujimura, M., Yamada, H. & Kumagai, H. (1975) *Eur. J. Biochem.* 51 155–164.
Fuji, T., Hanamitsu, K., Izumi, R. & Watanabe, T. (1973) Jap. Kokai 73,99393.

Fuji, T., Matsamoto, K. & Watanabe, T. (1976) *Proc. Biochem.* **11(8)**, 21.
Gassen, H. G. & Nolte, R. (1971) *Biochem. Biophys. Res. Common.* **44** 1410–1415.
Galbraith, W. (1981) US Patent 4,286,064.
Gerisch, G. (1974). U.S. Patent 3,810,822. Assigned to Max-Planck-Gessellschaft zur Forde-
    rung der Wissenschaften e.V., Germany.
Gestrelius, J. M. & Kjaer, J. H. (1983). U.S. Patent 4,380,552.
Givol, D., Weinstein, Y., Gorecki, M. & Wilchek, M. (1970). *Biochim. Biophys. Res. Comm.*
    38 825–830.
Godfrey, A. & Reschelt, J. (1983) *Ind. Enzymol.* Macmillan, London.
Grabner, R. W. (1975). U.S. Patent 3,897,309 assigned to Merck & Co. Inc. USA.
Gratzner, H. (1972) *J. Bacteriol.* **111** 443–446.
Gronow, M. (1984) *Trends in Biol. Sci.* Aug. 336–340.
Greenstein, J. P. (1954) *Adv. Prot. Chem.* **9** 122–202.
Gregoriadis, G. (1976). In: *Methods in Enzymology* **44** (Ed. Mosbach, K.) Academic Press
    698–708.
Guilbault, G. G. (1971) *Anal. Chim. Acta.* **56** 285–290.
Guilbault, G. G. (1980) *Enz. & Microb. Technol.* **2** 258–264.
Guilbault, G. G. & Montalvo, J. G. (1970). *J. Amer. Chem. Soc.* **92** 2533–2538.
Guilbault, G. G. & Nagy, G. (1973) *Anal. Letters* **6** 301–312.
Guilbault, G. G. & Stokbrow, W. (1975) *Anal Cheim. Acta* **76**, 237.
Gutcho, S. J. (1974) *Microbial Enzyme Production,* Noyes Data Corp., New Jersey, USA.
Hackel, U., Klein, J., Megnet, R. & Warner, F. (1975) *Eur. J. Applied Microbiol.* **1** 291–294.
Hahn, P. A., Hartdegen, F. J. & Espenshade, M. R. (1975). U.S. Patent 3,898,131, assigned
    to W. R. Grace & Co., USA.
Hajos, Z. G. & Parrish, D. R. (1974) *J. Org. Chem.* **39** 1615.
Halpern, M. G. (1981) *Industrial Enzymes from Microbial sources,* Noyes Data Corp. New
    Jersey, USA.
Hasegawa, S., Patel, M. N. & Snyder, R. C. (1982) *J. Agri. & Food Chem.* **30** 509–511.
Hasegawa, S. & Pelton, U. A. (1983) *J. Agric. & Food Chem.* **31** 178–180.
Haas, L. W. & Bohm, R. M. (1934) US Patents 1,957,333–337.
Heady, R. E, (1980) U.S. Patent 4,335,207 assigned to CPC International Inc.
Hicks, G. P. & Updike, S. J. (1966) *Anal. Chem.* **38** 726.
Higgins, J. J., Lewis, D. J., Daly, W. H., Mosquita, F. G., Dunnill, P. & Lilly, M. D. (1978)
    Biotechnol. Bioeng. **20** 159–182.
HMSO (1982) *Report on the review of enzyme preparations* by Food Additives and Con-
    taminants Committee of MAFF, HMSO.
Holfman, C. H., Harris, E., Chodroff, S., Micholson, S., Rothrock, J. W., Peterson, E. &
    Reuter, W. (1970) *Biochem. Biophys. Res. Commun.* **41** 710–714.
Hollo, J., Laszlo, E. & Hoschke, A. (1983) *Starch* **35** 169–175.
Homanberg, G. A., Komoriya, A. & Chaiken, I. M. (1982) *Biochem.* **21** 3385.
Horton, R. E. & Swaisgood, H. E. (1976). In: *Methods in Enzymol.* **44** (Ed. Mosbach, K.).
    Academic Press, pp. 517–526.
Hou, C. T., Patel, R., Laskin, A. I., Barnabe, N. & Barrist, I. (1983a) *Appl. Environ. Micro-
    biol.* **46** 171–177.
Hou, C. T., Patel, R., Laskin, A. I., Barnabe, N. & Barrist, I. (1983b) *Appl. Environ. Micro-
    biol.* **46** 178–184.
Huang, H. T. & Niemann, C. (1951) *J. Amer. Chem. Soc.* **73** 475–476.
Hung, P. P., Lee, S.-G., Roychoudhury, R., Ratzkin, B. J., Schrenck, W. J. & Chen, M. C.
    (1983) US Patent 4,370,417.
Hurst, T. L. (1975). U.S. Patent 3,897,305.
Isaac, O. (1977) US Patent 4,004,976.
Imamura, S., Matsumoto, T., Muto, N. & Misobi, H. (1982) US Patent 4,397,323.
Ishu, S., Kiho, K., Sugiyama, S. & Sugimoto, H. (1980). U.S. Patent 4,200,694 assigned to
    Kikkoman Shoyu Co. Ltd, Japan.
Isowa,, Y., Ohmori, M., Ichikawa, T., Kurita, M., Sato, M. & Mari, K. (1977) *Bull. Chem.
    Soc. Japan* **50** 2762.
Isowa, Y., Ohmori, M., Mori, K., Ichikawa, T., Nonaka, Y., Kihara, K., Oyama, K., Satoh, H.
    & Nishimura, S. (1979). U.S. Patent 4,165,311.
Jones, J. B. (1980) in *Enz. & Non-Enz. Cat.* (Ed. Dunnill, P. *et al.*) Ellis Horwood Ltd.
    pp. 55–81).
Jones, J. B. & Beck, J. F. (1976) *Tech. Chem.* **10** 107–401.
Jowitt, R. (ed.) (1980) *Hygienic Design and Operation of Food Plant.* Ellis Horwood Ltd.
Kainama, K., Kobayashi, S., Ito, T. & Suzuki, S. (1972) *FEBS Lett.* **26** 281–285.
Kamogashira, T., Kawaguchi, T., Miyazaki, W. & Doi, T. (1972) *Japan Kokai* **72** 28190.

Karube, I., Mitsuda, S. & Suzuki, S. (1979) *Eur. J. Microbiol. & Biotechnol* **7** 343–350.
Kato, T. & Homkoshi, K. (1984) *Biotechnol & Bioengin.* **26**, 595–598.
Kennedy, J. F. & White, C. A. (1979) *Starch* **31** 93–99.
Keyes, M. H. (1980). U.S. Patent 4,194,067 assigned to Technicon Instruments Corp, USA.
Kikuchi, M., Sakaguchi, K. & Nakano, E. (1974). U.S. Patent 3,796,633 assigned to Kikkoman Shoyu Co. Ltd, Japan.
Kirk, T. K. & Yang, M. H. (1979) *Biotech. letts.* 1 *347–352.*
Klibanov, A. M. & Morris, E. D. (1981) *Enz. Microbiol. Technol.* **3** 119–122.
Knorre, D. G., Melamed, N. U., Starostina, U. K. & Shubina, T. N. (1973) *Biokhima* **38** 121–123.
Koaze, Y., Nakajima, Y., Hidaka, H., Niwa, T., Adachi, K., Yoshida, J., Niida, T., Shomura, T. & Ueda, M. (1975). U.S. Patent 3,868, 464 assigned to Meiji Seika Kaisha Ltd, Japan.
Komoriya, A., Homanberg, G. A. & Chaiken, I. M. (1980) *Int. J. Peptide Protein Res.* **16** 433.
Kon, S. & Ohlson, A. C. (1973) *J. Food Sci.* **38** 215–217.
Korosi, M. (1976) US Patent 3,982,932.
Kulp, K. (1975). In: *Enzymes in Food Processing* (Ed. Reed, G.) 2nd edn. Academic Press, pp. 53–122.
Kusakabe, H., Kodama, K., Midorikawa, Y., Machida, H., Kuninaka, A. & Yoshino, H. (1976). U.S. Patent 3,930,955, assigned to Yamasa Shoyu KK, Japan.
Laboureur, P. & Villanon, M. (1972). U.S. Patent 3634191.
Lagerlof, E., Nathorst-Westfelt, L., Ekstrom, B. & Sjoberg, B. (1976). In: *Methods in Enzymol* Ed. Mosbach, K.) **44** Academic Press pp. 759–768.
Lambert, P. W. & Meers, J. L. (1983) *Phil. Trans; Royal Soc. B* **300** 263–281.
Law, B. A. & Wigmore, A. S. (1982) *J. Soc. Dairy Tech.* **35** 75–79.
Lee, D. D., Lee, Y. Y., Reilly, P. J., Collins, Jr., E. U. & Tsao, G. T. (1976) *Biotechnol. Bioeng.* **18** 253–267.
Liener, I. E. (1977) In: *Food improvements through chemical and enzymatic modification* (Feeney, R. E. & Whitaker, J. R. eds.), Advs. in Chem Series 160 pp. 283–300 Am. Chem. Soc.
Linden, J. C. (1982) *Enz. & Microb. Technol.* **4** 130–136.
Linko, Y., Saarinen, P. & Linko, M. (1975) *Biotechnol. Bioeng.* **17** 153–165.
Linko, Y-Y., Kautola, H. & Linko, P. (1981) *Advs. in Biotech.* (ed. Moo-Young, M.), Pergamon Press, 685–690.
Livernoch, D., Jurasek, L., Desrochers, M. & Veliky, I. A. (1981) *Biotech. letts.* **3** 701–706.
Macrae, A. R. (1983). In: *Microb. Enzs & Biotech.* Appl. Sci. publishers, pp. 225–250.
Maeda, H. & Tsao, G. T. (1979) *Process Biochem.* July, 2–5.
Marconi, W., Gulinelli, S. & Morisi, F. (1974) *Biotechnol. Bioeng.* **16** 501–511.
Marshall, R. O. & Kooi, E. T. (1957) *Science* **125** 648–649.
Mårtensson, K. (1974) *Biotechnol. Bioeng.* **16** 579–591.
Matsunaga, T., Karube, I. & Suzuki, S. (1980) *Biotechnol. Bioeng.* **22** 2607–2615.
Mattiasson, B., Mosbach, K. & Svenson, A. (1977) *Biotechnol. Bioeng.* **19** 1643–1651.
Mattiasson, B., Ramstorp, M., Nilsson, I. & Hahn-Hagerdal, B. (1981) *Biotech. letts.* **3** 561–566.
Matiasson. B. & Larsson, M. (1984) *Proc. Biotech. '84.* Online Pubs. pp. 683–691.
Mauer, A. M. & Simone, J. (1976) *Cancer Treat. Rev.* **3** 17–41.
Maul, S. B., Sinskey, A. J. & Tannenbaun, S. R. (1970) *Nature* **228** 181.
May, S. W. & Padgette, S. R. (1983) *Biotechnology Oct.* 677–686.
Mazur, R. H., Schlatter, J. M. & Goldkamp, A. H. (1969) *J. Amer. Chem. Soc.* **91** 2684.
Middlehoven, W. J. & Bakker, C. M. (1982) *Eur. J. Appl. Micro. Biotechnol.* **15** 214–217.
Miura, Y., Takamatsa, N. & Miyamoto, K. (1975), Jap. Patent 75148588.
Miura, Y., Yutani, K. & Izumi, Y. (1977). U.S. Patent 4,101,072 assigned to Tokuyama Soda K.K., Japan.
Misaki, H., Horiuchi, Y., Maluura, K. & Haroda, S. (1981) US Patent 4,275,161.
Mollenhauer, H. (1956) *Food Manufacture* **47** (Feb.).
Monsheimer, R. & Pfleiderer, E. (1981) US Patent 4,293,647.
Monti, J. C. & Jost, R. (1978) *Biotechnol. Bioeng.* **20** 1173–1185.
Morihara, K., Oka, T. & Tsuzuki, H. (1979) *Nature* **280** 5721.
Morikawa, U., Karube, I. & Suzuki, S. (1980) *Biotechnol. Bioeng.* **21** 261–270, and *ibid* (1980) **22** 1015–1023.
Munnecke, D. M. (1977) *Appl. & Envir. Microbiol.* **33** 503–507.
Munnecke, D. M. (1978) *Proc. Biochem.* **13** 14–17.
Nakanishi, T. & Machida, Y. (1981) US Patent 4,245,050.

Nakanishi, T. & Shigemasa, Y. (1981) US Patent 4,273,874.
Nakanishi, T. & Machida, Y. (1982) US Patent 4,341,868.
Neidelman, S. L., Amon, W. F. & Geigert, J. (1979). Eur. Patent Applic. 0 007 176.
Neidelman, S. L. & Geigert, J. (1983) Trends in Biotechnol. 1 21–25.
Nazaly, N. & Knowles, C. J. (1981) Biotechnol. lett. 3 363–368.
Nielsen. M. H. (1980). In: 13th Int. TNO Conf. Rotterdam, pp. 41–58.
Nielsen, G. C., Diers, I. V., Outtrup, N. & Norman, B. E. (1982). U.K. Patent Applic.
    GB 2 097 405 A, assigned to Novo Industri A/S.
Nielsen, G. C., Diers, V., Outtrup, H. & Norman, B. E. (1982) UK Pat. Applic. GB 2097
    405 A.
Nilsson, I., Ohlson, S., Haggstrom, L., Molin, N. & Mosbach, K. (1980) Eur. J. Appl. Micro-
    biol. & Biotech. 10 261–274.
Nobile, A. W., Charney, P. C., Perlman, H. L., Herzog, C. C., Payne, M. E., Tulley, M. E.,
    Jevnik, M. A. & Hershberg, E. B. (1955) J. Am. Chem. Soc. 77 4184.
Norman, B. E. (1981). In: Enzyme and Food Processing (Ed. Birch, G. G., Blakeborough, N.
    & Parker, K. J.), Applied Science Publishers pp.15–50.
Obara, J., Hashimoto, S. & Suzuki, H. (1976/7) Sugar Technol. Rev. 4 209.
Olson, N. F. & Richardson, T. (1974) J. Food Sci. 39 653–659.
Oyama, K., Kihara, K. I. & Nonaka, Y. (1981) J. Chem. Soc. Perkin 11 356–360.
Pastore, M., Morisi, F., Zaccardelli, D. (1974). In: Immobilized Enzymes (Eds. Salmona, M.,
    Saronio, C. & Garattini, S.), Raven Press, pp. 211–216.
Peterson, D. H. & Murray, H. C. (1952) J. Am. Chem. Soc. 74 1871–1872.
Pfisterer, E. & Wagner, H. (1975) J. Inst. Brew. 81 277–280.
Pilnik, W. & Romboutz, F. M. (1981) in Enzs. & Food Prod. Ed. Birch, G. G., Blakeborough,
    N. & Parker, K. J. Appl. Sci. Pub. pp. 105–128.
Pitcher, W. H., Ford, J. R. & Weetall, H. H. (1976). In: Methods in Enzymol. (Ed. Mos-
    bach, K.) 44 Academic Press 792–809.
Poulsen, P. B. & Zitton, L. (1976). In: Methods in Enzymol. 44 (Ed. Mosbach, K.) Aca-
    demic Press, 809–821.
Poulsen, P. B., Rugh, S. & Norman, B. E. (1980). U.S. Patent 4,206,285.
Prescott, S. C. & Dunn, C. G. (1959) Industrial Microbiology. McGraw-Hill.
Price, C. P. (1983) Phil. Trans. Royal Soc. B 300 411–422.
Priest, F. G. (1983). In: Microb. Enzs. & Biotechnol. Appl. Sci. publishers, pp. 319–366.
Ramaley, R. F. (1979). In: Advs. in Applied Microbiol. 25 37–55.
Reitz, B. & Guilboult, G. G. (1975) Anal. Chim. Acta 58 75.
Robyt, J. F. & Ackerman, R. J. (1971) Arch. Biochem. Biophys. 145 105–114.
Röder, A., Beaucamp, K., Weimann, G., Schreider, W. & Mühlegger, K. (1976). U.S. Patent
    3,951,744, assigned to Boehringer Mannheim GmbH, Germany.
Röder, A., Mollering, H., Graber, W., Beaucamp, K., Seidel, H., Stahl, P. & Hoerschelmann,
    D. von (1980). U.S. Patent 4,186,052, assigned to Boehringer Mannheim GmbH,
    Germany.
Roland, J. F. (1981) Enz. & Microbiol. Technol. 3 105–110.
Rose, A. H. (1961) Ind. Microbiol. Butterworths, pp. 264.
Rubenstein, K. E., Schneider, R. S. & Ullman, E. F. (1972) Biochem. Biophys. Res. Comm.
    47 846–851.
Ryu, D. D. Y. & Mandels, M. (1980) Enz. & Microb. Technol. 2 91–102.
Saimaru, H., Izumi, C., Narita, S. & Yamada, M. (1975), Ger. Offen. 2518280.
Saito, Y., Higachi, M. & Kobayashi, T. (1972) Arch. Biochem. Biophys. 153 150–187.
Sakaguichi, K. Uemeru, T. & Kinoshita, S. (1971). In: Biochemical & Industrial Aspects of
    Fermentation, Chapter 4, Kodansha, Tokyo.
Sakano, Y., Higachi, M. & Kobayashi, S. (1967) J. Ferm. Tech. 45 497–503.
Satoh, I., Karube, I. & Suzuki, S. (1976) Biotech. Bioeng. 18, 269–272.
Sato, T., Mori, T., Tosa, T., Chibata, I., Kurui, M., Yamashita, K. & Sumi, A. (1975) Bio-
    technol. Bioeng. 17 1797–1804.
Scheller, F. W., Schubert, F., Renneberg, R. & Kirstein, D. (1984) Proc. Biotech. '84 Online
    Publics. 367–378.
Scott, D. (1975). In: Enzymes in Food Processing (Ed. Reed, G.), 2nd edn., pp. 493–517.
Sealock, R. W. & Laskowski, Jr. M. (1969) Biochem. 8 3703.
Senn, D. R., Carr, P. W. & Klatt, L. N. (1976) Anal. Chem. 48 954–958.
Shimuzu, S., Morioka, H., Tani, Y. & Ogata, K. (1975) J. Ferm. Technol. 53 77–83.
Shinkarenko, L. N., Babenko, J. S. & Grigoriev, E. F. (1976). U.S. Patent 3,963,577.
Shiosaka, M. (1976). U.S. Patent 3,988,206, assigned to Hayashibara Biochem. Labs. Inc.
    Japan.
Sizer, I. W. (1972). Adv. Appl. Microbiol. 15 1–11.

Slater, H. J., Lovatt, D., Weightman, A. J., Senior, E. & Bull, A. T. (1979) *J. Gen. Microbiol.* **114** 125–136.
Snoke, R. E., Risley, H. A. & Goodhue, C. T. (1977). U.S. Patent 4,062,731, assigned to Eastman Kodak Co., USA.
Steere, N. V. (ed.) (1971) *CRC Handbook of Laboratory Safety* CRC Press.
Stone, J. V. & Townshend, A. (1973) *J. Chem. Soc.* **5** 495–501.
Storelli, G., Vogeli, H. & Semenza, G. (1972) *FEBS lett.* **24** 287–292.
Sugiura, M., Shimizu, H., Sugiyama, M., Kuratsu, T. & Hirata, F. (1977). U.S. Patent 4,003, 794 assigned to Ono Pharmaceutical Co., Japan.
Swaisgood, H. E. (1977) U.S. Patent 4,053,644.
Swaisgood, H. E. (1978) U.S. Patent 4,087,328.
Swaisgood, H. E. (1980) *Enz. & Microbiol. Technol.* **2** 265–272.
Swaisgood, H. E., Horton, R. E. & Mosbach, K. (1976). In: *Methods in Enzymol.* **44** (Ed. Mosbach, K.), pp. 504–516.
Swann, W. E. (1984) UK Patent Applic. GB 2 127 821 A.
Szalay, A. A., Mackay, C. J. & Langridge, W. H. R. (1979) *Enz. & Microbiol. Technol.* **1** 153–224.
Takada, N. & Misaki, H. (1982) US Patent 4,357,425.
Takamatsu, S., Umemura, I., Yamamoto, K., Sato, T., Tosa, T. & Chibata, I. (1982) *Eur. J. Appl. Microbiol. Biotechnol.* **15** 147–152.
Takasaki, Y. & Takahara, Y. (1974) *Japan Kokai* **76** 70875.
Takasaki, S., Itoh, M. & Kaneko, Y. (1981) *Eur. J. Appl. Microbiol. & Biotech.* **13** 175–178.
Terada, O. & Uwajima (1977). U.S. Patent 4,011,138 assigned to Kyowa Hakko Kogyo Co. Ltd, Japan.
Terado, O. & Aisaka, K. (1982) US Patent 4,335,213.
Terui, G., Okazaki, M. & Kinoshita, S. (1967) *J. Ferm. Tech.* **45** 497–503.
Thompson, J. W., Shovers, J. Sandine, W. E. & Elliker, P-R. (1970). *Appl. Microbiol.* **19** 883–889.
Thompson, K. N., Johnson, R. A. & Lloyd, N. E. (1974). U.S. Patent 3788945.
Townshend, A. (1973) *Process Biochem.* **8** 22–24.
van Beynum, G. M. A., Roels, J. A. & van Tilburg, R. (1980) *Biotechnol. Bioeng.* **22** 647–649.
Vandamme, E. J. (1983) *Enz. & Microb. Technol.* **5** 403–416.
Wallerstein, L. (1911). U.S. Patents 995,820 and 995, 823-6.
Walon, R. (1980). U.S. Patent 4,217.413.
Wandrey, C., Wichmann, R., Leuchtenberger, W., Kula,. M-R. & Buckmann, A. (1981). U.S. Patent 4,304,858.
Wang, D. I. C., Cooney, C. L., Demain, A. L., Dunnill, P., Humphrey, A. E. & Lilly, M. D. (1979) *Fermentation and Enzyme Technology*, J. Wiley & Sons.
Ward. O. P. (1983). In: *Microb. Enzs. & Biotechnol.* Appl. Sci. publishers, 251–317.
Watanabe, I., Satoh, Y. & Takano, T. (1981) US Patent 4,248,968, assigned to the Nitto Chem. Ind. Co.
Waterburg, J. B., Calloway, C. B. & Turner, R. D. (1983) *Science* **221** 1401.
Weetall, H. H. & Detar, C. C. (1974) *Biotechnol. Bioeng.* **16** 1095–1102.
Weil, T. A., Dzadzic, P. M., Shih, C.-CJ & Price, M. L. (1982) US Patent 4,338,399.
Whitaker, J. R. (1977). In: *Food Proteins: Improvements through chemical enzymatic modification.* A.C.S. Series 160, pp. 95–155. Am. Chem. Soc.
Wikstöm, P. Szajcer, E., Brodelius, P., Nilsson, K. & Mosbach, K. (1982) *Biotech. lett.* **4** 153–158.
Winter, G., Fersht, A. R., Wilkinson, A. J., Zoller, M. & Smith, M. (1982) *Nature* **299** 756–758.
Wolf, H. J. (1982) US Patent 4,535,987, assigned to Upjohn Co.
Wong, C-H., Gordon, J., Cooney, C. C. & Whitesides, G. M. (1981) *J. Org. Chem.* **46** 4676–4679.
Wong, C-H., Haynie, S. C. & Whitesides, G. M. (1982) *J. Org. Chem.* **47** 5418–5420.
Woodward, J. B. & Bennet, A. B. (1973). U.K. Pat. Spec. 1,421,955.
Yamada, H., Shimizu, S. & Tani, Y. (1981) US Patent 4,269,942.
Yamamoto, K., Sato, T., Tosa, T. & Chibata, I. (1974) *Biotechnol. Bioeng.* **16** 1601–1610.
Yamamoto, K., Tosa, T., Yamashita, K. & Chibata, I. (1976) *Eur. J. Appl. Microbiol.* **3** 169–183.
Yamamoto, K., Tosa, T., Yamashita, K. & Chibata, I. (1977) *Biotechnol. Bioeng.* **19** 1101–1114.
Yamashita, M., Arai, S., Kokubo, S., Aso, K. & Fujimaki, M. (1975) *J. Agric. Food Chem.*

23 27–30.
Yamashita, M., Arai, S. & Fujimaki, M. (1976) *J. Food Sci.* **41** 1029–1032.
Yoshimura, Y., Yokogawa, K. & Kawata, S. (1975). U.S. Patent 3,929,579 assigned to Dainippon Pharm. Co. Ltd, Japan.
Zingard, R. A. & Uziel, M. (1970) *Biochim. Biophys. Acta* **213** 371–379.

Additional information on the topics described in these chapters can be found in the following books:

Industrial Enzymology (Ed. Godfrey, T. & Reichelt, J.) Macmillan Publishers Ltd. (1983).
Characterization of Immobilized Biocatalysts (Ed. Buchoolz, K.) DECHEMA monograph vol. 84 (1979).
Principles of Biotechnology (Ed. Wiseman, A.) Surrey Univ. Press (1983).
Methods in Enzymology (Ed. Mosbach, K.) Academic Press, Vol. 44, (1976).
Fermentation & Enz. Technol. (Ed. Wang, D. I. C., *et al.*), J. Wiley Ltd. (1979).

and also in the many journals and series of monographs such as *Biotechnol. Bioengin.* (Wiley Interscience), *Enz. & Microb. Technol.* (IPC Magazines), *Advs. in Biochem. Engin.* (Springer Verlag), *Topics in Enzyme and Fermentation Biotechnology* (Ellis Horwood Ltd).

ACKNOWLEDGEMENTS

The author wishes to gratefully acknowledge the cooperation of the following publishers for granting permission for the reproduction of figures and tables in this chapter: Academic Press, The Biochemical Society, Butterworth & Company, Ellis Horwood, Macmillan Journals, Plenum Publishing Corp., Springer-Verlag, Surrey University Press (Blackie & Son Ltd.), Wheatland Journals, J. Wiley & Sons.

CHAPTER 4

Data on techniques of enzyme immobilization and bioaffinity procedures

Professor J. F. KENNEDY, Research Laboratory for the Chemistry of Bioactive Carbohydrate and Proteins, Department of Chemistry, University of Birmingham B15 2TT, England and The North East Wales Institute, Deeside, Clwyd CH5 4 BR, Wales and
Dr. C. A. WHITE, Vincent Kennedy Ltd, 47 Conchar Road, Sutton Coldfield, B72 1LL, England

4.1 INTRODUCTION

In this chapter attention is focussed on the chemical reactions involved, since primary losses of enzyme activity frequently occur by reaction at the active site or binding site of the enzyme. Secondary losses occur because of changes in the conformation of the enzyme during immobilization, and this is related to:

(a) the points of attachment in the enzyme,
(b) the new microenvironment on the surface of the support.

4.2 ENTRAPMENT

4.2.1 Gel entrapment

In Part A, the use of polymer gels to entrap enzymes was described, with the major gel being polyacrylamide which is obtained by the polymerization of acrylamide with a crosslinking agent such as N,N-methylenebis(acrylamide). This process requires an initiator such as potassium persulphate ($K_2S_2O_5$) or riboflavin and is catalyzed by 4-dimethylaminopropionitrile (DMAPN) or N,N,N',N'-tetramethylethylene diamine (TEMED) and results in a 3-dimensional structure as shown in Fig. 4.1. It is essential to use temperatures below 25°C and to carry out the reaction in the absence of oxygen to prevent thermal denaturation of the enzyme molecules, but as stated previously the generation of free radicals during the reaction can cause degradation of the enzyme.

When a gel is formed from a naturally occurring polymer such as gelatin, κ-carragenan, alginate, agar, etc., the polymer is dissolved in aqueous media and mixed into the enzyme solution. This enzyme—polymer mixture is then gelled by, in the case of gelatin, mixing with a solvent which is immiscible with water (frequently butanol) (van Velzen 1974), or by, in the case of κ-carragenan, cooling the solution in the presence of aqueous potassium chloride solution (Tosa *et al.* 1979). In the case of alginate gels, the aqueous solution of enzyme and sodium alginate is mixed with an aqueous solution of calcium chloride which results in the precipitation, in gel form, of calcium alginate-enzyme complex (Schovers & Sandine 1973). When organic solvents or high salt concentrations are used to form the gel network they can, in some cases, cause denaturation of the enzyme through permanent distortion of the enzyme's 3-dimensional tertiary structure which is important for enzyme activity.

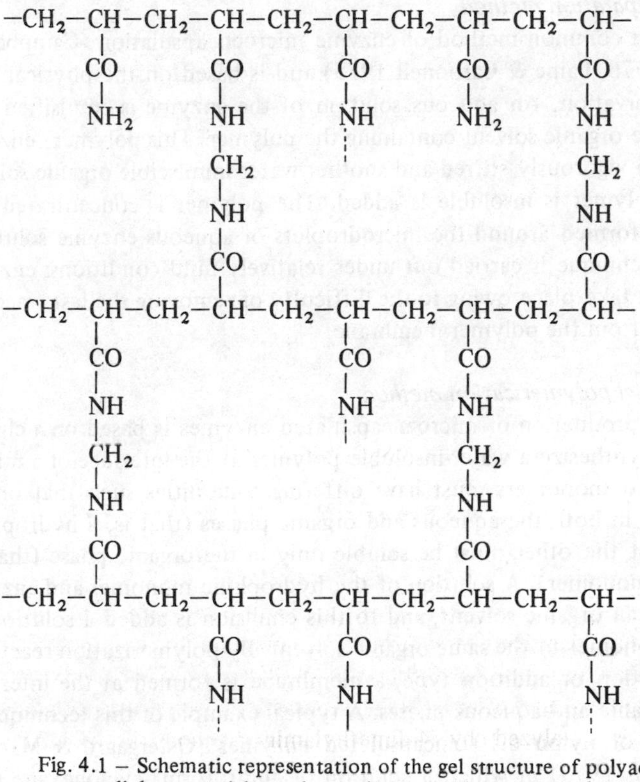

$$-CH_2-CH-CH_2-CH-CH_2-CH-CH_2-CH-CH_2-CH-$$

$$\begin{array}{ccccc}
| & | & | & | & | \\
CO & CO & CO & CO & CO \\
| & | & | & | & | \\
NH_2 & NH & NH & NH_2 & NH \\
 & | & | & & | \\
 & CH_2 & & & CH_2 \\
 & | & & & | \\
 & NH & & & NH \\
 & | & & & | \\
 & CO & & & CO
\end{array}$$

$$-CH_2-CH-CH_2-CH-CH_2-CH-CH_2-CH-CH_2-CH-$$

$$\begin{array}{ccc}
| & | & | \\
CO & CO & CO \\
| & | & | \\
NH & NH & NH \\
| & | & | \\
CH_2 & & CH_2 \\
| & & | \\
NH & & NH \\
| & & | \\
CO & & CO
\end{array}$$

$$-CH_2-CH-CH_2-CH-CH_2-CH-CH_2-CH-CH_2-CH-$$

$$\begin{array}{ccc}
| & | & | \\
CO & CO & CO \\
| & | & | \\
NH & NH & NH
\end{array}$$

Fig. 4.1 – Schematic representation of the gel structure of polyacrylamide.
(– – – represents linkages at right angles to the page which provide a 3-dimensional structure).

4.2.2 Fibre entrapment

With this technique the fibre-forming polymer is dissolved in an organic solvent which is immiscible with water and an emulsion formed with an aqueous solution of enzyme. This emulsion is extruded into a liquid coagulant (frequently toluene or petroleum ether) which causes the polymer to precipitate in the form of filaments with microdroplets of the enzyme solution entrapped within the microcavities of the polymer (Dinelli *et al.* 1978). The major source of deactivation of enzyme in this technique is the use of organic water immiscible solvents for polymer solubilization and dispersion.

4.2.3 Microencapsulation

The encapsulation of enzymes within spherical semipermeable polymer membranes (see Part A) can be brought about by a variety of methods which can be sub-classified into phase separation, interfacial polymerization, liquid—surfactant membrane, and liquid drying methods. Each type of method will be discussed in the subsequent sections with the relative merits and drawbacks being described.

4.2.3.1 *Phase separation method*

This is the most common method of enzyme microencapsulation (Campbell & Chang 1975, 1976, Paine & Carbonell 1975) and is based on the physical phenomenon, coacervation. An aqueous solution of the enzyme is emulsified in a water immiscible organic solvent containing the polymer. This polymer—enzyme emulsion is then vigorously stirred and another water immiscible organic solvent in which the polymer is insoluble is added. The polymer is concentrated and membranes are formed around the microdroplets of aqueous enzyme solution. Although the technique is carried out under relatively mild conditions, enzyme deactivation can take place owing to the difficulty of removing the last traces of organic solvents from the polymer membrane.

4.2.3.2 *Interfacial polymerization method*

This method of production of microencapsulated enzymes is based on a chemical reaction to synthesize a water-insoluble polymer at the interface of a microdroplet. The two monomers must have differing solubilities such that one is partially soluble in both the aqueous and organic phases (that is, a hydrophilic monomer) whilst the other must be soluble only in the organic phase (that is, a hydrophobic monomer). A solution of the hydrophilic monomer and enzyme is emulsified in an organic solvent, and to this emulsion is added a solution of hydrophobic monomer in the same organic solvent. By polymerization reactions of the condensation or addition type, a membrane is formed at the interface between the organic and aqueous phases. A typical example of this technique in the preparation of nylon 6,10 encapsulated enzymes (Ostergaard & Martiny 1973, Mori *et al.* 1973) in which a solution of chloroform—cyclohexane (1:4 by volume) to which is added the hydrophobic monomer 1,10-decanoyl chloride.

The major disadvantage of this procedure is that the hydrophilic monomer can cause deactivation of the enzyme before polymerization occurs.

4.2.3.3 *Liquid surfactant membrane method*

This procedure which provides immobilized enzymes within a nonpermanent microcapsule is based on the liquid surfactant membrane concept developed by Li (1971). Enzyme immobilization is brought about by emulsifying an aqueous enzyme solution with a surfactant to form the liquid membrane encapsulated enzymes.

The major advantages of this procedure are the nonchemical nature of the method and its reversibility of immobilization. However, the possibility of loss of enzyme by diffusion, and the same diffusion phenomenon controlling passage of substrates and products through the liquid membrane via a solubility mechanism rather than size exclusion mechanism, can be a disadvantage.

4.2.3.4 Liquid drying method
A process similar to the liquid surfactant membrane method has been developed which gives a permanent membrane (Kitajema et al. 1969). This, so-called, liquid drying method is carried out by emulsifying an aqueous solution of the enzyme in an organic solvent which has a boiling point lower than that of water (usually benzene, cyclohexane, or chloroform). The organic solvent also contains the membrane-forming polymer and surfactants. The emulsion is then dispersed in an aqueous medium, which contains protective colloidal substances (gelatin, polyvinylchloride) and surfactants, to give a second emulsion. Removal of the organic solvent by, for example, rotary evaporation from the second emulsion results in the formation of the microencapsulated enzymes.

The major advantage of this procedure is that no deactivation of the enzyme can occur as a result of monomer interactions due to the preformed polymer being used. Disadvantages of the procedure include the low yields of the microcapsules produced, due to problems of preparing the second emulsion, and the time required to completely remove the organic solvent which is essential for membrane production.

4.3 CARRIER BINDING
Since no chemical bonds are required to immobilize enzymes by physical adsorption or ionic binding, it is outside the scope of this chapter to discuss such methods of immobilization. There are references in Part A, Chapter 4, of this book to provide the reader with sufficient details to perform the necessary procedures to immobilize enzymes by either of these two methods.

4.3.1 Chelation or metal binding
Activation of support materials with transition metal compounds, usually titanium(IV) chloride, was originally thought to give supports containing active transition metal chelates which were able to bind enzyme molecules by replacement of, for example, metal—chloride bonds with groups present in the enzyme molecule. It has been shown (Kennedy 1979) that the washing stages involved in this process led to hydrolysis of the metal compound to give the corresponding hydrous metal oxide, and it is the replacement of hydroxyl ligands by amino, carboxyl, hydroxyl, etc., groups of the enzyme which is responsible for the immobilization of the enzyme.

Because of the variable results obtained using the contact/washing procedure when applied to inorganic supports, alternative procedures can be adopted. These include the activation of the inorganic support by coating it with a film of imperfectly crystallized titanium(IV) oxide obtained by refluxing the support with an acidic solution of titanium(IV) chloride (Cardoso et al. 1978). Owing to the relatively poor operational stability of immobilized enzymes, particularly with high molecular weight substrates (Flynn & Johnson 1978) produced by this procedure, other compounds such as tin(II) chloride (Messing 1976), 5-aminosalicylic acid (Kennedy & Chaplin 1979), or 1,6-diaminohexane and tannic acid or glutaraldehyde (Cabral et al. 1982a, b) have been used to coat the oxide layer (see Scheme 4.1). An alternative technique is to heat the support after contact with the titanium(IV) chloride at 80°C for 1 hour (Allen et al. 1979). Use of either of these techniques on the support, with the hydrous metal oxide showing a superior ability to bind biologically active materials including enzymes.

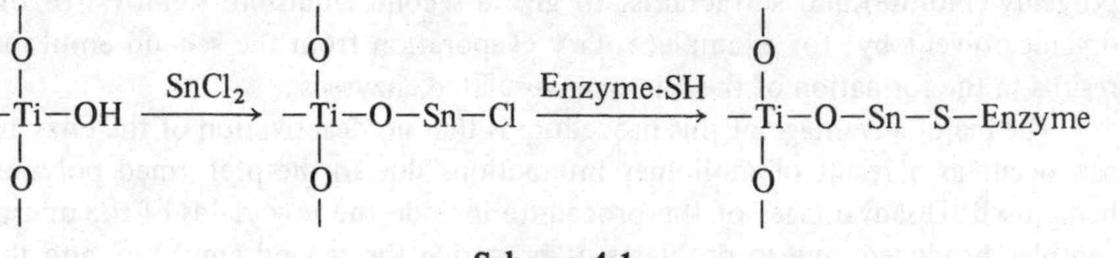

$$
\begin{array}{ccccc}
& \overset{|}{O} & & \overset{|}{O} & & \overset{|}{O} \\
& \overset{|}{} & \xrightarrow{SnCl_2} & \overset{|}{} & \xrightarrow{Enzyme\text{-}SH} & \overset{|}{} \\
-Ti-OH & & -Ti-O-Sn-Cl & & -Ti-O-Sn-S-Enzyme \\
& \overset{|}{O} & & \overset{|}{O} & & \overset{|}{O} \\
& \overset{|}{} & & \overset{|}{} & & \overset{|}{}
\end{array}
$$

Scheme 4.1

Preparation of hydrous oxide gels as supports, enzyme immobilization without additional support matrices, as an independent method to the titanium(IV) chloride activation method, provides a very simple and viable method. The support is prepared by the neutralization of transition metal salts, usually titanium-(IV) chloride, with ammonia to give polymeric insoluble gels (Figs. 4.2, 4.3) followed by chelation of the enzyme to the resulting precipitate. It is also possible to form a gel in the presence of the enzyme (Kennedy & Kay 1976) without loss of enzyme activity, whilst the use of controlled hydrolysis conditions in the presence of an insoluble matrix such as magnetic iron oxide can lead to the production of a hydrous oxide coating on the matrix (Kennedy et al. 1977a, b, Kennedy & White 1979) which is still able to bind enzymes etc.

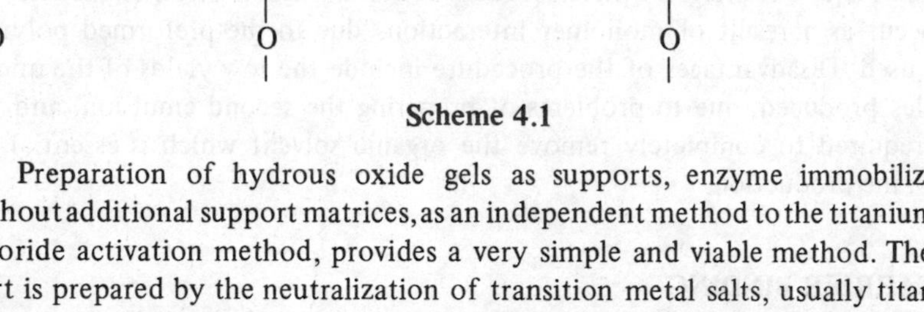

Fig. 4.2 – Schematic representation of hydrous titanium(IV) oxide.

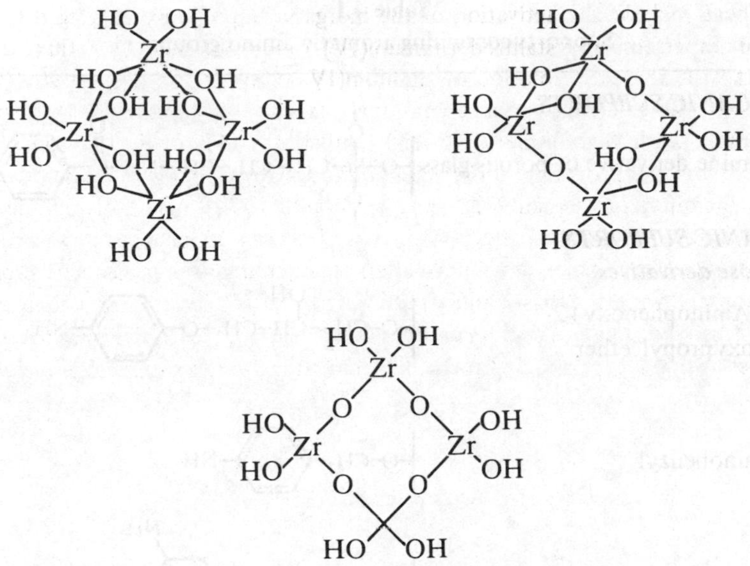

Fig. 4.3 − Structures of zirconium hydroxide.

4.3.2 Covalent binding

As described in Part A, Chapter 4, this method of attaching enzymes to a solid support is both the oldest and most widely used, but because the enzyme is in contact with highly active compounds, care must be taken to ensure that the conditions used are the mildest possible in order to prevent deactivation of the enzyme.

4.3.2.1 *Diazotization* (see Part A, p. 171)

This procedure is based on the formation of a diazo linkage between enzyme proteins and aryldiazonium electrophilic groups on the insoluble support. Supports containing aromatic amino groups are treated with sodium nitrite in acidic solutions (Scheme 4.2), and the resultant activated support reacts with aromatic residues such as the phenolic residue of L-tyrosine or the imidazole residue of L-histidine to give azo derivatives (Scheme 4.2). Free amino groups in the enzyme molecule (from L-lysine, L-arginine, and N-terminal residues) can also react (Scheme 4.3) to give disubstituted bisazo derivative. Some of the aromatic amino group-containing supports are shown in Table 4.1. Many other supports have been used in this procedure by prior incorporation of amino groups into their structure.

Table 4.1
Supports containing aromatic amino groups

INORGANIC SUPPORTS

Arylamine derivative of porous glass

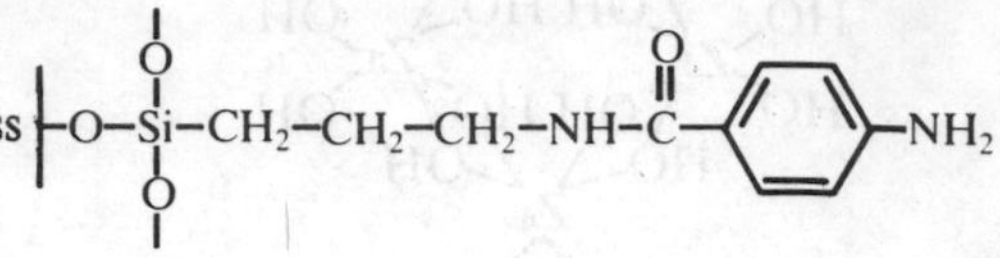

ORGANIC SUPPORTS
Cellulose derivatives

3-(4-Aminophenoxy)-2-hydroxypropyl ether

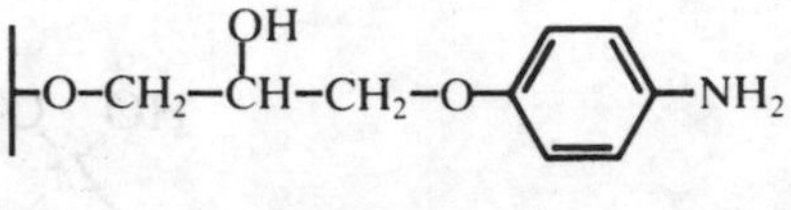

4-Aminobenzyl

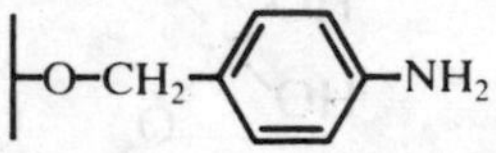

3-Aminobenzyloxymethyl ether

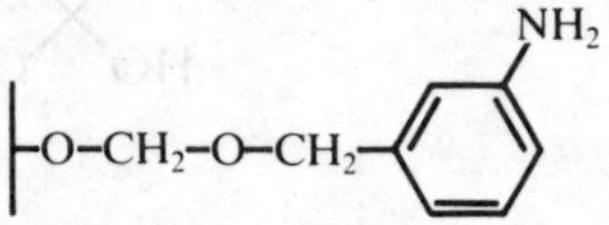

4-Aminobenzyl ester

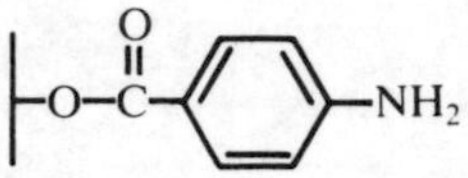

Benzidine derivative of carboxymethyl ether

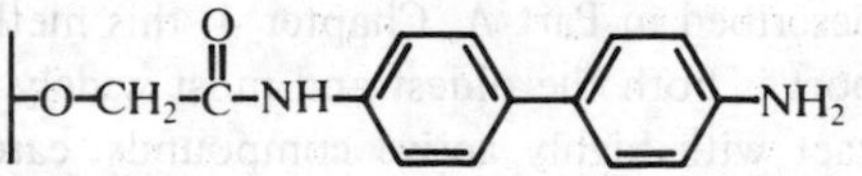

O-(3-Amino-4-methoxyphenyl) sulphonyl ethyl ether

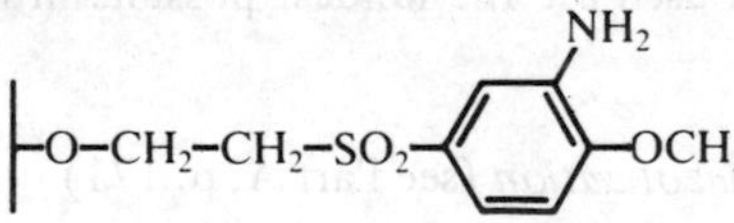

Other organic derivatives

Enacryl® AA

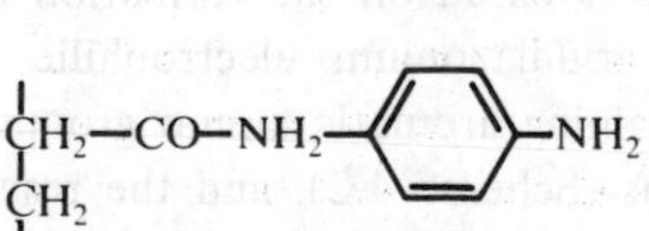

Polyaminostyrene

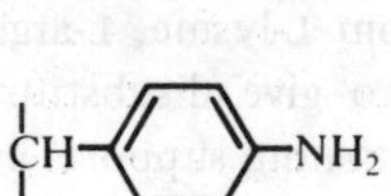

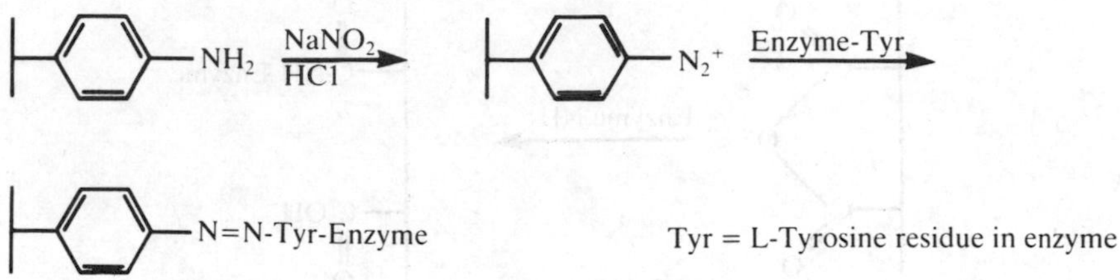

$$\left|\!\!-\!\!\text{C}_6\text{H}_4\!-\!\text{NH}_2 \xrightarrow[\text{HCl}]{\text{NaNO}_2} \left|\!\!-\!\!\text{C}_6\text{H}_4\!-\!\text{N}_2^+ \xrightarrow{\text{Enzyme-Tyr}}\right.$$

$$\left|\!\!-\!\!\text{C}_6\text{H}_4\!-\!\text{N=N-Tyr-Enzyme}\right.$$

Tyr = L-Tyrosine residue in enzyme

Scheme 4.2

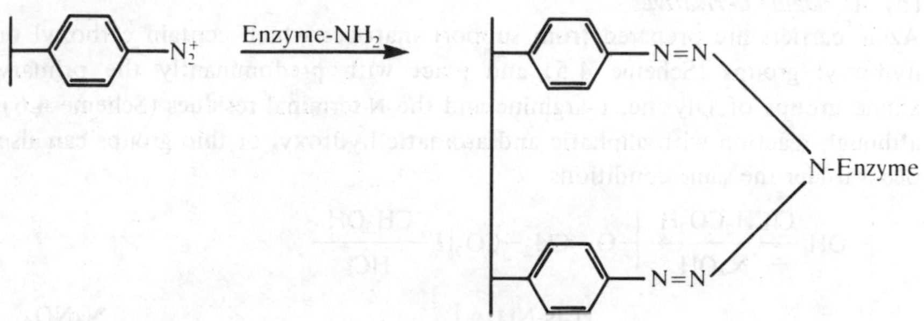

$$\left|\!\!-\!\!\text{C}_6\text{H}_4\!-\!\text{N}_2^+ \xrightarrow{\text{Enzyme-NH}_2}\right.$$

with products $-\text{N=N}$ and $-\text{N=N}$ linked to N-Enzyme

Scheme 4.3

4.3.2.2. *Amide bond formation*

This procedure comprises the formation of amide (peptide) bonds by the nucleophilic attack of such groups as amino, hydroxyl, or thiol groups present in the enzyme molecule on activated support matrices; but, owing to the irreversible denaturation of enzymes which occurs at high pH, the reactions are frequently carried out at near natural pH values rather than at high pH values at which the nucleophiles are most active. A number of different activated supports can be used in this procedure.

(a) *Acid anhydride derivatives*

Acid anhydride derivatives, which are frequently copolymers of maleic anhydride (see Table 4.13, Part A, Chapter 4) react with enzyme amino groups (Scheme 4.4) to produce, as a byproduct carboxyl groups by cleavage of the anhydride ring. These carboxyl groups do not react with enzyme functional groups, owing to the pH of the reaction medium which favours spontaneous ionization. This side reaction gives unfavourable ionic properties to the support, but can be overcome by the addition of diamines during the immobilization reaction. This addition also enhances the stability of the matrix by causing some degree of crosslinking.

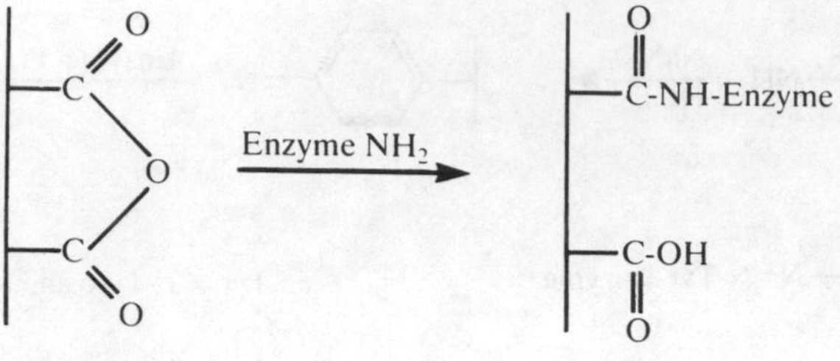

Scheme 4.4

(b) *Acylazide derivatives*
Azide carriers are prepared from support matrices which contain carboxyl or hydroxyl groups (Scheme 4.5) and react with predominantly the primary amine groups of L-lysine, L-arginine and the N-terminal residues (Scheme 4.6), although reaction with aliphatic and aromatic hydroxyl or thio groups can also occur under the same conditions.

$$|\!-OH \xrightarrow[NaOH]{ClCH_2CO_2H} |\!-O-CH_2-CO_2H \xrightarrow[HCl]{CH_3OH}$$

$$|\!-O-CH_2-CO_2-CH_3 \xrightarrow{H_2N\text{-}NH_2} |\!-O-CH_2-CO-NH-NH_2 \xrightarrow[NCl]{NaNO_2}$$

$$|\!-O-CH_2-CO-N_3$$

Scheme 4.5

$$|\!-O-CH_2-CO-N_3 \xrightarrow{Enzyme-NH_2} -O-CH_2-CO-NH-Enzyme$$

Scheme 4.6

(c) *Cyclic imidocarbonate derivatives*
For laboratory applications this method developed by Axén *et al.* (1967) is probably the one most commonly used. The hydroxyl groups of the support matrix (usually polysaccharides such as dextran, cellulose, or agarose) are reacted with cyanogen bromide to give the reactive cyclic imidocarbonate (Scheme 4.7). The product of this reaction reacts with primary amino groups under mildly basic conditions (pH 9–19) to give essentially (Wilchek *et al.* 1975, Kennedy *et al.* 1980a) the substituted isourea derivatives (Scheme 4.8). From evidence published in the last few years (Kohn & Wilchek 1982) it has been shown that the coupling of enzymes on agarose-based imidocarbonate, which explains the different behaviours of Sepharose (agarose-based) and Sephadex (dextran-based) gels during activation.

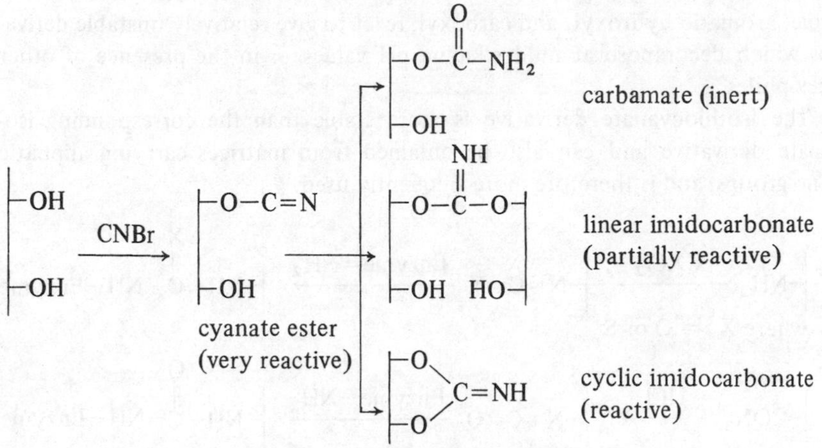

$$
\begin{array}{ccc}
\text{–OH} & \xrightarrow{\text{CNBr}} & \text{–O–C=N} \\
\text{–OH} & & \text{–OH}
\end{array}
$$

cyanate ester
(very reactive)

$$
\begin{array}{ll}
\text{–O–}\overset{\displaystyle O}{\overset{\|}{C}}\text{–NH}_2 \\
\text{–OH}
\end{array}
$$
carbamate (inert)

$$
\begin{array}{l}
\text{–O–}\overset{\displaystyle NH}{\overset{\|}{C}}\text{–O–} \\
\text{–OH   HO–}
\end{array}
$$
linear imidocarbonate
(partially reactive)

$$
\begin{array}{l}
\text{–O}\diagdown \\
\quad\;\; C=NH \\
\text{–O}\diagup
\end{array}
$$
cyclic imidocarbonate
(reactive)

Scheme 4.7

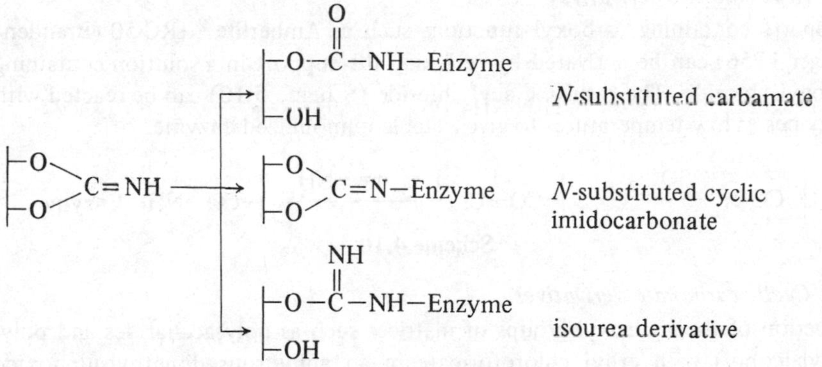

$$
\begin{array}{l}
\text{–O}\diagdown \\
\quad\;\; C=NH \\
\text{–O}\diagup
\end{array}
\longrightarrow
$$

$$
\begin{array}{l}
\text{–O–}\overset{\displaystyle O}{\overset{\|}{C}}\text{–NH–Enzyme} \\
\text{–OH}
\end{array}
$$
N-substituted carbamate

$$
\begin{array}{l}
\text{–O}\diagdown \\
\quad\;\; C=N\text{–Enzyme} \\
\text{–O}\diagup
\end{array}
$$
N-substituted cyclic
imidocarbonate

$$
\begin{array}{l}
\text{–O–}\overset{\displaystyle NH}{\overset{\|}{C}}\text{–NH–Enzyme} \\
\text{–OH}
\end{array}
$$
isourea derivative

Scheme 4.8

This simple method does suffer from a number of disadvantages such as the high toxicity of the cyanogen bromide, the high pH of the derivatizing and coupling reactions, and the presence of charged groups in the final product which cause variable nonspecific, interfering adsorptions. To offset some of these disadvantages, the commercial availability of cyanogen bromide activated polysaccharide supports is a big advantage to many workers.

(d) *Isocyanate and isothiocyanate derivatives*
Modification of matrices carrying aromatic amino and acyl azide groups with phosgene or thiophosgene under alkaline conditions leads to the formation of isocyanate and isothiocyanate derivatives respectively (Scheme 4.9) which react with primary amino groups in enzymes, etc. to form amide or thiomide linkages.

Other nucleophilic groups present in the protein molecule such as thiol, imidazole, aromatic hydroxyl, and carboxyl, react to give relatively unstable derivatives which decompose at mild alkaline pH values or in the presence of other nucleophiles.

The isothiocyanate derivative is more stable than the corresponding isocyanate derivative and can also be obtained from matrices carrying aliphatic amino groups, and is therefore more frequently used.

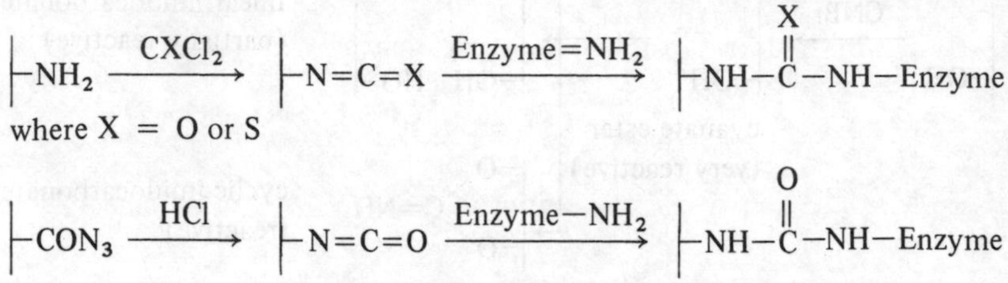

$$\left|-NH_2 \xrightarrow{CXCl_2} \right|-N=C=X \xrightarrow{Enzyme=NH_2} \left|-NH-\overset{\overset{\textstyle X}{\|}}{C}-NH-Enzyme\right.$$

where X = O or S

$$\left|-CON_3 \xrightarrow{HCl} \right|-N=C=O \xrightarrow{Enzyme-NH_2} \left|-NH-\overset{\overset{\textstyle O}{\|}}{C}-NH-Enzyme\right.$$

Scheme 4.9

(e) Acyl chloride derivatives

Supports containing carboxyl functions such as Amberlite® IRC-50 (Brandenberger 1956) can be activated by refluxing the support in a solution containing thionyl chloride. The resulting acyl chloride (Scheme 5.10) can be reacted with enzymes at low temperatures to give a stable immobilized enzyme.

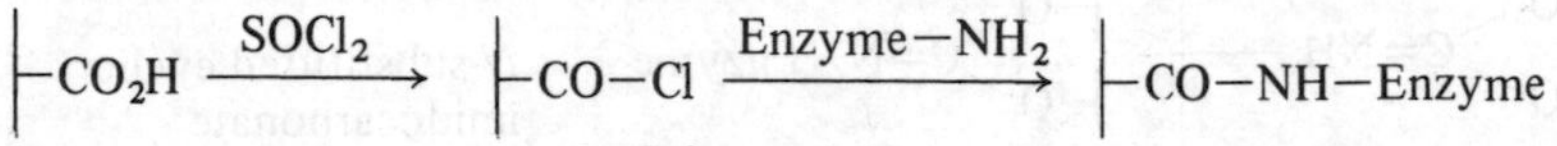

$$\left|-CO_2H \xrightarrow{SOCl_2} \right|-CO-Cl \xrightarrow{Enzyme-NH_2} \left|-CO-NH-Enzyme\right.$$

Scheme 4.10

(f) Cyclic carbonate derivatives

Reaction of the hydroxyl groups of matrices such as polysaccharides and poly-(allylalcohol) with ethyl chloroformate in an anhydrous dimethylsulphoxide solution containing dioxane and triethylamine at pH 7–8 (Barker et al. 1971) leads to the formation of cyclic carbonate derivatives (Scheme 4.11) which react, in an analogous manner to the cyclic imidocarbonate derivatives, with enzyme primary amino groups to give immobilized enzymes via formation of N-substituted imidocarbonates and N-substituted carbonates (see Scheme 4.11).

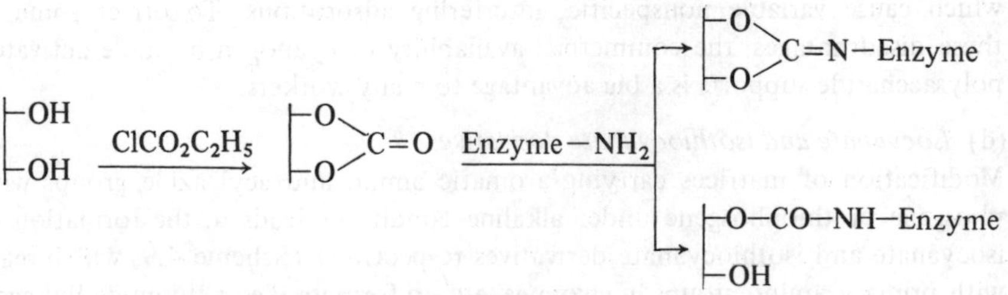

$$\left|\begin{matrix}-OH\\-OH\end{matrix}\right. \xrightarrow{ClCO_2C_2H_5} \left|\begin{matrix}-O\\-O\end{matrix}\right\rangle C=O \xrightarrow{Enzyme-NH_2} \begin{cases}\left|\begin{matrix}-O\\-O\end{matrix}\right\rangle C=N-Enzyme\\ \\ \left|\begin{matrix}-O-CO-NH-Enzyme\\-OH\end{matrix}\right.\end{cases}$$

Scheme 4.11

By comparison with the cyclic imidocarbonate method, this procedure is cheaper and more simple to perform, and does not involve the use of toxic chemicals, but, however, it has not as yet gained the popularity of the cyanogen bromide activation method.

(g) *Condensing reagents*

Activation of carboxylated supports can be brought about by use of carbodiimides and similar reagents. Carbodiimides react at slightly acid pH to give *O*-acylisourea derivatives which can couple enzymes via primary groups (Scheme 4.12) or with hydroxyl or thiol groups, but only at much lower rates of reaction than with amino groups.

The same type of reaction but with amino or hydrazine group-containing supports provides a method of binding enzymes via their carboxyl groups. In this process the carrier and condensing reagents are added simultaneously to the enzyme solution. The immobilization is brought about through amide bonds formed between amino groups of the support matrix and carboxyl groups of the enzyme.

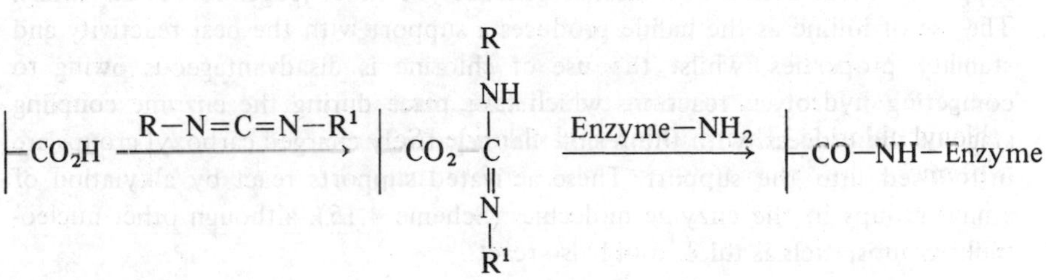

$$\begin{array}{c} R \\ | \\ NH \\ | \\ \end{array}$$

$$\left|-CO_2H \xrightarrow{\;R-N=C=N-R^1\;} CO_2-\begin{array}{c} R \\ | \\ NH \\ | \\ C \\ \| \\ N \\ | \\ R^1 \end{array} \xrightarrow{\;Enzyme-NH_2\;} \right|-CO-NH-Enzyme$$

Scheme 4.12

4.3.2.3 *Alkylation and arylation*

Reaction of amino, thiol, or aromatic hydroxyl groups in enzyme molecules with reactive suports containing active halides, oxirane, vinylsulphonyl, or vinylketo groups is the basis of the alkylation or arylation immobilization procedure. One of the most reactive supports used, which requires no prior activation, can be prepared by the copolymerization of methacrylic acid and methacrylic acid-5-fluoro-2,4dinitroanilide. This support, containing fluoro-substituted aromatic rings, reacts with enzyme primary amino groups, via arylation (Scheme 4.13).

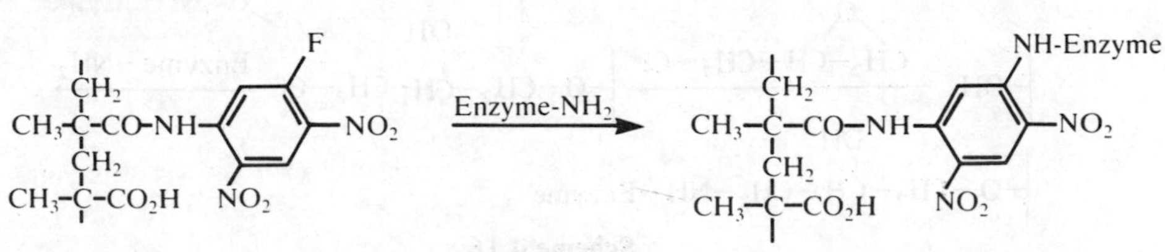

Scheme 4.13

Less reactive supports can be prepared by the reaction of 2,4,6-trichloro sym-triazine or its derivatives (such as 2-amino-4,6-dichloro-sym-triazine) with, for example, polysaccharides (such as cellulose) or minerals (such as bentonite) which again react with primary amino groups in enzyme molecules (Scheme 4.14).

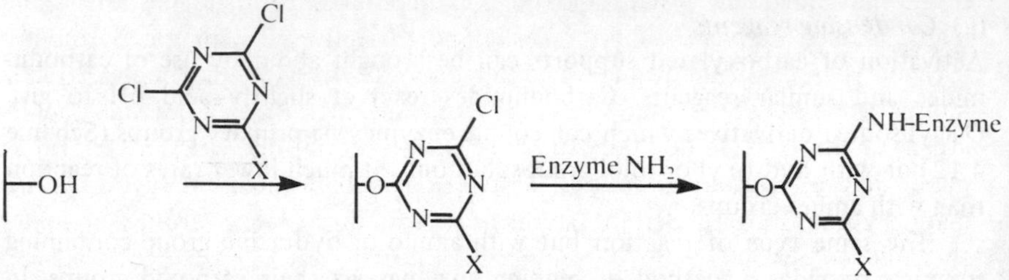

where X = Cl, NH$_2$ etc. **Scheme 4.14**

Active alkyl halide groups can be introduced into hydroxyl group-containing supports by the action of monohalogenoacetyl halides (Jagendorf *et al.* 1963). The use of iodine as the halide produces a support with the best reactivity and stability properties, whilst the use of chlorine is disadvantageous owing to competing hydrolysis reactions which take place during the enzyme coupling stage of the process, with the result that negatively charged carboxyl groups are introduced into the support. These activated supports react by alkylation of amino groups in the enzyme molecules (Scheme 4.15), although other nucleophilic groups, such as thiol, could also react.

$$\begin{array}{c} \text{—OH} \xrightarrow{\text{XCH}_2\text{COX}} \text{—O—CO—CH}_2\text{—X} \xrightarrow{\text{Enzyme—NH}_2} \\ \text{—O—CO—CH}_2\text{—NH—Enzyme} \end{array}$$

where X = I, Br or Cl

Scheme 4.15

The major disadvantage of this alkylation method is that the ester linkages, formed between the enzyme and carrier, are sometimes insufficiently stable to provide acceptable operational stability factors for large-scale usage. A more stable bond can be made if the support matrix is reacted with dihalogenoketones or epichlorohydrin in the presence of boron trifluoride etherate (Scheme 4.16).

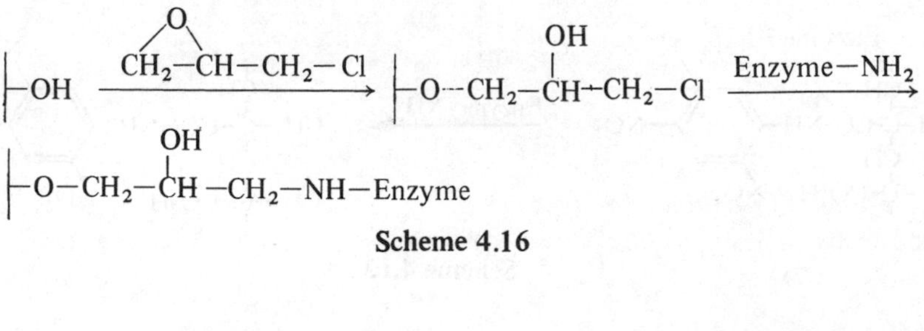

$$\begin{array}{c} \text{—OH} \xrightarrow{\overset{O}{\overset{\diagup\diagdown}{\text{CH}_2\text{—CH—CH}_2\text{—Cl}}}} \text{—O--CH}_2\text{—}\overset{\text{OH}}{\underset{|}{\text{CH}}}\text{--CH}_2\text{—Cl} \xrightarrow{\text{Enzyme—NH}_2} \\ \\ \text{—O—CH}_2\text{—}\overset{\text{OH}}{\underset{|}{\text{CH}}}\text{—CH}_2\text{—NH—Enzyme} \end{array}$$

Scheme 4.16

An alternative approach to the alkylation of enzymes is to use supports containing highly strained three-membered electrophilic ring systems containing oxygen (substituted oxiranes), sulphur (substituted ethylene sulphides), or nitrogen (substituted aziridines), with the reactivity of the epoxide ring making it the most favoured system and the sulphur-containing analogues least suited owing to their inherent stability. The support is activated with epichlorohydrin (Scheme 4.17), and the residual epoxide group reacts with enzyme nucleophilic groups such as thiol, amino, and hydroxyl in that order. The nucleophilic attack of the epoxide or thiol groups occurs at neutral pH, whilst attack on aliphatic hydroxyl group requires a more strongly basic medium (pH 11). Aromatic hydroxyl groups (from L-tyrosine), quanidino groups (from L-arginine), and imidazole groups from (L-histidine) also react. Although the bonds between the support matrix and enzyme are very stable and no unfavourable charged groups which could give rise to nonspecific interfering adsorptions are introduced as byproducts of the reaction (Murphy *et al.* 1977), the method is not strongly favoured, because of the extended reactions time (several days or weeks) which are required.

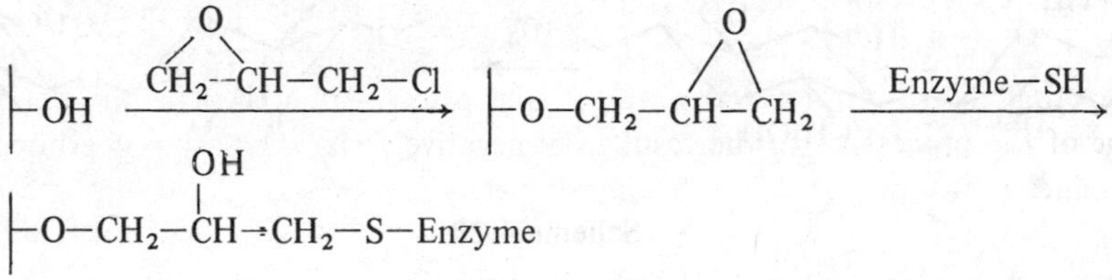

$$\text{—OH} \xrightarrow{\overset{\displaystyle O}{\overset{\diagup\diagdown}{CH_2\text{--}CH\text{--}CH_2\text{--}Cl}}} \text{—O—CH}_2\text{—}\overset{\displaystyle O}{\overset{\diagup\diagdown}{CH\text{—}CH_2}} \xrightarrow{\text{Enzyme—SH}}$$

$$\overset{\displaystyle OH}{\underset{|}{}}$$
$$\text{—O—CH}_2\text{—}\overset{|}{CH}\text{→}CH_2\text{—S—Enzyme}$$

Scheme 4.17

Activation of carrier hydroxyl groups with divinylsulphone in alkaline media (Porath 1979) renders the carrier susceptible to nucleophilic attack by the thiol, amino and hydroxyl groups in enzyme molecules (Scheme 4.18) in the same order of reactivity as for reaction with epoxides but with higher rates of reaction. The major disadvantage of this method is its unsuitability for use in solutions with pH values less than 8, owing to the slow release of the vinylsulphone ligand under these conditions. Provided that the material can be used in alkaline solutions, advantage can be taken of the simultaneous crosslinking reaction of the carrier which improves its properties (Porath *et al.* 1975).

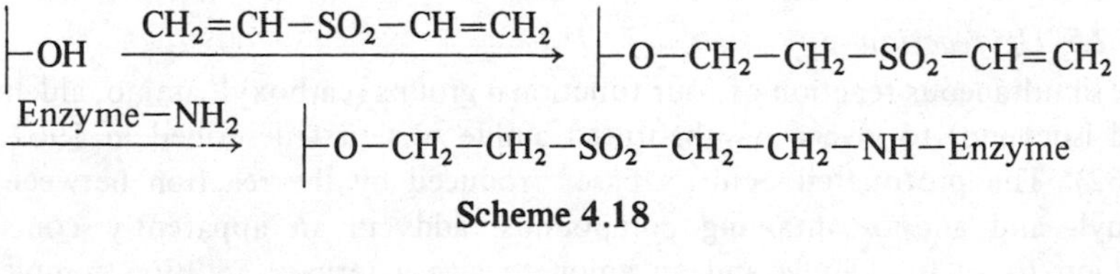

$$\text{—OH} \xrightarrow{CH_2=CH\text{—}SO_2\text{—}CH=CH_2} \text{—O—CH}_2\text{—CH}_2\text{—SO}_2\text{—CH}=CH_2$$

$$\xrightarrow{\text{Enzyme—NH}_2} \text{—O—CH}_2\text{—CH}_2\text{—SO}_2\text{—CH}_2\text{—CH}_2\text{—NH—Enzyme}$$

Scheme 4.18

4.3.2.4 *Schiff's base formation*
The formation of a Schiff's base (aldimine linkage) by the reaction between

carboxyl groups present in an activated support and free amino groups in enzyme molecules (Scheme 4.19) has been used as a method of enzyme immobilization. Supports containing aldehydo groups can be prepared from synthetic polymers such as polyacrylaylamino acetaldehyde (by the action of acid on Enzacryl® Polyacetal) (Epton et al. 1972), or the copolymer of allyl alcohol and vanillin methacrylate (Brown & Racois 1972), or by reaction of naturally occurring polysaccharide supports such as cellulose with periodic acid (Scheme 4.20), or dimethyl sulphoxide and acetic anhydride (Weakley & Mehltreatter 1973). Whilst oxidation of polysaccharides occurs rapidly in aqueous solution at normal temperatures, the close proximity of the two aldehyde functions can cause deactivation of the enzyme by twisting the molecule to allow two point binding.

$$\vdash CHO \xrightarrow{\text{Enzyme}-NH_2} \vdash CH=N-Enzyme$$

Scheme 4.19

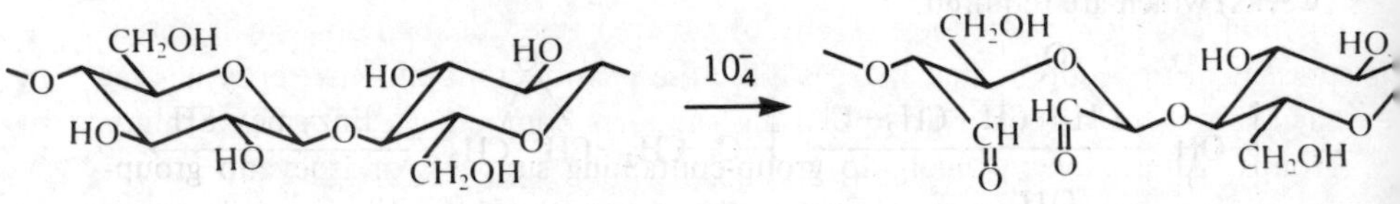

Scheme 4.20

Supports containing amino groups can be reacted with glutaraldehyde to yield an active aldehyde group-containing matrix. Although glutaraldehyde is normally used for crosslinking enzymes, its use here is based on the assumption that one aldehydo group reacts with the solid support, whilst the other, via Schiff's base formation, with the enzyme, although the nature of the reaction is not fully understood (see later, Section 4.4).

The major disadvantage of Schiff's base formation as an enzyme immobilization procedure, namely that of the reversibility of the reaction of low pH, can be overcome in some cases by reduction of the imine bond with sodium borohydride to give the more stable alkylamino linkage.

4.3.2.5 Ugi reaction

The simultaneous reaction of four functional groups (carboxyl, amino, aldehydo, and isocyano) to give a N-substituted amide was first described in 1962 (Ugi 1962). The protonated Schiff's base, produced by the reaction between carbonyl- and amino-containing compounds, adds in an apparently concerted fashion to an isocyanide and an anion to give a ternary addition complex. If the anion is carboxylate, an intermolecular rearrangememt occurs to give the N-substituted amide (Scheme 4.21).

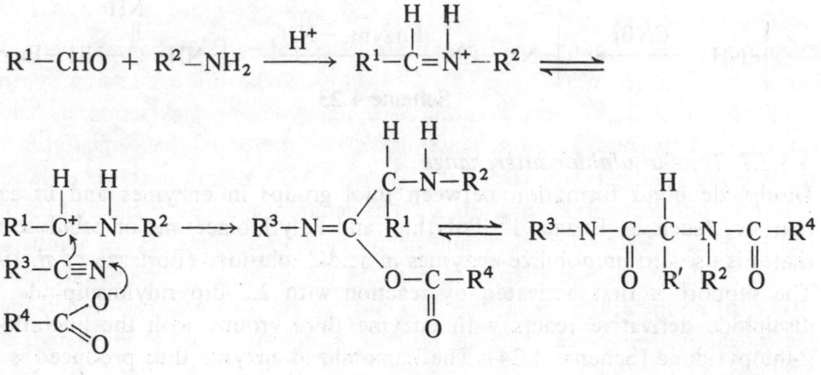

$$R^1-CHO + R^2-NH_2 \xrightarrow{H^+} \overset{H}{R^1-\overset{|}{C}}=\overset{H}{\overset{|}{N^+}}-R^2 \rightleftharpoons$$

$$\overset{H\ H}{R^1-\overset{|}{\underset{|}{\overset{+}{C}}}-N-R^2} \longrightarrow R^3-N=C\overset{\overset{H\ H}{C-N-R^2}}{\underset{O-C-R^4}{\underset{\|}{O}}}R^1 \rightleftharpoons R^3-N-\overset{H}{\underset{\|}{\overset{|}{C}}}-\overset{|}{\underset{R'}{C}}-\overset{H}{\underset{R^2}{N}}-\overset{|}{\underset{O}{C}}-R^4$$

$$\overset{R^3-C\equiv N^{\cdot}}{R^4-C\overset{O}{\underset{O}{\diagdown}}}$$

Scheme 4.21

This reaction, which takes place in an acidic medium, is sufficiently mild to be used for enzyme immobilization, and the method was optimized by Axén and his co-workers (Axén *et al.* 1971, Vretblad & Axén 1971) to give a very versatile method, owing to the choice of any of the four functional groups which can be present in the support matrix. It is therefore possible to immobilize enzymes via their amino or carboxyl groups, using supports containing carboxyl or amino groups. Alternatively, aldehydo group-containing supports or isocyano group-containing supports (in the presence of a water soluble aldehyde) can be used. The Ugi reaction therefore provides a versatile alternative to the use of active aldehydo carriers, although the requirements for the ternary addition complex can limit the yields obtained.

4.3.2.6 *Amidination reactions*
Supports containing cyano groups can be activated with dry hydrogen chloride in ethanol to give an imidoester derivative (Scheme 4.22) which is readily attacked by nucleophilic groups. At alkaline pH, the imidoester derivative reacts selectively with amino groups in enzyme molecules to give amidines which are stable in neutral or acidic solutions but are slowly hydrolyzed in high pH solutions.

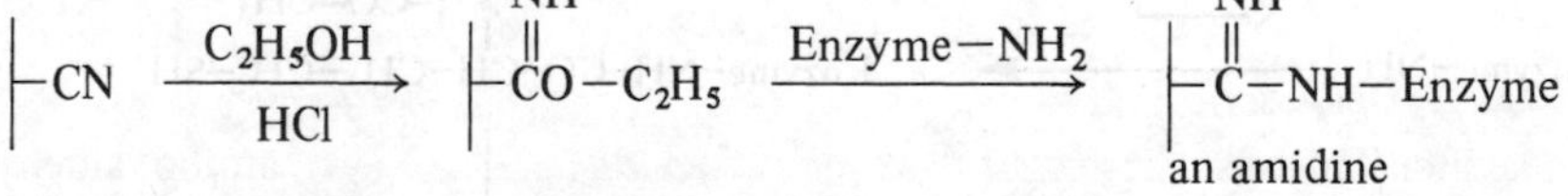

$$\left|-CN \xrightarrow[HCl]{C_2H_5OH} \left|-\overset{NH}{\overset{\|}{C}}O-C_2H_5 \xrightarrow{Enzyme-NH_2} \left|-\overset{NH}{\overset{\|}{C}}-NH-Enzyme\right.\right.\right.$$
an amidine

Scheme 4.22

Amidination reactions also occur between supports containing amino groups, which are activated with cyanogen bromide, and amino groups of enzymes leading to the formation of guanidino linkages between enzyme and carrier (Scheme 4.23).

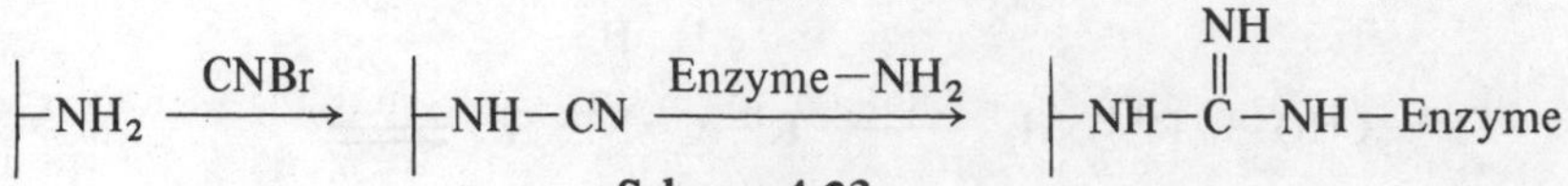

$$\text{---NH}_2 \xrightarrow{\text{CNBr}} \text{---NH--CN} \xrightarrow{\text{Enzyme--NH}_2} \text{---NH--}\overset{\displaystyle \overset{NH}{\|}}{C}\text{--NH--Enzyme}$$

Scheme 4.23

4.3.2.7 *Thiol-disulphide interchange*

Disulphide bond formation between thiol groups in enzymes and in carrier matrices (such as Enzacryl® Polythiol and Polythiolacetone or cellulose xanthate) is used to immobilize enzymes in acidic solutions (Borlazza *et al.* 1980). The support is first activated by reaction with 2,2'-dipyridyldisulphide. This disulphide derivative reacts with enzyme thiol groups with the liberation of 2-thiopyridone (Scheme 4.24). The immobilized enzyme thus produced is only stable under nonreducing conditions, and the immobilization can be reversed by low molecular weight thiol compounds.

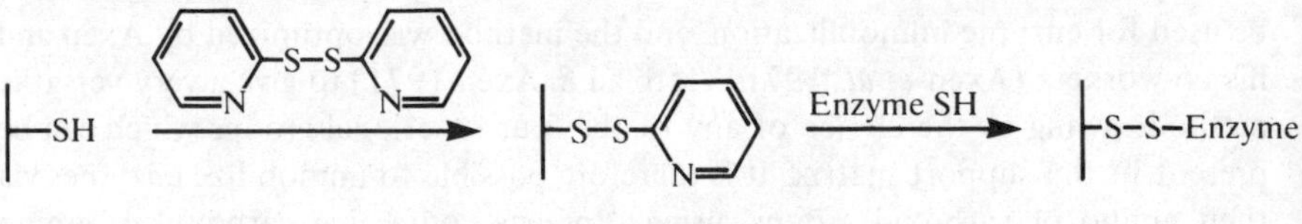

$$\text{---SH} \xrightarrow{\text{(pyridyl-S-S-pyridyl)}} \text{---S--S--(pyridyl)} \xrightarrow{\text{Enzyme SH}} \text{---S--S--Enzyme}$$

Scheme 4.2 4

There is an alternative procedure which is not only more versatile in that it uses the more readily available amino groups of an enzyme, but also introduces a spacer arm and thereby can cause less disruption to the tertiary structure of the enzyme on immobilization (Kennedy & Zamir 1975). The amino groups in the enzyme molecule are activated by reaction with *N*-acetylhomocysteine-thiolactone prior to coupling to the thiolated support under oxidizing conditions (Scheme 4.25). The reversible nature of the reaction under reducing conditions is exploited for regeneration of the carrier and reactivation of the immobilized enzyme

$$\text{Enzyme--NH}_2 \longrightarrow \text{Enzyme--NH--CO--CH(NH--CO--CH}_3\text{)--CH}_2\text{--CH}_2\text{--SH}$$

$$\downarrow \text{Cellulose--O--}\overset{\displaystyle \overset{S}{\|}}{C}\text{--SH}$$

$$\text{Enzyme--NH--CO--CH(NH--CO--CH}_3\text{)--CH}_2\text{--CH}_2\text{--S--S--}\overset{\displaystyle \overset{S}{\|}}{C}\text{--O--Cellulose}$$

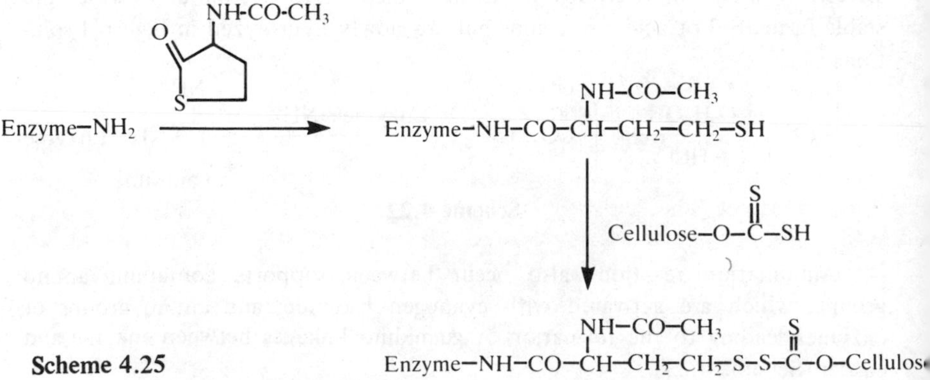

Scheme 4.25

4.3.2.8 *Mercury-enzyme interactions*

This method is based on the interaction of carriers containing mercury derivatives with thiol groups of the enzyme (Scheme 4.26). The bond which is formed
at pH 4–8 is not purely covalent, and this method is considered by some workers
as physical adsorption. Low molecular weight thiol agents can reverse this
immobilization.

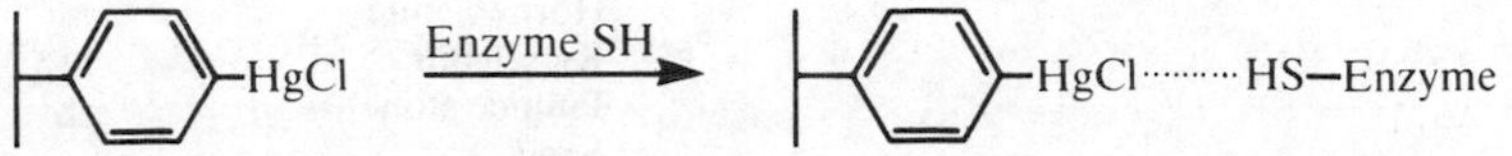

Scheme 4.26

4.3.2.9 *γ-Irradiation induced coupling*

γ-Irradiation of enzymes in the presence of such carriers as agarose and dextran
has been used as a means to immobilize enzymes (Brandt & Anderson 1976).
The radiation causes radicals to be formed from both enzyme and carrier, and
these radicals combine with the formation of a covalent bond (Scheme 4.27).

$$Matrix-H \longrightarrow Matrix^{\cdot} + H^{\cdot}$$
$$Enzyme-H \longrightarrow Enzyme^{\cdot} + H^{\cdot}$$
$$Matrix^{\cdot} + Enzyme^{\cdot} \longrightarrow Matrix-Enzyme$$

Scheme 4.27

The major disadvantages of the method include its nonspecificity, low
yields, and high loss of activity arising from radiation damage, whilst the method
has the advantage of being independent of temperature and pH.

4.3.3 Matrices for carrier binding

The solid supports used as matrices for enzyme immobilization can be classified
as shown in Table 4.2. As described in Part A, Chapter 4, the nature of the
support affects the activity, stability and kinetics of immobilized enzymes
and for these reasons reasearch work chooses the more specialized supports
specific for a given application, whilst industrial applications are limited by
economic consideration in terms of materials and techniques, with characteristics such as nontoxity, microbial resistance, reusability and ease of control
also being of major importance.

Table 4.2
Matrices for carrier binding methods of enzyme immobilization

Type		Example
INORGANIC		
	Naturally occurring	Attapulgite clays
		Bentonite
		Diatomaceous earth
		Horneblende
		Kieselguhr
		Pumice stone
		Sand
	Fabricated materials	Alumina
		Aluminosilicate
		Controlled pore glass
		Hydrous titanium oxide
		Magnetic iron oxide
		Nickel
		Silica
		Silochrome
		Stainless steel
		Titanium
		Zirconium hydroxide
ORGANIC		
	Natural polymers	
	Polysaccharides	Agarose
		Alginic acid
		Cellulose
		Chitin
		Chitosan
		Dextran
		Inulin
		Pectic acid
		Pectin
		Starch
	Proteins	Collagen
		Silk
	Carbon	Activated carbon
	Synthetic polymers	
	Polystyrenes	Polystyrene
	Polyacrylates	Polyacrylamide
		Polyacrylate
		Polyacrylonitrile
		Polyglycidylmethacrylate
		Polyhydroxyalkylmethacrylate
		Polymethacrylate
		Polymethacrylic acid anhydride
	Maleic anhydride based copolymers	Ethylene/maleic anhydride copolymer
	Polypeptides	4-Amino-DL-phenylalanine/L-leucine copolymer
	Vinyl polymers	Poly(vinyl alcohol)
		Poly(vinyl amine)
	Allyl polymers	Poly(allyl alcohol)
	Polyamides	Nylon

4.3.3.1 *Inorganic supports*

The earliest work on immobilized enzymes used inorganic ability to physically adsorb enzymes; but organic polymers used owing to the ease with which enzymes could be che them. Interest has now returned to inorganic supports, not economic considerations, but also because of their physica offer several advantages over organic supports:

high mechanical strength
thermal stability
resistance to organic solvents
resistance to microbial attack
ease of handling
long shelf life
facile regeneration via pyrolysis;

whilst many inorganic supports also have the advantage of not changing their structure over wide ranges of pH, temperature and pressure.

Porous materials have distinct advantages due to the high surface area to unit weight ratio which allows high enzyme loading with most of the enzyme being immobilized on internal surfaces which protects it from the high flow turbulence which exists around the external surfaces, particularly in industrial reactors. The size of the pores must be such that not only can the enzyme be accommodated, but there must also be sufficient space to allow facile access for the substrate and product molecules. For this reason, porous carriers with broad distributions of pore sizes have only a proportion of the pores of sufficient size, and the area which can effectively be used is only a small proportion of the total surface area of the support. The development of controlled pore supports has helped to overcome this problem by ensuring that the available surface is a maximum for a particular enzyme system using the different pore size supports which are produced. It must be noted that the larger the pore size the smaller is the available surface area.

(a) *Controlled pore supports*

Controlled pore glass is prepared by heating borosilicate glass to 500–700°C, when it undergoes a phase separation. Subsequent acid leaching of the borate compponent leaves a porous support with closely controlled pore sizes. However, controlled pore glass is not suitable for industrial purposes owing to its high price (Messing 1974) and lack of stability due to the continuous leaching of silica from the particles during prolonged usage in continuous flow systems, especially in alkaline media. Various methods have been employed to overcome the stability problem, including the deposition of a surface coating of zirconium oxide by impregnating the controlled pore glass with a zirconium salt under vacuum followed by calcination to the oxide (Tomb & Weetall 1979).

Controlled pore ceramics based on silica, alumina, and titania have been developed (Messing 1975) and are cheaper than controlled pore glass. However, their chemical durability demands attention in the choice of which support is chosen for a particular process. Thus titania and alumina are more suitable to applications in alkaline solutions, whilst silica is more suitable for acidic solutions.

(b) Other porous supports
Naturally occurring porous minerals such as attapulgite clays (Burns 1976), bentonite (Monsan & Durand 1971), kieselguhr (Greenfield & Lawrence 1975), and pumice stone (Cabral et al. 1981a) have been used as carriers for enzyme immobilization, but their broad pore distribution detracts from their general suitability.

(c) Non-porous supports
Non-porous supports have a major disadvantage of low surface area per unit weight, and whilst the use of fine particles can offset some of this disadvantage, it makes their use in continuous reactors unattractive owing to the high pressures required, whilst their difficult recovery from solution makes use in batch processes unattractive. However, a major advantage is the reduction or elimination of diffusional constraints with respect to the substrate, although this advantage does not effectively compete with the other advantages of using porous supports. There has recently been an increasing interest in the use of pure metals or metal oxides which exhibit ferromagnetic properties, including nickel (Chaplin & Kennedy 1976), stainless steel (Charles et al. 1975), and magnetic iron oxide (Kennedy et al. 1977b, Kennedy & White 1979), because of the ease of removal of enzymes immobilized on such supports from reaction mixtures which contain other insoluble material. Such magnetic materials also find applications in enzyme immunoassay systems.

Sand is a nonporous support which has gained some interest (Brotherton et al. 1976) owing mainly to economic reasons.

4.3.3.2 Coupling reactions for inorganic supports
As a consequence of their relative inertness, the covalent attachment to inorganic supports is difficult, and all methods involve the use of trialkoxysilane derivatives, which contain organic functional groups such as aliphatic or aromatic amino, halogeno, aldehydo or acetal groups, to form covalent bonds between the support and enzyme. The early work on silanization of inorganic supports was carried out by Weetall and co-workers (Weetall 1976) who introduced aminoalkyl groups into controlled pore glass using 3-aminopropyltriethoxysilane (Scheme 4.28) via a mechanism involving the displacement of the triethoxy residues of the silane by hydroxyl group on the oxidized surface of the glass.

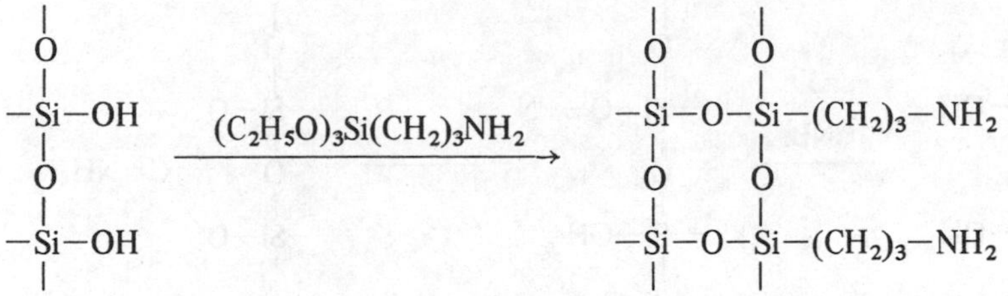

$$
\begin{array}{c}
\mid \\
O \\
\mid \\
-Si-OH \\
\mid \\
O \\
\mid \\
-Si-OH \\
\mid
\end{array}
\quad \xrightarrow{(C_2H_5O)_3Si(CH_2)_3NH_2} \quad
\begin{array}{c}
\mid \quad\quad \mid \\
O \quad\quad O \\
\mid \quad\quad \mid \\
-Si-O-Si-(CH_2)_3-NH_2 \\
\mid \quad\quad \mid \\
O \quad\quad O \\
\mid \quad\quad \mid \\
-Si-O-Si-(CH_2)_3-NH_2 \\
\mid \quad\quad \mid
\end{array}
$$

Scheme 4.28

Once the silane derivative of the matrix has been produced it can be used in any of the methods described in this chapter which use amino group-containing supports. Conversely, the amino group can be converted into other functionalities such as diazo, aldehydo, isothiocyano, carboxyl or alkylazide.

An alternative activation method of inorganic supports is to coat the glass, etc., with a titanium coating, which after drying can be reacted with a diamine to give an alkylamine derivative (Cabral *et al.* 1981b, 1982a) (Scheme 4.29). This method, which is simpler to perform, produces a derivative similar to that obtained by silanization, and the operational stabilities of the immobilized enzymes obtained compare favourably with those from the silanization method (Cabral *et al.* 1982b).

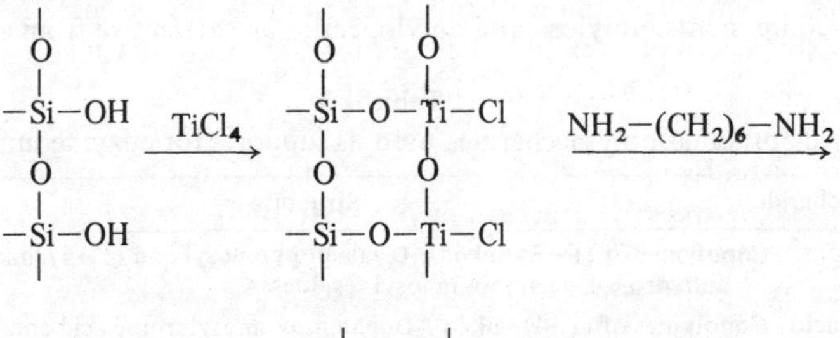

$$
\begin{array}{c}
\mid \\
O \\
\mid \\
-Si-OH \\
\mid \\
O \\
\mid \\
-Si-OH \\
\mid
\end{array}
\xrightarrow{TiCl_4}
\begin{array}{c}
\mid \quad\quad \mid \\
O \quad\quad O \\
\mid \quad\quad \mid \\
-Si-O-Ti-Cl \\
\mid \quad\quad \mid \\
O \quad\quad O \\
\mid \quad\quad \mid \\
-Si-O-Ti-Cl \\
\mid \quad\quad \mid
\end{array}
\xrightarrow{NH_2-(CH_2)_6-NH_2}
$$

$$
\begin{array}{c}
\mid \quad\quad\quad \mid \\
O \quad\quad\quad O \\
\mid \quad\quad\quad \mid \\
-Si-O-Ti-NH-(CH_2)_6-NH_2 \\
\mid \quad\quad\quad \mid \\
O \quad\quad\quad O \\
\mid \quad\quad\quad \mid \\
-Si-O-Ti-NH-(CH_2)_6-NH_2 \\
\mid \quad\quad\quad \mid
\end{array}
$$

Scheme 4.29

Cyclic imidocarbonate derivatives of glass and other inorganic supports have been prepared (Weetall & Detar 1975) using cyanogen bromide (Scheme 4.30), and chloride derivatives of brick, bentonite, and glass have been prepared (Monsan & Durand 1971) using thionyl or sulphonyl chloride as alternative methods of activation of inorganic supports to covalently bind enzymes.

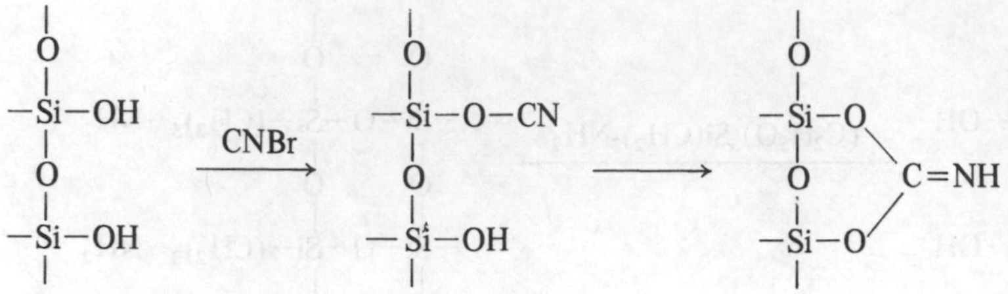

$$
\begin{array}{ccccc}
\overset{\displaystyle |}{O} & & \overset{\displaystyle |}{O} & & \overset{\displaystyle |}{O} \\
-Si-OH & & -Si-O-CN & & -Si-O \\
\overset{\displaystyle |}{O} & \xrightarrow{\text{CNBr}} & \overset{\displaystyle |}{O} & \longrightarrow & \overset{\displaystyle |}{O}\diagdown \\
-Si-OH & & -\overset{|}{Si}-OH & & -Si-O\diagup C=NH \\
| & & | & & |
\end{array}
$$

Scheme 4.30

4.3.3.3 *Organic supports*

In contrast to inorganic supports, organic supports can be obtained with a wide variety of functional groups. Organic supports can be classified, as in Table 4.2, as natural polymers, which include polysaccharides, proteins, and carbon, and synthetic polymers which can be prepared with hydrophilic or hydrophobic characteristics. Hydrophilic supports are to be preferred to hydrophobic supports owing to the potential destabilizing effects of the latter on enzymes, with synthetic polymers being preferred to naturally occurring materials because of their superior resistance to microbial attack.

(a) *Polysaccharides*

A number of polysaccharides have been used as supports for enzyme immobilization, the most important being cellulose, dextran, and agarose, whilst starch and its components, amylose and amylopectin, are attractive from an economic

Table 4.3

Structures of some polysaccharides used as supports for enzyme immobilization

Polysaccharide	Structure
Agarose	Copolymer of $(1\rightarrow3)$-linked β-D-galactopyranosyl and $(1\rightarrow4)$-linked 3,6-anhydro-α-L-galactopyranosyl residues
Alginic acid	Copolymer of $(1\rightarrow4)$-linked β-D-mannopyranosyluronic acid and $(1\rightarrow4)$-linked α-L-gulopyranosyluronic acid residues
Amylose	Polymer of $(1\rightarrow4)$-linked α-D-glucopyranosyl residues
Amylopectin	Polymer of $(1\rightarrow4)$-linked α-D-glucopyranosyl residues with branched structure via (1,4,6)-trisubstituted residues
Cellulose	Polymer of $(1\rightarrow4)$-linked β-D-glucopyranosyl residues
Chitin	Polymer of $(1\rightarrow4)$-linked 2-acetamido-2-deoxy-β-D-glucopyranosyl residues
Chitosan	Polymer of $(1\rightarrow4)$-linked 2-amino-2-deoxy-β-D-glucopyranosyl residues
Dextran	Polymer of $(1\rightarrow6)$-linked α-D-glucopyranosyl residues with branched sturcture via (1,3,6)-trisubstituted residues
Inulin	Polymer of $(2\rightarrow1)$-linked β-D-fructofuranosyl residues
Pectic acid	Polymer of $(1\rightarrow4)$-linked α-D-galactopyranosyluronic acid residues
Pectin	Polymer of $(1\rightarrow4)$-linked α-D-galactopyranosyluronic acid residues, many of which present as methyl esters
Starch	Mixture of amylose and amylopectin

point of view but not from its ease of microbial attack. In recent years, however, a number of other polysaccharides have been utilized, some of which contain amino or carboxyl groups as part of the polysaccharide structure. The structural features of these polysaccharides are given in Table 4.3. The hydroxyl groups of polysaccharides can be activated directly by introduction of an electrophilic group which is reactive towards enzyme molecules. The nucleophilic character of polysaccharides is so weak that pendant functional groups have to be introduced as the activation for coupling (direct coupling) or before activation for coupling (indirect coupling). A number of active derivatives have been prepared (see Table 4.4) using some of the hydroxyl groups of polysaccharides, whilst the residual hydroxyl groups provide a hydrophilic environment which protects the attached enzyme.

Table 4.4
Active polysaccharides for enzyme immobilization

Derivative	Group required in parent matrix	Activating agent	Reference
Direct coupling			
Bromoacetyl	Hydroxyl	Bromoacetylbromide	Maeda & Suzuki (1972)
Chloroalkyl	Hydroxyl	Epichlorohydrin/ boron trifluoride etherate	Almé & Nyström (1971)
Dialdehydo	Vicinal diol	Periodate	Glassmeyer & Ogle (1971)
Epoxy	Hydroxyl	Epichlorohydrin	Murphy *et al.* (1977)
Imidocarbonate	Vicinal diol	Cyanogen bromide	Meada *et al.* (1978)
Trans-2,3-carbonate	Vicinal diol	Ethylchloroformate	Kennedy & Zamir (1973)
Triazinyl	Hydroxyl	2,4,6-Trichloro-*sym*-triazine	Shimizu & Lenhoff (1979)
Vinylsulphone	Hydroxyl	Divinylsulphone	Porath (1979)
Indirect coupling			
Azide	Carboxyl	Hydrazine/nitrite/acid	Mitz & Summaria (1961)
Diazo	Amino	Nitrite/acid	Beddows *et al.* (1980)
Isocyanato	Amino	Phosgene	Maeda *et al.* (1978)

(b) *Proteins*

The main uses of nonenzymically active proteins as supports in enzyme immobilization have been in the complexation with collagen (Vieth & Venkatasubramanian 1976), the entrapment within a gelatin matrix (van Velzen 1974), or the crosslinking of enzymes together with inactive proteins such as albumin. There have, however, been a number of reports of the use of inactive proteins, such as collagen and silk *per se*, as supports (Coulet *et al.* 1975, Grasset *et al.* 1979).

Collagen is the most abundant protein constituent of higher vertebrates, and its ready availability from a large number of species ranging from fish to cattle makes it an expensive carrier. It can be isolated readily and reconstituted in various forms without loss of its native structure which, together with its inherent hydrophobicity, its open internal structure providing a high concentration of bind sites, and its fibrous structure with high swellability in aqueous solution, makes it an ideal support for enzyme immobilization. Recently a covalent technique was described (Coulet *et al.* 1980) in which azide derivation of collagen was used to immobilize a number of enzymes.

Silk, in its natural state, comprises two proteins; fibroin is water insoluble whilst sericin is soluble. When used as an enzyme support it is used in the form of industrial woven silk, which provides additional advantages of easy manipulation, low compaction, and abrasion resistance to its inherent advantages of good thermal stability, resistance to acid and alkali in this range pH 3–8 and resistance to microbial attack. Before silk can be used to covalently bind enzymes it must be activated, and some knowledge of the available functional groups is required. Whilst several methods have been reported the most effective is the azide derivation (Carlsson *et al.* 1974) which uses the carboxyl groups present in the fibroin protein molecules. By a process of acidic methylation followed by the action of hydrazine and nitrous acid, an active azide silk is obtained. Methylated silk itself has been used as a support for physical adsorption of enzymes.

(c) *Carbon*
Carbon materials are economically priced, mechanically strong supports for enzyme immobilization which can be obtained in a number of forms including high porous forms. The most used carbon material is activated carbon which is high porous, having three types of pores within the same structure (micropores, diameter 0–0.4 nm; transitional pores, 5–100 nm; and macropores, 100–4000 nm), and contains a number of active groups, which include aromatic hydroxyl and carboxyl groups, on its surface. Whilst many reports exist in which carbon materials are used to physically adsorb enzymes, the presence of these active groups does allow for reactions leading to covalent attachment of enzymes (Cho & Bailey 1978, 1979). The carboxyl groups can be activated by use of carbodiimides, whilst amino carbon can be obtained by vigorous oxidation and nitration followed by dithionate reduction. The amino groups in amino carbon can subsequently be converted to isothiocyanate groups (with thiophosgene) which, in turn, can be silanized with 3-aminopropyltriethioxysilane. Diazotized derivatives of carbon can be prepared from amino- and isothiocyanato-carbon.

(d) *Polystyrenes*
Polystyrene was the first synthetic polymer to be used for enzyme immobilization, owing to its ready availability and low cost (Brandenberger 1955). Polystyrene must be derivatized in order to allow covalent binding, the main intermediate being polyaminostyrene which is produced by nitration and reduction. The amino group has been activated by diazotization, or by reaction with thiophosgene or glutaraldehyde, but the inherent hydrophobicity of the result-

ing polymers results in poor yields of active immobilized enzymes. By using copolymers of styrene with hydrophilic monomers such as acrylic acid, methacrylic acid, 3-fluorostyrene and 3-isothiocyanatostyrene with lower amounts of divinyl benzene being used as a crosslinking agent (Manecke & Förster 1966, Manecke & Gunzel 1967), the problems of hydrophobicity have been overcome, resulting in an upsurge in the use of polystyrene.

(e) *Polyacrylates*

Acrylic polymers are about the most used synthetic polymers for enzyme immobilization being used for both entrapment and covalent binding of enzymes. Various derivatives can easily be prepared (see Table 4.5), some of which allow direct coupling of enzymes, and many of which, including polyacrylamides and polymethacrylates, are commercially available. These derivatives are used to introduce reactive groups into the polymer or to adjust the hydrophobicity. When the hydrophilic character is increased to a certain degree, the linear polymers are soluble in water and have to be rendered insoluble by crosslinking with such bifunctional groups as N,N'-methylene-bisacrylamide.

Table 4.5

Some acrylic polymers for enzyme immobilization

Polymer	Reactive groups	Reference
Acrylic acid/acrylamide	Amido, carboxyl	Torchilin *et al.* (1977)
Methacrylic acid/acrylamide	Amido, carboxyl	Torchilin *et al.* (1977)
Acrylic acid/acrolein	Aldehydo, carboxyl	Van Leemputten (1977)
Methacrylic acid anhydride	Acid anhydride	Conte & Lehmann (1971)
Copolymers of acrylamides	Various	Epton & Thomas (1971)

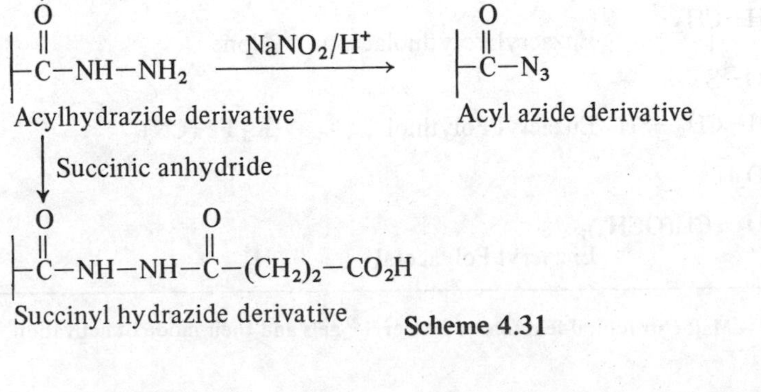

Scheme 4.31

Polyacrylamides can be activated to provide supports for direct coupling by a number of methods (Scheme 4.31) to give, for example, acylhydrazide-, aminoethyl- succinylhydrazide-, and acylazide-derivatives. The aminoethyl derivative and polyacrylamide *per se* can also be activated with glutaraldehyde, whilst succinyl hydrazide- and aminoethyl-derivatives react with enzymic amino and carboxyl groups respectively in the presence of carbodiimides.

A number of hydrophilic, electrically neutral acrylamide based polymers are commercially available under the trade name of Enzacryl® gels (Koch Light Labs), the uses of which and methods of activation have been described (White & Kennedy 1980). The major structural features of these supports are shown in Fig. 4.4. A number of similar copolymers have also been prepared (see Table 4.6). The copolymers containing 2-hydroxyethylmethacrylate correspond to

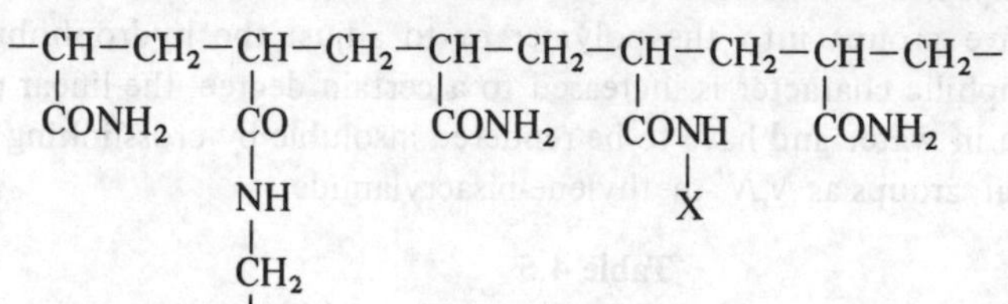

X = potentially active functional group

$-X$	Name	Activating agent
 $-NH-\langle\bigcirc\rangle-NH_2$	Enzacryl AA	Nitrite/H⁺ or thiophosgene
$-NH-NH_2$	Enzacryl AH	Nitrite/H⁺
$-NH-CH-CH_2$ $\quad\quad\mid\quad\mid$ $\quad\quad CO-S$	Enzacryl Polythiolactone	None
$-NH-CH-CH_2-SH$ $\quad\quad\mid$ $\quad\quad CO_2H$	Enzacryl Polythiol	$K_3 Fe (CN)_6$
$-NH-CH_2-CH(OCH_3)_2$	Enzacryl Polyacetal	H⁺

Fig. 4.4 — Major structural features of Enzacryl® gels and their mode of activation.

polysaccharide supports because of their hydroxyl content and are activated with cyanogen bromide (in an analogous process to cyclic imidocarbonate formation with polysaccharides), whilst copolymers of either acrylamide and acrylic acid or methacrylate (the latter being useful for reactions involving more hydrophobic substrates) are activated using carbodiimides. Poly(4- and 5-acrylamidosalicylic acids) are *N*-substituted amides which can be activated via metal ion chelation (Kennedy & Epton 1973).

Table 4.6

Some acrylamide copolymers for enzyme immobilization

Polymer	Structure
Acrylamide/2-hydroxyethyl methacrylate	$-CH-CH_2-C-CH_2-$ (with CH_3 above the C; $CONH_2$ and $CO_2-CH_2-CH_2-OH$ below)
Acrylamide/acrylic acid	$-CH-CH_2-CH-CH_2-$ ($CONH_2$ and CO_2H below)
Acrylamide/methacrylate	$-CH-CH_2-C-CH_2-$ (with CH_3 above the C; $CONH_2$ and CO_2-CH_3 below)
4-Acrylamidosalicyclic acid	$-CH-CH_2-CH-CH_2-$ (CONH groups attached to aromatic rings bearing OH and CO_2H)
Acryloyl-*N*,*N*′-bis (2,2′-dimethoxyethyl)amine	$-CH------CH_2------CH------CH_2-$ with CO and $N[(CH_2)_2-(OCH_3)_2]_2$ / $N[CH_2)_2-(OCH_3)_2]_2$
Acryloyl-*N*,*N*′-bis (2,2′-dimethoxyethyl) amine/acryloyl morpholine	$-CH------CH_2------CH------CH_2-$ with CO and $N[CH_2)_2-(OCH_3)_2]_2$ / N-morpholine

Closely related products are obtained by the polymerization of acrylo-N,N'-bis(2,2'-dimethoxyethyl)amine (Epton *et al.* 1975) or its copolymerization with N-acryloylmorpholine, the product of which is commercially available as Enacryl® Gel. (Epton *et al.* 1979), and these supports are activated by acid hydrolysis followed by reaction with sodium nitrite in acidic solution.

Polyglycidylmethacrylates are copolymers of glycidyl methacrylate and acrylamide (Krammer *et al.* 1976) which contain oxirane (epoxide) groups which react with free amine groups (and also with amide groups in the polymer unless stored in the cold). Similar products were obtained when ethylene dimethacrylate was used instead of acrylamide, and these materials have been shown to be very promising as supports for enzyme immobilization (Švec *et al.* 1978).

Polyacrylonitrile is activated by methanolic hydrogen chloride which generates an imidoester derivative capable of binding enzymes by covalent attachment via free amino groups.

(f) *Maleic anhydride based copolymers*

The most common example (Levin *et al.* 1964) of this group of supports is the copolymer of maleic anhydride and ethylene crosslinked with diamines (which also serve to neutralize the negatively charged groups produced during the immobilization reactions). The resulting copolymer can also be modified in a number of ways (see Scheme 4.32) to provide alternative methods of coupling with enzymes.

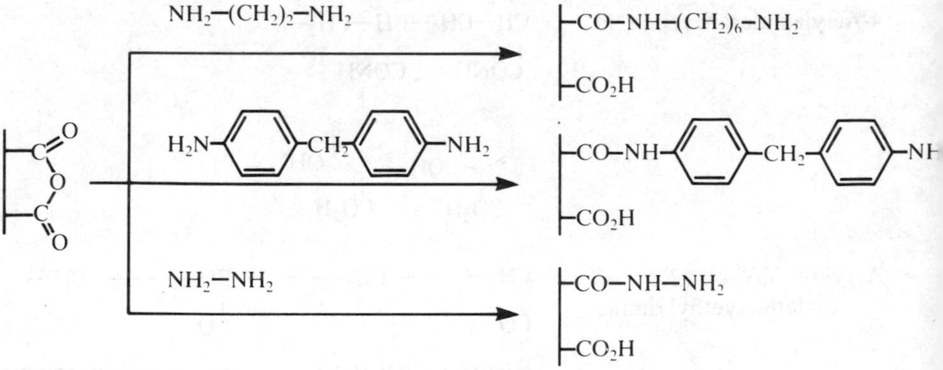

Scheme 4.32

(g) *Polypeptides*

Water insoluble polypeptides have been widely used for enzyme immobilization with the copolymer of L-leucine and 4-amino-DL-phenylalanine being the most commonly used (Engel & Alexander 1971). The polymer is prepared by the copolymerization of 4-N-carbobenzoxyamino-α-N-carboxy-DL-phenylalanine anhydride and N-carboxy-L-leucine anhydride (Scheme 4.33) and activated by acidic sodium nitrite.

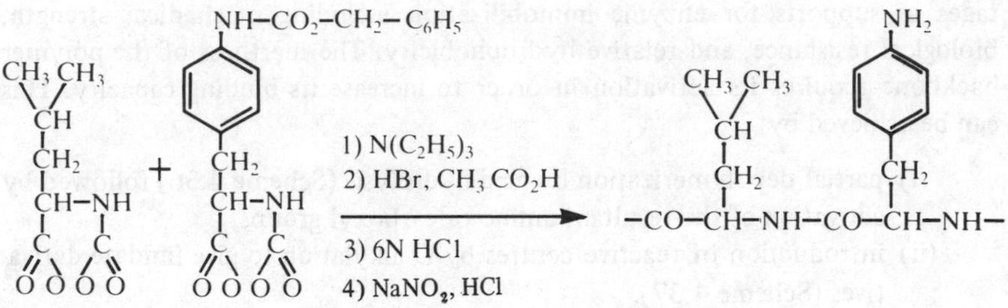

Scheme 4.33

(h) *Vinyl and allyl polymers*

Non-biodegradable supports, which have similar properties to polysaccharides, have been prepared from poly(vinyl alcohol), poly(vinyl amine), poly(vinyl ethers), or poly(allyl alcohol). The polymers are activated in similar ways as for polysaccharides with poly(vinyl alcohol) being activated (Scheme 4.34), with 2,4,6-trichloro-*sym*-triazine (Manecke & Vogt 1976) whilst poly(allyl alcohol) can be activated (Scheme 4.35) with ethyl chloroformate to give a cyclic carbonate derivative (Kennedy *et al.* 1972).

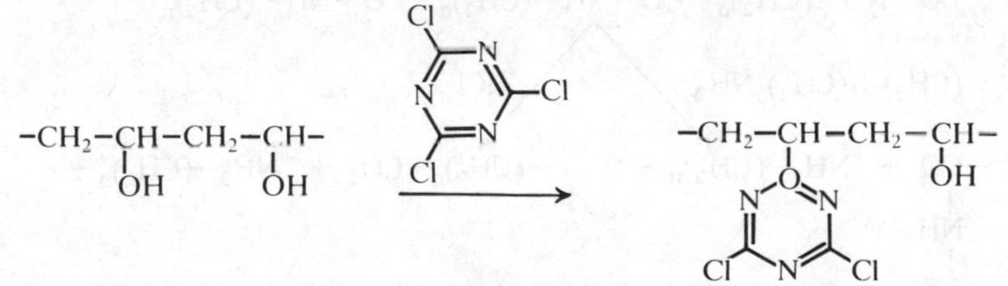

Scheme 4.34

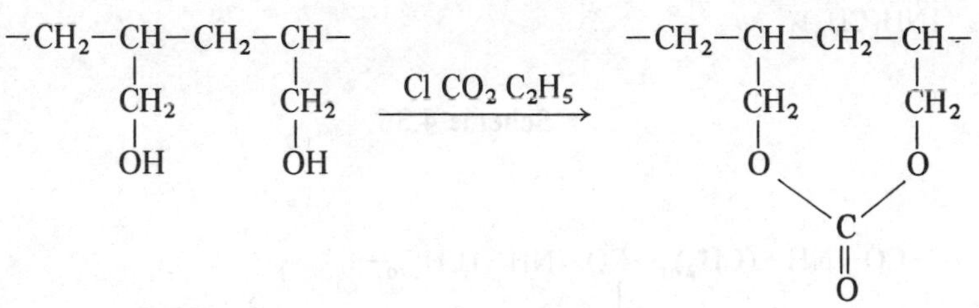

Scheme 4.35

(i) *Polyamides*

Synthetic polyamides produced as condensation polymers of α, ω-diamines and α, ω-dicarboxylic acids (or their diacid chlorides) are known generally as nylons. The different types of nylon differing only in the number of methylene groups in the repeating structure, are available in a number of forms including powders, fibres, hollow fibres, tubes and membranes, and they offer a number of advantages as supports for enzyme immobilization including mechanical strength, biological resistance, and relative hydrophobicity. The inertness of the polymer backbone requires its activation in order to increase its binding capacity. This can be achieved by:

(i) partial depolymerization by acid hydrolysis (Scheme 4.36) followed by activation of the resultant amino or carboxyl groups,

(ii) introduction of reactive centres by O-alkylation to give imidate derivatives (Scheme 4.37),

(iii) introduction of reactive centres by N-alkylation by mild acid depolymerization followed by condensation reactions (Scheme 4.38) of the amino and carboxyl groups with aldehydes and isocyanides to yield polyisonitrile nylon (Goldstein et $al.$ 1974). Enzymes can be bound to the isocyanide group via the Ugi reaction, or the isocyanide group can be modified to other functionalities.

The major disadvantage of using nylons is that the structure must be modified, and the surface areas of all activated forms are relatively small.

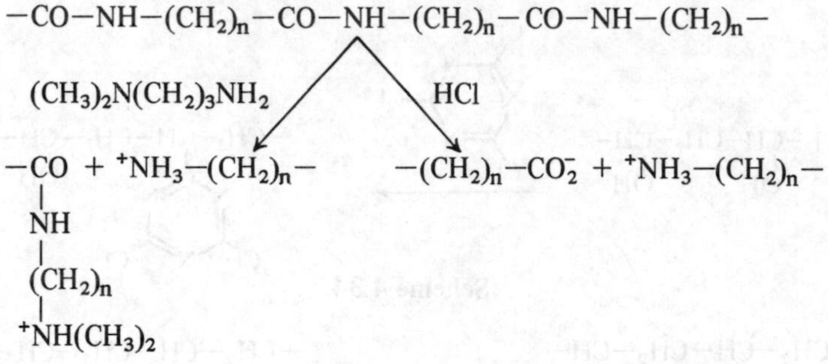

$$-CO-NH-(CH_2)_n-CO-NH-(CH_2)_n-CO-NH-(CH_2)_n-$$

$$(CH_3)_2N(CH_2)_3NH_2 \qquad\qquad HCl$$

$$-CO \; + \; {}^+NH_3-(CH_2)_n- \qquad -(CH_2)_n-CO_2^- \; + \; {}^+NH_3-(CH_2)_n-$$
$$|$$
$$NH$$
$$|$$
$$(CH_2)_n$$
$$|$$
$$^+NH(CH_3)_2$$

Scheme 4.36

$$-CO-NH-(CH_2)_n-CO-NH-(CH_2)_n-$$

$$(C_2H_5)_3O^+BF_4^-/CH_2Cl_2$$

$$-CO-NH-(CH_2)_n-C={}^+NH-(CH_2)_n-$$
$$|$$
$$O-C_2H_5$$

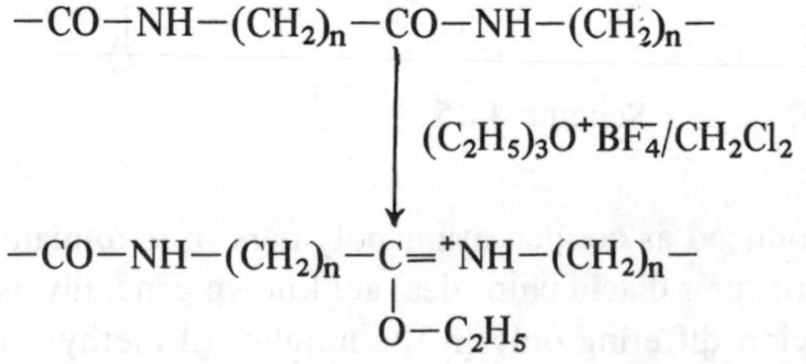

Scheme 4.37

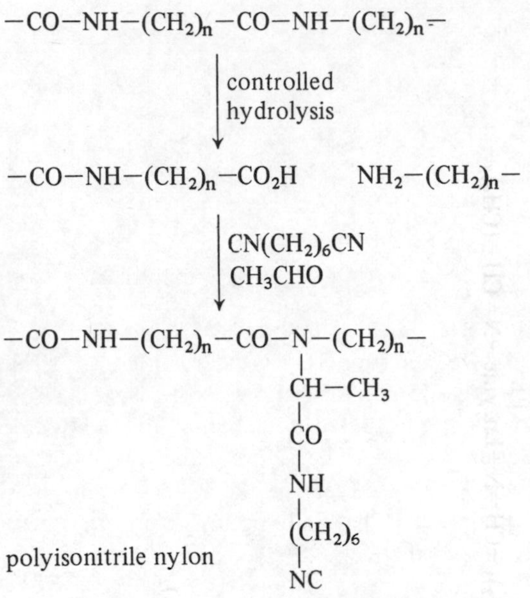

$$-CO-NH-(CH_2)_n-CO-NH-(CH_2)_n-$$

$\downarrow$ controlled hydrolysis

$$-CO-NH-(CH_2)_n-CO_2H \qquad NH_2-(CH_2)_n-$$

$\downarrow$ $\begin{array}{l} CN(CH_2)_6CN \\ CH_3CHO \end{array}$

$$-CO-NH-(CH_2)_n-CO-N-(CH_2)_n-$$
$$\qquad\qquad\qquad\qquad\qquad | $$
$$\qquad\qquad\qquad\qquad\qquad CH-CH_3$$
$$\qquad\qquad\qquad\qquad\qquad | $$
$$\qquad\qquad\qquad\qquad\qquad CO$$
$$\qquad\qquad\qquad\qquad\qquad | $$
$$\qquad\qquad\qquad\qquad\qquad NH$$
$$\qquad\qquad\qquad\qquad\qquad | $$
$$\qquad\qquad\qquad\qquad\qquad (CH_2)_6$$
polyisonitrile nylon $\qquad\qquad\quad | $
$$\qquad\qquad\qquad\qquad\qquad NC$$

Scheme 4.38

4.4 CROSSLINKING

The most common reagent used for intermolecular crosslinking is glutaraldehyde, and it was originally thought to involve Schiff's base formation between the reagent's aldehydo groups and amino groups from the enzyme molecules (Scheme 4.39), but the exceptional stability of the linkages to extremes of pH and temperatures has led to the suggestion (Richards & Knowles 1968) that the reaction involves the conjugate addition of enzyme amino groups to ethylenic double bonds of the α, β-unsaturated oligomers found in the commercial aqueous solutions of glutaraldehyde which are usually used (Scheme 4.40). The suggestion is further strengthened by the fact that the reactivity of freshly distilled glutaraldehyde is much lower towards proteins than is the commercial product. The formation of an insoluble complex is critically dependent on the relative concentrations of enzyme and reagent as well as on the pH and ionic strength of the solution (Brown 1976). Optimum concentrations for protein of between 50 and 200 mg/ml and between 0.3% and 0.6% for glutaraldehyde with a ratio of reagent to enzyme of 10% w/w will give a homogeneous insoluble product with optimum enzyme activity, but the effects of time, temperature, and ionic concentration require optimization for each enzyme (Janson *et al.* 1971).

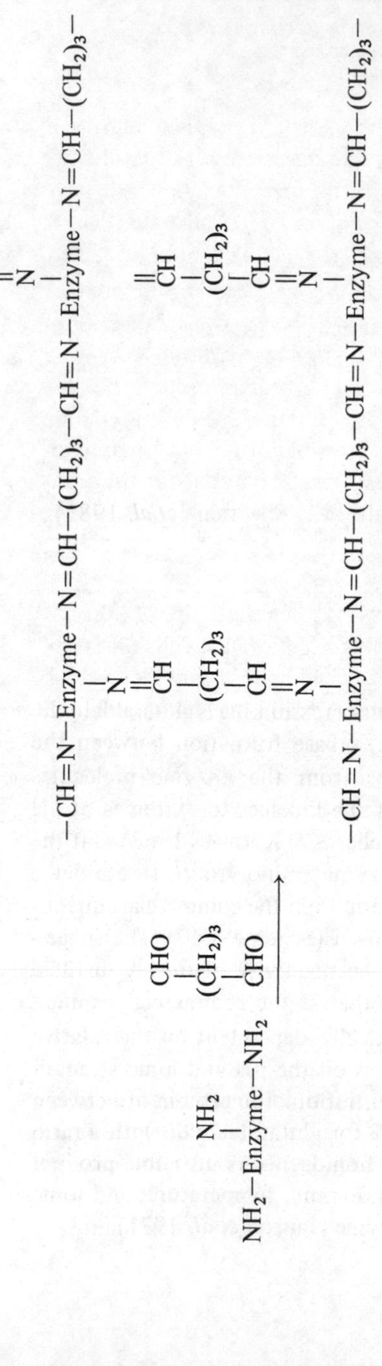

$$-CH=N-Enzyme-N=CH-(CH_2)_3-CH=N-Enzyme-N=CH-(CH_2)_3-$$

$$-N=CH-(CH_2)_3-CH=N-$$

$$-CH=N-Enzyme-N=CH-(CH_2)_3-CH=N-Enzyme-N=CH-(CH_2)_3-$$

Scheme 4.39

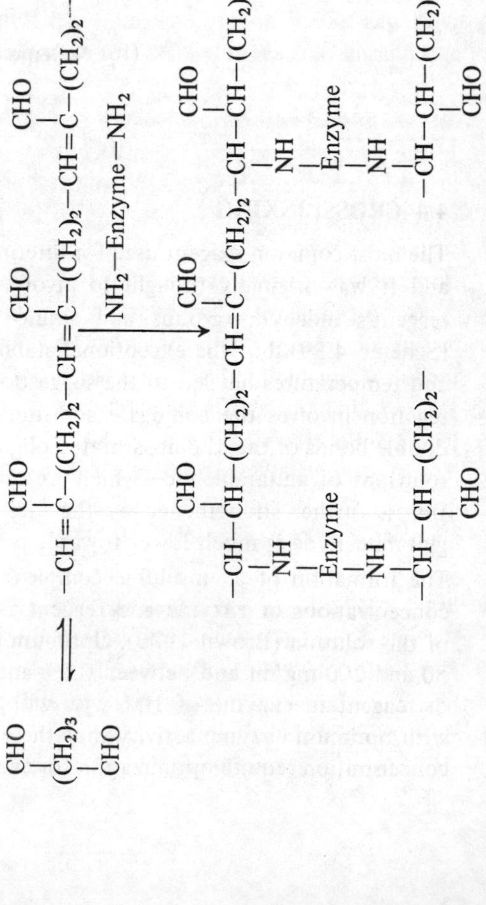

$$NH_2-Enzyme-NH_2$$

$$-CH=C-(CH_2)_2-CH=C-(CH_2)_2-CH=C-(CH_2)_2---$$

$$CHO \quad CHO \quad CHO$$

$$-CH-CH-(CH_2)_2-CH-CH-(CH_2)_2-$$

$$CHO \quad CHO$$

$$-NH \quad -NH$$

$$-Enzyme$$

$$-NH \quad -NH$$

$$-CH-CH-(CH_2)_2- \quad -CH-CH-(CH_2)_2-$$

$$CHO \quad CHO$$

$$NH_2-Enzyme-NH_2$$

$$NH_2$$

$$CHO-(CH_2)_3-CHO$$

4.5 IMMOBILIZED CELLS

Cells (either viable or nonviable) can be immobilized in much the same way as can enzyme, although many of the methods must be modified to accommodate the nature of cells. The methods used can be classified in a similar manner to the classification described in Part A, Chapter 4 (Kennedy & Cabral 1983).

4.5.1 Entrapment

The entrapment of cells within the interstitial spaces of a polymer matrix is one of the major methods employed and it uses the same methods of producing the polymer as described in Part A and B of this book. Polymers which have been used include polyacrylamide (D'Souza & Nadkarni 1980), alginate (Paul & Vignais 1980, Larreta Garde *et al.* 1979), silica hydrogel (Rouxhet *et al.* 1981), and gelatin (Park *et al.* 1980), with the polysaccharide gels being more popular owing to their formation under mild conditions. However, the method is of restricted use, owing to diffusion constraints on the final preparation to cellular enzyme reactions involving relatively small sized reactants and products; but the mild reaction conditions and the nontoxic stabilizing effects of the polysaccharide matrices have led to very stable immobilized cell systems which can operate on pilot plant scale to produce materials which have great potential for the food and pharmaceutical industries (for example isomaltulose, Cheetham *et al.* 1982).

4.5.2 Physical adsorption

In the natural state all microbial cells exist in an absorbed state, in soil, the silt in rivers, attached to skin, etc., and it is a property of microbial cells that they will attach to any surface with which they are in contact and, in time, render the surface biologically active. The adhesion of certain types of microorganisms can be attributed to the formation of stalks which attach to the surface, whilst other types of cells have slimy films or cement-like materials on the cell wall which causes adsorption. Many materials have been used for immobilization purposes including organic materials such as ion exchange materials (Hattori & Furusaka 1961, Gainer *et al.* 1981), poly(vinyl chloride) (Holló *et al.* 1980), polypropylene (Holló *et al.* 1979), coal (Ngian & Martin 1980) and wood chips (Moo-Young *et al.* 1980), and inorganic materials such as glass (Rouxhet *et al.* 1981), ceramics (Messing *et al.* 1979), diatomaceous earth (Grindbergs *et al.* 1977), and stainless steel (Atkinson *et al.* 1979).

4.5.3 Chelation or metal binding

Various cells have been immobilized using hydrous titanium(IV) oxide (Kennedy *et al.* 1980b) and zirconium hydroxide (Kennedy *et al.* 1976) *per se* or hydrous titanium oxide activated cellulose (Kennedy *et al.* 1980b) and used in pilot plant scale fermentors. The method is sufficiently mild as not to disrupt the cells' life processes, yet the cells have been shown to become firmly bound.

4.5.4 Covalent binding

Whilst this method is the most prevalent for the immobilization of enzymes, owing to the preparation of materials with high operational stabilities, it is not widely used for cell immobilization owing to the toxicity of the reagents used which can in some cases destroy both the enzyme activity and viability of the cell (Chipley 1974). In those cases where an active and viable immobilized cell is obtained (Table 4.7) the product is free from diffusion limitations and has minimal cell leakage over long periods of use.

Table 4.7
Principle examples of covalently attached whole cells

Support	Coupling reaction	Reference
Hydroxyalkylmethacrylate + epichchlorohydrin, diamine and glutaraldehyde	Schiff's base formation	Jirkú *et al.* 1980
Amine activated silica + glutaraldehyde	Schiff's base formation	Navarro & Durand 1977
Carboxymethylcellulose + carbodiimide	Peptide bond formation	Jack & Zajic 1977
Ethylene/maleic anhydride copolymer	Peptide bond formation	Shimizu *et al.* 1975
Glass + polyisocyanate	Peptide bond formation	Messing & Oppermann 1979, Messing *et al.* 1979
Zirconia ceramics + polyisocyanate	Peptide bond formation	Messing & Oppermann 1979, Messing *et al.* 1979
Silochrome + toluene isocyanate	Peptide bond formation	Romanov-Skaya *et al.* 1980
Cellulose + triazine	Alkylation	Gainer *et al.* 1981

4.5.5 Crosslinking

The crosslinking method, though relatively mild in character, does suffer from the drawback that the toxicity of the bifunctional reagents used for enzyme immobilization towards a number of cells is such that the application of the method is severely limited. The method has been applied to cells which catalyze single reactions (Chibata *et al.* 1974) and employs glutaraldehyde, diazotized diamines, and toluene diisocyanate which react with the free amino groups of the cell wall peptidoglycans. Crosslinking and entrapment have been used as a combined method of immobilization using glutaraldehyde and alginate or polyacrylamide (Ziomek *et al.* 1982)

A milder, less toxic method of crosslinking has been developed based on the physical crosslinking by flocculation, which leads to high cell concentrations per unit volume. A variety of flocculating agents have been used (Lee & Long 1974, Long 1976) including: cationic polyelectrolytes, such as polyamines, polyethyleneimine and cationic polyacrylamides; anionic polyelectrolytes such as carboxyl-substituted polyacrylamides, polystyrene sulphonates and polycarboxylic acids; and metal salts such as the oxides, hydroxides, sulphates and phosphates of magnesium(II), calcium(II), iron(II) and manganese(II).

4.6 OTHER IMMOBILIZED BIOLOGICALLY ACTIVE MOLECULES

Many other biologically active molecules have been immobilized using the technology developed for enzyme immobilization and the products can be classified into immunoadsorbents, affinity chromoatographic media and immobilized antibiotics.

4.6.1 Immunoadsorbents

Immobilized antibodies are prepared principally for the purification of antigens by a type of column chromatography, referred to as immunoadsorption, in which the solution containing impure antigen is passed through a bed of immobilized antibody. The specific antibody is adsorbed via the antibody—antigen interaction, whilst all other components of the solution are washed through the column. By changing the eluant the antibody—antigen interaction can be eliminated, resulting in the elution of purified antigen (Kennedy *et al.* 1980c), thereby short-circuiting the more conventional lengthy chromatographic techniques. Immunoadsorption can of course be applied to the purification of antibodies by use of immobilized antigens, or to the isolation of specific components of immunoglobulin by use of, for example, immobilized antibodies with high binding activity for allotypic determinants (Kennedy *et al.* 1982), thereby allowing opportunities to study the minor allotypic components of rarely expressed immunoglobulin. Immobilized antibodies and antigens have also been used in radioimmunoassay techniques (Kennedy & Cho Tun 1973a). The immobilization methods most used for the preparation of immunoadsorbents are the cyanogen bromide activation of polysaccharides such as agarose and the cyclic carbonate derivatives of cellulose.

4.6.2 Affinity chromatography media

Affinity chromatography is a technique by which substances can be separated on the basis of their differing strengths of interactions with a support material which has been modified to contain groups with specific interactions. The use of biological rather than chemical interactions is of direct relevance to this chapter, and the molecules which are immobilized to modify the solid support can be classified into: lectins; amino acids and peptides; carbohydrates; and nucleosides, nucleotides and nucleic acids. With all these materials the mode of use is essentially the same. A solution of an impure material is allowed to contact the immobilized material in a column or as an additive to the solution. The solution is separated from the solid support, and after washing, the molecule to be purified is removed from the solid support by disruption of the specific interaction.

5.6.2.1 *Immobilized lectins*
Lectins are a family of proteins and glycoproteins which have a specific affinity
for given carbohydrate residues which can be in the form of monosaccharides or
combined with other monosaccharides to give polysaccharides or carbohydrate-
containing macromolecules such as glycoproteins. Immobilization of lectins
has led to the development of specific methods of separation and purification of
carbohydrate materials from similar materials for which the particular lectin has
no affinity (Kennedy & White 1983). The supports commonly used for lectin
immobilization are the polymeric carbohydrates including agarose (Kennedy &
Rosevear 1973), and a number of immobilized lectins are commercially available
(for example ConA-Sepharose® from Pharmacia and Glycaminosylex® from
Miles).

4.6.2.2 *Immobilized amino acids and peptides*
Immobilized amino acids and their polymers (peptides and proteins) have uses
in affinity chromatography in addition to the catalytic properties of some
immobilized proteins (namely immobilized enzymes). Immobilized enzymes
have been used for the affinity chromatographic purification of their substrates
using conditions under which the protein can bind its substrate but not react
with it. However, examples of the reverse situation in which a substrate is
immobilized and used to purify an enzyme are more common. For example,
L-lysine has been immobilized on cyanogen bromide activated agarose and used
for the purification of plasmin and plasminogen (Burge *et al.* 1981). It is also
possible to immobilize inhibitors to protein action and use the product to
purify enzymes as shown by Slater & Strout (1981) who immobilized the
synthetic D-leucine-containing octapeptide inhibitor for the purification of
renin.

4.6.2.3 *Immobilized carbohydrates*
Carbohydrates in the form of monosaccharides, oligosaccharides, and polysaccharides have been used:

(a) to study the interaction between carbohydrates and their enzymes or
    between carbohydrate-containing macromolecules and other compounds
    as shown by the immobilization of heparin to study the interaction of
    glycosaminoglycans with proteases (Marossy 1981) or the interaction
    between heparin and thrombin (Larsson *et al.* 1980).
(b) to purify carbohydrate-directed lectins or enzymes as shown by the immo-
    bilization of 2-amino-2-deoxy-D-glucose for the purification of yeast hexo-
    kinase isoenzymes (Kopetzki & Entian 1982), or the purification of α-1,4-
    D-glucan phosphorylase using immobilized starch (Steup 1981).
(c) to prepare solid phase substrates for the estimation of carbohydrate-direc-
    ted enzymes as typified by the diepoxide crosslinking of starch for the
    determination of α-amylase (Ceska 1972).

4.6.2.4 *Immobilized nucleosides, nucleotides and nucleic acids*
Insoluble derivatives of nucleic acids and their components have been used:

(a) for the fractionation and purification of other nucleic acids, nucleotides etc. via the mechanism of base pairing (Lee *et al.* 1970);

(b) for synthesis by the multiplication of single strands of nucleic acids via the base pairing mechanism (Wagner *et al.* 1971);

(c) as insoluble substrates for the affinity chromatography or estimation of nucleic acid-directed enzymes such as the DNA glycosylases (Thomas *et el.* 1982);

(d) as affinity chromatography materials for the purification of nucleic acid-binding proteins (Trevillyan & Paul 1982).

5.6.3 Immobilized antibiotics
Immobilization of antibiotics with retention of antimicrobial activity has been achieved using a variety of solid supports including poly(*N*-acryloyl-4- and -5-aminosalicyclic acids) (Kennedy *et al.* 1973), cellulose carbonate (Kennedy & Cho Tun 1973b), and hydrous metal oxides (Kennedy & Humphreys 1976, Kennedy *et al.* 1981), and has provided the potential to protect biodegradable substances (including immobilized enzymes) from microbial attack and for slow-release formulation in the medical and industrial fields. Immobilized antibiotics can also be used as affinity chromatography media for the purification of antibiotic-directed enzymes (Coombe & George 1982) and antibiotic-binding proteins (Tamura *et al.* 1980).

REFERENCES
Allen, B. R., Charles, M. & Coughlin, R. W. (1979) *Biotechnol. Bioeng.* **21** 689–706.
Almé, B. & Nyström, E. (1971) *J. Chromatogr.* **59** 45–52.
Atkinson, B., Black, G. M., Lewis, P. J. S. & Pinches, A. (1979) *Biotechnol. Bioeng.* **21** 193–200.
Axén, R., Porath, J. & Ernback, S. (1967) *Nature* **214** 1302–1304.
Axén, R., Vretblad, P. & Porath, J. (1971) *Acta Chem. Scand.* **25** 1129–1132.
Barker, S. A., Doss, S. H., Gray, C. J., Kennedy, J. F., Stacey, M. & Yeo, T. H. (1971) *Carbohydr. Res.* **20** 1–7.
Beddows, C. G., Mirauer, R. A. & Guthrie, J. T. (1980) *Biotechnol. Bioeng.* **22** 311–321.
Borlazza, V. C., Cheetham, N. W. H. & Sowthwell-Kelly, P. T. (1980) *Carbohydr. Res.* **79** 125–132.
Brandenberger, H. (1955) *Angew. Chem.* **67** 661.
Brandenberger, H. (1956) *Rev. Ferment. Ind. Aliment* **11** 237–241.
Brandt, J. & Anderson, L.-O. (1976) *Acta Chem. Scand.* **B30** 815–819.
Brotherton, J. E., Emery, A. & Rodwell, V. W. (1976) *Biotechnol. Bioeng.* **18** 527–543.
Broun, G. B. (1976) In: *Methods in Enzymology* Vol. 44, p. 263–280. Ed. by Mosbach, K. Academic Press, New York.
Brown, E, & Racois, A. (1972) *Tetrahedron Lett.* 5077–5080.
Burge, J., Nicholson-Weller, A. & Austin, K. F. (1981) *Mol. Immunol.* **18** 47–54.
Burns, R. A. (1976) US Patent 3,953,292.
Cabral, J. M. S., Novais, J. M. & Cardoso, J. P. (1981a) In: *3rd Internat. Chem. Eng. Symp.* Chempor '81, Porto
Cabral, J. M. S., Novais, J. M. & Cardoso, J. P. (1981b) *Biotechnol. Bioeng.* **23** 2083–2092.
Cabral, J. M. S., Kennedy, J. F. & Novais, J. M. (1982a) *Enzyme Microb. Technol.* **4** 337–342.

Cabral, J. M. S., Kennedy, J. F. & Novais, J. M. (1982b) *Enzyme Microb. Technol.* 4 343–348.
Campbell, J. & Chang, T. M. S. (1975) *Biochim. Biophys. Acta* 397 101–109.
Campbell, J. & Chang, T. M. S. (1976) *Biochem. Biophys. Res. Comm.* 69 562–569.
Cardoso, J. P., Chaplin, M. F., Emery, A. N., Kennedy, J. F. & Revel-Chion, L. P. (1978) *J. Appl. Chem. Biotechnol.* 28 775–785.
Carlsson, J., Axén, R., Brocklehurst, K. & Crook, E. M. (1974) *Eur. J. Biochem.* 44 189–194.
Ceska, M. (1972) *Clin. Chem. Acta* 36 453–461.
Chaplin, M. F. & Kennedy, J. F. (1976) *Carbohydr. Res.* 50 267–274.
Charles, M., Couglin, R. W., Parachuri, E. K., Allen, B. R. & Hasselberger, F. X. (1975) *Biotechnol. Bioeng.* 17 203–210.
Cheetham, P. S. J., Imber, C. E. & Isherwood, J. (1982) *Nature* 299 628–631.
Chibata, I., Tosa, T. & Sato, T. (1974) *Appl. Microbiol.* 27 878–885.
Chipley, J. R. (1974) *Microbios* 10 115–120.
Cho, Y. K. & Bailey, J. E. (1978) *Biotechnol. Bioeng.* 20 1651–1665.
Cho, Y. K. & Bailey, J. E. (1979) *Biotechnol. Bioeng.* 21 461–476.
Conte, A. & Lehmann, K. (1971) *Z. Physiol. Chem.* 352 533–541.
Coombe, R. G. & George, A. M. (1982) *Biochemistry* 21 871–875.
Coulet, P. R., Godinot, C. & Gautheron, D. C. (1975) *Biochim. Biophys. Acta* 391 272–281.
Coulet, P. R., Sternberg, R. & Thévenot, D. R. (1980) *Biochim. Biophys. Acta* 612 317–327.
Dinelli, D., Marconi, W., Cecere, F., Galli, G. & Morisi, F. (1978) In: *Enzyme Engineering* Vol. 3, p. 477–481. Ed. by Pye, E. K. & Weetall, H. H. Plenum Press, New York.
D'Souza, S. F. & Nadkarni, G. B. (1980) *Enzyme Microb. Technol.* 2 217–222.
Engel, A. M. & Alexander, B. (1971) *J. Biol. Chem.* 246 1213–1221.
Epton, R. & Thomas, T. H. (1971) In: *An Introduction to Water Insoluble Enzymes,* Koch-Light Labs. Ltd, England.
Epton, R., McLaren, J. V. & Thomas, T. H. (1972) *Carbohydr. Res.* 22 301–306.
Epton, R., Hibbert, B. L. & Marr, G. (1975) *Polymer* 16 314–320.
Epton, R., Hobson, M. E. & Marr, G. (1979) *Enzyme Microb. Technol.* 1 37–40.
Flynn, A. & Johnson, D. B. (1978) *Biotechnol. Bioeng.* 20 1445–1454.
Gainer, J. L., Kirwan, D. J., Foster, J. A. & Seylan, E. (1981) *Biotechnol. Bioeng. Symp.* 10 35–42.
Glassmeyer, C. K. & Ogle, J. D. (1971) *Biochemistry* 10 786–792.
Goldstein, L., Freeman, A. & Sokolovsky, M. (1974) *Biochem. J.* 143 497–509.
Grasset, L., Cordier, D. & Ville, A. (1979) *Process Biochem.* 14 2–5.
Greenfield, P. F. & Laurence, R. L. (1975) *J. Food Sci.* 40 906–910.
Grindbergs, M., Hildebrand, R. P. & Clarke, B. J. (1977) *J. Inst. Brew.* 83 25–29.
Hattori, T. & Furusaka, C. (1961) *J. Biochem. (Tokyo)* 50 312–315.
Holló, J. To'th, J., Tengerdy, R. P. & Johnson, J. E. (1979) In: *Immobilized Microbial Cells,* p. 73–86. Ed. by Venkatsubramanian, K. ACS Symp. Ser. 106, Washington.
Holló, J. Weinbrenner, Z., Czakó, L. & To'th, J. (1980) *Biotechnol. Lett.* 2 87–92.
Jack, T. R. & Zajic, J. E. (1977) *Biotechnol. Bioeng.* 19 631–648.
Jagendorf, A. T., Patchornik, A., & Sela, M. (1963) *Biochim. Biophys. Acta* 78 516–528.
Jansen, E. F., Tomimatsu, Y. & Olson, A. C. (1971) *Arch. Biochem. Biophys.* 144 394–400.
Jirkú, V., Turková, J. & Krumphanzl, V. (1980) *Biotechnol. Lett.* 2 509–513.
Kennedy, J. F. (1979) *Chem. Soc. Rev.* 8 221–257.
Kennedy, J. F. & Cabral, J. M. S. (1983) In: *Applied Biochemistry and Bioengineering,* Vol. 4, p. 189–280, Ed. by Wingard, L. B. & Chibata, I. Academic Press, New York.
Kennedy, J. F. & Chaplin, M. F. (1979) *Enzyme Microb. Technol.* 1 197–200.
Kennedy, J. F. & Cho Tun, H. (1973a) *Carbohydr. Res.* 30 11–19.
Kennedy, J. F. & Cho. Tun, H. (1973b) *Antimicrob. Agents Chemother.* 3 575–579.
Kennedy, J. F. & Epton, J. (1973) *Carbohydr. Res.* 27 11–20.
Kennedy, J. F. & Humphreys, J. D. (1976) *Antimicrob. Agents Chemother.* 9 766–770.
Kennedy, J. F. & Kay, I. M. (1976) *J. Chem. Soc. Perkin Trans.* 1 329–335.
Kennedy, J. F. & Rosevear, A. (1973) *J. Chem. Soc., Perkin Trans.* 1 2041–2046.
Kennedy, J. F. & White, C. A. (1979) *Stärke* 31 375–381.
Kennedy, J. F. & White, C. A. (1983) *Bioactive Carbohydrates: in Chemistry, Biochemistry and Biology,* Ellis Horwood, Chichester.
Kennedy, J. F. & Zamir, A. (1973) *Carbohydr. Res.* 29 497–501.
Kennedy, J. F. & Zamir, A. (1975) *Carbohydr. Res.* 41 227–233.
Kennedy, J. F., Barker, S. A. & Rosevear, A. (1972) *J. Chem. Soc. Perkin Trans.* 1 2568–2573.

Kennedy, J. F., Epton, J. & Kennedy, G. R. (1973) *Antimicrob. Agents Chemother.* **3** 29–32.
Kennedy, J. F., Barker, S. A. & Humphreys, J. D. (1976) *Nature* **261** 242–244.
Kennedy, J. F., Barker, S. A. & White, C. A. (1977a) *Carbohydr. Res.* **54** 1–12.
Kennedy, J. F., Barker, S. A. & White, C. A. (1977b) *Stärke* **29** 240–243.
Kennedy, J. F., Barnes, J. A. & Matthews, J. B. (1980a) *J. Chromatogr.* **196** 379–389.
Kennedy, J. F., Humphreys, J. D., Barker, S. A. & Greenshileds, R. N. (1980b) *Enzyme Microb. Technol.* **2** 209–216.
Kennedy, J. F., Catty, D. & Keep, P. A. (1980c) *Int. J. Biol. Macromol.* **2** 137–142.
Kennedy, J. F., Humphreys, J. D. & Barker, S. A. (1981) *Enzyme Microb. Technol.* **3** 129–136.
Kennedy, J. F., Keep, P. A. & Catty, D. (1982) *J. Immunol. Methods* **50** 57–75.
Kitajema, M., Miyano, S. & Kondo, A. (1969) *Kogyo Kageku Zasshi* **72** 493–499.
Kohn, J. & Wilchek, M. (1982) *Enzyme Microb. Technol.* **4** 161–163.
Kopetzki, E. & Entian, K. D. (1982) *Analyt. Biochem.* **121** 181–185.
Krämmer, D. M., Lehman, K., Pennerviss, H. & Plainer, H. (1976) *XXIII Coll. "Part. Biol. Fluids"* **23** 505–513.
Larreta Garde, V., Thomasset, B. & Barbotin, J.-N. (1981) *Enzyme Microb. Technol.* **3** 216–218.
Larsson, R., Olsson, P. & Lindahl, U. (1980) *Thromb. Res.* **19** 43–54.
Lee, C. K. & Long, M. E. (1974) US Patent 3,821,086.
Lee, J. C., Weith, H. L. & Gilham, P. T. (1970) *Biochemistry* **9** 113–118.
Levin, Y., Pecht, M., Goldstein, L. & Katchalski, E. (1964) *Biochemistry* **3** 1905–1913.
Li, N. N. (1971) *Ind. Eng. Chem. Process Res. Develop.* **10** 215–221.
Linko, Y.-Y., Poutanen, K., Weckeström, L. & Linko, P. (1979) *Enzyme Microb. Technol.* **1** 26–30.
Long, M. E. (1976) US Patent 3,935,069.
Maeda, H. & Suzuki, H. (1972) *Agric. Biol. Chem.* **36** 1581–1593.
Maeda, H., Tsao, G. T. & Chen, L. T. (1978) *Biotechnol. Bioeng.* **20** 383–402.
Manecke, G. & Förster, H. J. (1966) *Makromol. Chem.* **91** 136–154.
Manecke, G. & Gunzel, G. (1967) *Naturwissenschafen* **54** 531–533.
Manecke, G. & Vogt, H. G. (1976) *Makromol. Chem.* **177** 725–739.
Maressy, K. (1981) *Biochim. Biophys. Acta* **659** 351–361.
Messing, R. A. (1974) *Process Biochem.* **9** (11) 26–28.
Messing, R. A. (1975) US Patent 3,910,851.
Messing, R. A. (1976). In: *Methods in Enzymology* Vol. 44, p. 148–169. Ed. by Mosbach, K., Academic Press, New York.
Messing, R. A. & Opperman, R. A. (1979) *Biotechnol. Bioeng.* **21** 49–58.
Messing, R. A., Oppermann, R. A. & Kolot, F. (1979). In: *Immobilized Microbial Cells*, p. 13–28. Ed. by Venkatsubramanian, K. ACS Symp. Ser. 106, Washington.
Mitz, M. A. & Summaria, L. J. (1961) *Nature* **189** 576–577.
Monsan, P. & Durand, G. (1971) *C.R. Acad. Sci., Ser. C* **273** 33–36.
Mori, T., Tosa, T. & Chibata, T. (1973) *Biochim. Biophys. Acta* **321** 653–661.
Moo-Young, M., Lamptey, J. & Robinson, C. W. (1980) *Biotechnol. Lett* **2** 541–548.
Murphy, R. F., Conlon, J. M., Iman, A. & Kelly, G. J. C. (1977) *J. Chromatogr.* **135** 427–433.
Navarro, J. M. & Durand, G. (1977) *Eur. J. Appl. Microbiol. Biotechnol.* **4** 243–254.
Ngian, K. F. & Martin, W. R. B. (1980). *Biotechnol. Bioeng.* **22** 1007–1014.
Ostergaard, J. C. W. & Martiny, S. C. (1973) *Biotechnol. Bioeng.* **15** 561–564.
Paine, M. A. & Carbonell, R. G. (1975) *Biotechnol. Bioeng.* **17** 617–619.
Park, Y. H., Chung, T. W. & Moon, H H. (1980) *Enzyme Microb. Technol.* **2** 227–233.
Paul, F. & Vignais, P. M. (1980) *Enzyme Microb. Technol.* **2** 281–287.
Porath, J. Låås, T. & Janson, J. C. (1975) *J. Chromatogr.* **103** 49–62.
Wilchek, M., Academic Press, New York.
Porath, J. Laas, T. & Janson, J. C. (1975) *J. Chromatogr.* **103** 49–62.
Richards, F. M. & Knowles, J. R. (1968) *J. Mol. Biol.* **37** 231–233.
Romanov-Skaya, V. A., Karpenko, V. I., Pantskhava, E. S., Grinberg, T. A. & Malashenko, Y. R. (1980) *6th Int. Ferment. Symp. Abstracrs* F-12.1.14, p. 123, Ottawa.
Rouxhet, P. G., Van Haecht, J. L., Didelez, J., Gerard, P. & Briquet, M. (1981) *Enzyme Microb. Technol.* **3** 49–54.
Schovers, J. & Sandine, W. E. (1973) US Patent 3,733,205.
Shimizu, S. Y. & Lenhoff, H. M. (1979) *J. Solid-Phase Biochem.* **4** 75–94.
Shimizu, S., Moriok, H., Tani, Y. & Ogata, K. (1975) *J. Ferment. Technol.* **53** 77–83.
Slater, E. E. & Strout, H. V. (1981) *J. Biol. Chem.* **256** 8164–8171.

Steup, M. (1981) *Biochim. Biophys. Acta* **659** 123–131.
Švec, F., Kálal, J., Menyailova, I. I. & Nakhapteyan, L. A. (1978) *Biotechnol. Bioeng.* **20** 1319–1328.
Szwajcer, E., Brodelius, P. & Mosbach, K. (1982) *Enzyme Microb. Technol.* **4** 409–413.
Tamura, T., Suzuki, H., Nishimura, Y., Mizoguchi, J. & Hirato, Y. (1980) *Proc. Natl. Acad. Sci. USA* **77** 4499–4503.
Thomas, L., Yang, C. H. & Goldthwaite, D. A. (1982) *Biochemistry* **21** 1162–1169.
Tomb, W. H. & Weetall, H. H. (1979) US Patent 3,783,101.
Torchilin, V. P., Tischenko, E. G. & Smirnov, V. N. (1977) *J. Solid-Phase Biochem.* **2** 19–29.
Tosa, T., Sato, T., Mori, T., Yamamoto, K., Takata, I., Nishida, Y. & Chibata, I. (1979) *Biotechnol. Bioeng.* **21** 1697–1709.
Trevillyan, J. M. & Paul, M. L. (1982) *J. Biol. Chem.* **257** 3978–3986.
Ugi, I. (1962) *Angew. Chem. Int. Ed.* **1** 8–21.
Van Leemputten, E. (1977) US Patent 4,017,364.
van Velzen, A. G. (1974) US Patent 3,838,007.
Vieth, W. R. & Venkatasubramanian, K. (1976). In: *Methods in Enzymology,* Vol. 44, pp. 243–263. Ed. by Mosbach, K., Academic Press, New York.
Vretblad, P. & Axén, P. (1971) *FEBS Letts.* **18** 254–256.
Wagner, A. F., Bugianesi, R. L. & Shen, T. Y. (1971) *Biochem. Biophys. Res. Comm.* **45** 184–189.
Weakley, F. B. & Mahltretter, C. L. (1973) *Biotechnol. Bioeng.* **15** 1189–1192.
Weetall, H. H. (1976). In: *Methods in Enzymology* Vol. 44, p. 134–148. Ed. by Mosbach, K. Academic Press, New York.
Weetall, H. H. & Detar, C. C. (1975) *Biotechnol. Bioeng.* **17** 295–299.
White, C. A. & Kennedy, J. F. (1980) *Enzyme Microb. Technol.* **2** 82–90.
Wilchek, M., Oka, T. & Topper, Y. J. (1975) *Proc. Nat. Acad. Sci. USA* **72** 1055–1058.
Ziomek, E., Martin, W. G., Veliky, I. A. & Williams, R. E. (1982) *Enzyme Microb. Technol.* **4** 405–408.

Enzymes in clinical analyses – data

Dr. B. J. GOULD, Department of Biochemistry, University of Surrey, Guildford, England and **Dr. B. F. ROCKS,** Department of Pathology, The Royal Sussex County Hospital, Brighton, England

5.1 INTRODUCTION AND ACKNOWLEDGEMENT

The information gathered together in this chapter comes from literature written by the manufacturers and suppliers in the UK who are listed in Table 5.5 at the end of this chapter. Many of these are internationally known firms whose products are widely available. Some, however, are small firms, particularly those involved in EIA, which is the newest of the technologies described in Part A, Chapter 5. We would like to thank all those firms who sent us information, several of whom went to considerable trouble to find the required information and then to keep us up-to-date. Szabo & Örs (1983) produced two comprehensive tables listing (i) substrates measured by enzymic methods in clinical chemical analysis, and (ii) the most frequently used EIA methods. Their tables include many references to materials for which reagent kits are not currently available.

5.2 SUBSTRATES MEASURED ENZYMICALLY IN CLINICAL LABORATORIES

Substrates are measured in blood, plasma, serum and urine to help with the diagnosis and monitoring of many disease states. It is essential to collect data on the normal healthy population for comparison. Further information on the interpretation and usefulness of these analyses can be obtained from specialist publications (Zilva & Pannall 1979, Whitby, Percy-Robb & Smith 1980).

Table 5.1 shows the substrates that are analysed using enzymes as analytical tools in clinical laboratories. Only those substrates for which pre-packed kits of reagents are available have been listed in Table 5.1. Single enzyme systems are available for ammonia, α_1-antitrypsin, ethanol, galactose, glucose(vi), lactate, α_2-macroglobulin, pyruvate, urea(i), and uric acid(i). For the other substrates, and also for galactose, glucose, urea and uric acid, there are systems which involve more than one enzyme. The most common end point is the conversion between NAD^+ and NADH. Only with glucose(v) and triglycerides(ii) is NADH measured by the formation of a formazan. The measurement of O_2 or H_2O_2 by

Table 5.1   Use of enzymes as analyti

SUBSTRATE		REACTIONS
Adenosine-5′ monophosphate (AMP)	ADP + Phosphoenolpyruvate	Pyruvate kinase
Adenosine-5′ diphosphate (ADP)	Pyruvate + NADH + H⁺	Lactate dehydrogenase
	AMP + ATP	Myokinase
Adenosine-5′-triphosphate (ATP)	ATP + 3-Phosphoglycerate	Phosphoglycerate kinase
	1,3-Bisphosphoglycerate + NADH + H⁺	Glyceraldehyde phosphate dehydrogena
Ammonia	2-Oxoglutarate + NH₄⁺ + NAD(P)H	Glutamate dehydrogenase
α₁-Antitrypsin	Trypsin + α₁-Antitrypsin	
	N-Benzoylarginine	Trypsin (residu
2,3 Bisphospho-glycerate (formerly 2,3-diphospho-glycerate, DPG)	2,3-Bisphosphoglycerate	2,3-Bisphosphoglycera Phosphatase
	3-Phosphoglycerate + ATP	3-Phosphoglycerate kinase
	1,3-Bisphosphoglycerate + NADH + H⁺	Glyceraldehyde 3-phosphate dehydrogen
Carbon dioxide	CO₂ + Phosphoenolpyruvate	Phosphoenolpyruvate carboxylase
	Oxaloacetate + NADH + H⁺	Malate dehydrogenase
Cholesterol (i)	Cholesterol esters + H₂O	Cholesterol esterase
	Cholesterol + O₂	Cholesterol oxidase
	H₂O₂ + Chromogen	Peroxidase
(ii)	Cholesterol esters + H₂O	Cholesterol esterase
	Cholesterol + O₂	Cholesterol oxidase
	H₂O₂ + CH₃OH	Catalase

ols in the measurement of substrates

	END POINT	KIT† SUPPLIERS
TP + Pyruvate	Loss of NADH. Measure ADP, then add myokinase and measure AMP	5
Lactate + NAD⁺		
ADP		
DP + 1,3-Bisphosphoglycerate	Loss of NADH	5, 10, 22
lyceraldehyde 3-phosphate + P$_i$ + NAD⁺	Total conversion by by enzymic removal of glyceraldehyde 3-phosphate using (TIM and GDH)	
Glutamate + NAD(P)⁺ + H$_2$O	Loss of NAD(P)H at 340 nm	5, 7, 22
₁-Antitrypsin-Trypsin] + Trypsin (residual)	Formation of p-nitroaniline	5
ɛnzylarginine + p-Nitro-aniline	Initial incubation partially inactivates the trypsin.	
Phosphoglycerate + P$_i$	Loss of NADH	22
3-Bisphosphoglycerate + ADP		
lyceraldehyde-3-phosphate + NAD⁺		
xaloacetate + P$_i$	Loss of NADH	3, 7, 17
-Malate + NAD⁺		
holesterol + fatty acids	Follow formation of dye	1, 3, 4, 5, 6, 7, 8, 14, 16, 18, 20, 22
holest-4-en-3-one + H$_2$O$_2$	Can determine free cholesterol if cholesterol esterase is omitted	
ye + H$_2$O		
holesterol + fatty acids	Monitor formation of 3,5-diacetyl-1,4-dihydro-lutidine at 410 nm	5
holest-4-en-3-one + H$_2$O$_2$		
CHO + 2H$_2$O		

Table

SUBSTRATE		REACTIONS
	HCHO + 2 Acetylacetone + NH$_3$	
Citrate	Citrate	Citrate lyase
	Oxaloacetate + NADH + H$^+$	Malate dehydrogenase
	Pyruvate + NADH + H$^+$	Lactate dehydrogenase
Creatinine	Creatinine + H$_2$O	Creatininase
	Creatine + ATP	Creatine kinase
	ADP + Phosphoenolpyruvate	Pyruvate kinase
	Pyruvate + NADH + H$^+$	Lactate dehydrogenase
Ethanol	Ethanol + NAD$^+$	Alcohol dehydrogenase
Formimino-L-glutamic acid (FIGLU) in urine	FIGLU + Tetrahydrofilic acid	FIGLU transferase
	Formimino-tetrahydrofolate	Formiminotetrahydrofo cyclodeaminase
Galactose (i)	Galactose	Galactose oxidase
	H$_2$O$_2$ + Chromogen	Peroxidase
(ii)	Galactose + NAD$^+$	Galactose dehydrogena
Glucose (i)	Glucose + O$_2$ + H$_2$O	Glucose oxidase
	H$_2$O$_2$ + Chromogen	Peroxidase
(ii)	Glucose + O$_2$ + H$_2$O	Glucose oxidase
(iii)	Glucose + O$_2$ + H$_2$O	Glucose oxidase
(iv)	Glucose + ATP	Hexokinase
	Glucose-6-Phosphate + NAD(P)$^+$	Glucose-6-Phosphate dehydrogenase
(v)	Glucose + ATP	Hexokinase

tinued

	END POINT	KIT SUPPLIERS†
Diacetyl-1,4-dihydrolutidine + $3H_2O$		
loacetate + Acetate	Loss of NADH	5
alate + NAD^+	Oxaloacetate spontaneouse decarboxylates to pyruvate – hence need need for third reaction	
actate + NAD^+		
atine	Loss of NADH	5
atine phosphate + ADP		
+ Pyruvate		
ctate + NAD^+		
taldehyde + NADH + H^+	Formation of NADH	5, 10, 22
mimino-tetrahydrofolate + glutamate	Increase in absorbance at 365 nm due to formation of 5,10-methenyl tetrahydrofolate	22
)-Methenyl-tetrahydrofolate + NH_4^-		
actolactone + H_2O_2	Formation of dye	11
+ H_2O		
actolactone + NADH + H^+	Formation of NADH	5
conic acid + H_2O_2	Formation of dye	5, 6, 7, 11, 14, 17, 18, 22
+ H_2O		
conic acid + H_2O_2	Use of O_2 with oxygen electrode.	3
conic acid + H_2O_2	Formation of H_2O_2 with peroxide electrode	6
cose-6-phosphate + ADP	Formation of NAD(P)H	1, 3, 5, 7, 8, 10, 16, 19, 20, 22
hosphogluconate + NAD(P)H + H^+		
cose-6-phosphate + ADP	Follow formation of NADPH	22

Tabl

SUBSTRATE		REACTION
		Glucose-6-phosphate + NADP⁺
		$$\text{Glucose-6-phosphate + NADP}^+ \xrightarrow{\text{Glucose-6-phosphate dehydrogen}}$$

SUBSTRATE		REACTION
		Glucose-6-phosphate + NADP⁺ —— Glucose-6-phosphate dehydrogen
		NAPDH + H⁺ + Phenazine methosulphate ——
		Phenazine methosulphate (reduced + Iodonitrotetrazolium violet ——
	(vi)	β-D-Glucose + NAD⁺ —— Glucose dehydrogenase and Muta
Lactate		L-Lactate + NAD⁺ —— Lactate dehydrogenase
Lecithin		Lecithin + H_2O —— Phospholipase C
		Phosphorylcholine + H_2O —— Alkaline phosphatase
		Choline + ATP —— Choline kinase
		ADP + Phosphoenolpyruvate —— Pyruvate kinase
		Pyruvate + NADH + H⁺ —— Lactate dehydrogenase
α₂-Macroglobulin		Trypsin + α₂-Macroglobulin ——
		Trypsin (residual) + Aprotinin ——
		Z-val-gly-arg-p-nitroaniline + H_2O —— [α₂-Macroglobulin-Tryp
Pyruvate		Pyruvate + NADH + H⁺ —— Lactate dehydrogenase
Triglycerides	(i)	Triglycerides + $3H_2O$ —— Lipase / Esterase / Glycolipase
		Glycerol + ATP ——
		ADP + Phosphoenolpyruvate —— Pyruvate kinase
		Pyruvate + NADH + H⁺ —— Lactate dehydrogenase

ntinued

	END POINT	KIT SUPPLIERS†
hosphogluconate + NADPH + H⁺	due to formazan production	
DP⁺ + Phenazine methosul- te (reduced)		
nazine methosulphate + onitrotetrazolium violet duced)		
Gluconolactone + NADH + H⁺	Formation of NADH is followed at 340 nm	3, 7, 14
ruvate + NADH + H⁺	Formation of NADH at 340 nm. Use pH 9–10. Can trap pyruvate with hydrazine or remove enzymically with alanine transaminase	5, 10, 17, 22
lyceride + Phosphorylcholine	First two reactions com- pleted and stopped by boiling. Follow loss of	5
oline + Phosphate	NADH at 340 nm in second phase	
osphorylcholine + ADP		
ruvate + ATP		
actate + NAD⁺		
-Macroglobulin-Trypsin] + Trypsin (residual)	Formation of p-nitro- aniline produced by [α₂-Macroglobulin- Trypsin] after inactivation	5
protinin-Trypsin] inactive	of excess trypsin	
al-gly-arg-OH + p-Nitro-aniline		
actate + NAD⁺	Loss of NADH. pH 7 to force equilibrium to lactate formation	4, 5, 10, 17, 22
cerol + 3 fatty acids	Follow loss of NADH at 340 nm. Also first sapon- ification reaction can be	1, 3, 5, 7, 10, 14, 19, 22
cerol 3-phosphate + ADP	done in ethanolic KOH	
ruvate + ATP		
ctate + NAD⁺ + H₂O		

SUBSTRATE			REACTIO
	(ii)	Triglycerides + $3H_2O$	Lipase/Esterase
		Glycerol + NAD^+	Glycerol dehydrogenase
		Tetrazolium salt + NADH	Diaphorase
Urea	(i)	Urea + H_2O	Urease
	(ii)	Urea + $2H_2O$	Urease
		2-Oxoglutarate + 2NADH + $2NH_4^+$	Glutamate dehydrogenase
Uric acid	(i)	Uric acid + $2H_2O + O_2$	Uricase
	(ii)	Uric acid + $2H_2O + O_2$	Uricase
		$H_2O_2 + C_2H_5OH$	Catalase
		$CH_3CHO + NADP^+$	Aldehyde dehydrogenase
	(iii)	Uric acid + $2H_2O + O_2$	Uricase
		$2H_2O$ + Chromogens	Peroxidase

†See Table 5.5 for names and addresses of kit suppliers.

the methods discussed in Part A (5.2.4.2) are also important for cholesterol, glucose and uric acid. The conversion between $NADP^+$/NADPH is used for the estimation of ammonia and glucose(iv). Only the estimation of α_1-antitrypsin, formimino-L-glutamic acid (FIGLU), α_2-macroglobulin, urea(i) and uric acid(i) have different end points. It is apparent that there are alternative enzymic assays for cholesterol, galactose, glucose, triglycerides, urea and uric acid. The assays where an enzymic kinetic method can be used are glucose(vi), lactate, triglycerides(i), urea(ii), and uric acid(ii). There are also other analytical methods, which do not involve enzymes, for most of the substrates listed in Table 5.1.

Adenosine-5'-diphosphate (ADP) and adenosine-5'-monophosphate (AMP) are measured on the same sample. First, ADP is measured with a coupled system that utilizes NADH. Only when this reaction is complete is the third enzyme, myokinase, added. This converts AMP to ADP and enables AMP to be measured using the same enzymes as were previously used to measure ADP.

Two proteins are included in Table 5.1. α_1-Antitrypsin is measured after a

tinued

	END POINT	KIT SUPPLIERS†
cerol + 3 fatty acids	Formation of NADH linked to production of formazan	3, 5, 6, 7
ydroxyacetone + NADH + H⁺		
mazan + NAD⁺		
$I_3 + CO_2$	Measure NH_3 with phenol/ hypochlorite	5, 8, 14, 17, 22
$I_4^+ + HCO_3^-$	Loss of NADH at 340 nm	3, 5, 7, 10, 17, 19, 20, 22
·Glutamate + $2NAD^+ + 2H_2O$		
antoin + $CO_2 + H_2O_2$	Measure loss of uric acid at 293 nm	5, 11, 14, 17, 22
ntoin + $CO_2 + H_2O_2$	Formation of NADPH	5, 14
CHO + $2H_2O$		
etate + NADPH + H⁺		
antoin + $CO_2 + H_2O_2$	Formation of dye	3, 4, 5, 6, 7, 18
e + $4H_2O$		

preliminary incubation with trypsin. Only uninhibited trypsin reacts (Zazgornik *et al.* 1980). α_2-Macroglobulin also combines with trypsin, but the complex formed is active in this case. Therefore excess trypsin is reacted with aprotinin before measurement of the activity of the α_2-macroglobulin-trypsin complex.

The most common substrate analyses done using enzymes are urea, glucose, cholesterol, triglycerides and uric acid. A kinetic methodology is available for all of them.

The suppliers produce leaflets for each test. These describe the principle of the test together with references. Particular attention should be given to the sample that can be used and to any pretreatment, sometimes protein precipitation, which may be required, and to make up of reagents and their stability. There is a detailed description of the test procedure together with methods of calibration and calculation of results. There should also be a note of common interfering substances. The normal range, although quoted, should be determined by each laboratory. A guide to the interpretation of results may be given, but this is the responsibility of the clinician.

Table 5.2   Use of enzymes as analyt

ENZYME		REACTIONS
Alanine aminotransferase	L-Alanine + 2-Oxoglutarate	$\dfrac{\text{Alanine}}{\text{aminotransferase}}$
	Pyruvate + NADH + H$^+$	$\dfrac{\text{Lactate}}{\text{dehydrogenase}}$
Aldolase	Fructose-1,6-bisphosphate	$\dfrac{\text{Aldolase}}{}$
	Glyceraldehyde-3-phosphate	$\dfrac{\text{Triosephosphate}}{\text{isomerase}}$
	2 Dihydroxyacetate phosphate + 2NADH + 2H+1	$\dfrac{\text{Glycerol-1-phosphat}}{\text{dehydrogenase}}$
α-Amylase (i)	Maltoheptaose + H$_2$O	$\dfrac{\text{α-Amylase}}{}$
	Maltotriose + 2H$_2$O	$\dfrac{\text{α-Glucosidase}}{}$
	3 Glucose + 3ATP	$\dfrac{\text{Hexokinase}}{}$
	3 Glucose-6-phosphate + 3NAD$^+$	$\dfrac{\text{Glucose-6-phosphate}}{\text{dehydrogenase}}$
(ii)	p-Nitrophenylmaltopentaoside	$\dfrac{\text{α-Amylase}}{}$
	p-Nitrophenylmaltoside	$\dfrac{\text{α-Glucosidase}}{}$
(iii)	Maltotetraose + H$_2$O	$\dfrac{\text{α-Amylase}}{}$
	Maltose + phosphate	$\dfrac{\text{Maltose}}{\text{phosphorylase}}$
	Glucose-1-phosphate	$\dfrac{\text{Phosphoglucomutas}}{}$
	Glucose-6-phosphate + NAD$^+$	$\dfrac{\text{Glucose-6-phosphate}}{\text{dehydrogenase}}$
Aspartate aminotransferase	L-Aspartate + 2-Oxoglutarate	$\dfrac{\text{Aspartate}}{\text{aminotransferase}}$
	Oxaloacetate + NADH + H$^+$	$\dfrac{\text{Malate}}{\text{dehydrogenase}}$
	(In some assays lactate dehydrogenase is also added to rem	
Creatine phosphokinase (and isoenzymes)	Creatine phosphate + ADP	$\dfrac{\text{Creatine}}{\text{kinase}}$
	ATP + Glucose	$\dfrac{\text{Hexokinase}}{}$
	Glucose-6-phosphate + NADP$^+$	$\dfrac{\text{Glucose-6-phosphat}}{\text{dehydrogenase}}$

ols in the measurement of other enzymes

	ENDPOINT	KIT SUPPLIERS†
ruvate + L-Glutamate	Follow loss of NADH	1, 3, 4, 5, 7, 8, 10, 14, 16, 17, 19, 20, 22
ᴖactate + NAD$^+$		
ᵧceraldehyde-3-phosphate + Dihydroxyacetone-phosphate	Follow loss of NADH	5, 10
ᴖydroxyacetone phosphate		
Glycerol-1-phosphate + 2NAD$^+$		
ᴖltotriose + Maltotriose	Follow formation of NADH	3, 5
ᴖlucose		
ᴖlucose-6-phosphate + 3ADP		
ᴖluconate-6-phosphate + 3NADH + 3H$^+$		
Nitrophenylmaltoside + Maltotriose	Follow formation of p-nitrophenol	10
Nitrophenol + 2 Glucose		
Maltose	Follow formation of NADH	
ᴖucose + Glucose-1-phosphate		
ᴖucose-6-phosphate		
Phosphogluconate + NADH + H$^+$		
ᴖaloacetate + L-Glutamate	Follow loss of NADH	1, 3, 4, 5, 7, 8, 10, 14, 16, 17, 19, 20, 22
Malate + NAD$^+$		
ᴖgenous pyruvate before the enzyme assay is started)		
ᴖeatine + ATP	Follow formation of NADPH	1, 3, 5, 6, 7, 8, 10, 14, 18, 19, 20, 22
ᴖucose-6-phosphate + ADP		
ᴖuconate-6-phosphate + NADPH + H$^+$		

Table

SUBSTRATE	REACTIONS
Galactose-1-phosphate uridyl transferase	Galactose-1-phosphate + UDPG $\xrightarrow{\text{Galactose-1-phosphate uridyl transferase}}$
	UDPG + 2NAD$^+$ $\xrightarrow{\text{UDPG dehydrogenase}}$
Lipase	H$_2$O + Vinyl 8-phenyloctanoate $\xrightarrow{\text{Lipase}}$
	Acetaldehyde + NADH + H$^+$ $\xrightarrow{\text{Alcohol dehydrogenase}}$
Malate dehydrogenase	L-Aspartate + 2-Oxoglutarate $\xrightarrow{\text{Aspartate aminotransferase}}$
	Oxaloacetate + NADH + H$^+$ $\xrightarrow{\text{Malate dehydrogenase}}$
5′-Nucleotidase	AMP $\xrightarrow{\text{5′-Nucleotidase}}$
	Adenosine $\xrightarrow{\text{Adenosine deaminase}}$
	NH$_4^+$ + 2-Oxoglutarate + NADH + H$^+$ $\xrightarrow{\text{Glutamate dehydrogenase}}$
Phosphohexose isomerase	Fructose-6-phosphate $\xrightarrow{\text{Phosphohexose isomerase}}$
	Glucose-6-phosphate + NADP$^+$ $\xrightarrow{\text{Glucose-6-phosphate dehydrogenase}}$
Pyruvate kinase	Phosphoenolpyruvate + ADP $\xrightarrow{\text{Pyruvate kinase}}$
	Pyruvate + NADH + H$^+$ $\xrightarrow{\text{Lactate dehydrogenase}}$

† See Table 5.5 for names and addresses of kit suppliers.

5.3 ENZYMES MEASURED BY COUPLED ENZYME SYSTEMS IN CLINICAL LABORATORIES

The data collected in Table 5.2 lists those commercially available kits which use a coupled enzyme for the measurement of other enzymes. Only two of the assays require more than one coupled enzyme in order to reach an NAD$^+$/NADH end point; they are creatine kinase and α-amylase(i). The only assay that does not use this end point is α-amylase(ii) where the production of yellow p-nitrophenol is followed (Wallenfels *et al.* 1978).

tinued

	END POINT	KIT SUPPLIERS†
ose-1-phosphate + UDP-Galactose	Determine difference in NADH used in presence and absence of galactose-1-phosphate	22
-Glucoronic acid + NADH + 2H$^+$		
enyloctanoic acid + Acetaldehyde	Follow loss of NADH	10
nol + NAD$^+$		
utamate + Oxaloacetate	Follow loss of NADH	5
late + NAD$^+$		
osine + P$_i$	Follow loss of NADH	22
ne + NH$_4^+$		
utamate + NAD$^+$ + H$_2$O		
ose-6-phosphate	Follow formation of NADPH	22
osphogluconate + NADPH H$^+$		
vate + ATP	Follow loss of NADH	5
ctate + NAD$^+$		

An unusual use of a coupled enzyme is in the assay of malate dehydrogenase. In this case the first enzyme, aspartate aminotransferase, is used to generate oxaloacetate which is the substrate for the second reaction, malate dehydrogenase. The reason is that oxaloacetate is unstable and tends to decarboxylate to form pyruvate.

One of the enzymes listed in Table 5.2 is measured to ascertain whether it is absent or deficient in children. This is galactose-l-phosphate uridyltransferase, an enzyme required in the overall conversion of galactose to glucose. The enzyme is measured in erythrocytes by a discontinuous assay. An haemolysate is prepared and incubated with galactose-1-phosphate for 15 min. Then the reaction is stopped and the amount of uridine diphosphoglucose (UDPG) remaining is determined in the second reaction by following the formation of NADH. The amount of UDPG in the original reaction mixture is detemined in another cuvette where the galactose-1-phosphate is omitted. Blood taken from a normal person uses up nearly all of the UDPG. In contrast, blood from a child with galactosaemia uses very little of this compound.

The other enzymes in Table 5.2 are measured in serum or plasma to investigate specific tissue damage. The enzymes most commonly assayed are alanine aminotransferase, aspartate aminotransferase, and creatine kinase. There are alternative assays, for these enzymes and for most of the others shown in Table 5.2, which do not involve the use of coupled enzymes.

Table 5.3

Commercially available diagnostic products using Immobilized Enzymes

Substrate	Immobilized enzymes	Kit suppliers†††
(a) Enzymes attached to coils		
Creatinine	Creatininase	9
Glucose	Glucose oxidase	2, 9
Glucose	Hexokinase/Glucose-6-phosphate- dehydrogenase	24
Urea	Urease	2, 9
Uric acid	Uricase	9, 24
(b) Bioanalytical probe		
Glucose	Glucose oxidase	6
(c) Dry reagent strips		
Cholesterol††	Cholesterol oxidase/cholesterol esterase/ peroxidase	2
Glucose†	Glucose oxidase/peroxidase	2,5
Triglycerides††	Lipase/Glycerol kinase/α-glycerol phosphate dehydrogenase/diaphorase	2
Uric acid	Uricase/peroxidase	2
(d) Multilayer slides		
Cholesterol††	Cholesterol ester hydrolase/cholesterol oxidase/peroxidase	12
Creatinine	Creatinine iminohydrolase	12
Glucose	Glucose oxidase/peroxidase	12
Triglycerides††	Lipase/glycerol kinase/L-α-glycerophosphate oxidase/peroxidase	12
Urea	Urease	12
Uric acid	Uricase/peroxidase	12

 † Reflectance photometers for use with blood glucose test strips are available from several manufacturers including 2, 5.

 †† At the time of writing the methods for the assay of cholesterol and triglycerides are not commercially available, but they are at a late stage of development and are undergoing evaluations.

††† See Table 5.5 for names and addresses of kit suppliers.

5.4 IMMOBILIZED ENZYMES FOR MEASURING SUBSTRATES

Various types of immobilized enzyme systems are listed in Table 5.3. In general the chemical reactions involved are covered in Table 5.1. This particular area of the application of enzymes in clinical analysis is dominated by the firms who produce the enzyme reagents in immobilized form. Some of the early work on enzymes attached to coils and to bioanalytical probes was done prior to involvement of commercial organisations. But once the potential application of this form of enzyme was appreciated, it was also realized that considerable capital investment was required, not just to produce suitable immobilized enzymes but also to develop the whole technology which generally includes the dedicated equipment which is necessary for quantitation. It is apparent from Table 5.3 which areas these firms have specialized in.

5.5 ENZYMES USED IN ENZYME IMMUNOASSAY (EIA)

In general, different enzymes are used in heterogeneous EIA and in homogeneous EIA (Table 5.4). Enzymes commonly used for heterogeneous assays have high turnover numbers and form a coloured product. This is particularly useful when

Table 5.4
Enzymes commonly used for Enzyme Immunoassay (EIA)

Enzyme	Source	End point	Kit suppliers†
(a) Heterogeneous assays			
Alkaline phosphatase	Calf intestine	Formation of p-nitrophenol p-nitrophenylphosphate	10, 15, 25, 26
β-D-Galactosidase	*Escherichia coli*	Formation of o-nitrophenol or 4-methylumbelliferone from corresponding β-galactoside substrate	25
Peroxidase	Horseradish	H_2O_2/Chromagen	1, 5, 7, 10, 13, 15, 21, 25
Urease	Jack beans	NH_3/Chromogen	21
(b) Homogeneous assays			
Acetylcholinesterase	*Electrophorus electricus*	Colorimetric	1
β-D-Galactosidase	*Escherichia coli*	Formation of fluorescent 4-methylumbelliferone	15
Glucose-6-phosphate dehydrogenase	*Leuconostoc mesenteroides*	Formation of NADH	23
Lysozyme	Chicken egg-white	Formation of cell wall fragments of *Micrococcus luteus*	23
Malate dehydrogenase	Pig heart mitochondria	Formation of NADH	23

†See Table 5.5 for names and addresses of kit suppliers.

only qualitative results are required, for example in the screening of infectious diseases. However, with these enzymes very specific and sensitive quantitative assays are also possible. With alkaline phosphatase and β-galactosidase it is also possible to perform fluorimetric assays if improved sensitivity is needed.

In homogeneous EIA, continuous assays are done. The enzymes most frequently used are glucose 6-phosphate dehydrogenase and malate dehydrogenase, both of which are easily measured by following the increase in absorbance at 340 nm. However, this change can also be monitored fluorimetrically if greater sensitivity is required. The commercial assays that use β-galactosidase do not allow the normal amplification effect found in other types of EIA. In this case a fluorimetric assay is essential to give the required sensitivity.

Table 5.5
The Names and Addresses of Reagent Kit Suppliers and Instrument Manufacturers in the United Kingdom

For cross reference with Tables 5.1, 5.2, 5.3 and 5.4, the number given below is used. Most companies have agencies in other countries.

Company name	Address and telephone number	Number
Abbott Laboratories Ltd, Diagnostics Division	Brighton Hill Parade, Basingstoke, Hampshire RG22 4EH 0256–54051	1
Ames Division Miles Laboratories Ltd.	P.O. Box 37, Stoke Court, Stoke Poges, Buckinghamshire, SL2 4LY 028–14–2151	2
Beckman-RIIC Ltd, Analytical Instruments Sales and Service Operation	Progress Road, Sands Industrial Estate, High Wycombe, Buckinghamshire, HP12 4JL 0494–41181	3
BDH Chemicals Ltd	Poole, Dorset, BH12 4NN 0202–745520	4
BCL, Boehringer Corporation (London) Ltd.	Bell Lane, Lewes, East Sussex, BN7 1LG 07916–71611	5
Clandon Scientific Ltd	Lysons Avenue, Ash Vale, Aldershot, Hampshire GU12 5QR 0252–514711–5	6
Corning Medical and Scientific Corning Ltd.	Halstead, Essex, CO9 2DX 0787–472461	7
Diamed Diagnostics Ltd	Mast House, Derby Road, Bootle, Merseyside, L20 1EA 051–933–7277	8
Erba Science (UK) Ltd	14 Bath Road, Swindon, SN1 4BA 0793–33551	9
Hoechst UK Ltd Pharmaceutical Division	Hoechst House, Salisbury Road, Hounslow, Middlesex, TW4 6JH 01–570–7712	10
Hughes & Hughes Ltd	Elms Industrial Estate, Church Road, Harold Wood, Romford, Essex, RM3 0HR 040–23–49017	11

Table 5.5 — *continued*

Company name	Address and telephone number	Number
Kodak Ltd	Station Road, Hemel Hempstead, Hertfordshire, HP1 1JU 0442–61122	12
Kone Instruments	Regent House, Heaton Lane, Stockport, Cheshire, SK4 1BS 061–477–0662	13
Merk EM Diagnostics	169 Oldfield Lane, Greenford, Middlesex, UB6 8PN 01–575–5228	14
Miles Laboratories Ltd.	P.O. Box 37, Stoke Court, Stoke Poges, Slough, SL2 4LY 02814–5151	15
Monitor International	Robell Way, Storrington, Pulborough, West Sussex RH20 3DW 090–66–5311	16
Norris Biomedical Ltd	P.O. Box 25, Basingstoke, Hampshire (0)256–881877	17
Pierce & Warriner (UK) Ltd	44 Upper Northgate Street, Chester, CH1 4EF 0244–382525	18
Roche Products Ltd Diagnostics Division	P.O. Box 8, Welwyn Garden City, Hertfordshire, AL7 3AY 07073–28128	19
Ross Lab. G. S. Ross Ltd.	Unit 13, Fence Avenue Industrial Estate, Macclesfield, Cheshire, SK11 1LT 0625–610077	20
Sera-Lab Ltd	Crawley Down, Sussex, RH10 4FF 0342–716366	21
Sigma London Chemical Co. Ltd	Fancy Road, Poole, Dorset, BH17 7NH 0202–733114	22
Syva UK	Syntex House, St Ives Road, Maidenhead, Berkshire, SL6 1RD 0628–70969	23
Technicon Instruments Co. Ltd	Evans House, Hamilton Close, Basingstoke, Hampshire, RG21 2YE 0256–29181	24
Townson & Mercer Ltd	93–96 Chadwick Road, Astmoor, Runcorn, Cheshire WA7 1PR 09285–76245	25
Don Whitley Scientific Ltd	Green Lane, Baildon, Shipley, West Yorkshire, BD17 5JS 0274–595728	

REFERENCES

Szabó, A. & Örs, E. (1983) *J. Clin. Chem. Biochem.* **21** 209–215.
Wallenfels, K., Földi, P., Niermann, N., Bender, H. & Linder, D. (1978) *Carbohydrate Research* **61** 359–368.
Whitby, L. G., Percy-Robb, I. W. & Smith, A. F. (1984) *Lecture Notes on Clinical Chemistry,* 3rd ed., Blackwell Scientific Publications, Oxford, London, Edinburgh, and Melbourne.
Zazgornik, J., Kopsa, H., Schmidt, P., Pils, P., Balcke, P., Hysek, H. & Deutsch, E. (1980) *J. Clin. Chem. Clin. Biochem.* **18** 241–244.
Zilva, J. F. & Pannall, P. R. (1984) *Clinical Chemistry in Diagnosis and Treatment* 3rd ed., Lloyd-Luke (Medical Books) Ltd, London.

Index